7th International Conference on Nanomaterials by Severe Plastic Deformation 2017 (NanoSPD7)

IOP Conference Series: Materials Science and Engineering
Volume 194

Sydney, Australia
2 – 7 July 2017

Editors:

Yuri Estrin
Xiaozhou Liao

ISBN: 978-1-5108-4029-4
ISSN: 1757-8981

Printed from e-media with permission by:

Curran Associates, Inc.
57 Morehouse Lane
Red Hook, NY 12571

Some format issues inherent in the e-media version may also appear in this print version.

Printed by Curran Associates, Inc. (2017)

For permission requests, please contact the Institute of Physics
at the address below.

Institute of Physics
Dirac House, Temple Back
Bristol BS1 6BE UK

Phone: 44 1 17 929 7481
Fax: 44 1 17 920 0979

techtracking@iop.org

Additional copies of this publication are available from:

Curran Associates, Inc.
57 Morehouse Lane
Red Hook, NY 12571 USA
Phone: 845-758-0400
Fax: 845-758-2633
Email: curran@proceedings.com
Web: www.proceedings.com

Table of contents

Volume 194

7th International Conference on Nanomaterials by Severe Plastic Deformation
2–7 July 2017, Sydney, Australia

Accepted papers received: 24 April 2017
Published online: 8 May 2017

Preface

Papers

IOP Conf. Series: Materials Science and Engineering **194** (2017) 011001 doi:10.1088/1757-899X/194/1/011001

The 7th International Conference on Nanomaterials by Severe Plastic Deformation – Preface

X Z Liao[1] and Y Estrin[2,3]

[1] School of Aerospace, Mechanical and Mechatronic Engineering, The University of Sydney, Sydney, NSW 2006, Australia
[2] Department of Materials Science and Engineering, Monash University, Clayton, VIC 3800, Australia
[3] Laboratory of Hybrid Nanostructured Materials, NUST MISIS, Moscow 119049, Russia

E-mail: xiaozhou.liao@usyd.edu.au; yuri.estrin@monash.edu

Severe plastic deformation (SPD) has arguably been the most effective top-down approach to producing bulk nanostructured materials from more conventional, coarse-grained ones. During SPD processing, a giant shear deformation of a material is combined with very high hydrostatic pressure to produce ultrafine (< 1 μm) or nanocrystalline (< 100 nm) grain structure. Despite the huge strain involved, popular SPD techniques practically do not change the dimensions of a work-piece, which makes them very attractive for future industry-scale applications. By appropriately choosing SPD processing parameters that give rise to desirable microstructure, one can produce materials with exceptional mechanical properties and in many cases unusual functional characteristics.

A series of international events briefly referred to as NanoSPD conferences have become major forums for the growing community of researchers exploring the potential of severe plastic deformation as a means of producing bulk ultrafine-grained or nanostructured materials. The 7th International Conference on Nanomaterials by Severe Plastic Deformation in Sydney (2-7 July 2017) is the seventh in this series of conferences held on a triennial basis. It follows the conferences that took place in Moscow (1999), Vienna (2002), Fukuoka (2005), Goslar (2008), Nanjing (2011), and Metz (2014). NanoSPD conferences have provided an important platform for discussion of a wealth of challenging scientific problems and new technological developments in the NanoSPD area. These Proceedings contain a selection of peer-reviewed papers presented at NanoSPD7 that cover the major themes of the conference. These include SPD processing techniques as well as microstructures, deformation mechanisms and properties of SPD-processed materials. Naturally, mechanical properties figure in the Proceedings most prominently, but functional properties, including biodegradation and biocompatibility, are represented as well. A number of articles are dedicated to modelling of SPD processes and the mechanical behaviour of bulk nanostructured materials. A historical perspective and a survey of the most important developments in the field since the NanoSPD6 conference in Metz as well as the emerging industry-scale applications of SPD-modified materials are given in an introductory article by R.Z. Valiev and T.G. Langdon.

We would like to thank all those who contributed to the success of NanoSPD7. These include all conference participants, and especially the authors of the papers collected in these Proceedings, the International NanoSPD Steering Committee (www.nanospd.org), the Local Organising Committee (www.nanospd7.com), staff from ICMS Australasia Pty Ltd, and student helpers. We also appreciate the generous financial support from the conference sponsors.

IOP Conf. Series: Materials Science and Engineering **194** (2017) 011002 doi:10.1088/1757-899X/194/1/011002

Peer review statement

All papers published in this volume of *IOP Conference Series: Materials Science and Engineering* have been peer reviewed through processes administered by the proceedings Editors. Reviews were conducted by expert referees to the professional and scientific standards expected of a proceedings journal published by IOP Publishing.

The 7th International Conference on Nanomaterials by Severe Plastic Deformation: a report of the International NanoSPD Steering Committee

R Z Valiev[1,3*] and T G Langdon[2,3]**

[1] Institute of Physics of Advanced Materials, Ufa State Aviation Technical University, Ufa, Russia and Saint Petersburg State University, 28 Universitetskiy pr., Saint Petersburg 198504 Russia

[2] Materials Research Group, Faculty of Engineering and the Environment, University of Southampton, Southampton SO17 1BJ, UK

[3] International NanoSPD Steering Committee (www.nanospd.org)

[*] ruslan.valiev@ugatu.su, [**] langdon@soton.ac.uk

Abstract. The 7th International Conference on Nanomaterials by Severe Plastic Deformation (NanoSPD7) is hosted by the University of Sydney (Australia) following a series of earlier conferences: in Moscow (1999), Vienna (2002), Fukuoka (2005), Goslar (2008), Nanjing (2011) and Metz (2014). This introductory paper reports on several major developments in NanoSPD activities as well as on recent NanoSPD citation data which illustrate the growth and expansion of this important research area for the time period following the conference in Metz. Close attention is given to topics of nanostructuring of metals by SPD processing for advanced properties and on new trends in developing SPD techniques for practical applications. A special concern of the committee is the appropriate terminology that is used in this new field of science and engineering as well as the innovation potential of recent applied studies and developments.

1. Introduction

In recent years there has been growing interest in bulk nanostructured materials produced by severe plastic deformation (SPD) processing, as demonstrated by the increasing numbers of publications with high citations as well as numerous specialized conferences, workshops and symposia on the subject (www.nanospd.org). The conferences of the NanoSPD series serve as a live-to-live platform allowing the attendees and contributors to present and share the most cutting-edge research and receive feedback from colleagues, attend numerous exciting sessions and become aware of the innovative studies in particular areas of research. These conferences have been held all over the world, starting in 1999 (Moscow, Russia) and then continuing in 2002 (Vienna, Austria), 2005 (Fukuoka, Japan), 2008 (Goslar, Germany), 2011 (Nanjing, China), 2014 (Metz, France) and now in Sydney, Australia. This latest meeting offers even more stimulus for scientific and applied research and summarizes the important events related to NanoSPD and to the activities of the International NanoSPD Steering Committee.

This introductory report is designed to present a summary of the main results that have occurred in NanoSPD over the last three years since our meeting in Metz, France, in 2014 with its record-breaking 428 participants from 42 different countries and to discuss the important achievements and provide a performance summary. In addition, statistics on publication citation data are presented to illustrate the major impact of NanoSPD within the broad field of Materials Science.

2. Main events and publications as a platform for the exchange of ideas and research in the NanoSPD field

Major NanoSPD related conferences and symposia over the period 2014-2017

The last three years since the 6th International Conference on Nanomaterials by Severe Plastic Deformation (NanoSPD6) in Metz, France, have witnessed extensive and profound research in the NanoSPD area with a number of international conferences and meetings, thus providing a sustainable content for thorough exchange within the research community, industry representatives and emerging experts to bring the ideas from a laboratory scale to the mass production of products as well as to offer the participants an opportunity for networking with new people within the field, cultural activities and receptions to establish new connections and ties. Of the primary events aimed at enhancing research in these areas, the authors would like to emphasize the following:

The 69th ABM International Annual Congress, the annual meeting of the Brazilian Metallurgical Association, was held in Sao Paulo, Brazil, on July 21-24, 2014. Incorporated within the meeting was the Second Pam American Materials Conference co-sponsored by TMS and consisting of several English-language symposia on topics of current interest. One of these symposia was a one-day meeting on Ultrafine-Grained Materials organized by Roberto Figueiredo and Terence Langdon. This symposium was well-attended and contained several presentations of interest to members of the NanoSPD community.

The 22nd International Conference on Metastable and Nano Materials (ISMANAM) held on July 13-17, 2015 in Paris, France, was a success as a multidisciplinary forum promoting international scientific and technological exchanges on all aspects related to metastable, amorphous and nanostructured materials, in particular, produced by severe plastic deformation.

The 5th International Symposium on Bulk Nanostructured materials: from fundamentals to innovations (BNM-2015) were held on August 26-28, 2015 in Ufa, Russia, by the Institute of Physics of Advanced Materials of Ufa State Aviation Technical University (IPAM USATU). The symposium was dedicated to the 20th anniversary of IPAM USATU, headed by Prof. Ruslan Valiev. The symposium covered the following topics: New trends in SPD, Physical and mechanical properties of BNM, Microstructure evolution and phase transformations, Grain boundary engineering of UFG materials, Computer modeling and texture analysis. The specific feature of this symposium was its special focus on practical applications of nanoSPD materials. That is the reason for organizing a round-table on "Innovation trends and mechanisms & commercialization of BNM".

International Workshops on Giant Straining Process for Advanced Materials (GSAM 2015, GSAM2016) were held on September 3-6, 2015, and July 28-31, 2016, in Fukuoka, Japan, by the International Research Center on Giant Straining for Advanced Materials (IRC-GSAM) at Kyushu University to discuss the topics from structural properties as well as from functional properties of the SPD-processed materials. The workshop proceedings were published under the title «Exploring SPD Potential: Innovative Approach for Production of High-Performance Materials» and «Promoting Functionality by Severe Plastic Deformation: Significance of Lattice Defects and Phase Transformation» (The proceedings are available upon request to zenjihorita@gmail.com). As a GSAM series of workshops, GSAM2017 is planned on September 2-5, 2017, in Fukuoka, Japan, to promote exchanging idea of different fields and pursue new emerging research topics (visit http://irc-gsam.kyushu-u.ac.jp/index-e.html for more information).

The International Conference on Superplasticity in Advanced Materials (ICSAM 2015), including a session on Severe Plastic Deformation was held on September 7-9, 2015, in Tokyo, Japan, with approximately 130 presentations, more than 160 participants and 3 exhibition-making companies from over 20 countries. Many presentations were focused on the superplasticity of ultrafine-grained alloys from SPD processing, their unique behavior and potential for applications.

The 14th International Union of Materials Research Societies-International Conference on Advanced Materials (IURMS-ICAM 2015) was held on October 25-29, 2015, at the Jeju International Convention Center in Jeju, Korea. IUMRS-ICAM, a bi-annual conference, once again provided great opportunities for scientists, researchers, engineers and business executives from all parts of the world not only to exchange the latest knowledge but to share the state-of-art techniques in the field of material sciences.
During the event, Prof. Ruslan Valiev (Russia) and Prof. Yuntian Zhu (U.S.A.) received the 2015 Somiya Award (presented biennially by the International Union of Materials Research Societies (IUMRS) for the most significant research conducted by an interactive group or team whose members are drawn from at least two continents) for resolving the paradox of superior strength and ductility of ultrafine-grained metals through microstructural design (Fig. 1). This is the second time this award has been given to members of the NanoSPD International Steering Committee because in 2005 the award was presented to Prof. Zenji Horita (Japan) and Prof. Terry Langdon (U.S.A.).

Figure 1. Prof. Valiev receiving the awards on behalf of both recipients presented by General Secretary of IUMRS, Prof. Robert P.H. Chang (center) and the Conference Chair, Prof. Soo Wohn Lee (right).

EUROMAT 2015 including a session on Severe Plastic Deformation and Nanostructuring was held on September 21-24, 2015, in Warsaw, Poland, and hosted by the Warsaw University of Technology, the oldest and the largest technical university in the country. The event witnessed an increased record of bringing together up to 2000 researchers, scientists, trainees, and students from both academia and industry to discuss critical developments and perspectives in the field of materials science and technology and their applications.
In 2015 Warsaw University of Technology celebrated its 100th anniversary of the university and the introduction, 45 years ago, of classes in metallurgy and materials science. In a special award ceremony conducted in the University on September 25, 2015, seven senior professors were presented with Friend of the Faculty Awards to mark their long-term collaborations with the Faculty of Material Science and Engineering at Warsaw University of Technology. Two of the awardees were Prof.

Michael Zehetbauer (University of Vienna) and Prof. Terry Langdon (University of Southampton) who were recognized specifically for their collaborations in the field of NanoSPD.

The International Biennial Conference on Ultrafine Grained and Nanostructured Materials (UFGNSM 2015) was held on November 11-12, 2015, in Tehran, Iran, following the success of a series of biennial conferences focused on ultrafine-grained and nanostructured materials. The event was jointly organized by University of Tehran, Iran, and University of Trento, Italy. The conference topics were fully covered in Procedia Materials Science with 100 papers devoted to bulk nanostructured materials, advanced characterization techniques, nanobiomaterials, nanoelectronic and magnetic materials, synthesis and nanofabrication, modeling and simulation of UFGNSM, nanocoatings and thin films, and nanocomposites.

The 2016 TMS Annual Meeting with the 9th International Symposium on Ultrafine Grained Materials (UFG IX) was held on February 14-18, 2016, in Nashville, Tennessee, USA. This symposium, as with previous symposia in the series, attracted a wide audience and many topical papers describing recent developments in NanoSPD.

THERMEC 2016 was held from May 30 to June 3, 2016 in Graz, Austria and covered all aspects of processing, fabrication, structure/property relationships and applications of various metallic materials. In the framework of the conference, Professor Reinhard Pippan of the Erich Schmid Institute of Materials Science in Leoben, Austria, received a THERMEC Distinguished Award for pioneering research in the field of fatigue/fracture and severe plastic deformation of advanced materials and for leadership in materials education in Austria (Fig. 2).

Figure 2. Professor Reinhard Pippan of the Erich Schmid Institute of Materials Science in Leoben, Austria (on right) receiving THERMEC Distinguished Award from Terry Langdon (left).

The 2016 Engineering Mechanics Institute International Conference (2016-EMI-IC), including a mini-symposium on Severe Plastic Deformation was held on October 25-27, 2016 in Metz, France. The symposium was organized by Prof. Laszlo Toth and devoted to priority trends in NanoSPD through 28 lectures.
An Internet discussion forum has been created by the Metz research group on SPD issues:

www.lem3-univ-lorraine.fr/forumspd/phpBB3. After inscription to the forum, one can make questions on research topics which can be answered by others and the dialogue can be followed by the members of the forum.

The 2017 TMS Annual Meeting was held from February 26 to March 2, 2017, in San Diego, California, USA. This comprehensive annual meeting was dedicated to fostering the exchange of learning and ideas across the entire range of materials science and engineering, from minerals processing and primary metals production, to basic research and the advanced applications of materials and the meeting included the following SPD-related symposia: Nanocrystalline and Ultrafine Grained Materials and Bulk Metallic Glasses, Multiscale Architectured Materials (MAMII): Tailoring Mechanical Incompatibility for Superior Properties Mechanical Behavior of Advanced Materials.

Major NanoSPD related reviews, special issues and books on NanoSPD research over 2014-2017

A number of new reviews and special issues on the results of recent NanoSPD materials-related studies and developments were published following the NanoSPD6 conference in Metz, France, in 2014. The authors would like to draw attention to some significant key publications during the period under review and these are now summarized:

Nanofabrication using Nanomaterials, edited by J. Ebothé, W. Ahmed, One Central Press (OCP) 2014, The book consists of seven chapters considering nanofabrication as a technical procedure for the preparation or the synthesis of a nanomaterial as well as a project of building a particular system wherein the nanomaterial plays a significant role.

K. Edalati, Z. Horita, A review on high-pressure torsion (HPT) from 1935 to 1988, Mater Sci Eng A 652 (2016) 325-352 reviewing the findings of Bridgman and his successors from 1935 to 1988 using the HPT method and summarizing their historical importance in recent advancement of materials, properties, phase transformations and HPT machine designs.

I. Sabirov, N.A. Enikeev, M.Y. Murashkin, R.Z. Valiev, Bulk nanostructured materials with multifunctional properties, Series: SpringerBriefs in Materials, Springer, 2015, IX, 161 p written by researchers at the forefront of this new materials design approach, the present volume provides a comprehensive introduction to multifunctional design of bulk nanostructured materials, with applications ranging from hydrogen storage to construction engineering.

R.A. Andrievsky, A.V. Khatchoyan, Nanomaterials in Extreme Environments: Fundamentals and Applications, Springer Series in Materials Science, vol. 230, Springer 2016. This book focuses on the behaviour of nanomaterials under extreme conditions of high temperature, irradiation by electron/ions and neutrons as well as in mechanical and corrosion extremes. The theoretical approaches and modeling are presented with numerous results of experimental studies as well as the examples of many applications and some unresolved problems.

R. Z. Valiev, Y. Estrin, Z. Horita, T. G. Langdon, M. J. Zehetbauer, Y.T. Zhu, Producing bulk ultrafine-grained materials by severe plastic deformation: ten years later, JOM, vol. 68, No. 4 (2016) pp. 1216-1226 to provide, exactly ten years after the first article with a similar title was published in JOM in the issue dated April 2006, an update to the fast-moving and challenging field of the fabrication of bulk ultrafine-grained (UFG) materials through the application of severe plastic deformation (SPD) and to give a description of new developments in the production and properties of UFG materials.

Severe Plastic Deformation Technology, edited by A. Rosochowski, Whittles Publishing, Dunbeath, Scotland, UK, 2016, covering the most popular SPD process of equal-channel angular pressing, ECAP and its incremental version, I-ECAP. A separate chapter is devoted to tooling used in ECAP/IECAP. Another popular SPD process is high-pressure torsion (HPT) which produces very good results in terms of refining grain structure but faces some interesting challenges. A less known SPD process is cyclic extrusion compression (CEC), which is thoroughly explained in the book as is twist extrusion (TE) which is a relatively new process that is showing good potential. Finally, an original SPD process of accumulated roll bonding (ARB), capable of refining grain structure in sheets, is discussed.

Defect Structure and Properties of Nanomaterials, 2nd Ed, J. Gubicza. Elsevier, 2017, provides a detailed overview of processing methods, defect structure, and defect-related mechanical and physical properties of nanomaterials and includes new chapters on recent advances in both processing techniques and methods for the study of lattice defects.

3. Citation statistics for publications in NanoSPD

Over the last years several special articles have been published dealing with the historical background of the NanoSPD research [1,2]. From an historical viewpoint, the processing of metals through the application of severe plastic deformation has a long history dating back to the fabrication of steels in ancient China more than 2000 years ago [3] and the subsequent development of Wootz steel in India [4]. Nevertheless, the first scientific reports of processing by SPD appeared about eighty years ago with the work of Bridgman at Harvard University in the United States [5]. These early reports were devoted exclusively to the mechanical characteristics associated with the processing operation and it was only later, in 1988, that the first reports appeared in the scientific literature documenting the ability to achieve ultrafine-grained (UFG) microstructures through SPD and the consequent exceptional properties that may be attained in these materials [6] using HPT and later by ECAP [7]. This was subsequently followed, in 1993, by the first publication from an international team describing the properties of an aluminum alloy processed by SPD with an emphasis on the flow properties and the dependence on strain rate at elevated temperatures [8]. In practice, these early reports led to further publications from several different laboratories and gradually to the development of a worldwide interest in the processing and properties of UFG metals.

It is apparent from this brief summary that NanoSPD is a relatively new field within Materials Science having a scientific history of less than 25 years. Nevertheless, the impact of NanoSPD has been considerable to the extent that this topic now competes easily with more conventional and well-established topics in Materials Science such as flow behavior and crystalline plasticity, the effects of alloying, phase transformations and traditional subjects such as creep, fracture and fatigue. Three earlier reports examined the impact of NanoSPD using citation data that was current at the time of publication [9-11] and this paper follows this earlier tradition by reporting citation data collected from Thomson Reuters Web of Science in March 2017.

The citation data are collected in five Tables covering five of the major journals currently published in the field of Materials Science. Each journal is contained within a different listing in Tables 1 to 5 and for each journal there is information on the total numbers of papers published to date and the Impact factor (IF) for the journal. Within each Table, papers relevant to the field of NanoSPD are listed where these are the papers contained within the top ten all-time most cited articles appearing in each journal. Thus, Table 1 is for *Progress in Materials Science* - a major review journal devoted exclusively to publishing long review articles on selected topics. Inspection of Table 1 shows that this journal has published a total of 501 papers and the IF is exceptionally high at >36. It is remarkable that four of the all-time top ten papers in this journal are from NanoSPD and this includes the papers in the 1st, 3rd, 4th and 5th positions. Data for *Acta Materialia* are given in Table 2 where again there are four papers in the top ten and in Table 3 for *Scripta Materialia* with five NanoSPD papers in the top ten, including the papers lying in the 1st, 2nd and 3rd positions. Table 4 lists results for *Materials Science and Engineering A* where there are more than 28,000 published papers and yet four of the top ten are again from the field of NanoSPD. Finally, Table 5 contains data for *JOM* which is a U.S. journal primarily publishing invited papers and distributed as a house magazine to all members of TMS. It is apparent that *JOM* contains two NanoSPD papers within the top ten.

As already noted, the numbers collected in Tables 1-5 correspond to data available on the internet in Web of Science as of March 2017. These numbers will gradually increase with time as more citations become available but the overall trends will remain reasonably unchanged. A good indication of the changes occurring over a long period of time may be gained by comparing these data with those published earlier for the NanoSPD6 conference [11]. The important conclusion from this analysis is that reports on NanoSPD have had a major impact within the field of Materials Science and they account for four or five of the all-time most-cited papers in several major journals.

Table 1. All-time ranking for *Progress in Materials Science* (501 papers; IF = 36.083)

Ranking	Author(s)	Year	No. of citations	Reference
1	Valiev *et al.*	2000	4157	[12]
3	Gleiter	1989	3102	[13]
4	Valiev and Langdon	2006	2139	[14]
5	Meyers *et al.*	2006	1976	[15]

Table 2. All-time ranking for *Acta Materialia* (12,518 papers; IF = 5.058)

Ranking	Author(s)	Year	No. of citations	Reference
2	Gleiter	2000	1743	[16]
4	Kumar *et al.*	2003	1272	[17]
5	Saito *et al.*	1999	1232	[18]
6	Iwahashi *et al.*	1998	989	[19]

Table 3. All-time ranking for *Scripta Materialia* (10,863 papers; IF = 3.305)

Ranking	Author(s)	Year	No. of citations	Reference
1	Iwahashi *et al*	1996	1426	[20]
2	Saito *et al.*	1998	810	[21]
3	Tsuji *et al.*	2002	641	[22]
5	Mukai *et al.*	2001	586	[23]
8	Nieh and Wadsworth.	1991	443	[24]

Table 4. All-time ranking for *Materials Science and Engineering A* (28,458 papers; IF = 2.647)

Ranking	Author(s)	Year	No. of citations	Reference
2	Segal	1995	1779	[25]
4	Valiev *et al.*	1993	1016	[26]
5	Furukawa *et al.*	1998	836	[27]
6	Valiev *et al.*	1991	798	[7]

Table 5. All-time ranking for *JOM* (1890 papers; IF = 1.05)

Ranking	Authors(s)	Year	No. of citations	Reference
2	Valiev *et al.*	2006	798	[28]
9	Ma	2006	232	[29]

4. Innovation activity

Over the last few years, studies of bulk nanostructured materials tend to be oriented more toward the development of their advanced and superior properties and in this context the concept of nanostructural design plays an important role. In addition to grain refinement down to the nanometer range, grain boundary structure engineering is also important because boundaries having different structures can exhibit specific transport mechanisms, in terms of deformation and diffusion, and this can be used to control the properties [28,30]. This opens up the potential for developing new ways for improving the properties of various ultrafine-grained materials.

Attention to commercial production of nanostructured metals is being motivated by the increasing maturity of SPD technology and the ever-growing number of published reports on the notable properties achieved in a widening range of alloys. Companies are seeking to evaluate or develop nanostructured alloys to use in existing and prospective new products, often working with the assistance of university labs or research institutes that conduct SPD research. For example, the Ufa State Aviation Technical University established a spin-off company NanoMet Ltd. already in 2009. The company's objective was to organize in Ufa the first in the world pilot-commercial production of semi-products (rods) from nanostructured titanium for medical applications. Since 2010, NanoMeT has been commercially producing semi-products (rods) for nanostructured titanium for medical applications. In 2012 NanoMeT received the international ISO 9001:2008 certificate for its quality management system. To date, NanoMeT has mastered the production of long-length rods (over 2.5 m in length) with a diameter of 3-8 mm. The estimated production capacity is 2 tons per year [31].

The development of commercial production methods is being also systematically addressed at the Nanostructured Metals Manufacturing Testbed facility established in Golden, Colorado, in association with the Colorado School of Mines. This facility has multiple SPD machines networked together to transition research scale SPD processing into pilot scale production. Scale up of production rates and billet sizes have been demonstrated for alloys of magnesium, aluminium, titanium, copper, and iron [32]. Product development effects within this pilot scale manufacturing environment show the roles of enabling technologies such as on-machine process monitoring, design of tooling for high pressure, specialized lubrication schemes, and non-isothermal forming processes.

Acknowledgements

The authors express their sincere gratitude to colleagues and members of the International NanoSPD Steering Committee – Professors Yuri Estrin, Zenji Horita, Michael J. Zehetbauer, Yuntian T. Zhu,

Roberto B. Figueiredo, Hyoung Seop Kim, Laszlo Toth, Gerhard Wilde, Terry Lowe – for their kind support and assistance in writing this paper. Special thanks are due to Prof. Yuntian Zhu for his dedicated work in maintaining and providing technical support for the www.nanospd.org website. Also, the authors are grateful to the Russian Ministry of Education and Science for support through Grant 14.B25.31.0017 (RZV) and the European Research Council for support through Grant Agreement No. 267464-SPDMETALS (TGL).

References

[1] Langdon TG 2013 Twenty five years of ultrafine-grained materials: Achieving exceptional properties through grain refinement *Acta Mater* **61** 7035

[2] Edalati K, Horita Z 2016 A review on high-pressure torsion (HPT) from 1935 to 1988 *Mater Sci Eng A* **652** 325

[3] Wang JT 2006 Historic retrospection and present ststus of severe plastic deformation in China *Mater Sci Forum* **503-504** 363

[4] Srinivasan S and Ranganathan S (2004) *India's Legendary Wootz Steel: An Advanced Material of the Ancient World* (Bangalore, India: National Institute of Advanced Studies and IISc).

[5] Bridgman PW 1935 Effects of high shearing stress combined with high hydrostatic pressure *Phys Rev* **48b 825**

[6] Valiev RZ, Kaibyshev OA, Kuznetsov RI, Musalimov RSh and Tsenev NK 1988 Low-temperature superplasticity of metallic materials *Dokl Akad Nauk SSSR* **301** 864.

[7] Valiev RZ, Krasilnikov NA and Tsenev NK 1991 Plastic-deformation of alloys with submicron-grained structure *Mater Sci Eng A* **137** 35

[8] Wang J, Horita Z, Furukawa M, Nemoto M, Tsenev NK, Valiev RZ, Ma Y and Langdon TG 1993 An investigation of ductility and microstructural evolution in an Al-3% Mg alloy with submicron grain size *J Mater Res* **8** 2810

[9] Langdon TG 2010 The impact of bulk nanostructured materials in modern research *Rev Adv Mater Sci* **25** 11

[10] Langdon TG 2012 The current status of bulk nanostructured materials *Rev Adv Mater Sci* **31** 1

[11] Valiev RZ and Langdon TG 2014 Report of international NanoSPD steering committee and statistics on recent NanoSPD activities *IOP Conf. Series: Mater Sci Eng* **63** 011002

[12] Valiev RZ, Islamgaliev RK, Alexandrov IV 2000 Bulk nanostructured materials from severe plastic deformation *Prog Mater Sci* **45** 103

[13] Gleiter H 1989 Nanocrystalline materials *Prog Mater Sci* **33** 223

[14] Valiev RZ and Langdon TG 2006 Principles of equal-channel angular pressing as a processing tool for grain refinement *Prog Mater Sci* **51** 881

[15] Meyers MA, Mishra A and Benson DJ 2006 Mechanical properties of nanocrystalline materials *Prog Mater Sci* **51** 427

[16] Gleiter H 2000 Nanostructured materials: Basic concepts and microstructure *Acta Mater* **48** 1

[17] Kumar KS, Van Swygenhoven H, Suresh S 2003 Mechanical behavior of nanocrystalline metals and alloys *Acta Mater* **51** 5743

[18] Saito Y, Utsunomiya H, Tsuji N and Sakai T 1999 Novel ultra-high straining process for bulk materials – Development of the accumulative roll-bonding (ARB) process *Acta Mater* **47** 579

[19] Iwahashi Y, Horita Z, Nemoto M and Langdon TG 1998 The process of grain refinement in equal-channel angular pressing *Acta Mater* **46** 3317

[20] Iwahashi Y, Wang J, Horita Z, Nemoto M and Langdon TG 1996 Principle of equal-channel angular pressing for the processing of ultra-fine grained materials *Scripta Mater* **35** 143

[21] Saito Y, Tsuji N, Utsunomiya H, Sakai T and Hong RG 1998 Ultra-fine grained bulk aluminum produced by accumulative roll-bonding (ARB) process *Scripta Mater* **39** 122

[22] Tsuji N, Ito Y, Saito Y and Minamino Y 2002 Strength and ductility of ultrafine grained

aluminum and iron produced by ARB and annealing *Scripta Mater* **47** 893
[23] Mukai T, Yamanoi M, Watanabe H and Higashi K 2001 Ductility enhancement in AZ31 magnesium alloy by controlling its grain structure *Scripta Mater* **45** 89
[24] Nieh TG and Wadsworth J 1991 Hall-Petch relation in nanocrystalline solids *Scripta Metall Mater* **25** 955
[25] Segal VM 1995 Materials processing by simple shear *Mater Sci Eng A* **197** 157
[26] Valiev RZ, Korznikov AV and Mulyukov RR 1993 Structures and properties of ultrafine-grained materials produced by severe plastic-deformation *Mater Sci Eng A* **168** 141
[27] Furukawa M, Iwahashi Y, Horita Z, Nemoto M and Langdon TG 1998 The shearing characteristics associated with equal-channel angular pressing *Mater Sci Eng A* **257** 328
[28] Valiev RZ, Estrin Y, Horita Z, Langdon TG, Zehetbauer MJ and Zhu YT 2006 Producing bulk ultrafine-grained materials by severe plastic deformation *JOM* **58**(4) 33
[29] Ma E 2006 Eight routes to improve the tensile ductility of bulk nanostructured metals and alloys *JOM* **58**(4) 49
[30] Valiev RZ, Estrin Y, Horita Z, Langdon TG, Zehetbauer MJ, Zhu YT 2016 Fundamentals of superior properties in bulk nanoSPD materials *Mater Res Lett* **4** 1
[31] Reshetnikova T, Shcherbakov A, Salakhova M, Khakimova L 2011 Towards Perspective Applications of Nanostructured Ti in Medicine *Mater Sci Forum* ***667-669*** *1201*
[32] Lowe TC, Davis CF, Rovira PM, Hayne ML, Campbell GS, Grzenia JE, Stock PJ, Meagher RC, Rack HJ 2017 Scientific and technological foundations for scaling production of nanostructured metals *IOP Conf. Series, Mater Sci Eng.* (in press for nanoSPD7)

IOP Conf. Series: Materials Science and Engineering **194** (2017) 012002 doi:10.1088/1757-899X/194/1/012002

Fabrication of hybrid metal systems through the application of high-pressure torsion

Megumi Kawasaki[1,2,*], Dae-Kuen Han[1], Jae-Kyung Han[1], Jae-il Jang[1], Terence G Langdon[2,3]

[1]Division of Materials Science and Engineering, Hanyang University, Seoul 133-791, South Korea
[2]Departments of Aerospace & Mechanical Engineering and Materials Science, University of Southern California, Los Angeles, CA 90089-1453, U.S.A.
[3]Materials Research Group, Faculty of Engineering and the Environment, University of Southampton, Southampton SO17 1BJ, U.K.

*megumi@hanyang.ac.kr

Abstract. This presentation demonstrates a simple and very rapid synthesis of metal matrix nanocomposites (MMNCs) in Al-based hybrid systems which are achieved by processing stacked disks of two pure metals through the application of high-pressure torsion (HPT) at ambient temperature. These synthesized hybrid systems exhibit exceptionally high hardness through rapid deformation-induced diffusion and the simultaneous formation of different intermetallic compounds. The experiments also show further improvement in mechanical properties is available by increasing the numbers of HPT turns and additional short-term annealing. Thus, by demonstrating the fabrication of the Al-Mg and Al-Cu hybrid systems, the present paper suggests a potential for simply and expeditiously fabricating a wide range of MMNCs through HPT.

1. Introduction

The light-weight metals including aluminum, magnesium and cupper are widely used for structural applications in the automotive, aerospace and electronic industries but improvements in the mechanical properties of these metals would be attractive for enhancing their future use. In practice, an earlier study on an aerospace-grade Al-7075 alloy reported an increase in the strength limit of the alloy by processing using high-pressure torsion (HPT) while maintaining reasonable formability [1]. Nevertheless, it is reasonable to anticipate that there is probably an upper limit in the maximum achievable mechanical properties when the processing is conducted directly on the alloy. This suggests a new strategy for achieving superior properties of materials by bonding dissimilar metals and synthesizing new metal systems.

Bulk nanostructured materials with ultrafine grains (UFG) can be fabricated through the application of severe plastic deformation (SPD) which is a promising technique for achieving significant grain refinement in bulk metals [2]. Among the reported SPD techniques, one of the most attractive methods refers to processing by HPT where this type of processing leads to exceptional grain refinement that is not generally achieved using other procedures [3]. The processed metals generally demonstrate an enhancement in the physical and mechanical characteristics through significant grain refinement at room temperature (RT) and the intensive introduction of point and line defects [4]. Because of the introduction of intense plastic straining during processing, HPT has been applied also for the bonding of machining chips [5,6] and the consolidation of metallic powders [7-9].

Accordingly, a new approach of applying conventional HPT processing was studied for the formation of Al-Mg and Al-Cu hybrid systems and ultimately attaining a metal matrix nanocomposite (MMNC) from separate Al, Mg and Cu disks through diffusion bonding at RT under high pressure. Specifically, the present reports describe the unique microstructure, hardness distributions and the micro-mechanical response for an Al-Mg hybrid system synthesized by HPT. A study of post-deformation annealing (PDA) was applied to demonstrate the improvement in plasticity by an increase in the strain rate sensitivity, *m*, in the synthesized alloy system. Moreover, the feasibility of the HPT processing in the preparation of new alloy systems is demonstrated by forming an Al-Cu alloy system recording very high hardness due to significant grain refinement during HPT.

2. Fabrication of an Al-Mg hybrid system

A commercial purity Al-1050 alloy and a ZK60 magnesium alloy were used for the experiments. The extruded bars of the alloys having a diameter of 10 mm were cut into billets with lengths of ~65 mm and a number of disks was sliced from the billets and polished to achieve uniform thicknesses of ~0.83 mm. The direct bonding of the Al and Mg disks was performed through the conventional HPT procedure at RT using a quasi-constrained facility under a hydraulic pressure of 6.0 GPa for 1, 5, 10 and 20 turns at a rotational speed of 1 rpm. In particular, a set of the separate Al and Mg disks were placed in the depression on the lower anvil in the order of Al/Mg/Al where the Mg disk was positioned between the two Al disks but without using any glue or metal brushing treatment. A schematic illustration of the sample set-up between the conventional HPT anvils is shown in Figure 1(a) [10,11].

Figure 1(b) shows overviews of the microstructure taken at the cross-sections of the Al-Mg disks by optical microscopy (OM) after HPT for 1, 5 and 10 turns from the top, respectively, where the bright regions denote the Al-rich phase and the dark regions correspond to the Mg-rich phase in these micrographs [10-12].

A disk after 1 turn by HPT showed a multi-layered structure with fragmented Mg layers with thicknesses of ~200 μm without any segregation of Al and Mg phases throughout the disk diameter. A similar microstructure consisting of multi-layers of the Al and Mg phases was observed at the central regions at r <2.0 and <1.0 mm of the disks after 5 and 10 turns, respectively, where r denotes the radius of the HPT disk. However, the disk peripheries at r >2.5 mm after 5 turns demonstrated a unique microstructure involving a homogeneous distribution of very fine Mg phases having thicknesses of ~5-10 μm to even true nano-scale sizes of ~100-500 nm within the Al matrix. Furthermore, the Mg phases disappeared at the disk edge and thus there was no evidence of visible Mg phases at ~3< r <5 mm after 10 turns.

The distributions of Vickers microhardness was examined over the vertical cross-sections of the processed disks and the data set was visualized by constructing color-coded hardness contour maps as shown in Figure 1(c) for 1, 5 and 10 turns from the top, respectively, where the hardness values are indicated in the key on the right. For reference, the Al-1050 alloy and the ZK60 alloy show a saturation hardness of Hv ≈ 65 and ~110 across the disk diameters after HPT for 5 turns providing sufficient torsional straining.

The total cross-section after HPT for 1 turn shows an average microhardness value of ~70. This is similar to the value of ~65 for the base material of the Al-1050 alloy after HPT for 5 turns and this hardness value remains constant at the centers at r <2.5 mm of the Al-Mg disks up to 10 turns. However, additional HPT turns to 5 introduced high hardness with a maximum of Hv ≈ 130 in the peripheral region where the fine Mg phase is homogeneously distributed within the Al matrix. Moreover, there is a significant increase in Hv after 10 turns where a maximum hardness of ~270 was recorded at the peripheral region at r >3.0 mm. These high hardness values measured in the Al-Mg system after HPT are much higher than the base alloys of Al and Mg after HPT and a detailed microstructural analysis is necessary to understand the hardening mechanism of this unique Al-Mg system produced by HPT processing.

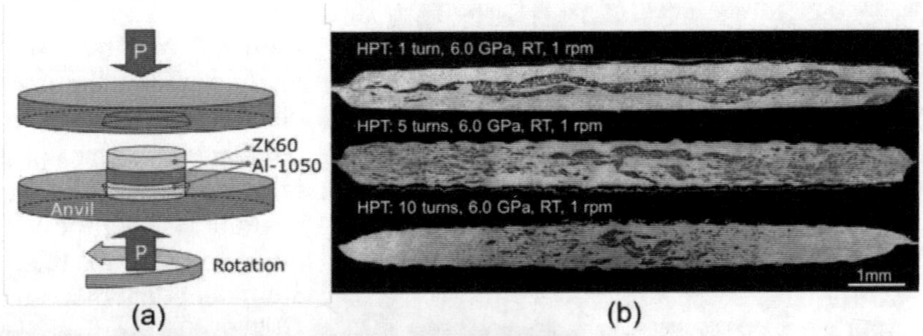

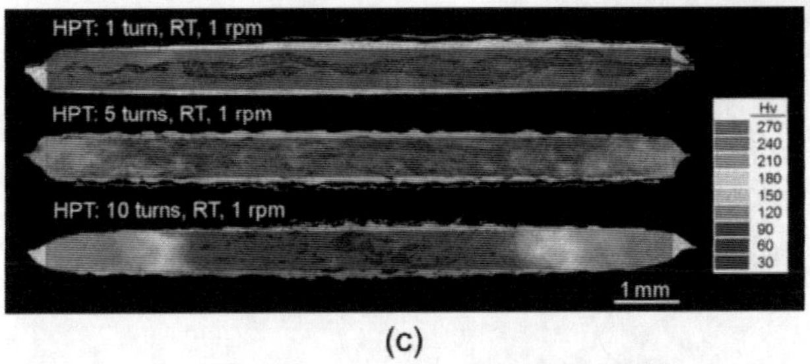

Figure 1. (a) Schematic illustration of the sample set-up for HPT processing and (b) an overview of the microstructure and (c) color-coded hardness contour maps taken at the vertical cross-sections along the disk diameters after HPT for 1, 5 and 10 turns, respectively [10-12].

Processing by HPT was continued up to 20 turns to evaluate the possibility of achieving further improvement in hardness. Moreover, PDA at 573 K for 1 hour was applied to investigate the change in microstructure and the enhancement in mechanical properties in the Al-Mg alloy system. There are several recent reports demonstrating the significance of PDA on the mechanical properties at RT of HPT-processed materials [13-15]. Microstructures taken at the vertical cross-sections and the color-coded hardness contour maps are shown in Figure 2(a) and (b), respectively, for the Al-Mg disks after HPT at RT for 20 turns (upper) and after HPT followed by PDA at 573 K for 1 hour (lower) [16].

It is apparent that the deformed microstructure after HPT for 20 turns is reasonably similar to the sample conditions in the Al-Mg system after HPT for 10 turns as shown in Figure 1(b) where there are multi-layered Al and Mg phases at the disk center but there is a complete dissolution of Mg in the Al matrix. A similar microstructure was observed after PDA whereas the outer region involving the complete mixture of Mg within Al was reduced. Accordingly, the synthesized Al-Mg system after HPT and after PDA consists of gradient-type microstructures across the disk diameters.

There is a drastic increase in hardness in the Al-Mg disk after HPT for 20 turns. Specifically, a Vickers microhardness of Hv $\approx$ 60 and slightly higher was observed at the central region of the disk at $r < 2.5$ mm and then a transition in hardness to ~150-240 was measured at $r \approx 2.5$-3.5 mm followed by an exceptionally high hardness of Hv $\approx$ 330 at $r \approx 4.0$-5.0 mm. This high hardness is significantly higher than the highest value of Hv $\approx$ 270 observed in the Al-Mg disk processed by HPT for 10 turns as shown in Figure 1(c). After PDA, there was a slight reduction in hardness to Hv $\approx$ 30 in the central regions up to $r \approx 3.0$ mm and a reduction to Hv $\approx$ 220 at the outer region of the disk at $r \approx 4.0$-5.0 mm. Thus, the changes in microstructure correlate directly with the values of hardness in the Al-Mg system after HPT and there is a significant hardness variation in the synthesized Al-Mg through the HPT processing. Accordingly, the detailed microstructures at the disk peripheries demonstrating exceptional hardness were examined by TEM and the microstructures are displayed in Figure 3 for the samples after (a)-(b) 20 HPT turns and (c) HPT followed by PDA. In addition, the XRD profiles for these two samples are demonstrated in Figure 3(d) and (e), respectively.

As seen in Figure 3(a) and (b), there is a mixture of regions with a layered structure and an equiaxed microstructure at the disk edge immediately after HPT for 20 turns. In practice, the layered microstructure has an average thickness of ~20 nm and these layers contain numerous dislocations which subdivide the layers in a vertical sense. The equiaxed grains showed an average grain size of $d \approx 60$ nm. By contrast, it is apparent from Figure 3(c) that after PDA the HPT-processed Al-Mg system contained a homogeneous equiaxed microstructure with an average grain size of $d \approx 380$ nm.

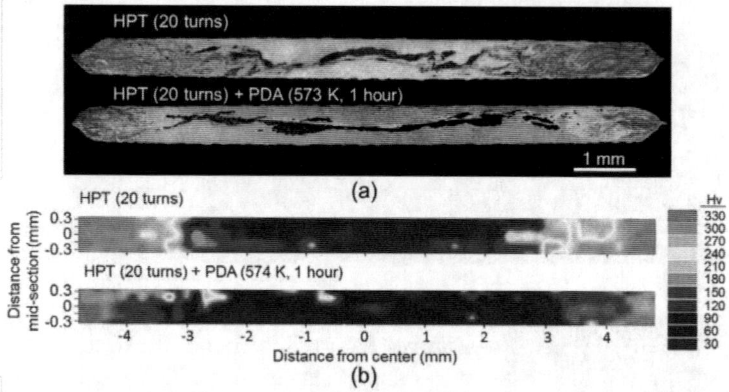

Figure 2. (a) The OM micrographs of the Al-Mg disks after HPT at room temperature under a pressure of 6.0 GPa and (b) color-coded contour maps of the Vickers microhardness for the vertical cross-sectional planes after HPT for 20 turns (upper) and after HPT and PDA (lower), respectively. [16]

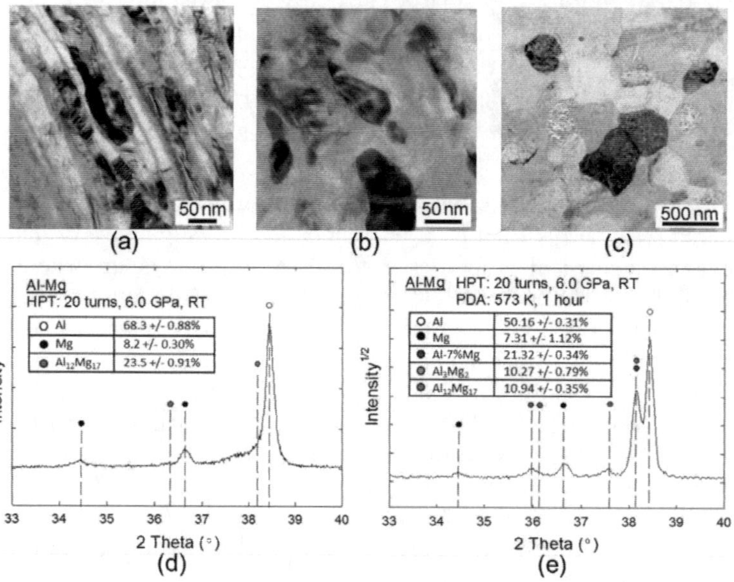

Figure 3. Representative TEM bright-field images taken at the disk edges after (a)-(b) HPT for 20 turns and (c) HPT followed by PDA in the Al-Mg system and the X-ray diffraction profiles for the disk edges of the Al-Mg system after (d) HPT for 20 turns and (e) HPT and PDA [16].

The results of the X-ray analysis are shown in Figure 3 (d) and (e) where additional compositional analysis based on the X-ray profile through MOUD was displayed as a table in each plot. It should be noted that the samples at the disk edges were carefully prepared by removing the central regions so that the measurements focused only on the peripheral regions where Mg was completely dissolved into the Al matrix after HPT for 20 turns and after PDA as seen in Figure 2 whereas the inevitable concentrations of the Mg-rich phase existed close to the mid-radius of the processed samples.

The disk edge immediately after HPT for 20 turns showed there is evidence of a γ-$Al_{12}Mg_{17}$ intermetallic compound in the Al matrix whereas after PDA there is an Al-7% Mg solid solution phase with two different β-Al_3Mg_2 and γ-$Al_{12}Mg_{17}$ intermetallic compounds. Thus, it is established that processing by HPT for 20 turns and additional PDA produced two different types of deformation-induced MMNCs containing intermetallic compounds at the disk edges of the Al-Mg system.

The synthesis of intermetallic compounds at RT is feasible due to the accelerated diffusivity of Mg atoms into the Al matrix that is attributed to SPD under high pressure [10]. Several recent studies demonstrated experimental evidence for enhanced atomic diffusion in nanostructured materials processed by ECAP [17] and HPT [10,18]. Moreover, a recent review describes the significance of the fast atomic mobility during SPD by recognizing the significant increase in the vacancy concentration through SPD processing [19]. Accordingly, the present experimental results also anticipate that the formation of these intermetallic compounds provides an excellent potential for reinforcing the Al matrix by improving the hardness and strength.

3. Micro-mechanical response of the MMNCs in the Al-Mg system

The micro-mechanical response was examined by applying the nanoindentation technique for the disk edges consisting of MMNCs in the Al-Mg system after HPT for 20 turns and subsequent PDA. The analysis used a nanoindentation facility, Nanoindenter-XP (formerly MTS; now Agilent, Oak Ridge, TN) with a three-sided pyramidal Berkovich indenter having a centerline-to-face angle of 65.3°. More than 15 indentations were conducted at each specific phase at the measured locations to provide statistically valid data. All measurements were conducted under a predetermined peak applied load of P_{max} = 50 mN at constant indentation strain rates of 0.0125, 0.025, 0.05 and 0.1 s^{-1} which are equivalent to general strain rates of 1.25×10^{-4}, 2.5×10^{-4}, 5.0×10^{-4} and 1.0×10^{-3} s^{-1} calculated through an empirical relationship [20].

Figure 4 shows representative load-displacement curves for the Al-Mg disk edges after (a) HPT for 20 turns and (b) HPT and PDA when measuring at four strain rates [16]. The disk edge after HPT for 20 turns demonstrated no strain rate dependency where all four load-displacement curves placed in reasonably consistent locations as shown in Figure 4(a). However, there is a slight tendency of less displacement when the strain rate is slower, thereby indicating a negative strain rate dependency of plasticity. By contrast, the Al-Mg system after HPT followed by PDA demonstrated an apparent positive strain rate dependency where increasing displacements were achieved at slower strain rates of nanoindentation as shown in Figure 4(b). It is also apparent by comparing the load-displacement curves between the two samples that the MMNC immediately after HPT shows much lower displacements than the MMNC after HPT and PDA at all strain rates, thereby demonstrating the high hardness of the MMNC in the Al-Mg disk edge immediately after HPT for 20 turns. This result is fully consistent with the Vickers hardness measurements as shown in Figure 2(b).

Considering Tabor's empirical prediction where the flow stress is equivalent to one-third of hardness, H, measured through nanoindentaion testing for fully-plastic deformation at a constant strain rate, the strain rate sensitivity, *m*, was determined from the slope of the line for each examined sample in a logarithmic plot of H/3 against the strain rate as shown in Figure 4(c) for the disk edges of the Al-Mg system after HPT for 20 turns and after HPT followed by PDA [16].

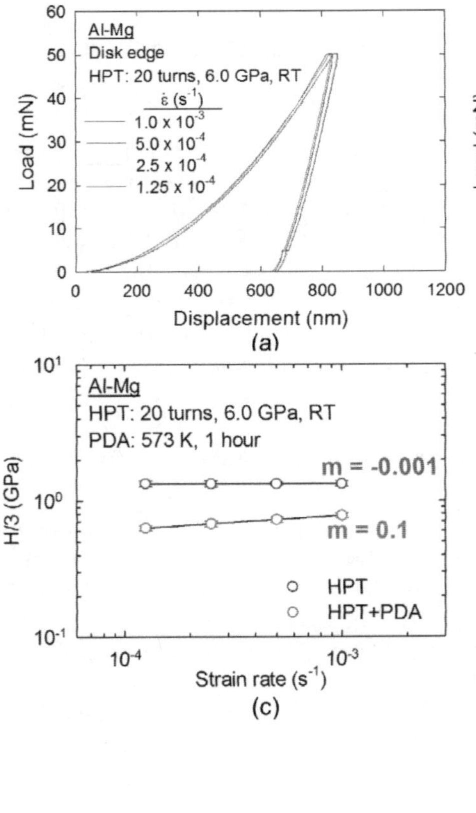

Figure 4. Representative load-displacement curves for the Al-Mg disk edges after (a) HPT for 20 turns and (b) HPT and PDA when measuring at four strain rates at 1.0×10^{-3} - 1.25×10^{-4} s^{-1} under a predetermined maximum peak load of 50 mN and (c) variations in the strain rate sensitivity with increasing strain rate for the disk edges of the Al-Mg system after HPT for 20 turns and after HPT and PDA [16].

The analysis estimated the *m* values of –0.001 and 0.1 for the MMNCs at the Al-Mg disk edges immediately after HPT and after HPT followed by PDA, respectively. It has been well described that there is generally a significant loss of ductility in bulk nanostructured materials where a marked increase in hardness and strength is achieved by grain refinement through SPD [21-23]. Specifically, the lower overall ductility in the UFG materials is attributed to an inter-relationship between the lower strain hardening and an increasing strain rate sensitivity [24,25]. Nevertheless, the present nanoindentation analysis shows that there is a significant increase in the strain rate sensitivity in the HPT-induced MMNC after PDA. Thus, a PDA treatment is a feasible procedure for enhancing the RT plasticity in the MMNC in the Al-Mg hybrid system synthesized by HPT while maintaining reasonably high hardness as shown in Figure 2(b).

The significance of PDA was demonstrated earlier for improving the overall ductility of nanostructured Ti after HPT [13]. A recent review demonstrated that a PDA treatment produces an ordering of the defect structures within the grain boundaries leading to an equilibrium state without any significant grain growth [23]. In addition, short-term annealing reduces the dislocation density in the grain interior of the UFG material after SPD so that the dislocation storage capability may increase and thus the strain hardening capability is enhanced leading to the possibility of high ductility in the SPD-processed material. Accordingly, by the addition of a PDA treatment, the MMNC in the Al-Mg system produced by HPT has a great potential for demonstrating both high hardness and superior ductility.

4. Future potential of the HPT technique for synthesizing new metal systems

There are only limited numbers of demonstrations to date of the fabrication of hybrid material showing high mechanical performance by bonding dissimilar bulk metals to form new metal systems through HPT. A first report demonstrated a solid-state reaction in an Al-Cu system through the bonding of semi-circular half-disks of commercial purity Al and Cu using HPT at ambient temperature for up to 100 turns [26]. Thereafter, a similar approach was applied for forming a spiral texture by processing of an Al-Cu hybrid material though HPT where four quarter-disks, including two of pure Cu and two of an Al-6061 alloy, were positioned to make a complete disk and then processed by HPT at RT for 1 turn [27]. However, this study described only the computational calculation of the distribution of equivalent stress in the processed Al-Cu disk using a finite element method and there was no detailed microstructural analysis of the processed disk.

Accordingly, a simpler approach for utilizing conventional HPT processing for the bonding of separate Al and Mg disks was processed by stacking a set of two disks for up to 20 turns [28] and stacking three disks for 5-10 turns [10-12] and for 20 turns [16] for producing bi-layered and multi-layered structures, respectively, in an Al-Mg system through HPT at RT. It should be emphasized that the MMNCs synthesized at the disk edges of the Al-Mg system receive very high straining through HPT but in the present study the final bulk metal contains a gradient-type microstructure or a heterogeneous nanostructure which involves gradations of not only grain size but also compositions [29]. This is a new category of bulk structure in engineering materials and it is expected to lead to a significant potential for exhibiting excellent mechanical properties and functionalities which extend the fields of applications of the HPT-induced hybrid materials.

For examining the feasibility of the HPT processing with the simple sample set-up for fabricating new metal systems, a further examination was conducted for producing an Al-Cu system through 10, 20 and 60 turns by HPT under 6.0 GPa at RT from the commercial purity Al and Cu. Overviews of the HPT-processed Al-Cu metal disks are shown in Figure 5(a) for 10 and 60 turns. It should be noted that at the central regions in both disks the bright regions denote the Al-rich phase and the dark regions correspond to the Cu-rich phase whereas the grey color at the overall disk edges may describe a complete mixture of Al and Cu as was demonstrated in the Al-Mg system.

There is a consistent trend of microstructural evolution between the Al-Cu and the Al-Mg systems during HPT with increasing numbers of turns. Thus, due to diffusion bonding of the Al and Cu phases, the Al-Cu disk obtained a multi-layered microstructure in a wide region from the disk center towards *r*

IOP Conf. Series: Materials Science and Engineering **194** (2017) 012002 doi:10.1088/1757-899X/194/1/012002

$\approx$ 3-4 mm and a mixture of very fine Cu phases within the Al matrix in the remainder of the region at the disk edge after 10 HPT turns. Additional HPT through 60 turns demonstrated a significant influence to reduce the multi-layered region at the disk center to r <2 mm and the wide peripheral region showed a complete mixture of Al and Cu.

Representative TEM micrographs are shown in Figure 5(b) and (c) for the Al-Cu disk edges after HPT for 20 and 60 turns, respectively. Processing by HPT for 20 turns produced a layered nanostructure with an average thickness of ~20 nm where the microstructure consists of a simple mixture of the Al-rich phase with a brighter color and the Cu-rich phase with a darker color as shown in Figure 5(b). By contrast, after 60 HPT turns the Cu is completely dissolved into the Al matrix and there was no evidence of the Cu-rich phase at the disk edge. Significant grain refinement was achieved so that true nano-scale grains with an average grain size of ~30 nm were observed in an equiaxed microstructure as shown in Figure 5(c).

The Vickers microhardness was recorded at the mid-height along the disk diameters of the Al-Cu disks after HPT for 10, 20 and 60 turns and the results are shown in Figure 5(d) where the hardness of the base materials of Al and Cu after 10 HPT turns are denoted by the dashed lines at Hv $\approx$ 65 and ~150, respectively. The central regions of r <3-4 mm after 10 turns and r <1 mm after 60 turns where the areas hold the layered microstructure as seen in Figure 5(a) demonstrated the lower hardness values which are consistent with the hardness of the base material of Cu when processed separately by HPT for 10 turns. The hardness at the peripheral regions recorded exceptionally high hardness values of Hv $\approx$ 250, 400 and 500 with increasing numbers of turns to 10, 20 and 60, respectively.

These results on the Al-Cu system were obtained from preliminary examinations and further studies are necessary to investigate the detailed compositional changes through the diffusion bonding during HPT, the strengthening mechanism and the functionalities of the synthesized system after HPT. Nevertheless, the exceptionally high hardness is greater than the hardness which is achievable by grain refinement on the separate base metals and thereby it indicates the synthesis of new phases exhibiting excellent hardness in the HPT-processed Al-Cu system. Thus, the present study demonstrates a considerable potential for applying HPT processing for the synthesis of new alloy systems containing gradient-type microstructures, especially by simply and expeditiously fabricating a wide range of MMNCs from simple metals and alloys and by incorporating an additional treatment of PDA to further improve the mechanical properties of these HPT-processed materials.

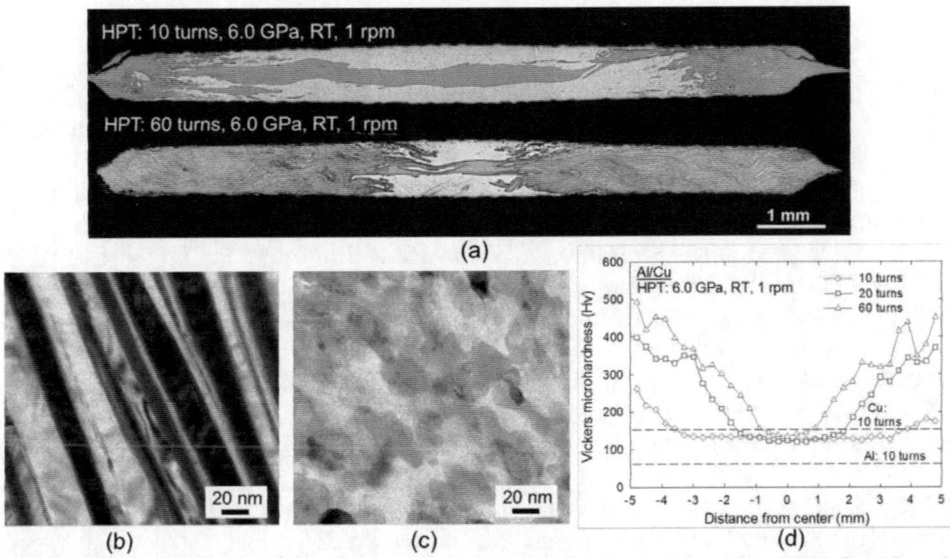

Figure 5. (a) an overview of the microstructure of the Al-Cu system after HPT for 10 and 60 turns, TEM micrographs at the disk edges of the Al-Cu system after HPT for (b) 20 turns and (c) 60 turns and (d) Vickers microhardness variations of the Al-Cu disks after HPT.

5. Summary and conclusions

The synthesis of an MMNC by forming intermetallic compounds was demonstrated in an Al-Mg system using conventional HPT processing for 20 turns at RT through the diffusion bonding. The significance of applying PDA was demonstrated by improving the strain rate sensitivity leading to an excellent potential for achieving high plasticity in the Al-Mg system. Moreover, a potential for making use of HPT at ambient temperature was demonstrated for synthesizing an Al-Cu system exhibiting exceptionally high strength after HPT through 60 turns. Additional results are now needed to confirm the feasibility of using this approach for other metallic systems.

Acknowledgements

This work was supported by the NRF Korea funded by MoE under Grant No. NRF-2016R1A6A1A03013422 and by MSIP under Grant No. NRF-2016K1A4A3914691 (MK); the NRF Korea funded by MSIP under Grant No. NRF-2015R1A5A1037627 (JIJ); and the European Research Council under ERC Grant Agreement No. 267464-SPDMETALS (TGL).

References

[1] Liddicoat PV, Liao XZ, Zhu YT, Zhao YH, Lavernia EJ, Murashkin MY, Valiev RZ and Ringer SP 2010 *Nature Commun.* **1** 63/1.
[2] Valiev RZ, Islamgaliev RK and Alexandrov IV 2000 *Prog. Mater. Sci.* **45** 103.
[3] Zhilyaev AP and Langdon TG 2008 *Prog. Mater. Sci.* **53** 893
[4] Langdon TG 2013 *Acta Mater.* **61** 7035.
[5] Zhilyaev AP, Gimazov AA, Raab GI and Langdon TG 2008 *Mater. Sci. Eng. A* **486** 123.
[6] Edalati K, Yokoyama Y and Horita Z 2010 *Mater. Trans.* **51** 23.
[7] Korznikov AV, Safarov IM, Laptionok DV and Valiev RZ 1991 *Acta Metall. Mater.* **39** 3193.
[8] Edalati K, Horita Z, Fujiwara H and Ameyama K 2010 *Metall. Mater. Trans. A* **41A** 3308.
[9] Zhilyaev AP, Ringot G, Huang Y, Cabrera JM and Langdon TG 2017 *Mater. Sci. Eng. A* **688** 498.
[10] Ahn B, Zhilyaev AP, Lee H-J, Kawasaki M and Langdon TG 2015 *Mater. Sci. Eng. A* **635** 109.
[11] Kawasaki M, Ahn B, Lee H-J, Zhilyaev AP and Langdon TG 2015 *J. Mater. Res.* **31** 88.
[12] Ahn B, Lee H-J, Choi I-C, Kawasaki M, Jang J-I and Langdon TG 2016 *Adv. Eng. Mater.* **18** 1001.
[13] Valiev RZ, Sergueeva AV and Mukherjee AK 2003 *Scripta Mater.* **49** 669.
[14] Huang Y, Lemang M, Zhang NX, Pereira PHR and Langdon TG 2016 *Mater. Sci. Eng. A* **655** 60.
[15] Shahmir H, He J, Lu Z, Kawasaki M and Langdon TG 2016 *Mater. Sci. Eng. A* **676** 294.
[16] Han J-K, Lee H-J, Jang J-i, Kawasaki M and Langdon TG 2017 *Mater. Sci. Eng. A* **684** 318.
[17] Divinski SV, Reglitz G, Rösner H, Estrin Y and Wilde G 2011 *Mater. Sci. Eng. A* **59** 1974.
[18] Lee D-H, Choi I-C, Seok M-Y, He J, Lu Z, Suh J-Y, Kawasaki M, Langdon TG and Jang J-i 2015 *J. Mater. Res.* **30** 2804.
[19] Sauvage X, Wilde G, Divinski SV, Horita Z, Valiev RZ 2012 *Mater. Sci. Eng. A* **540** 1.
[20] Lucas BN and Oliver WC 1999 *Metall. Mater. Trans. A* **30A** 601.
[21] Valiev RZ, Alexandrov IV, Zhu YT and Lowe TC 2002 *J. Mater. Res.* **17** 5.
[22] Valiev R 2002 *Nature* **419** 887.
[23] Kumar P, Kawasaki M and Langdon TG 2016 *J. Mater. Sci.* **51** 7.
[24] Zhu YT and Liao XZ 2004 *Nat. Mater.* **3** 351.
[25] Valiev RZ, Estrin Y, Horita Z, Langdon TG, Zehetbauer MJ and Zhu YT 2016 *Mater. Res. Lett.* **4** 1.
[26] Oh-ishi K, Edalati K, Kim HS, Hono K and Horita Z 2013 *Acta Mater.* **61** 3482.
[27] Bouaziz O, Kim HS and Estrin Y 2013 *Adv. Eng. Mater.* **15** 336.
[28] Qiao X, Li X, Zhang X, Chen Y, Zheng M, Golovin IS, Gao N and Starink MJ 2016 *Mater. Lett.* **181** 187.
[29] Kang JY, Kim JG, Park HW and Kim HS 2016 *Sci. Rep.* **6** 26590/1.

IOP Conf. Series: Materials Science and Engineering **194** (2017) 012003 doi:10.1088/1757-899X/194/1/012003

Characterization of UFG materials by X-ray and Neutron diffraction

T Ungár[1,2], G Ribárik[1]

[2] Department of Materials Physics, Eötvös University, H-1117 Budapest, Hungary
[3] Materials Performance Centre, School of Materials, The University of Manchester, Manchester M13 9PL, UK

Corresponding author: ungar@ludens.elte.hu

Abstract. Grain size, subgrain size, microstrain, dislocation density and dislocation types and intergranular strains determine the mechanical, chemical and other physical properties in materials of submicron grain size. Electron microscopy is a straightforward method to determine the substructure. X-ray and neutron diffraction methods supplement dynamic properties, especially strains and stresses and number densities of substructure elements. The synergy of electron microscopy and diffraction patterns provides a more comprehensive characterization of the substructure and a better understanding of the performance of ultra-fine-grain (UFG) materials.

1. Introduction

X-ray and neutron diffraction line-profile-analysis (LPA) proved to be a powerful tool for determining the sub-structure of crystalline materials in terms of (i) grain size or subgrain size, (ii) dislocation density and dislocation type, (iii) twinning and faulting and (iv) other types of lattice defects. The convolutional multiple whole profile (CMWP) method of LPA procedure is based on physically modeled profile functions of the different microstructural elements, and is available as a free software package. In the present work, a few of applications to determine the dislocation structure in severe plastic deformed bulk nanocrystalline materials will follow a brief summary of the diffraction procedure. At the end, the paper provides a comprehensive list of relevant references.

2. Basic principles of line profile analysis

Within the frame of the kinematical scattering theory, the intensity distribution diffracted from a large defect-free crystal is similar to a δ function, i.e. its breadth is zero, and however its maximum is finite. When (i) the scattering crystal becomes smaller than about a micron [1-4], or when it contains (ii) dislocations [5-8], (iii) planar defects [9-12], (iv) chemical inhomogeneities [13], (v) lattice strains [14] or other types of lattice defects [15], the diffraction peaks broaden. There are two fundamentally different broadening types: (1) size-broadening and (2) strain broadening. Size broadening is diffraction order independent. The broadening of diffraction peaks stays globally constant in reciprocal space. Strain broadening, however, increases, also in a global manner as a function of distance from the origin of the reciprocal space. Both types of broadening can vary for different *hkl*-s. In the case of size broadening the *hkl* variation is caused by shape anisotropy [16,17], whereas in the case of strain broadening the *hkl* dependence is produced by strain anisotropy [7,18]. Strain anisotropy in line broadening means that neither the breadths in a Williamson-Hall plot [19] nor the Fourier coefficients in the Warren-Averbach method [2] increase or decrease monotonously as a function of g or g^2, respectively, where g is the absolute value of the diffraction vector. A large variety of lattice defects cause strain in a crystal. There is a strong correlation between the space dependence of strain fields and the dimensionality of the different lattice defects [5]. Point defects, GP-zones, small precipitates, small dislocation loops or small second phase particles have zero dimension and the strain field is of short range character, i.e. $\varepsilon(r)\cong 1/r^2$. Dislocations are one dimensional defects and the strain fields are of long range character, i.e. $\varepsilon(r)\cong 1/r$. Different types of planar defects are two dimensional defects and the strain fields are homogeneous, i.e. $\varepsilon(r)\cong$constant. Since there is reciprocity in the distance relation between crystal and reciprocal space, short-range strain fields will cause diffraction far from the Bragg peaks, whereas long-range strain fields will cause diffraction clustered around the Bragg peaks. The former is slowly varying diffuse scattering below the Bragg peaks, whereas latter is line broadening. Homogeneous strain fields, corresponding to two dimensional lattice defects, like planar defects, cause peak shifts [2,9-12]. This relatively simple dimensional consideration is the dislocation theorem of line broadening [5,6,18]. It means that strain broadening always indicates the presence of linear type lattice defects, i.e. the presence of dislocations.

The Fourier transform of the strain profile is [2]:

$$A^D_{hkl}(\mathrm{L}) \cong \exp[-\,2\pi^2 g^2 L^2 \langle \varepsilon^2_{g,L}\rangle]\,, \qquad (1)$$

where L is the Fourier variable and $\langle \varepsilon^2_{g,L}\rangle$ is the mean square strain. Krivoglaz [5] and Wilkens [6] derived the mean square strain for dislocated crystals:

$$\langle \varepsilon^2_{g,L}\rangle = \frac{\rho C b^2}{4\pi} f(\eta)\,, \qquad (2)$$

where ρ, C and b are the density, the contrast factor and the Burgers vector absolute values of dislocations. The function $f(\eta)$ describes the L dependence of the mean-square-strain, $\eta=L/R_e$, where R_e is the effective outer cut-off radius of dislocations. R_e in eq. (2) is the same physical quantity as the effective outer cut-off radius of dislocations in the elastic stored energy of dislocated crystals [20]. R_e only makes sense in relation to the average distance of dislocations: $d=1/\sqrt{\rho}$ [6]. If R_e is small or large relative to d, the dislocations are in a stronger or weaker dipole configuration. The dislocation arrangement parameter $M=R_e\sqrt{\rho}$ gives the dipole character, or the effectiveness of strain field screening [6]. If M is smaller or larger than unity the dipole character, or the screening of strain fields, of dislocations is stronger or weaker, respectively. The M parameter is in direct correlation with the profile shape in the lowest intensity range. For stronger or weaker dipole character, the tails of the diffraction peaks will be longer or shorter, respectively. The full-width-at-half-maximum (FWHM) of diffraction peaks is only one part of line broadening. The other, more decisive part is the tail region [5,6]. Fig. 1a shows the 200 diffraction profiles measured at the surface (solid line) and in the bulk (dashed line) of a Cu single crystal oriented for single slip and tensile deformed to a shear strain of $\gamma=0.43$ at RT [21]. The two profiles were measured on the crystal surface where the unpaired edge components of the dislocation dipoles escaped from the near surface region. In this or in the bulk region the dislocation arrangement has a weaker or stronger dipole character, respectively. The dislocation density in the bulk is larger than what the increase of FWHM would predict alone. Figure 1b shows the changing of dislocation density and arrangement parameter *vs.* the distance from the surface in the primary slip direction. Fig. 1c shows the same profile shape behavior in a high-pressure torsion (HPT) deformed Cu-10%Zn alloy [22]. The 8 mm diameter diskette was HPT deformed by 6 GPa pressure and 5 rotations. The figure shows the 220 reflections for the center (blue line) and the edge (red line) regions of the specimen, where the M parameter values are 2.6 and 1.3, respectively. The ratio of the FWHM values of the 220 reflections is $R_{\mathrm{FWHM}}=1.57$, whereas the ratio of the dislocation densities is R_ρ=5.9. The value of R_ρ is by far larger than R_{FWHM}, in correlation that, at larger dislocation densities the dipole character is usually stronger.

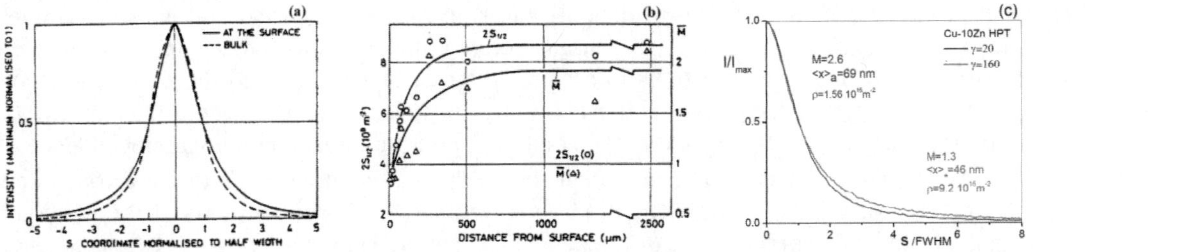

Figure 1. (a) The 200 peak profiles measured on the surface and in the bulk, (b) the FWHM ($2S_{1/2}$) and the M parameter vs. the distance from the surface in a tensile deformed Cu single crystal [21]. (c) The profile shape of 220 reflections for the center (blue line) and the edge (red line) regions of a HPT deformed Cu-10%Zn alloy [22]. (*By courtesy of [21,22]*).

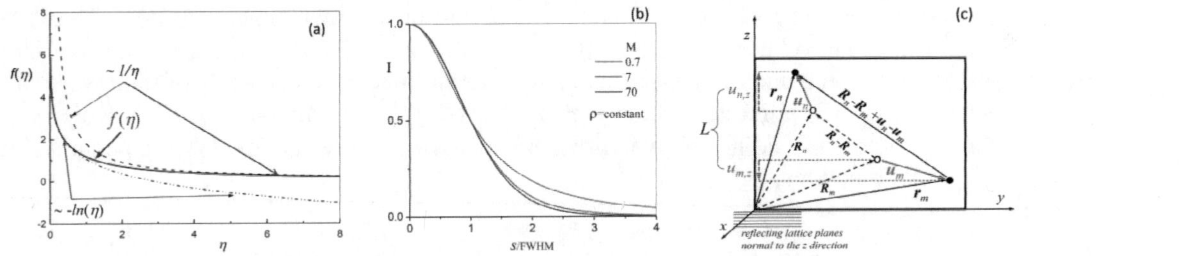

Figure 2. (a) The dash double-dot and the dashed line are logarithmic and the hyperbolic asymptotes at small and large η values, respectively. The blue solid line is the numerical matching of the two asymptotic functions having smooth derivatives.

Krivoglaz derived eq. (2) for small L values where $\langle \varepsilon^2_{g,L} \rangle$ has a logarithmic singularity [5]. Wilkens derived $\langle \varepsilon^2_{g,L} \rangle$ in the entire L range and showed that at large L values it is hyperbolic. Fig. 2a shows the $f(\eta)$ function [6]. Consistently with experiment, the average density, ρ and the arrangement parameter, M describe the dislocation structure. The replacement of the $\langle \varepsilon^2_{g,L} \rangle$ function by a single number, e.g. in the Williamson-Hall method, automatically assumes that the strain profile is a pure Gaussian function. The experimental Figs. 1a and 1c show that this is almost never the case. A single number cannot replace the $\langle \varepsilon^2_{g,L} \rangle$ function.

The term "mean square strain" is rather misleading. The scattered intensity, $I(K)$, is the square of the scattered amplitudes of waves. Fig. 2c shows schematically the relation between $\langle \varepsilon^2_{g,L} \rangle$ and lattice distortions, $\boldsymbol{u}$. Line broadening is determined by the pair-correlation of atomic displacements, $\boldsymbol{u}_n$ and $\boldsymbol{u}_m$ at distances between the two positions, $\boldsymbol{r}_n$-$\boldsymbol{r}_m$. Quantitatively the pair-correlation of atomic displacements is given by the non-local distortion tensor, β [5,6,23]: $\boldsymbol{r}_n-\boldsymbol{r}_m=\boldsymbol{R}_n-\boldsymbol{R}_m+\boldsymbol{u}_n-\boldsymbol{u}_m=[\underline{\underline{1}}+\underline{\underline{\beta}}(\boldsymbol{R}_n,\boldsymbol{R}_m)](\boldsymbol{R}_n-\boldsymbol{R}_m)$, and the correspondence between β, and the infinitesimal strain tensor, $\underline{\underline{\hat{\beta}}}=\mathrm{grad}\boldsymbol{u}$ is: $\boldsymbol{u}_n-\boldsymbol{u}_m=\underline{\underline{\beta}}(\boldsymbol{R}_n,\boldsymbol{R}_m)\cdot(\boldsymbol{R}_n-\boldsymbol{R}_m)=\int_{\mathbf{R}_m}^{\mathbf{R}_n}\underline{\underline{\hat{\beta}}}d\boldsymbol{R}$ [6,23]. The distance L is the projection of $\boldsymbol{R}_n$-$\boldsymbol{R}_m$ in the diffraction direction, $\boldsymbol{g}$. $\langle \varepsilon^2_{g,L} \rangle$ is the square of the pair-correlation of lattice distortion in the $\boldsymbol{g}$ direction. The more appropriate term for $\langle \varepsilon^2_{g,L} \rangle$ would be, 'mean square strain-correlation distribution', indicating that this is a function, irreplaceable by a shear number.

Size broadening is an optical scattering effect. When the crystallites become small, the diffraction peaks blow up uniformly in reciprocal space [24]. Assuming equiaxed shape and log-normal size distribution the size profile can be expressed in a close form [25,26]. Line profiles of faulted and twinned crystals consist of several sub-profiles, where their numbers, shifts and breadths depend on the *hkl* indices of the fundamental Bragg reflection [27-31]. It was shown by numerical [32] and theoretical [30,31] methods that sub-profiles are the sum of a symmetrical and an anti-symmetrical Lorentzian profile function. The anti-symmetrical Lorentz function is the consequence of interference between two overlapping sub-reflections in reciprocal space, where one and the other correspond to the parent and twin crystal, respectively [30,31].

The contrast factor, C in $\langle\varepsilon_{g,L}^2\rangle$, in eq. (2), is taking into account the directional dependence of the strength of strain. The *hkl* dependence of C depends on the relative orientation of the Burgers and line vectors of dislocations, $\boldsymbol{b}$ and $\boldsymbol{l}$, the diffraction vector $\boldsymbol{g}$ and the elastic constants of the material [33-36]. In polycrystalline materials, the *hkl* dependent contrast factors have to be averaged over the permutations of the *hkl* indexes [34,35,37,38]. The averaged contrasts are functions of the fourth order independents of *hkl*-s [37]. In cubic and hexagonal crystals, they are [8,37]:

$$\overline{C}=\overline{C}_{h00}(1-qH^2)\,,\quad \overline{C}_{hk.l}=\overline{C}_{hk.0}\,[1+q_1x+q_2\,x^2]\,, \tag{3}$$

where $H^2=(h^2k^2+h^2l^2+k^2l^2)/(h^2+k^2+k^2)^2$, $\overline{C}_{h00}$ and $\overline{C}_{hk.0}$ are the average contrast factor of the *h00* and *hk.0* reflections in cubic and hexagonal crystals, $x=(2/3)(l/ga)^2$ and, l and a are the lattice constants of a hexagonal crystal. The q and q_1, q_2 parameters depend on the elastic constants of the crystal and the type of dislocations [36,37]. In ref. [39] the averaged contrasts were evaluated for all crystal systems.

Based on the principles outlined above, diffraction patterns, $I^{PM}(2\theta)$ are modelled using physically established profile functions:

$$I^{PM}(2\theta)=\sum_{hkl} I_{hkl}^{S} * I_{hkl}^{D} * I_{hkl}^{PD} * I_{hkl}^{INST} + I_{BG}\,. \tag{4}$$

The defect related profile functions are the size, I^{S}_{hkl}, the strain I^{D}_{hkl} and the planar defect profiles, I^{PD}_{hkl}. The instrumental pattern I_{hkl}^{INST} is measured and the background, I_{BG} added to the model-calculated pattern. We developed the convolutional multiple whole profile (CMWP) software package to simulate the measured diffraction pattern using the physically modelled diffraction pattern in eq. (4) [25,40].

3. Laboratory X-ray diffraction experiments

A large number of laboratory X-ray diffraction experiments used line profile analysis (LPA) to characterize the sub-structure of bulk UFG materials [7,38-64]. Fig. 3a shows the thermal stability of the dislocation density and long-range internal stresses in Ni deformed by HPT to ε=58 applying 8 GPa pressure and 4 rotations [47].

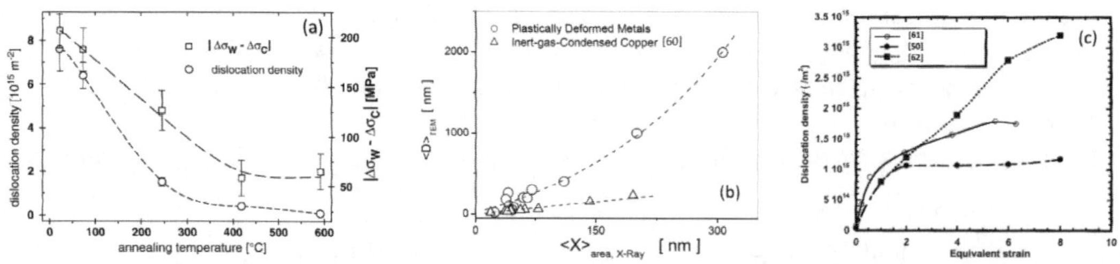

Figure 3. (a) Thermal stability of the dislocation density and long-range internal stresses [47]. (b) Comparison of X-ray crystallite size with TEM grain size [50]. Comparison of X-ray and TEM dislocation densities [52]. (*By courtesy of [47,50,52]*).

The long-range internal stresses were determined from the peak asymmetries. The grain size was determined by SEM. Both, the dislocation density and the grain size started recovering at about 300 °C. Fig. 3b shows the X-ray and TEM grain size in different UFG materials deformed either by HPT or ECAP along with the grain size in inert-gas condensed Cu measured in the as-received state and after long term natural aging at RT [50]. The figure indicates that in the smallest grain-size region the X-ray and TEM data are identical, whereas when the TEM grain size exceeds about 100 nm the X-ray grain size values tend to show smaller values. TEM grain size gives the true grain-size values, while above about 100 nm X-rays tend to give the sub-grain size. Fig. 3c shows the comparison of dislocation densities in cold rolled Cu determined by TEM [62] or X-ray LPA [61] and ECAP deformed Cu determined by TEM [52]. The figure indicates a good correlation between the data despite the differences in methods, samples and investigators.

4. Synchrotron X-ray diffraction experiments

A relatively large number of synchrotron X-ray diffraction experiments used LPA to characterize the dislocation structure and grain size in bulk UFG materials [65-74]. Lee at al. [70] measured the dislocation density evolution in HPT processed Cu by using synchrotron X-ray diffraction. Fig. 4a shows a typical synchrotron X-ray diffraction pattern. Fig. 4b shows the dislocation density as a function of strain. Authors developed a dislocation-density based constitutive model embedded in a finite element code (FEC). Fig. 4c shows the model calculated and the measured dislocation densities along with data from the literature. The average FEC calculated values (solid red line in Fig. 4c) are in good agreement with the synchrotron XRD values in Fig. 4b and the results in ref. [64] (open squares). The open triangles are TEM calculated dislocation density values taken from ref. [52]. Fig. 4b shows that the dislocation density drops at the largest deformation. The drop at very large HPT deformation values relates most likely to the increased excess vacancy production during HPT processing [76]. Releasing the specimen from the HPT machine, the excess vacancies become free and annihilate some of the stored dislocations, as shown in ref. [76].

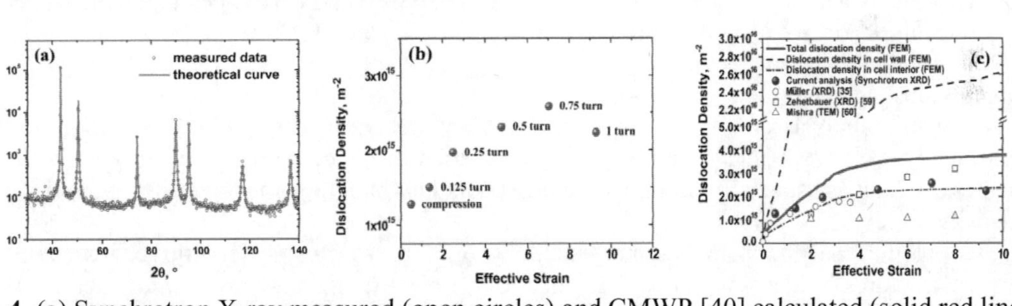

Figure 4. (a) Synchrotron X-ray measured (open circles) and CMWP [40] calculated (solid red line) of an ECAP deformed Cu specimen [72]. (b) Dislocation density values obtained by using the CMWP procedure [40,72]. (c) Finite-element calculated dislocation densities in the cell walls (dashed line), in the cell interiors (dash dot lines) and the total density (solid red line). The full symbols are the same values as in (b). The open symbols are taken from the literature. (*By courtesy of Lee et al. [72]*)

5. Neutron diffraction experiments

The angular resolution of neutron diffractometers is usually substantially lower than that of X-ray diffractometers. During the last two decades spallation neutron sources with considerably improved angular resolution have become perspective equipment for LPA [77-84]. A fully transformed lath-martensite specimen was tensile deformed in-situ in the high resolution and high intensity neutron diffractometer at the TAKUMI beamline of the spallation neutron source J-PARC [84]. In the as-quenched state, the initial crystallite size determined by LPA was 60 nm in good correlation with lath thickness, obtained by TEM. The initial dislocation density was 4.5×10^{15} m^{-2}. The diffraction peaks in the as-quenched state are perfectly symmetrical, see in Fig. 5a. During tensile deformation, the diffraction peaks become characteristically asymmetric. Peaks with diffraction vectors parallel (see Fig. 5b) or perpendicular to the tensile axis have a longer tail in the smaller or larger d* direction, where d* is the reciprocal space coordinate. Similarly, as in the case of tensile deformed Cu single crystals with dislocation cell structures, see refs. [85,86]. Characteristically asymmetric peak broadening indicates the presence of long-range internal stresses between coherent domains within grains. Either these are the lath packets oriented with lath-planes parallel or oblique to the largest Schmid factor directions. In the lath packets oriented parallel to the largest Schmid factor, dislocations glide easily with long mean-free paths. In lath packets oriented oblique to the largest Schmid factor, dislocations are stuck in the lath boundaries and will have short mean-free paths. In the former case, the initial dislocation density is larger than what can carry the strain, therefore it decreases and these packets soften (see the open triangles in Fig. 5c and ref. [57]). In the latter case, even the large initial dislocation density is not sufficient to carry the strain, therefore it increases and these packets harden (see the open circles in Fig. 5c). As a result, plastic deformation divides the lath packets, initially statistically identical, into hard and soft packets and the material behaves like a composite [84].

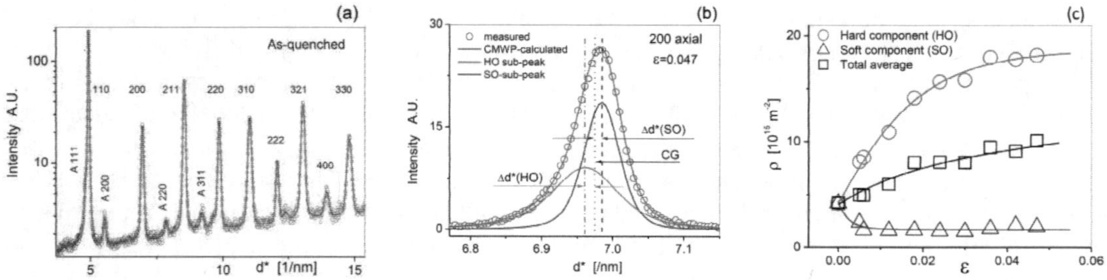

Figure 5. (a) Measured (open circles) and CMWP [40] calculated neutron diffraction pattern of the as-quenched lath martensite. (b) Asymmetric 200 peak when the diffraction vector and the tensile axis are parallel. The figure also shows the long-range internal stresses between soft and hard packets. (c) Evolution of the dislocation densities in the soft (open triangles) and hard (open circles) packets and the average values (open squares) (*Figures (b) and (c) by courtesy of [84]*).

6. Conclusions

(1) Line profile analysis developed during the last few decades, being able to provide reliable and physically well-established quantities to characterize the microstructure of plastically deformed materials. This is especially true for UFG materials in which the grain size is small and the defect density large.

(2) Line profile analysis provides an excellent opportunity for synergetic characterization of microstructures by the simultaneous application of electron microscopy and diffraction. Whether the application of laboratory X-ray, synchrotron X-ray or neutron diffraction, the synergy works.

(3) Diffraction experiments provide of all kinds internal stresses not easily accessible by electron microscopy methods.

References

[1] Stokes A R, Wilson A J C 1942 *Proc Camb Phil Soc* **38** 313

[2] Warren B E 1959 *Progr Metal Phys* **8** 147

[3] Warren B E 1969 *X-ray Diffraction* New York; Dover.

[4] Rebuffi L, Troian A, Ciancio R, Carlino E, Amimi A, Leonardi A, Scardi P 2016 *Sci Report* **6** 20712

[5] Krivoglaz M A 1996 in *Theory of X-ray and Thermal Neutron Scattering by Real Crystals* Plenum Press, New York

[6] Wilkens M 1970 in *Fundamental Aspects of Dislocation Theory* J.A. Simmons, R. de Wit, R. Bullough (Eds.), vol II, Nat Bur Stand (US) Spec Publ No 317, Washington, DC, USA, p. 1195.

[7] Ungár T, Borbély A 1996 *Appl Phys Lett* **69** 3173
[8] Ungár T, Dragomir I, Révész Á, Borbély A 1999 *J Appl Cryst* **32** 992
[9] Estevez-Rams E, Leoni M, Scardi P, Aragon-Fernandez B, Fuess H 2003 *Philos Mag* **83** (2003) 4045
[10] Velterop L, Delhez R, de Keijser Th H, Mittemeijer E J, Reefman D J 2000 *Appl Cryst* **33** 296
[11] Balogh L, Ribárik G, Ungár T *J Appl Phys* **100** 023512
[12] Balogh L, Tichy G, Ungár T 2009 *J Appl Cryst* **42** 580
[13] Leineweber A, Mittemeijer E 2006 *Z Kristallographie* S23 117
[14] Leineweber A 2016 *J Appl Cryst* **49** 1632
[15] Akhlaghi M, Steiner T, Meka S R, Leineweber A, Mittemeijer E J 2015 *Acta Mater* **98** 254
[16] Langford J I, Boultif A, Auffrédic J P, Louër D 1993 *J Appl Cryst* **26** 22
[17] Scardi P, Leoni M *Acta Cryst* **A57** 604
[18] Ungár T 2008 *J Powdwr Diffr* **23** 125
[19] Williamson G K, Hall W H 1953 *Acta metall* **1** 22
[20] Bailey J E, Hirsch P B 1960 *Phil Mag* **5** 485
[21] Ungár T, Mughrabi H, Wilkens M 1982 *Acta Metall* **30** 1861
[22] Balogh L, Ungár T, Zhao Y, Zhu Y T, Horita Z, Xu C, Langdon T G 2008 *Acta Mater* **56** 809
[23] Zilahi Gy, Ungár T, Tichy G 2015 *J Appl Cryst* **48** 418
[24] James, R W 1965 in *The optical principles of the diffraction of X-ray* Bell G & Sons, Ltd, London
[25] Ungár T, Gubicza J, Ribárik G, Borbély A 2001 *J Appl Cryst* **34** 298
[26] Scardi P, Leoni M 2002 *Acta Cryst* **A58** 190
[27] Balogh L, Ribárik G, Ungár T 2006 *J Appl Phys* **100** 023512-10.
[28] Balogh L, Tichy G, Ungár T 2009 *J. Appl. Cryst* **42** 580
[29] Velterop L, Delhez R, de Keijser Th H, Mittemeijera E J, Reefman D 2000 *J Appl Cryst* **33** 296
[30] Estevez-Rams R, Penton-Madrigal A, Lora-Serrano R, Martinez-Garcia J 2001 *J Appl Cryst* **34** 730
[31] Estevez-Rams R, Leoni M, Scardi P, Aragon-Fernandez B, Fuess F 2003 *Philosophical Magazine* **83** 4045
[32] Treacy M M J, Newsam J M, Deem M W 1991 *Proc Roy Soc London* **A433** 499
[33] Wilkens M 1970 *Phys Stat Sol* **A2** 359
[34] Klimanek P J, Kuzel R L Jr 1988 *J Appl Cryst* **21** 59
[35] Kuzel R L Jr, Klimanek P J 1989 *J Appl Cryst* **22** 299
[36] Borbély A, Dragomir I C, Ribárik G, Ungár T, *J Appl Cryst* **36** 160
[37] Ungár T, Tichy G 1999 *Phys Stat Sol* **A147** 425
[38] Révész Á, Ungár T, Borbély A, Lendvai J 1996 *Nanostr Mater* **7** 779
[39] Leoni M, Marinez-Garcia J, Scardi P 2007 *J Appl Cryst* **40** 719
[40] Ribárik G, Ungár T 2010 *Mater Sci Eng* **A528** 112
[41] Islamgaliev R K, Chmelik F, Kuzel R 1997 *Mater Sci Eng* **A234-236** 335
[42] Islamgaliev R K, Chmelik F, Kuzel R 1997 *Mater Sci Eng* **A27** 43
[43] Fátay D, E. Bastarash, Nyilas K, Dobatkin S, Gubicza J, Ungár T 2003 Z Kristllgr 94 1
[44] Zhu Y T, Huang J Y, Gubicza J, Ungár T, Wang Y M, Ma E, Valiev R Z 2003 *J Mater Res* **18** 1908
[45] Vives S, Gaffet E, Meunier C 2004 *Mater Sci Eng* **A366** 229
[46] Gubicza J, Nam N H, Balogh L, Hellmig R J, Stolyarov V V, Estrin Y, Ungár T 2004 *Mater Sci Eng* **A378** 248
[47] Schafler E, Pippan R 2004 *Mater Sci Eng* **A387-389** 799
[48] Gubicza J, Balogh L, Hellmig R J, Estrin Y, c 2005 *Mater Sci Eng* **A400-401** 334
[49] Schafler E, Steiner G, Korznikova E, Kerber M, Zehetbauer M J 2005 *Mater Sci Eng* **A410-411** 169
[50] Ungár T, Tichy G, Gubicza J, Hellmig R J 2005 *Powder Diffr* **55** 366
[51] Gubicza J, Chinh N G, Langdon T G, Ungár T 2006 *Proc Ultraf Grain Mater IV* Ed Zhu Y T et

al pp 231

[52] Mishra A, Kad B K, Gregori F, Meyers M A 2007 *Acta Mater* **55** 13

[53] Ungár T, Schafler E, Hanák P, Bernstorff S, Zehetbauer M J 2007 *Mater Sci Eng* **A462** 398

[54] Sarkar A, Bhowmik A, Suwas S 2009 *Appl Phys* **A94** 943

[55] Liu M, Roven H J, Liu X, Murashkin M, Valiev R Z, Ungár T, Balogh L 2010 *J Mater Sci* **45** 4659

[56] Kerber M B, Zehetbauer M J, Schafler E, Spieckermann F C, Bernstorff S, Ungár T 2011 *JOM* **63** 61

[57] Ungár T, Li L, Tichy G, Pantleon W, Choo H, Liaw P K 2011 *Scripta Mater* **64** 876

[58] Yoon E Y, Lee, D J, Kim T S, Chae H J, Jenei P, Gubicza J, Ungár T, Janecek M, Vratna J, Lee S, Kim H S 2012 *J Mater Sci* **47** 7117

[59] Jóni B, Schafler E, Zehetbauer M J, Tichy G, Ungár T 2013 *Acta Mater* **61** 632

[60] Skrotzki W, Eschke A, Jóni B, Ungár T, Toth L S, Ivanisenko Yu, Kurmanaeva L 2013 *Acta Mater* **61** 7271

[61] Bahl S, Suwas S, Ungár T, Chatterjee K 2017 *Acta Mater* **122** 138

[62] Ungár T, Ott S, Sanders P G, Borbély A, Weertman J R 1998 *Acta Mater* **46** 3693

[63] Ungár T, Zehetbauer M 1996 *Scripta Mater* **35** 1467

[64] Zehetbauer M, Seumer V 1993 *Acta Mater* **41** 577

[65] Zehetbauer M J, Steiner G, Schafler E, Korznikov A, Korznikova E 2006 *Mater Sci Forum* **503-504** 57

[66] Schafler E, Nyilas K, Bernstorff S, Zeipper L, Zehetbauer M J, Ungár T 2006 *Z Kristallogr Suppl* **23** 129

[67] Fan G J, Fu L F, Wang Y D, Ren Y, Choo H, Liaw P K, Wang G Y, Browning N D 2006 *Appl Phys Lett* **89** 101918-3

[68] Li L, Ungár T, Wang Y D, Fan G J, Yang Y L, Jia N, Ren Y, Tichy G, Lendvai J, Choo H, Liaw P K 2009 *Scripta Mater* **60** 317

[69] Li L, Ungár T, Wang Y D, Morris J R, Tichy G, Lendvai J, Yang Y L, Ren Y, Choo H, Liaw P K 2009 *Acta Mater* **57** 4988

[70] Barabash R I, Ice G E, Liu W, Barabash O M 2009 *Micron* **40** 28

[71] Kilmametov A R, Vaughan G, Yavari A R, LeMoulec A, Botta W J, Valiev R Z 2009 *Mater Sci Eng* **A503** 10

[72] Lee D J, Yoon E Y, Ahn D-H, Park B H, Park H W, Park L J, Estrin Y, Kim H S 2014 *Acta Mater* **76** 281

[73] Zhang H, Fong K S, Song X, Ying S, Salvati E, Sui T, Korsunsky A M 2016 Proc Int MConf Eng Comp ISBN 978-988-14047-6-3

[74] Miyajima Y, Okubo S, Miyazawa T, Adachi H, Fujii T 2016 *Phil Mag* **96** 294

[75] Li L, Ungár T, Toth L S, Skrotzki W, Wang J D, Ren J, Choo H, Fogarassy Z, Zhou X Z and Liaw P K 2016 *Metall Mat Trans* A **47** 6632

[76] Dubravina I, Zehetbauer M, Schafler E, Alexandrov I 2004 *Mater Sci Eng* **A387–389** 817

[77] Woo W, Choo H, Brown D W, Bourke M A M, Feng Z, David S A, Hubbard C R, Liaw P K 2005 *Appl Phys Lett* **86** 231902-3

[78] Tao K, Choo H, Li H, Jin J E, Lee Y K 2007 *Appl Phys Lett* **90** 101911-3

[79] Lukas P, Strunz P, Davydov V, Kuzel R 2009 *ICDD* **ISSN 1097-0002** 421

[80] Woo W, Ungar T, Feng Z, Kenik E, Clausen B 2010 *Metall Mater Transact* **A41** 1210

[81] González Crespo P A, Luis Pérez C, Hughes D, Turillas X 2013 *J Nanomater* **2013** 404903-6

[82] Ahmmed K F, Balogh L, Idrees Y, Yu H, Long F, Daymond M R 2016 *J Appl Cryst* **49** 1609

[83] Harjo S, Kawasaki T, Gong W, Aizawa K 2016 J Phys Conf Ser **746** 012046-7

[84] Ungár T, Harjo S, Kawasaki T, Tomota Y, Ribárik G, Shi Z 2017 *Metall Mater Transact* **A48** 159

[85] Ungár T, Mughrabi H, Ronnpagel D, Wilkens M 1984 *Acta Metall* **32** 333

[86] Mughrabi H, Ungár T, Kienle W, Wilkens M 1986 *Phil Mag* **53** 793

IOP Conf. Series: Materials Science and Engineering **194** (2017) 012004 doi:10.1088/1757-899X/194/1/012004

Strength, corrosion resistance, and biocompatibility of ultrafine-grained Mg alloys after different modes of severe plastic deformation

S V Dobatkin[1,2*], E A Lukyanova[1,2], N S Martynenko[1], N Yu Anisimova[3], M V Kiselevskiy[3], M V Gorshenkov[1], N Yu Yurchenko[4], G I Raab[5], V S Yusupov[2], N Birbilis[6], G A Salishchev[4], Y Z Estrin[1,6]

[1] Laboratory of Hybrid Nanostructured Materials, National University of Science and Technology "MISIS", Moscow, 119991, Russia
[2] A.A. Baikov Institute of Metallurgy and Materials Science of the Russian Academy of Sciences, Moscow, 119334, Russia
[3] N.N. Blokhin Russian Cancer Research Center, Moscow, 115478, Russia
[4] Belgorod National Research University, Belgorod, 308015, Russia
[5] Ufa State Aviation Technical University, Ufa, 45008, Russia
[6] Department of Materials Science and Engineering, Monash University, Melbourne, VIC 3800, Australia

*E-mail: dobatkin.sergey@gmail.com

Abstract. The effect of severe plastic deformation on the structure, mechanical properties, corrosion resistance, and biocompatibility of the WE43 (Mg-Y-Nd-Zr) alloy earmarked for applications as bioresorbable material has been studied. The alloy was deformed by rotary swaging (RS), equal channel angular pressing (ECAP), and multiaxial deformation (MAD). The microstructure examination by transmission electron microscopy showed that all SPD modes lead to the formation of ultrafine-grained structure with a structural element size of 0.5-1 μm and the $Mg_{12}Nd$ phase particles 0.3 μm in size. The microstructure refinement by all three treatments resulted in strengthening of the alloy. ECAP and MAD also raised ductility to up to 12-17%, while RS increased the ultimate tensile strength to up to 415 MPa. The study of the corrosion properties showed that SPD does not affect the electrochemical corrosion of the alloy. Its biocompatibility *in vitro* was estimated after incubation of the samples with red blood cells (hemolysis study), white blood cells (cell viability assay), and mesenchymal stromal cells (cell proliferation analysis). The biodegradation rate in fetal bovine serum was also evaluated. ECAP and MAD were found to cause some deceleration of biodegradation by slowing down the gas formation in the biological fluid and, compared to MSC, to improve the biocompatibility of the WE43 alloy.

1. Introduction

Over the past decade, magnesium and its alloys have attracted a great deal of attention as possible medical implant materials [1, 2]. Their potential for medical applications, particularly in bioresorbable implants, owes to their low density, high specific strength, and good biocompatibility. However, because of its high chemical activity, magnesium is characterized by a high corrosion rate, which is accompanied by the active evolution of hydrogen gas [3]. Such disadvantages limit the use of magnesium alloys as materials for bioresorbable implants. The WE43 alloy containing rare earth metals (REMs) is one of the most popular medical magnesium alloys [4]. Alloying with REMs improves the corrosion resistance of magnesium [5] and also enhances its strength. Another method of increasing the strength characteristics is severe plastic deformation (SPD), which leads to the formation of ultrafine-grained (UFG) structure in magnesium and its alloys [6]. Such structure provides substantial strengthening [6-9] and also leads to improved corrosion resistance and reduced gas evolution. The reduction of gas evolution, in turn, gives rise to improved biocompatibility, which expands the suitability of WE43 alloy as a biodegradable implant material.

In this work, the WE43 alloy was processed by rotary swaging (RS), equal-channel angular pressing (ECAP), and multiaxial deformation (MAD). The effect of deformation on the microstructure, mechanical properties and corrosion resistance of the alloy has been studied, and its biocompatibility *in vitro* has been evaluated in the initial state and after the deformation by the three SPD modes.

2. Experimental

The material studied in this work was a Mg-3.56% Y-2.20% Nd-0.47% Zr (wt %) alloy. In the initial state, the alloy was homogenized at a temperature of 525°C for 8 hours. This was followed by deformation by various SPD methods, including rotary swaging, multiaxial deformation, and equal-channel angular pressing. The processing schemes are shown in Fig. 1. Deformation was performed at gradually decreasing temperature and increasing degree of strain. In the case of RS, temperature was decreased from 425°C to 350°C in 25°C steps, and the cumulative equivalent strain reached a level of 1.02. MAD was performed on workpieces 25 mm in diameter and 40 mm in length in an Instron 300LX universal hydraulic testing machine at a low rate of 1 mm/min. Again, deformation was performed in several steps, with successive drops in temperature of 25°C from 450°C to 300°C. The total number of deformation cycles was 28. ECAP was performed using a die with an angle of 120° between the channels with stepwise drops in temperature from 425°C to 300°C at the final 12th pass. The cumulative equivalent strain in the ECAP billets (10 mm in diameter and 80 mm in length) was about 7.6.

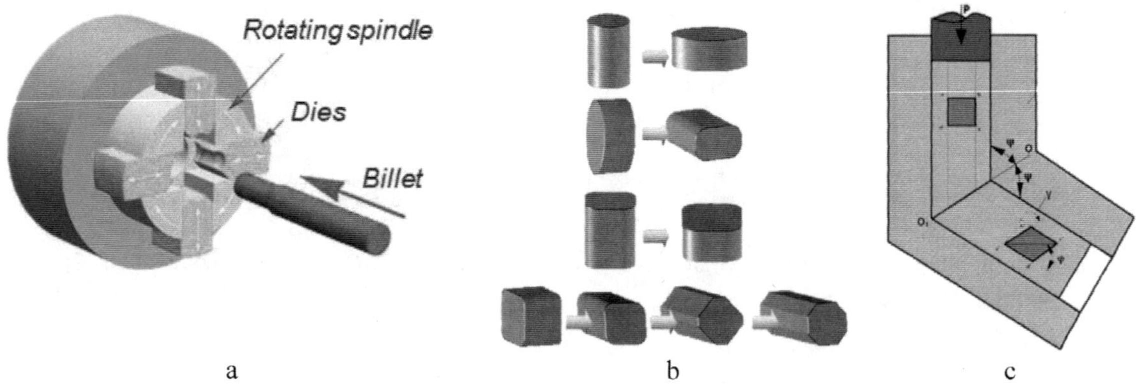

a b c

Figure 1. Schemes of deformation processing of the alloy: (a) RS, (b) MAD, and (c) ECAP.

The structure of the alloy after homogenization was examined with a Reichert MeF optical microscope. The TEM analysis of the alloy after SPD was carried out with a JEM 1400 transmission electron microscope operating at a voltage of 120 kV. The room temperature mechanical properties before and after deformation were measured in tension on cylindrical samples with a 3 mm diameter and 15 mm gage length in an Instron 1096 machine at an extension rate of 1.5 mm/min.

The potentiodynamic polarization (PDP) method was used to evaluate the corrosion resistance of the alloy. The PDP tests were performed with a VMP potentiostat under the control of EC-Lab specialized software (Bio-logic). The unit used a standard flat cell with three electrodes (working electrode, saturated calomel reference electrode, and Pt-mesh counter electrode). The corrosive medium was 0.9% NaCl solution in water. Before testing, all surfaces were ground to 2000 grit by sandpaper.

To study biocompatibility and biodegradation, samples 2 mm thick and 8 mm in diameter were sanded on sandpaper (P2000) and then polished on a wet cloth. Prior to testing, the samples were sterilized by steam in an autoclave at 120°C for 30 minutes and then washed three times with RPMI-1640 (PanEco, Russia) sterile medium.

For hemolysis studies, anticoagulated CBA mouse whole blood was centrifuged at 1000 rpm for 10 min at 20°C for removal of plasma. Then red blood cells (RBCs) were washed for additional three times with phosphate buffered saline (PBS), pH 7.4, for 5 min. Washed RBCs were mixed with PBS (1:9) and used within 2 h for the hemolysis assay as described by A.V. Maximkin et al. [10]. 1000 μL of RBCs suspension were introduced into a well of a 24-well plate with the sample. Positive and negative controls were produced by adding 100 μL of Triton-X-100 (Sigma-Aldrich, USA) or PBS, respectively. The plate with RBCs and alloy samples was incubated at 37°C. Probes for analysis were collected 1, 2, 3, 4, 5, 6, 24 h after start of the test and centrifuged at 300 rpm for 3 min. Optical density (OD) of the supernatant was measured at 540 nm using a Labsystems Multiscan MS plate reader. The percentage hemolysis was calculated from the calibration curve.

For evaluation of the effect of alloys on cell vitality, 1 ml of mouse white blood cells (WBCs) in RPMI-1640 medium supplemented with 10% fetal bovine serum (Thermo Scientific, USA), 100 μg/ml penicillin, 100 μg/mL streptomycin, and 2 mM L-glutamine (all-Gibco, USA) at a density of 5×10^5 cells/cm^2 were cultivated with alloy samples in a 24-well plate (Nunc, USA) at 37°C in a humidified atmosphere with 5% CO_2. The control WBCs were treated only with RPMI 1640. After incubation for 2 and 24 h, 20 μL 3- (4.5-dimethylthiazol-2-yl)-2.5-diphenyltetrazolium bromide (MTT) (Sigma) with a concentration of 5 mg·mL^{-1}, was added into each well. The plates were incubated for 4 h under cell culture conditions. Subsequently, 500 μL dimethyl sulphoxide (PanEco, Russia) was added to the cells by replacing the medium, to dissolve the formazan crystals. OD of the solution (100 μL) was measured at 540 nm in 96-well plates. The vitality of WBCs was calculated using the following formula:

$$\textit{Vitality, \%} = \textit{OD samples/OD control} \times 100.$$

Mesenchymal stromal/stem cells (MSCs) were isolated according to the standard techniques for the isolation of mononuclear cells from bone marrow of CBA mouse femurs using density gradient centrifugation [11]. The cells were cultured in Dulbecco's modified Eagle's medium (DMEM, high glucose) (Gibco), supplemented with 10% fetal bovine serum (FBS) in a humidified incubator at 95% relative humidity and 5% CO_2 at 37°C. Cells were passaged at about 80%-90% confluency. For the experiments, cells after the third passage were used. Cells were tested with flow cytometry for the presence of characteristic markers on BD FACSCanto II Cells Analyzer (Becton Dickinson). We used the specific CD105, CD90, CD45 and CD34 antibodies (Becton Dickinson or Bioscience). MSCs were trypsinized by TryplE (Life Technologies, USA) and seeded in wells at a density of 1×10^5 cells. The 24-wells plate containing tested alloy samples was incubated at 37°C in a humidified atmosphere for 5 days. At the end of the assay, MTT and dimethyl sulphoxide were added to each well as described earlier. The solution of each sample was placed in a microtiter plate, and the absorbance was measured in triplicate at 540 nm. OD baseline was measured during the first day of co-incubation of MSCs with alloy samples. The percentage of MSCs proliferation was calculated using the following formula:

$$\textit{Proliferation, \%} = \textit{(Mean OD sample - Mean OD baseline)/Mean OD baseline} \times 100.$$

In vitro degradation of the WE43 alloy after various SPD treatments (ECAP, MAD, RS) was evaluated after holding samples at 37°C in fetal bovine serum (FBS). To estimate the biodegradation rate, each WE43 alloy of (W_0) mg mass was immersed into 2 mL of FBS (HyClone UK Ltd., Thermo Scientific). The samples were placed into a CO_2 incubator (AutoFlow NU-4750 Water Jacket CO_2 Incubator, NuAire) at 37°C and 5% CO_2 for 1, 2, 3 and 4 weeks. Every week a sample of each type

IOP Conf. Series: Materials Science and Engineering **194** (2017) 012004 doi:10.1088/1757-899X/194/1/012004

was removed and washed ten times with distilled water. Then it was dried at a temperature of 20°C to a constant weight (W_{FBS}). The weight loss was defined as $W_{FBS}/W_0 \times 100\%$.
In order to assess the effect of the degree of surface preparation on the corrosion properties and biocompatibility of the alloy, we measured roughness parameters such as the arithmetic average roughness (R_a), the root mean square roughness (R_q), the average maximum peak-to-valley value (R_z), and the average maximum height of the profile (R_t). The kurtosis (R_{ku}) and skewness (R_{sk}) characteristics describing the shape and distribution of peaks [12] were also determined. The roughness measurements were carried out using a WYKO NT 1100 optical profilometer (Veeco Instruments, USA). The measurements were performed by vertical scanning interferometry. The average roughness was measured on a surface area of 900 μm x 1200 μm for each state of the alloy.

3. Results and Discussion

Upon homogenization, a uniform grain structure with an average grain size of 70 μm is formed in the WE43 alloy. In addition, the structure contains a small amount of the $Mg_{12}Nd$ phase, which was not dissolved upon homogenization (Fig. 2a). All three types of SPD result in similar microstructures of the alloy with close average grain sizes of 0.73 ± 0.21 μm after ECAP, 0.61 ± 0.19 μm after RS, and 0.93 ± 0.29 μm after MAD (Figs. 2b, 2c, and 2d). The size of the $Mg_{12}Nd$ particles precipitated upon heating for processing and upon deformation is 0.3 μm for all three types of SPD. It should also be noted that the microstructure of the samples that underwent RS exhibited an increased dislocation density and deformation twins. In this case, $Mg_{24}Y_5$ phase platelets are observed in addition to the globular $Mg_{12}Nd$ phase particles.

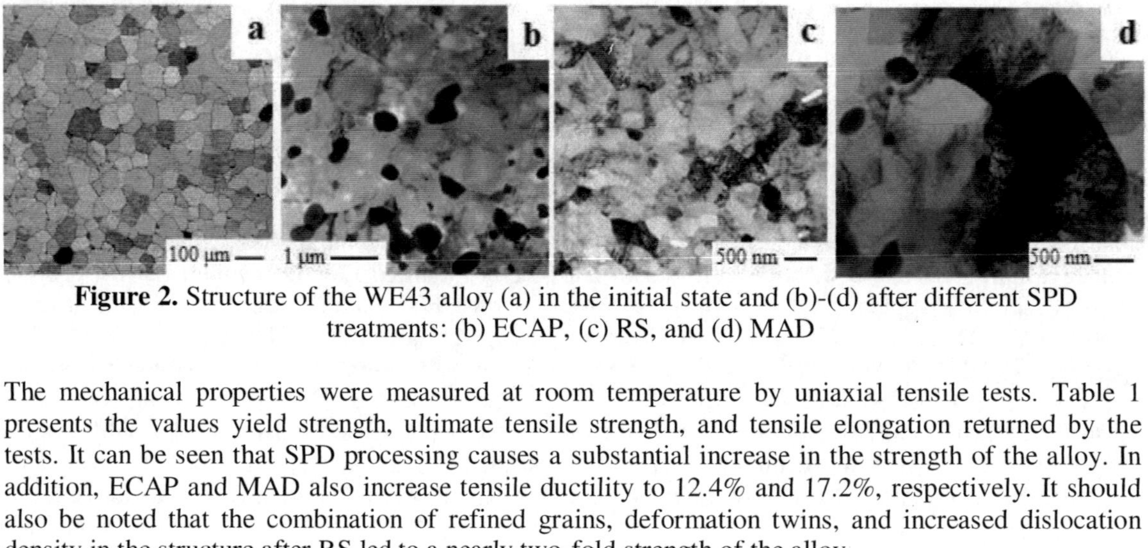

Figure 2. Structure of the WE43 alloy (a) in the initial state and (b)-(d) after different SPD treatments: (b) ECAP, (c) RS, and (d) MAD

The mechanical properties were measured at room temperature by uniaxial tensile tests. Table 1 presents the values yield strength, ultimate tensile strength, and tensile elongation returned by the tests. It can be seen that SPD processing causes a substantial increase in the strength of the alloy. In addition, ECAP and MAD also increase tensile ductility to 12.4% and 17.2%, respectively. It should also be noted that the combination of refined grains, deformation twins, and increased dislocation density in the structure after RS led to a nearly two-fold strength of the alloy.

Table 1. Mechanical properties of WE43 alloy in the initial state and after different SPD treatments

Treatment	UTS, MPa	YS, MPa	El, %
Homogenization at 525°C, 8 h	220	150	10.2
Rotary swaging (RS)	415	285	7.0
Multiaxial deformation (MAD)	300	210	17.2
Equal channel angular pressing (ECAP)	300	260	12.4

The resistance to electrochemical corrosion was estimated at room temperature by potentiodynamic tests. Figure 3 shows the polarization curves after various treatments of the alloy in comparison with the initial state. It can be seen that the corrosion resistance is not compromised by the severe deformation of the alloy. Table 2 summarizes the data on the corrosion potential E_{corr} and the corrosion current density i_{corr}, as determined from the Tafel plots, for the alloy in the initial state and after various SPD treatments. The data obtained show that the best corrosion resistance of the alloy is achieved by MAD. In this case, a slight increase in the corrosion potential is observed, which indicates an increase in the corrosion resistance. At the same time, the corrosion current density remains at the level of the initial, undeformed state.

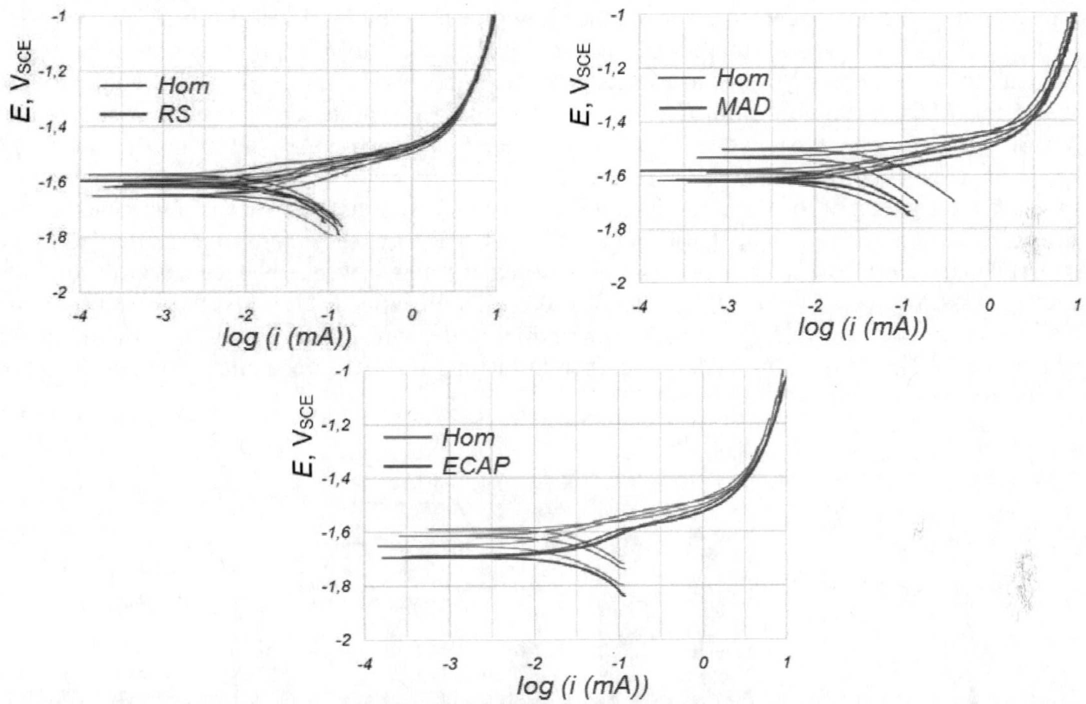

Figure 3. Polarization curves (potential *E* in volts with respect to a saturated calomel electrode (SCE) *vs.* current *i*) for the WE43 alloy in the initial homogenized state ("Hom") and after various SPD treatments.

Table 2. Results of PDP tests of the WE43 alloy in the initial state and after different SPD treatments

Treatment	E_{corr}, mV_{SCE}	i_{corr}, $\mu A/cm^2$
Homogenization at 525°C, 8 h	-1630 ± 34	21.3 ± 4.5
Rotary swaging (RS)	-1686 ± 8	29.8 ± 6.5
Multiaxial deformation (MAD)	-1595 ± 38	21.9 ± 8.9
Equal channel angular pressing (ECAP)	-1608 ± 27	30.0 ± 11.6

The hemolysis characteristics were tested for 1, 2, 3, 4, 5, and 24 hours after the incubation of the WE43 alloy samples with RBC (red blood cells) (Fig. 4a). It should be noted that all SPD treatments of the WE43 alloy decelerate the hemolysis to some extent as compared to the initial undeformed state. The lowest value of the hemolysis after the incubation time of 1 hour is reached for the alloy after ECAP and MAD: 4-5% and 8-9% compared to 80% in the undeformed state. However, with increasing incubation time, the sample processed by ECAP shows a rapid increase of the hemolysis. In the case of MAD, the hemolysis level increased more slowly than in the case of other treatments. After 5 hours of incubation, the hemolysis level of the samples after MAD was substantially lower than in other test samples.

Figure 4b presents the viability of WBCs evaluated before and after the SPD treatments. A sharp decrease in the viability of WBCs is observed already after 2 hours of incubation for a homogenized sample and an RS treated one. By contrast, samples of ECAP and MAD processed material had virtually no effect on the cell viability. After 24 hours, a significant decrease in the viability of WBCs was recorded for all SPD treatments. Note that the lowest (30%) and the highest (60%) viability levels were exhibited by the homogenized alloy and the alloy processed by MAD, respectively.

According to Fig. 4c, all WE43 samples cause a strong inhibition of MSCs proliferation. The greatest cytotoxicity to bone marrow MSCs was exhibited by the samples of homogenized alloy and those after RS treatment, whilst no significant difference was observed between these two conditions: -22% and -18% MSC proliferation, respectively. The samples after ECAP and MAD did exhibit a lower MSC proliferation: they showed 17% and 2%, respectively.

Analysis of biodegradation of the alloy in embryonic bovine serum after various treatments showed the greatest mass loss for a sample that underwent RS. By contrast, the alloy samples after ECAP and MAD exhibited a very low degradation rate. In particular, after 4 weeks of incubation in the blood serum, the mass loss was 51.1 ± 0.07% after RS; 0.2 ± 0.04% after ECAP; and 0.02 ± 0.006% after MAD – to be compared with 5.5 ± 0.16% in the initial undeformed state (Fig. 4d). It should also be noted that the WE43 alloy in the undeformed condition and after RS processing exhibited long-term gas evolution upon incubation.

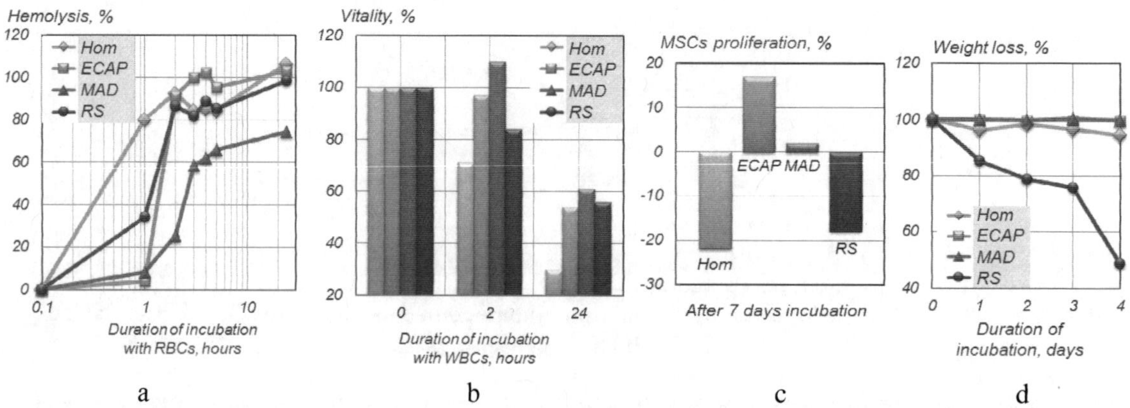

Figure 4. Results of biocompatibility tests of the WE43 alloy in the initial state and after different SPD treatments.

The results of the surface roughness evaluation for the WE43 alloy after various treatments are presented in Fig. 5.

IOP Conf. Series: Materials Science and Engineering **194** (2017) 012004 doi:10.1088/1757-899X/194/1/012004

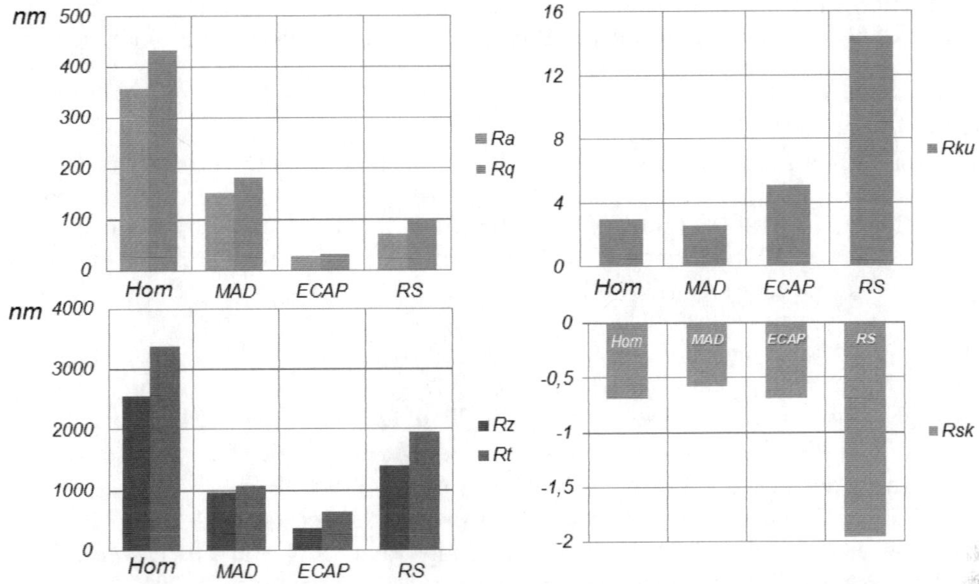

Figure 5. Surface roughness parameters of the WE43 alloy in the initial state and after different SPD treatments.

The surface roughness characteristics obtained in the work correlate with the results of the biocompatibility study. The R_a, R_t, R_z, and R_q parameters decreased after SPD in a way similar to a decrease in the hemolysis and an increase in the WBC viability and MSC proliferation relative to those of the undeformed state.

A closer look at kurtosis and skewness shows that these parameters are the largest for the samples after RS. This encourages the assumption that these parameters are the ones responsible for the extremely high biodegradation rate in the fetal serum even by comparison with that of the undeformed condition. Based on this correlation it can be surmised that the effect of SPD processing on biocompatibility of alloy WE43 can be associated with the differences in their surface morphology. These surface features are believed to be associated with the differences in the microstructures caused by SPD, particularly with regard to the grain size, dislocation density and the occurrence of deformation twins. Further studies substantiating this conjecture are certainly necessary.

4. Conclusions

1. Severe plastic deformation of the WE43 alloy by all three schemes leads to the formation of similar ultrafine-grained (UFG) structures characterized by a grain size of 0.5-1.0 μm and the presence of $Mg_{12}Nd$ phase particles 0.3 μm in size. Only the structure formed upon rotary swaging (RS) exhibited an increased dislocation density and deformation twins.

2. The UFG structure provides a high level of mechanical properties of the WE43 alloy. ECAP and MAD increase strength to 300 MPa to be compared with 220 MPa in the initial state. They also raise tensile ductility to a level of 12-17%. RS increases the ultimate tensile strength to 415 MPa, but at the cost of ductility.

3. The evaluation of the corrosion properties showed that SPD does not deteriorate the resistance to electrochemical corrosion.

4. Grain refinement by SPD predominantly improves the biocompatibility of the WE43 alloy.

As a general conclusion, it can be stated that SPD processing trialed in this study imparts to alloy WE43 a range of beneficial properties that qualify it as a candidate material for bioresorbable medical implants.

Acknowledgments
Part of this work relating to studies of biocompatibility was funded by the Ministry of Education and Science of the Russian Federation (grant #14.A12.31.0001) and the Increased Competitiveness Program of NUST «MISiS» (grant #K2-2016-062). Funding support of investigations of microstructure, mechanical performance, and corrosion properties was provided by the Russian Science Foundation (project #17-13-01488).

References

[1] Witte F, Zheng Y F and Gu X N 2014 *Mater. Sci.Eng. R* **77** p 1
[2] Hort N, Huang Y, Fechner D, Störmer M et al. 2010 *Acta Biomater.* **6** 1714
[3] Kirkland N T, Birbilis N and Staiger M P 2012 *Acta Biomater.* **8** 925
[4] Liu Y, Zheng S, Li N, Guo H et al. 2016 *Mater. Lett.* **V179** 100
[5] Li N, Guo C, Wu Y H, Zheng Y F et al. 2012 *Corros. Eng. Sci. Techn.* **V47(5)** 346
[6] Lukyanova E A, Martynenko N S, Shakhova I E, Belyakov A N et al. 2016 *Mater. Lett.* **170** 5
[7] Salandari-Rabori A, Zarei-Hanzaki A, Fatemi S M, Ghambari M et al. 2011 J. *Alloys Compd.* **693** 409
[8] Diez M, Kim H-E, Serebryany V, Dobatkin S et al. 2014 *Mater. Sci.Eng. A* **V612** 287
[9] Zhang F, Zhang K-x, Tan C-w, YU X-d et al. 2011 *Trans. Nonferrous Met. Soc. China* **21** 2140
[10] Maksimkin AV, Senatov FS, Anisimova NYu et al. 2017 *Mater.Sci.Eng.* C **73** 366
[11] Yaochite J N, De Lima K W, Caliari-Oliveira C, Palma P V et al. 2016 *Stem Cell Res Ther.* **7(1)** 14.
[12] Gadelmawla E S, Koura M M, Maksoud T M A, Elewa I M et al. 2002 *J. Mater. Process. Technol.* **123** 133

IOP Conf. Series: Materials Science and Engineering **194** (2017) 012005 doi:10.1088/1757-899X/194/1/012005

Scientific and Technological Foundations for Scaling Production of Nanostructured Metals

Terry C. Lowe[1], Casey F. Davis[1], Peter M. Rovira[1], Mathew L. Hayne[1], Gordon S. Campbell[1], Joel E. Grzenia[1], Paige J. Stock[1], Rilee C. Meagher[1], Henry J. Rack[2]

[1]George S. Ansell Department of Metallurgical and Materials Engineering, Colorado School of Mines, Golden, Colorado 80401 USA
[2]School of Materials Science and Engineering, Clemson University, Clemson SC 29634 USA

lowe@mines.edu

Abstract. Severe Plastic Deformation (SPD) has been explored in a wide range of metals and alloys. However, there are only a few industrial scale implementations of SPD for commercial alloys. To demonstrate and evolve technology for producing ultrafine grain metals by SPD, a Nanostructured Metals Manufacturing Testbed (NMMT) has been established in Golden, Colorado. Machines for research scale and pilot scale Equal Channel Angular Pressing-Conform (ECAP-C) technology have been configured in the NMMT to systematically evaluate and evolve SPD processing and advance the foundational science and technology for manufacturing. We highlight the scientific and technological areas that are critical for scale up of continuous SPD of aluminum, copper, magnesium, titanium, and iron-based alloys. Key areas that we will address in this presentation include the need for comprehensive analysis of starting microstructures, data on operating deformation mechanisms, high pressure thermodynamics and phase transformation kinetics, tribological behaviors, temperature dependence of lubricant properties, adaptation of tolerances and shear intensity to match viscoplastic behaviors, real-time process monitoring, and mechanics of billet/tooling interactions.

1. Introduction

The foundational principles underlying Severe Plastic Deformation (SPD) and Ultrafine Grained (UFG) metals and alloys have been developed over the past 25 years, as highlighted by multiple excellent overviews [1,2]. Alloy systems based upon aluminum, copper, magnesium, iron, titanium have been thoroughly explored to elucidate the mechanisms by which SPD imparts desirable UFG microstructures and associated properties, including high strength and ductility [3,4]. Computational models of thermomechanical deformation, some including microstructure evolution, have been developed to guide SPD process specification to produce UFG metals and alloys [5–8]. Over 60 variants of SPD processing methods have been published or patented [9]. These advancements support the prospect of industrial production of UFG metals and alloys. Commercial offerings of UFG metal products are presently few in number. Notable examples include the use of ECAE® by Honeywell in the USA to produce aluminum and copper sputtering targets [10]. Another example is production of Grade 4 titanium rods by NanoMet in Ufa, Russia for the fabrication of Nanoimplant® dental implants by Timplant in the Czech Republic [11].

The advantages of UFG metals and alloys merit manufacture for additional applications. However, for each prospective product application, appropriate SPD processing must be developed. For the examples of aluminum, copper, and titanium products cited above, several years of process development, equipment optimization, and product inspection technologies were undertaken [10]. However, with the larger knowledge base now available on microstructures and techniques for SPD, the time needed to develop new products can be shortened. Nevertheless, systematic progression through sequential manufacturing readiness levels (MRLs), illustrated in Figure 1, is needed.

Manufacturing Readiness Level (MRL)		
Phase	MRL	State of Development
Phase 3: Production Implementation	9	Full production qualified against performance metrics
	8	Production qualified for full range of products
	7	Capability & rate confirmed
Phase 2: Pre-production	6	Process optimized for production rate/equipment
	5	Basic capability demonstrated
Phase 1: Technology Assessment	4	Production validation in lab environment
	3	Experimental proof of concept complete
	2	Application and validity of concept demonstrated
	1	Concept proposed with scientific validation

Figure 1. Manufacturing readiness evolves through three phases and nine distinct levels.

The majority of contributions to the archival literature on SPD appear to pertain to aspects of UFG metals outside the scope of manufacturing. Even as new SPD methods are proposed, their development and publication generally falls within MRLs 1, 2, and 3. Only in a few instances, for example, Levashov et al. [12] are developments above MRL 4 reported. Yet, for reproducible manufacturing of UFG metals, SPD methods and their ancillary processes must all be developed through level 9.

A Nanostructured Metals Manufacturing Testbed (NMMT) has been established to enable research and development up to MRL 8. The NMMT contains four progressively larger and increasingly sophisticated machines for SPD. These machines allow development of SPD parameters and demonstration of pilot scale production for a wide range of alloys. The sections that follow highlight areas that are being addressed in the NMMT for pre-commercial development and pilot-scale production of iron, titanium, aluminum, magnesium, copper, and zirconium based alloys.

2. Achieving process efficiency through high shear deformation

Manufacturing demands the highest possible efficiency of all production sub-processes. One measure of process efficiency is the degree of enhancement of product value per unit cost. For bulk metals this includes achieving the greatest increments in properties with the smallest possible investment in capital resources and processing time. While many factors influence this measure, the efficiency with which strength-enhancing microstructures can form during SPD of billets is increased by utilizing techniques that maximize the shear components of the strain tensor and maintain high lattice rotation rates via intermittent strain path changes. The manner in which shear-dominated deformation maximizes microstructure evolution rate has been analyzed and summarized by Gu et al. [13]. They showed that shear deformation modes are most effective for grain refinement, particularly when combined with non-monotonic strain paths. The rate of dislocation storage and subsequent rearrangement into sub-boundaries and grain boundaries has been correlated with rotation-induced lattice curvature and associated curvature-induced dislocations [14]. Equal Channel Angular Pressing (ECAP) and one of its variants capable of producing long rods or coil, Equal Channel Angular Pressing-Conform (ECAP-C), impart lattice rotation rates that are greater than can be achieved by conventional rolling, drawing, extrusion, and forging. Thus, the efficiency of grain refinement is intrinsically higher than achievable by conventional metal forming methods. Others, for example, Kocich et al. [15] have focused on innovative variants of ECAP that further enhance the degree of shearing and lattice rotation to maximize microstructure refinement. To emphasize the importance of shearing, we introduce the term High Shear Deformation (HSD) to describe a subset of SPD for which one or more of the shear components of the Green strain tensor E_{ij} are greater than 4, and also much larger than the diagonal components E_{ii}. ECAP and High Pressure Torsion (HPT) are examples of SPD processes which are also HSD processes.

3. Starting microstructures for commercial HSD processing

For most thermomechanical processing, the starting condition of billets influences the subsequent evolution of microstructure. For commercial HSD processing the starting microstructures, including starting grain size and initial texture, must be well characterized. HSD results in the formation of grain boundaries having desirable features that are not present in boundaries created by crystallization during solidification, static recrystallization or most other primary or secondary deformation processes. Specifically, grain boundaries formed by HSD possess various degrees of disorder that enhance their effectiveness as barriers to dislocations and are also resistant to migration at high temperatures [16]. With this in mind, higher strength and microstructural stability can be achieved by maximizing the area of grain boundaries formed by HSD. Consequently, initiating HSD processing with fully annealed coarse grain samples is desirable. However, the number of HSD steps to achieve sub-micron ultrafine grain sizes may be reduced if the starting grain size is small, despite having boundaries with characteristics that are not as desirable. Thus, there is a tradeoff between manufacturing efficiency and achieving optimal properties by HSD. The choice of initial microstructural condition for beginning HSD depends ultimately upon the economic value of achieving specific increments in properties for specific product applications.

Similar considerations pertain to the effects of initial texture. HSD processing of rod, bar, wire or plate generally begins with product forms that contain classical textures from prior rolling, drawing or extrusion. Random or weak initial textures provide the widest range of grain orientations for subsequent HSD. During HSD distinct textures form [17], eventually replacing most attributes of any prior texture. However, the presence of strong initial texture can influence the deformation mechanisms that are active during HSD, particularly for alloys that undergo significant deformation twinning. For example, the presence of specific textures can encourage the activation of $\{10\bar{1}2\}\langle\bar{1}011\rangle$ and $\{11\bar{2}2\}\langle\bar{1}\bar{1}23\rangle$ twins in titanium at an early stage of HSD processing [18]. This affects grain refinement and the textures that ultimately form by HSD.

4. Deformation mechanism analysis

The design and implementation of HSD processing significantly depends upon the anticipated deformation mechanisms. Intrinsic alloy characteristics including stacking fault energy, second phase morphology, solute concentrations, initial grain size, and initial texture determine the relative prominence of slip, twinning, solid state diffusionless transformations, grain boundary sliding, and diffusional deformation mechanisms. These characteristics influence the selection of temperature and strain rate for commercial HSD.

IOP Conf. Series: Materials Science and Engineering **194** (2017) 012005 doi:10.1088/1757-899X/194/1/012005

An excellent example of the interplay of deformation mechanisms during HSD can be seen in ECAP processing of austenitic stainless steels. Ueno et al. [19] studied the prospect of enhancing microstructure refinement through selection of processing temperatures during ECAP that promote twinning and strain-induced phase transformation. They selected ECAP temperatures to achieve optimum strength, recognizing the grain size and temperature dependencies of mechanical twinning. By choosing conditions to maximize twinning during the first ECAP pass they were able to enhance the rate of microstructure change during further ECAP.

5. Plastic and elastic billet response

Because of the very large strains imposed during HSD, most attention is focused on the plastic response of billets. However, for continuous HSD methods that involve multiple process passes, such as ECAP-C, the elastic properties of billets are important. Billets for ECAP-C are elastically and plastically deformed as they are introduced into and constrained by ECAP-C tooling. The relative compliances of the billet and the tooling, combined with the billet flow stress, affect the magnitude of shear tractions present at the tooling-billet interface. The ECAP-C method relies upon billet-die friction within the billet channel to provide the force to cause intense shear at the die channel intersection. Alloys with high elastic moduli such as steels attain high levels of force with small elastic deflections, while alloys with low elastic moduli such as magnesium undergo larger elastic deflections to attain a comparable level of shear traction during HSD. Consequently, control of forces for continuous feeding of billets must consider elastic modulus. Furthermore, elastic properties also influence the choice of type and degree of lubrication used with ECAP-C billets.

Billets exiting from HSD tooling expand elastically as the constraint of the confining dies is removed. This results in geometrical changes in the billets, which increase in magnitude following successive passes of multistage HSD. For example, during ECAP-C the geometric properties of the ECAP-C billet are determined by dimensional settings, elastic properties, flexural constraints provided by the tooling. Billets undergo elastic and plastic deformation within the die channel. However, the dimension of the exiting billet is determined by the combination of the yield stress and the elastic moduli of the billet material. After each pass of ECAP-C, the yield strength is increased, while the elastic properties remain approximately constant. Therefore, for fixed die channel dimensions, the size of the exiting billet increases with increasing number of passes and increasing billet strength. For non-continuous multi-pass HSD methods, the practice of reshaping or machining billets to desired dimensions between passes is common. However, for continuous multistage HSD the tooling and process parameters must anticipate evolving billet dimensions and flow characteristics, without the option of halting processing between passes to reshape billets.

6. Tribology, lubricants, surface strains, and homogeneity

The interface between billets and tooling during HSD can affect homogeneity of deformation in the billet, surface quality, tool wear, and overall effectiveness of microstructure refinement. For ECAP-C, billet-tooling interactions also influence the forces available for imposing intense shear at the die channel intersection. Selection of lubricants is determined by these factors. Because the properties of lubricants depend upon temperature, pressure, and billet/tooling surface condition, it is essential to specify lubricant composition and application for every commercial HSD process. Lubricants can also be optimized for specific HSD tooling materials.

Lubricants for ECAP-C processing have been proposed for specific temperature ranges (77°K – 293° K, 20°C-200°C, 200°C-400°C, 400°C-650°C, and greater than 650°C. For the range from 20°C-200°C lubricants for pressures up to 800 MPa have been tested and found to produce dynamic coefficients of friction between 0.07 and 0.90 for total sliding distances up to 45 meters. In practice, intermediate coefficients of friction can be obtained by mixing of multiple lubricants.

7. Tooling and tolerances

Selection of tooling materials and tolerances for HSD dies requires consideration of billet properties, operating temperatures, pressures, and production rates. For difficult-to-deform alloys such as Co-Cr and some titanium alloys tooling wear resistance and fracture toughness are the leading factors governing the economics of production. For ductile, low melting point alloys such as aluminum or magnesium, tooling tolerances are among the most important factor limiting production efficiency and multi-pass operations.

Tooling for processes such as extrusion, drawing, or rolling possess simple, axisymmetric geometries. For HSD processes such as ECAP-C, complex asymmetric tooling components are employed, requiring special fabrication methods and in some cases, limiting the choice of tooling materials that be used.

8. Process monitoring

The quality of billets produced by ECAP-C can measured and adjusted in real-time by in-process monitoring of temperatures, surface condition, and billet dimensions. In-process monitoring increases the complexity and cost of HSD operations, but provides essential benefits. Because the properties of billets evolve so significantly during HSD, real-time feedback is needed to ensure uniformity of properties along the entire length of continuously fed billets.

9. High pressures effect phase transformation

Severe Plastic Deformation processing requires the presence of high pressure to ensure stability of shearing. The presence of high pressure introduces the possibility of pressure-induced phase transformation. For alloys, we rely upon equilibrium phase diagrams to establish the temperatures and compositions needed to obtain specific phases. Virtually all phase diagrams are published based on experiments or calculations for ambient pressure. However, for SPD and HSD processing, consideration must be given to the presence of pressures on the order of 1 GPa and higher.

HSD of titanium provides an important example of the importance of high pressure. Studies of high pressure torsion and ECAP documented the formation of the high pressure omega phase in commercial purity titanium[20,21]. Similar behavior has also been reported for zirconium [22] .

10. Summary

Scaling of Severe Plastic Deformation and High Shear Deformation processing for full scale production requires attention to overall process efficiencies, details of microstructure, deformation mechanisms, alloy elastic properties, high pressure thermodynamics, and ancillary technologies. Tribological phenomena are particularly important for continuous High Shear Deformation processes such as ECAP-C that critically depend upon billet-die friction. Die-billet interactions influence the cost of production and the homogeneity of microstructures that can be achieved. Tooling characteristics must be matched to specific alloys to achieve efficient and reliable operations. Furthermore, in-process monitoring is essential to mitigate the effects of variation in billet characteristics during sequential processing passes. Finally, because of the high pressures that are essential for reliable High Shear Deformation processing, it can be important to consider the possibility of high pressure phases being introduced in some alloys.

11. References

[1] Valiev R Z, Estrin Y, Horita Z, Langdon T G, Zehetbauer M J and Zhu Y T 2016 Fundamentals of Superior Properties in Bulk NanoSPD Materials *Mater. Res. Lett.* **4** 1–21

[2] Valiev R Z and Langdon T G 2006 Principles of equal-channel angular pressing as a processing tool for grain refinement *Prog. Mater. Sci.* **51** 881–981

[3] Valiev R Z, Murashkin M Y, Kilmametov A, Straumal B, Chinh N Q and Langdon T G 2010 Unusual super-ductility at room temperature in an ultrafine-grained aluminum alloy *J. Mater. Sci.* **45** 4718–24

[4] Zhao Y H, Liao X Z, Cheng S, Ma E and Zhu Y T 2006 Simultaneously Increasing the

Ductility and Strength of Nanostructured Alloys *Adv. Mater.* **18** 2280–3

[5] Kocich R, Greger M and Machackova A 2007 FEM simulation of extrusion by ECAP method on magnesium alloy AZ91 *Int. J. Comput. Mater. Sci. Surf. Eng.* **1** 438–51

[6] Yang C, An Y, Tort M and Hodgson P D 2014 Fabrication, modelling and evaluation of microstructured materials in a digital framework *Comput. Mater. Sci.* **81** 89–97

[7] Stops A, Wilcox R and Jin Z 2012 Computational modelling of the natural hip: a review of finite element and multibody simulations *Comput. Methods Biomech. Biomed. Engin.* **15** 963–79

[8] Mishnaevsky Jr L, Levashov E, Valiev R Z, Segurado J, Sabirov I, Enikeev N, Prokoshkin S, Solov'yov A V, Korotitskiy A and Gutmanas E 2014 Nanostructured titanium-based materials for medical implants: Modeling and development *Mater. Sci. Eng. R Reports* **81** 1–19

[9] Lowe TC and Valiev RZ 2014 Frontiers for bulk nanostructured metals in biomedical applications *Biomaterials and Biodevices* (Wiley Scrivener) pp 3–52

[10] Valiev R Z (Ruslan Z, Zhilyaev A P and Langdon T G *Bulk nanostructured materials : fundamentals and applications*

[11] Valiev R Z Z, Semenova I P P, Latysh V V V., Rack H, Lowe T C C, Petruzelka J, Dluhos L, Hrusak D and Sochova J 2008 Nanostructured Titanium for Biomedical Applications *Adv. Eng. Mater.* **10** B15--B17

[12] Levashov E A, Petrzhik M I, Shtansky D V, Kiryukhantsev-Korneev P V, Sheveyko A N, Valiev R Z, Gunderov D V, Prokoshkin S D, Korotitskiy A V and Smolin A Y 2013 Nanostructured titanium alloys and multicomponent bioactive films: Mechanical behavior at indentation *Mater. Sci. Eng. a-Structural Mater. Prop. Microstruct. Process.* **570** 51–62

[13] Gu C F, Tóth L S, Arzaghi M and Davies C H J 2011 Effect of strain path on grain refinement in severely plastically deformed copper *Scr. Mater.* **64** 284–7

[14] Tóth L S, Estrin Y, Lapovok R and Gu C 2010 A model of grain fragmentation based on lattice curvature *Acta Mater.* **58** 1782–94

[15] Kocich R, Kunčická L and Macháčková A 2014 Twist Channel Multi-Angular Pressing (TCMAP) as a method for increasing the efficiency of SPD *IOP Conf. Ser. Mater. Sci. Eng.* **63** 12006

[16] Hasnaoui A, Van Swygenhoven H and Derlet P . 2002 On non-equilibrium grain boundaries and their effect on thermal and mechanical behaviour: a molecular dynamics computer simulation *Acta Mater.* **50** 3927–39

[17] Li S, Gazder A A, Beyerlein I J, Davies C H J and Pereloma E V 2007 Microstructure and texture evolution during equal channel angular extrusion of interstitial-free steel: Effects of die angle and processing route *Acta Mater.* **55** 1017–32

[18] Sun J L, Trimby P W, Yan F K, Liao X Z, Tao N R and Wang J T 2013 Grain size effect on deformation twinning propensity in ultrafine-grained hexagonal close-packed titanium *Scr. Mater.* **69** 428–31

[19] Ueno H, Kakihata K, Kaneko Y, Hashimoto S and Vinogradov A 2011 Nanostructurization assisted by twinning during equal channel angular pressing of metastable 316L stainless steel *J. Mater. Sci.* **46** 4276–83

[20] Edalati K, Daio T, Arita M, Lee S, Horita Z, Togo A and Tanaka I 2014 High-pressure torsion of titanium at cryogenic and room temperatures: Grain size effect on allotropic phase transformations *Acta Mater.* **68** 207–13

[21] Dyakonov G S, Zemtsova E, Mironov S, Semenova I P, Valiev R Z and Semiatin S L 2015 An EBSD investigation of ultrafine-grain titanium for biomedical applications *Mater. Sci. Eng. A* **648** 305–10

[22] Pérez-Prado M T, Gimazov A A, Ruano O A, Kassner M E and Zhilyaev A P 2008 Bulk nanocrystalline ω-Zr by high-pressure torsion *Scr. Mater.* **58** 219–22

IOP Conf. Series: Materials Science and Engineering **194** (2017) 012006 doi:10.1088/1757-899X/194/1/012006

Tribological and Geometrical Analysis of the Friction Forces of Continuous Severe Plastic Deformation

T Donič[1,3], B Hadzima[1] and G Raab[2]

[1] Research Centre of University of Žilina, University of Žilina, Žilina, Slovak Republic
[2] Institute for Physics of Advanced Materials, Ufa State Aviation Technical University, Ufa, Russia

E-mail: tibor.donic@gmail.com

Abstract. Equal Channel Angular Rolling (ECAR), a severe plastic deformation process, is suitable for the shear deforming a long and thin sheet continuously. An interesting issue is that thickness of a sheet is not reduced during ECAR. Although the shear texture and fine grain structure are easily obtained by ECAR, yet the ECAR process's difficulties in terms of technical control, such as surface defect, low ductility and low processing speed, still remain. The surface defects and processing speed are partially improved by applying a series deformation of rolling and ECAR. The article presents methodological and analytical procedures, adaptation of a conventional rolling mill Duo-system for processes of a SPD-representative which are represented by the method ECAR. They specify in detail the technical requirements whose new design solutions should be implemented. They present in detail the mechanical load drive shafts, as well as the calculation of compression forces with consideration of the friction forces. The geometric analyses of localisation positions of the outlet channel in terms of size and orientation of the dominant shear plane are also presented.

1. Introduction

In the framework of solving the project No. 0393-2016 – Investigation of Combined Techniques of Severe Deformation by Shear, Promising for Industrial Application – Creation of Nanostructures in Metal Materials through Intensive Plastic Deformations – a partial task developed – to make suitable technical equipment for realising experiments oriented on the continual processes. Therefore a team of designers entered the investigation group and their task is to prepare design documentation of this equipment.

From the design point of view it is a new task whose functional core cannot be taken over from other design patterns [1] [2] [3]. The process of the intensive plastic deformation has been experimentally verified at the University of Žilina, however, only with small samples in short-term processes during rectilinear motion to the place of the plastic deformation.

As several meetings of the scientific and research teams have shown the tasks of the aforementioned groups are as follows:

1. To make (design, manufacture and put into operation) equipment for realising the experiments for creating the nanostructures in the metal materials through intensive plastic deformations by the ECAR process which will be an extending module in the experimental rolling mill which is available

in the labs of the Department of Forming at the Centre of Theoretical and Technological Plastometry of Construction Materials.

2. To invent a design of equipment for carrying out the experiments of creating the nanostructures in the metal materials through intensive plastic deformations for long samples with a possibility of realising the continual process [4]. This procedure will be realised at the University of Žilina in Žilina. The working title is ECAR 1 Žilina.

As a matter of fact there are two design tasks which distinguish from each other by the assignment and realisation time deadlines. While the first task is connected with utilising the existing equipment (the rolling mill) and has strictly monitored deadlines, the second task has not that strict deadlines and is freer also from the point of view of its solution.

2. ECAR 1 Žilina

2.1. Assignment

The table 1 shows and defines the conditions and requirements of the solution. The list of requirements is still not finished. In this solution phase (the solution design) it is necessary - not only for the interested parties (designers) but especially for the users of the equipment in the future- to confirm these requirements and if there are any new ones to present all of them without any delay. This is the way how to complete the assignment for the designers to be appropriately informed and could take into account these requirements during designing the equipment.

These requirements can be divided into the general and specific ones. The designers are informed about the general requirements; they are also valid in other design tasks. They are, e.g. requirements on the strength, toughness, manufacturability, safety, etc. The designer has to learn about the specific requirements from those who assign the task, from the user of the device in the future because they are based on special requirements of the technological process, operating conditions, the available space, the goals of the ordering party in the area of its utilisation, the life span and other necessary properties but also economic possibilities in which the designed equipment will be produced and utilised.

2.2. Concept

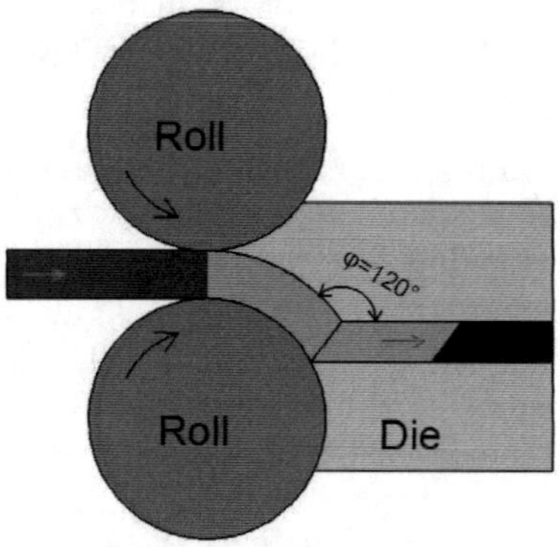

Figure 1. The solution concept is based on the kinematic scheme of the process presented as ECAR 1. The specialist in the area of forming materials will account for its design. So we know the physical process and its effects. The figure 1 already roughly defines the solution concept. The designer's task is to find and define such technical means that will realise this process.

Table 1. Specific requirements for device properties ECAR 1 Žilina

Requirements, wishes	Source	Notes, giving reasons
The main requirement: realising ECAR process	Task	The purpose of this task
The maximal permissible forces of the rolling mill DUO-210 in Žilina for the design calculations are: **2 x 150 KN** (the force under the left and right adjusting screw).	Minutes from the meeting held on 8th December 2016	

The maximal strength of the deformation material: **Rm = 900 MPa**	Minutes from the meeting held on 8th December 2016	
Utilising the existing rolling mill in Žilina	Discussion	Economic reason, valid for the variant
Possibilities of adjustments for the ECAR 1 needs - angles	Discussion	
The maximal width of the cross-section 100 mm	Discussion	According to the Žilina rolling mill
The rolling speed 25 mm/s	Bibliography	Approximate
Modular design for replacing the tool	Discussion	Variability
Increase of the thrust of the rollers – grooved rollers, other pairs of driven rollers at the input - removable	Discussion	One pair of rollers will be probably not enough
Simple and quick disassembly	Discussion	Taking out the sample, cleaning
Modular design from the point of view of the tool variability	Discussion	Variability of experiments
Modular design from the point of view of extending the system	Discussion	Possibility to extend
Necessity of implementing scanners for measuring the forces and speeds (revolutions)	Discussion	Regulation of forces and speeds. Side results of the experiments
Safety and protection of health, technical safety	Discussion	A cover?
Possibility of cooling the tools	Research of patents	Deformation heat
Using the mechanical systems, not hydraulics	Discussion	Based on the experience

The utilisation of the existing rolling mill is the most important requirement. It means to look for the methods of connection, the shapes, arrangement and dimensioning for the requirements presented to designers to be fulfilled.

2.3. Mechanical Load of ECAR 1 Drive – Estimation

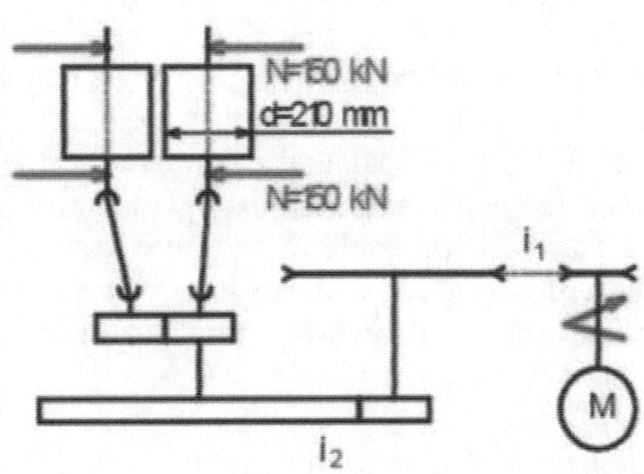

Figure 2. Rolling mill drive

Assumptions:
- Normal contact force on the rollers $N = 2{\cdot}150$ kN
- R_m of the sample up to 900 MPa
- Factor of increasing the pressure $k_{90} = 5$

k_{90} is experimentally verified value of the pressure necessary for extruding a sample of the AlMg alloy through a 90° angle in the R_m ratio during experiments in Žilina

Given facts:
Engine output $P_0 = 9.5$ kW at the speed of $n_0 = 550$ min^{-1} (data from the machine label)

Gears:	Belt transmission	$i_1 = 460/875$ mm
	Gearing	$i_2 = 17/100$
	Roller diameter	$d = 210$ mm

Calculated:

Roller speed	$n = n_0 \cdot i_1 \cdot i_2 = 550 \cdot (460 / 875) \cdot (17 / 100) = 49$ min^{-1}
Roller peripheral velocity	$v = n \cdot d \cdot \pi \cdot 60^{-1} = 49 \cdot 210 \cdot \pi \cdot 60^{-1} = 538$ mm·s^{-1}

The force F by which the rollers are able to press the sample to the tool (horizontal) is calculated from the output of the drive and the roller peripheral velocity (the effectiveness of the gear, dynamic effects and the possibility of overloading the electric motor are not taken into account) is approximately:

$$F = P/v = 9500/0.538 = 17\,658 \text{ N} = 17 \text{ kN}$$

For the ECAR 1 purposes this force seems to be small and the speed too high.
To achieve the ECAR 1 parameters, i.e. the force on the roller circumference 120 kN, there are the following possibilities:

- An ideal solution: to insert another gear to multiple the torque approximately 1:7, then the force would be $F = 120$ kN; the peripheral velocity $v=76$ mm.s^{-1}, however, this possibility would mean a principal reconstruction of the drive,
- Partial (compromise) measures,
  - to increase the gear ratio of the belt transmission by exchanging the small belt,
  - to increase the torque of the driving engine by changing the drive regulation,
  - to utilise the dynamic effects (inertia) of the drive,
  - to combine all the aforementioned measures.

A probable solution could be also reducing the required output parameters – i.e. to deal only with materials of lower strengths and samples with smaller cross-sections.
The frame of the rolling mill is designed for transmitting normal forces to the surface of the rolled materials, not for achieving such large tangential forces on the roller circumference required by the ECAR 1. This fact logically results from the function of the rolling mill. The stability of the transmission of the output to the rollers is not guaranteed too.

There is an assumption that the rollers (also the frame, bearings, shafts) are dimensioned for the thrust of 2 x 150 MPa. We can assume that the design is over-dimensioned and is able to transmit larger forces than the adjusting thrust screws. If there is no structure disruption during the first overload, it is probable that the life span of the bearings and other components will be reduced after repeated overloads. Due to the cyclic strain at higher stresses there is a higher probability of the fatigue defects. However, this fact regarding the experimental character of the equipment is not essential..

2.4. Check of the Input Shafts
This is only an approximate calculation because we do not know the exact values of the mechanic load, our calculations are based on estimated values. The shafts at the input are loaded only by rotation.

The torque on the shaft of the roller develops by the force on the roller circumference, on one roller there is half of the necessary force 120 kN; i.e. F/2=60 kN (see the figure 4). The shaft at the input has a diameter of 120 mm and is loaded only by the torque. The bending moments on the input shaft can be derived only from its own weight.

$$Mk = (F/2) \cdot r = 60\,000 \cdot 0.105 = 6\,300 \text{ Nm}$$

where r is the roller radius.

Figure 3. Shafts of the roller drives

Check of the roller – rotation:

$$W_k = 0.2 \cdot D^3 = 0.2 \cdot 120^3 = 345\ 600\ \text{mm}^3$$

$$\tau_k = M_k/W_k = 6\ 300\ 000/345\ 600 = 18.23\ \text{MPa}$$

where D is the roller diameter at the input.

If we assume the roller is made of common structural steel where R_e = 300 MPa, the roller shaft is suitable and has high safety.

Check of the propeller shaft with square section 82 x 82:

$$W_k = 0.208 \cdot h^3 = 0.208 \cdot 823^3 = 114\ 684\ \text{mm}^3$$

$$\sigma_k = M_k/W_k = 6\ 300\ 000/114\ 684 = 54.93\ \text{MPa}$$

Also these shafts are suitable.

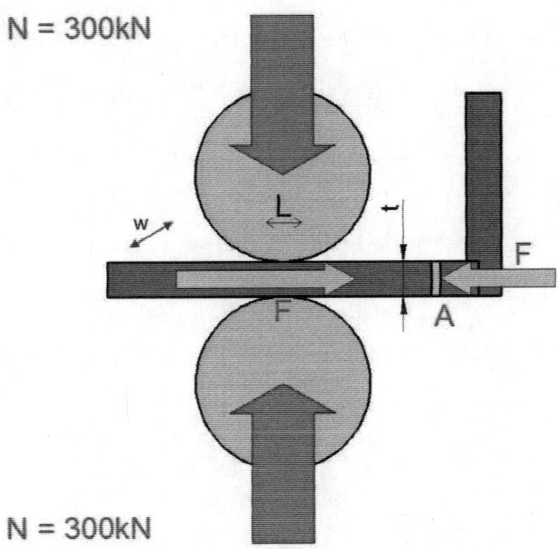

Figure 4. Scheme for the calculation of compression forces on the basis of friction

3. Conclusion

This article documents the complicatedness of the area – design of experimental equipment for creating nanostructures by continual method of intensive plastic deformation. The most important area of the design will be to cope with large forces and pressures which have to be generated high above the threshold of the material strengths of the samples being processed.

The design activity during the next solution stage will be concentrated on the task ECAR 1 Žilina by selecting a variant, making the building structure through necessary dimensional calculations and analyses more accurate as well working out the design documentation.

References

[1] Valiev R 2009 *Int. Journal of Materials Research* **100** 757

[2] Zhu Y T , Han B Q, Lavernia E J 2009 *Deformation mechanisms of nanostructure materials, In: Bulk Nanostructured Materials*, ed. M J Zehetbauer, Y T Zhu, (Weinheim: WILEY-VCH Verlag) pp 89-108

[3] Raab G J, Valiev R Z, Lowe T C, Zhu Y T 2004 *Materials Science and Engineering* A vol 382 pp 30-34

[4] Valiev R Z, et al 2000 *Bulk nanostructured materials from severe plastic deformation, Progress in Materials Science* vol 45 pp 103-189

Acknowledgments

The authors gratefully acknowledge the help rendered by Schneider Electric, Ltd – system integrator of Schneider Electric for expert consultation and hardware support of electric subsystems. Sincere gratitude also goes to Martin Žarnay and Ronald Baštovanský who participated in the early stage of the tooling design. The authors gratefully acknowledge Miroslav Oleš for his help with designs in the article.

The results were supported by Project of international cooperation between the Slovak Republic and the Russian federation, No. 0393-2016 – Investigation of Combined Techniques of Severe Deformation by Shear, Promising for Industrial Application and by the Ministry of Education and Science of the Russian Federation under the Grant agreement No. 14.586.21.0025 (unique identification number RFMEFI58616X0025) and by ERDF and State budget of the Slovak Republic by the project ITMS 313011D011.

Improving hydro-formability of stainless steel tubes by tube channel pressing

Y Kitano[1], M Yuasa[2], H Miyamoto[2], M H Farshidi[3], and E Bagherpour[4]
[1]Graduate School of Science and Engineering, Doshisha University, Kyoto, Japan.
[2]Department of Mechanical Engineering, Doshisha University, Kyoto, Japan.
[3]Department of Materials Engineering, Ferdowsi University of Mashhad, Razavi Khorasan, Iran,
[4]Faculty of Metallurgical and Materials Engineering, Semnan University, Semnan, Iran.

E-mail: duq0532@mail4.doshisha.ac.jp

Abstract. Tube channel pressing (TCP), which is one of the severe plastic deformation (SPD) technologies to refine grain size into submicron size for tubular materials, have been applied to ferritic stainless steel tubes for one pass, in order to alleviate ridging and enhance the hydro-formability. It was found that grain-scale shear bands were introduced by one-pass TCP, and texture and microstructure was successfully modified by promoting recrystallization of deformation microstructure, which is otherwise hard-to-recrystallize, in the post-TCP annealing. Elongation to failure, strain-hardening exponent (n-value) and Lankford values of both longitudinal and circumferential directions increased in comparison to with the tube fabricated by conventional process.
Keywords: ridging, formability, recrystallization, randomize, tube channel pressing

1. Introduction

Ferritic stainless steels (FSS) have high corrosion resistance, heat resistance and press workability. Therefore, they have been applied to various industrial applications, mostly as sheets or tubes. For example, tubes of FSS has been applied to exhaust tubes of automobile, which is frequently deformed into complicated shapes by hydro-forming. However, FSS exhibit ridging when they are subjected to tensile plastic strain in the rolling direction leading to reduction of the formability and the early fracture [1]. The ridging is a kind of rumple and it is caused by plastic anisotropy of the so-called colony with similar crystallographic orientations and is a feature of ferritic stainless steel. Therefore, it is important to promote recrystallization and randomize the texture to alleviate ridging.

In our previous study, we demonstrated that the ridging can be alleviated and the formability is enhanced in FSS sheets by equal-channel angular pressing (ECAP) [2], which is one of severe plastic deformation (SPD) methods. It was found that <100>//ND grains effectively recrystallized and changed into other orientations leading to randomization of texture in the post-ECAP annealing. Dense grain-scale shear bands were introduced by ECAP in large grains of <100>//ND which is otherwise hard-to-recrystallize, and promote recrystallization. There are some processes of SPD about tube materials, for examples, tube twisting is effective to impart strains [3]. Recently, modified ECAP was invented for tube materials and called tube channel pressing [4] and TCP which have been successfully applied to aluminium alloys tube to obtain ultra-fine grain (UFG) structure [4]. In the present study, we applied one-pass TCP to FSS tube in order to control texture and alleviate ridging and enhance the formability.

2. Experimental Procedure

2.1. Entire Process

Figure 1 shows entire flow of conventional process and TCP process. FSS (AISI409L) tubes of 42.7 mm in diameter and 1.0 mm in thickness, 25 mm in length were deformed by TCP process for one pass. Before and after TCP process, tubes were annealed. For comparison, the tubes were annealed before TCP as conventional process in order to relieve the plastic strain of as-received state. Tensile test was carried out to evaluate *r*-value, *n*-value and ridging. An optical microscope was used to see microstructure and a laser microscope was used to measure the surface roughness.

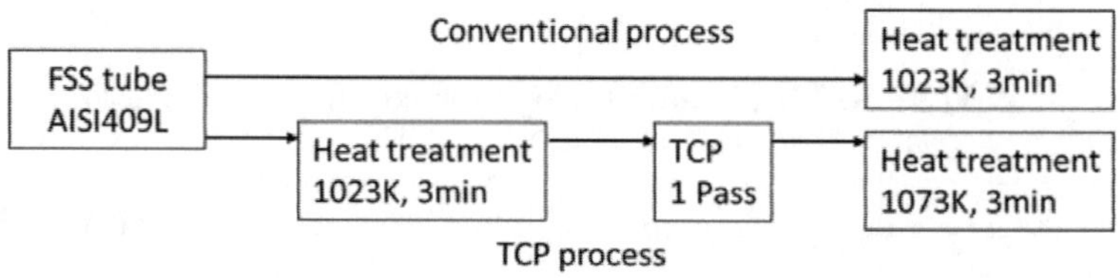

Fig.1 Entire flows of both conventional and TCP process.

2.2. TCP process

TCP is one of SPD process developed by Iranian researchers in order to refine grain size of tube materials [5]. TCP is a modified SPD of ECAP, and intensive shear deformation is given at the bended part of the circular channel. As shown in Figure 2, tube sample is pressed into the die by a tube plunger. In this way, the sample can be pressed for several times until UFG formation. Although, the sample is commonly deformed by simple shear in both ECAP and TCP, there are two important different points. First, there are three to four bended parts where the material is subjected to shear strain in TCP, whereas the typical ECAP has only one, so that billets are subjected to shear strain for several times according to die design in one pass.

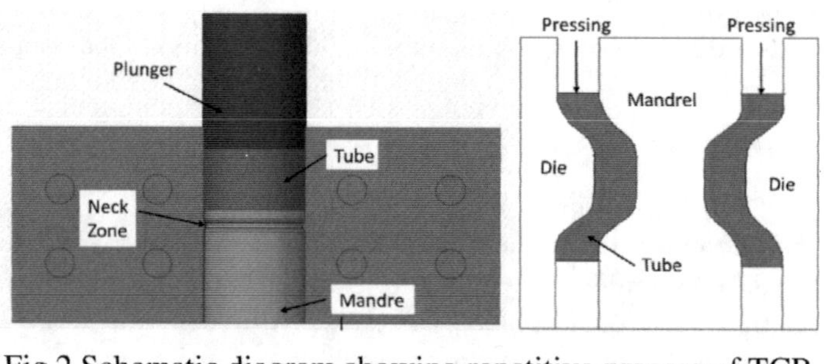

Fig.2 Schematic diagram showing repetitive process of TCP.

3. Results and discussion

3.1. Microstructure

Figure 3 shows optical micrographs before and after TCP. Equiaxial grains before TCP was deformed to elongated shape by shear deformation by one-pass TCP. After TCP, small shear bands are visible inside some grains as shown in Figure 3 (b) and (d). As discussed previous study, such small shear bands are observed in BCC metals such as stainless steels[6], carbon steels[7-10] and IF steels[7,8,11,12], and FCC metals with high stacking fault energy such as aluminum alloys[13-15]. This small shear bands are alternatively called in-grain shear bands [7-9, 11], grain-scale shear bands [13] or micro shear bands [15] by different researchers, and in this paper they are called grain-scale shear bands. It can be confirmed there are shear bands in two directions. This is because there are several shearing points through 1 pass TCP. The tube is squeezed in the radial direction accompanied by thickening of tube wall. Thus, shear strain is given in not only through-thickness direction as in ECAP, but also circumferential direction as shown in Figure 3 (e). Vickers hardness and microstructures of neck zone after TCP are shown in Figure 4. They have three shear parts. It is confirmed that Vickers hardness increase as much as every shear part and shear bands increase gradually.

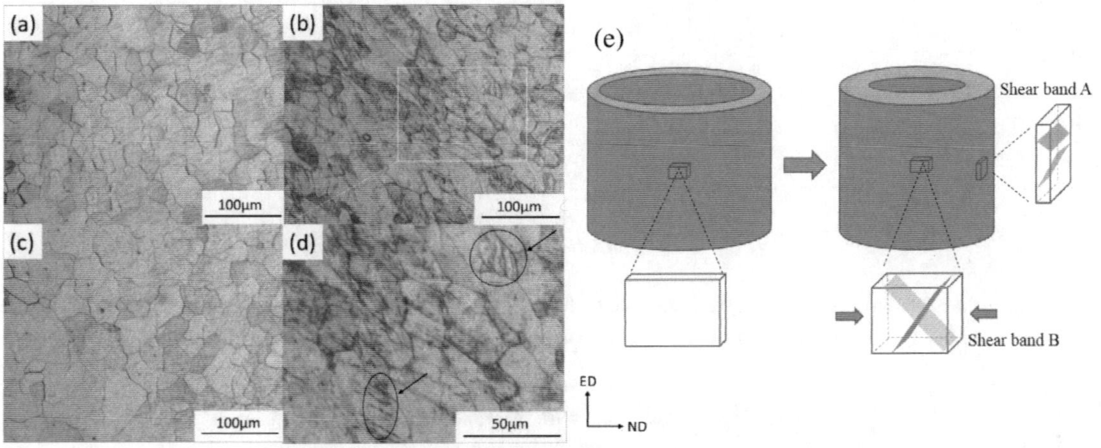

Fig.3 Microstructures of (a) before TCP, (b) after TCP, (c) after final heat treatment and (d) Magnified photo of (b). Arrows in (d) indicate grain scale shear bands and geometry (e) of shear bands A and B formed by shear strain by channel angle and squeezing (reduction of diameter), respectively.

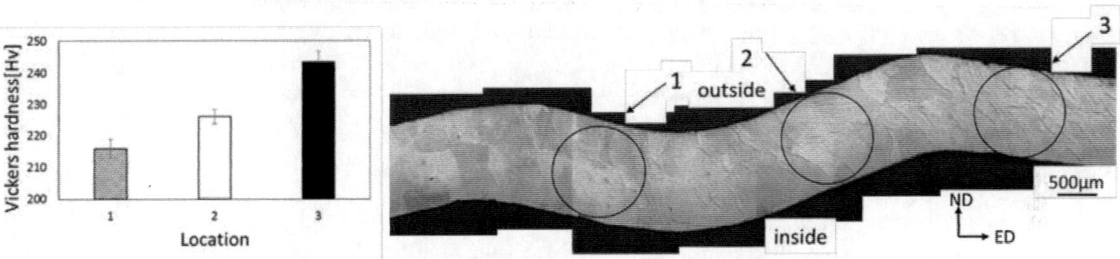

Fig.4 Vickers hardness and microstructures of neck zone after TCP.

3.2. Tensile test

Figure 5 shows the stress-strain curves of axial and circumferential directions. The latter data were obtained by tensile test of ring specimens. Higher elongation to failure in TCP process is evident as shown in stress-strain curve and the photos (Figure 6). Figure 7 shows the relationship between work hardening ratio and true strain of the axial direction, it is evident that the strain hardening rate is higher in TCP process throughout the strain. The strain-hardening exponent (n-value) of conventional and TCP processes are 0.22 and 0.26, respectively. Thus, the tube in TCP process is expected to have higher hydro-formability. Considerable effect is manifest in Lankford value both in axial and circumferential directions as shown in Table 1. Ridging evaluated by tensile test of 15% as shown in Figure 8. Ridging is clearly visible in the conventional process whereas it is mostly alleviated in TCP process. Figure 9 shows {100}-pole figures after the final annealing. Since the tubes are fabricated by electric resistance welding (ERW) of cold-rolled sheets, typical cold-rolled and annealed texture can be recognized in the conventional processing with main components of {111}<112> and {322}<236> orientation. It is known that {111}<112> orientations tends to split into these {322}<236> orientations after cold rolling with very high rolling reduction. In contrast, texture is randomly oriented with no clear peak orientations as shown in TCP process. Thus, TCP and the final annealing randomized texture leading to alleviation of ridging and enhanced formability.

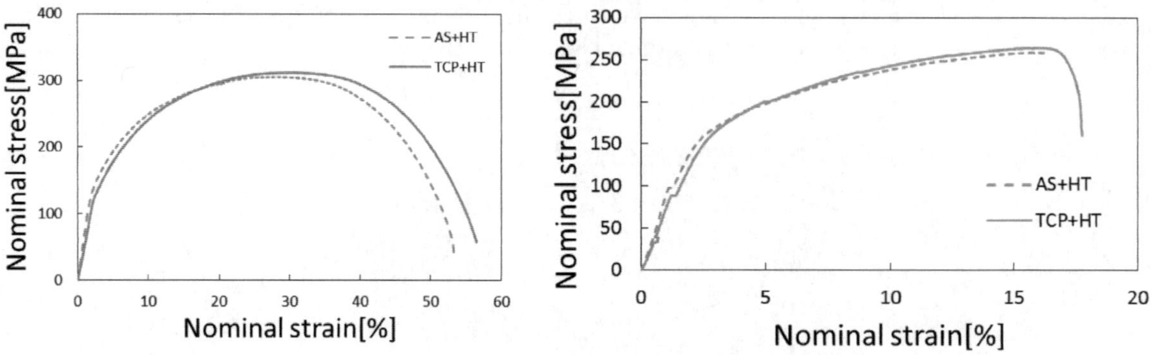

Fig. 5 Stress-strain curves of ring specimens to evaluate tensile behavior of axial (left) and circumferential (right) direction of conventional process and TCP process.

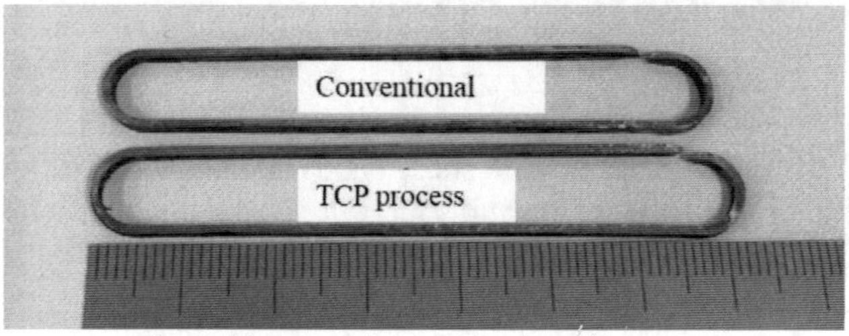

Fig.6 Ring specimen after tensile tests.

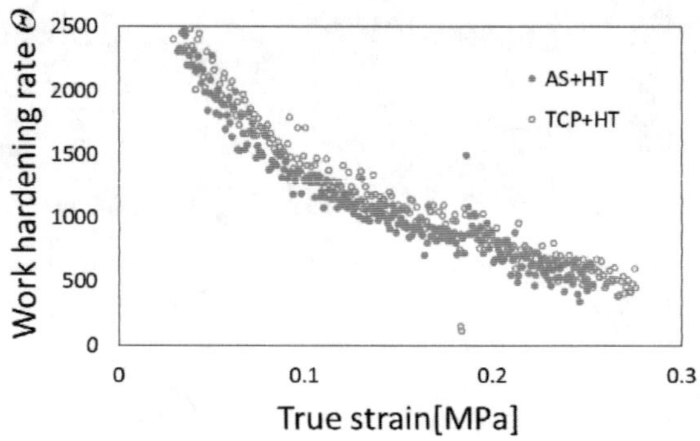

Fig. 7 Relationship between the work hardening rate and true stress of as annealed and 1 Pass of TCP annealed

Table 1 Lankford values of axial and circumferential directions

	axial	circumferential
AS+HT	1.14	0.90
TCP+HT	1.73	1.64

Fig.8 Ridging after 15% tensile elongation of (a) conventional and (b) TCP processing

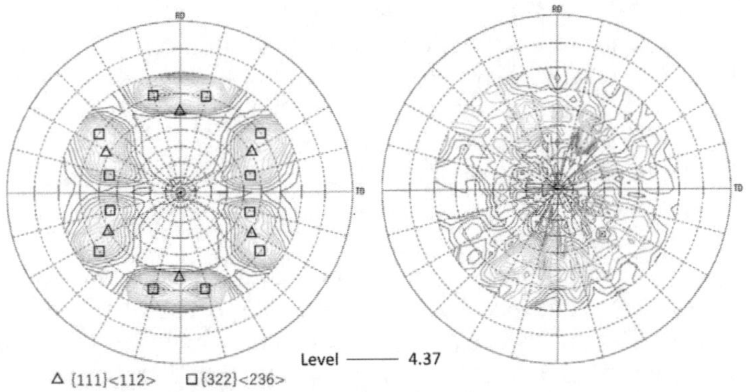

Fig.9 {100} pole-figures after the final heat treatment of conventional and TCP process

4. Conclusions

Tube channel pressing was applied to ferritic stainless steel (Fe-10%Cr) tubes for one pass to examine the effect of TCP on recrystallization microstructure and texture and formability. After one pass TCP, grain-scale shear bands were visible in several grains. These shear bands facilitate recrystallization and randomize the texture. Ridging was alleviated, and both strain-hardening capability and *r*-value was enhanced by one-pass TCP. It can be expected that the hydro-fomability is also enhanced.

Acknowledgement
This work was financially supported by Japan Science and Technology Agency (JST) matching planner program and this support is gratefully appreciated.

References

[1] Chao H-C1973 Recent studies into the mechanism of ridging in ferritic stainless steels *Metallurgical Transactions* **4** 1183-1186

[2] Miyamoto H, Xiao T, Uenoya T, Hatano M 2010 Effect of simple shear deformation prior to cold rolling on texture and ridging of 16% Cr ferritic stainless steel sheets *ISIJ International* **50(11)** 1653-1659

[3] Pougis A, Toth L S, Bouaziz O, Fundenberger J J, Barbier D, Arruffat R 2012 Stress and strain gradients in high-pressure tube twisting *Scripta Materialia* **66(10)** 773-776

[4] Farshidi M H, Kazeminezhad M 2016 The effects of die geometry in tube channel pressing: Severe plastic deformation. *Proceedings of the Institution of Mechanical Engineers, Part L: Journal of Materials: Design and Applications* **230(1)** 263-272

[5] Farshidi M H, Kazeminezhad M, Miyamoto H 2015 Microstructure and mechanical properties of an Al–Mg–Si tube processed by severe plastic deformation and subsequent annealing. *Materials Science and Engineering*: A **640** 42-50

[6] Zhang C, Liu Z, Wang G 2011 Effects of hot rolled shear bands on formability and surface ridging of an ultra purified 21%Cr ferritic stainless steel *Journal of Materials Processing Technology* **211(6)** 1051-1059

[7] Barnett M R, Jonas J J 1997 Influence of ferrite rolling temperature on microstructure and texture in deformed low C and IF steels *ISIJ International* **37(7)** 697-705

[8] Barnett M R, Jonas J J 1997 Influence of ferrite rolling temperature on grain size and texture in annealed low C and IF steels *ISIJ International* **37(7)** 706-714

[9] Barnett M R 1998 Role of in-grain shear bands in the nucleation of <111>//ND recrystallization textures in warm rolled steel *ISIJ International* **38(1)** 78-85

[10] Nave M D, Barnett M R, Beladi H 2004 The influence of solute carbon in cold-rolled steels on shear band formation and recrystallization texture *ISIJ International* **44(6)** 1072-1078

[11] Jonas J J 2001 Effects of shear band formation on texture development in warm-rolled IF steels *Journal of Materials Processing Technology* **117(3)** 293-299

[12] Chen Q Z, Quadir M Z, Duggan B J 2006 Shear band formation in IF steel during cold rolling at medium reduction levels, *Philosophical Magazine* **86(23)** 3633-3646

[13] Korbel A, Embury J D, Hatherly M, Martin P L, Erbsloh H M 1986 Microstructural aspects of strain localization in Al-Mg alloys *Acta Metallurgica* **34(10)** 1999-2009

[14] Gasperini M, Pinna C, Swiatnicki W 1996 Microstructure evolution and strain localization during shear deformation of an aluminium alloy *Acta Materialia* **44(10)** 4195-4208

[15] Hurley P J, Humphreys F J 2003 The application of EBSD to the study of substructural development in a cold rolled single-phase aluminium alloy, *Acta Materialia* **51** 1087-1102

Processing and characterization of Al-Al_3Nb prepared by mechanical alloying and equal channel angular pressing

P Chandran, A Zafari, E W Lui and K Xia

Department of Mechanical Engineering, The University of Melbourne, Victoria 3010, Australia

Corresponding author's email: prathapc@student.unimelb.edu.au

Abstract. Mechanically alloyed Al with immiscible elements such as Nb can lead to a uniform distribution of nanoscaled precipitates which are highly stable compared to conventional alloying and with excellent interface, resulting in significant increase in strength without problems associated with nano ceramic particles in metal matrix composites. Although immiscible, Nb can be alloyed with Al through mechanical milling, forming trialuminide (Al_3Nb), either directly or upon subsequent precipitation, which possesses high strength, stiffness and stability at elevated temperatures. In the present study, Al-5 at.% Nb supersaturated solid solution was achieved after prolonged ball milling and nano Al_3Nb precipitates were formed during subsequent ageing at 530°C. The Al-Al_3Nb powder was consolidated by equal channel angular pressing (ECAP) at 400°C, resulting in a fully dense material with a uniform distribution of nanoscaled Al_3Nb precipitates in the Al matrix.

1. Introduction

Al-Nb is a promising system for high strength aluminium alloys thanks to the possible formation of hard Al_3Nb intermetallic precipitates. Al_3Nb possesses high strength and stiffness, and excellent mechanical properties at elevated temperatures [1, 2]. However, the equilibrium solubility of Nb in Al is zero at room temperature and very low (~0.065 at.%) at the maximum solubility temperature [3], making it impossible to follow the conventional solution and aging treatment to achieve precipitation strengthening. High energy mechanical alloying as a non-equilibrium processing method has been successfully employed to extend the solubility well above the equilibrium values in several Al binary systems [4-6]. One study on mechanical alloying of Al and Nb has shown that a maximum of 15 at.% Nb can be dissolved in Al [1]. Depending on the composition and milling conditions, different structures including solid solution, intermetallic or amorphous can be obtained in the Al-Nb system [1, 2, 7]. In addition, mechanical milling leads to the formation of nanostructures with improved mechanical properties [8].

The objectives of the present study are to first achieve a supersaturated solid solution of Nb in Al by mechanical alloying and to bring about nano precipitates by ageing, and then to consolidate the powder into bulk material by a severe plastic deformation (SPD) method.

The SPD processes such as equal channel angular pressing (ECAP) and high pressure torsion (HPT) have been developed to consolidate particulate materials to form bulk nanostructured materials [9, 10]. Deformation of the particles during SPD, rather than diffusion, causes their bonding [11]. a number of ultrafine and nanostructured bulk materials have been prepared by SPD consolidation including several based on Al and Ti, leading to significantly improved mechanical properties [9, 12]. Unlike the conventional sintering process at high temperatures, which tends to destroy the nanostructures achieved by milling, ECAP enables the particles to consolidate at much lower temperatures with little porosity [13]. The preliminary results from the present investigation into the Al-Nb system demonstrated that it is possible to achieve high solubility by mechanical milling and high strength by subsequent ageing and ECAP consolidation.

2. Experiments

Elemental powders of Al (-325 mesh, 99.8% purity) and Nb (-325 mesh, 99.8% purity) were used. The powder mixture of Al-5 at.% Nb in composition was placed in a stainless steel jar filled with stainless steel balls of 10 mm in diameter with a ball to powder ratio (BPR) of 20:1. A process control agent (PCA) of stearic acid (1 wt.%) was added. High energy ball milling was performed at 400 rpm under vacuum at room temperature in a planetary mill until complete solid solution of Al-Nb was achieved. The solid solution powder was annealed at 530°C for an hour in vacuum to cause precipitation. The annealed powder was then encapsulated in a thin copper can and placed in an ECAP die with channels of 12 mm in diameter intersecting at 90°. The consolidation was carried out at 400 °C for 4 passes following Route C with a back pressure of 50 MPa. XRD (Cu Kα) was carried out to identify the phases in the powder after different milling durations. The microstructure was observed by SEM on polished surfaces and TEM on samples prepared using focused ion beam (FIB). Compression tests were performed on cylindrical samples of dimensions of 6 mm in diameter and 9 mm in height at a strain rate of 0.001/s.

3. Results and Discussion

The XRD patterns of the Al-5 at.% Nb powder after milling for 1, 45 and 90 hours are shown in figure 1, revealing the phase transformation, and the corresponding back scattered SEM images in figure 2, displaying the morphological evolution of the particles. It is clear from the XRD patterns that the peaks of the elemental Al and Nb still existed after one hour of milling, and the peaks of Nb disappeared after 45 hours, indicating the apparent formation of complete solid solution. There seems to be no further phase changes during further milling and only fcc Al peaks were observed after 90 hours of milling. The broadening of the fcc Al peaks was observed with increasing milling time as a result of the refinement of the crystallite size by heavy plastic deformation [14]. In fact, the starting grain size of ~1.3 μm in the elemental Al powder was reduced to ~23 nm after 90 hours. The Al (grey) and Nb (white) particles can be clearly distinguished after 1 hour (figure 2a), although some Al particles were elongated. With increasing milling time, the Nb particles were severely fragmented to become smaller particles and be entrapped and cold welded in the Al particles. Despite the disappearance of the Nb peaks in the XRD pattern after 45 hours, closer inspection of the corresponding SEM image (figure 2c) reveals ultrafine Nb particles of < 1 μm embedded uniformly in the Al matrix (the amount is too small for XRD to pick up). In contrast, no Nb particles were detected by the SEM images for the powder milled for 90 hours (figures 2d and e), indicating the true formation of a complete solid solution between Al and Nb. The average particle size was ~21 μm after 90 hours. The particle morphology remained nearly equiaxed with irregular boundaries after 45 and 90 hours. EDS analysis on several particles after 90 hours confirmed that they all have the same composition as the overall composition (5 at.% Nb). As expected, the lattice parameters remained constant at all the times thanks to similar atomic radii of Al (0.1430 nm) and Nb (0.1432 nm) [1].

Figures 3 a and b show the bright field TEM image after 90 hours and the corresponding selected area electron diffraction (SAED) pattern, respectively. The SAED pattern identifies fcc Al rings only, i.e. the formation of the complete solid solution, and the complete rings confirm that the prolonged milling had resulted in the formation of a nanocrystalline structure. During the early stages of milling the Al and Nb particles are subjected to high energy collisions from the impacting balls. As milling continues, fracturing and cold welding occur repeatedly, leading to severe plastic deformation. As a result, a large number of crystal defects are introduced and the amount of grain boundaries significantly increases, leading to rapid diffusion for Nb and the eventual formation of a uniform solid solution of Al-Nb [3, 15].

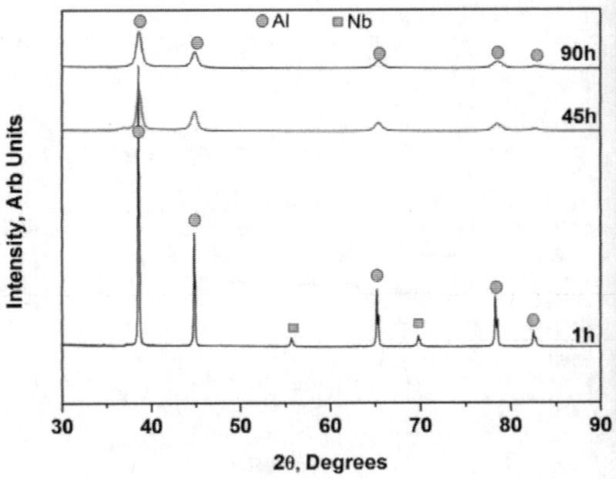

Figure 1. XRD patterns of Al-5 at.% Nb after milling for 1, 45 and 90 hours.

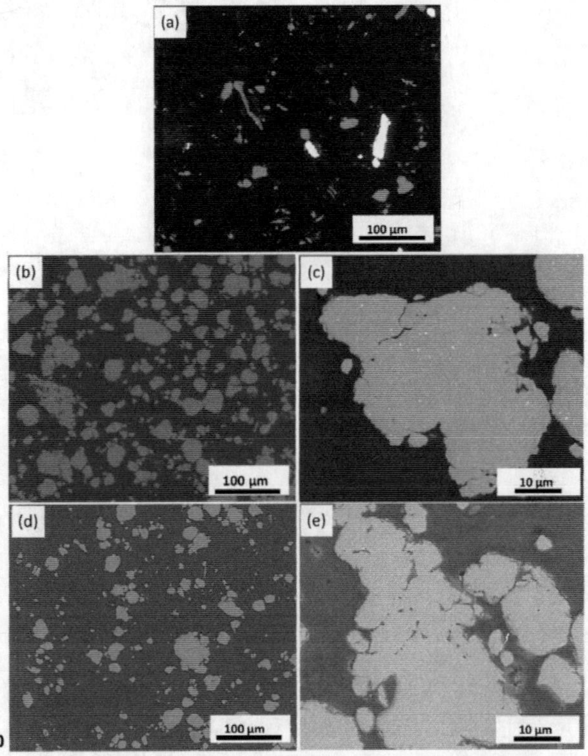

Figure 2. Backscattered SEM micrographs showing Al-5 at.% Nb particles created after milling for (a) 1 hour, (b) and (c) 45 hours, (d) and (e) 90 hours at two different magnifications, respectively.

Following mechanical milling, the solid solution powder was annealed at 530°C for one hour in vacuum to bring about the Al_3Nb precipitates. The XRD pattern of the annealed powder is shown in figure 4, confirming the formation of ~20 at.% of Al_3Nb which is the maximum amount that can result from an Al-5 at.% Nb supersaturated solid solution. It is also observed that the fcc Al peaks has become shaper compared to the solid solution peaks before annealing, indicating coarsening of the crystallites. In fact, the Al grains are now ~46 nm in size based on the peak broadening measurement, still in the nanoscale. On the other hand, the precipitates are measured to be ~29 nm in size. No other intermetallic phases are detected after the annealing process.

After ECAP consolidation of the annealed powder, a fully dense bulk material within ±0.1% of the theoretical density (3.05 g/cm^3) was obtained. The application of a back pressure ensures the shearing of the particles during ECAP, resulting in well bonded particles with no pores [12, 16]. The cross-sectional SEM microstructures (low and high magnifications) of the consolidated material are displayed in figures 5 a and b with two distinct regions, the lighter grey and the darker region between. The grey region is Al with numerous nano Al_3Nb precipitates as revealed by TEM, from the framed area in figure 5 b, in figures 5 c and d, and the darker region consist of mostly Al with a very

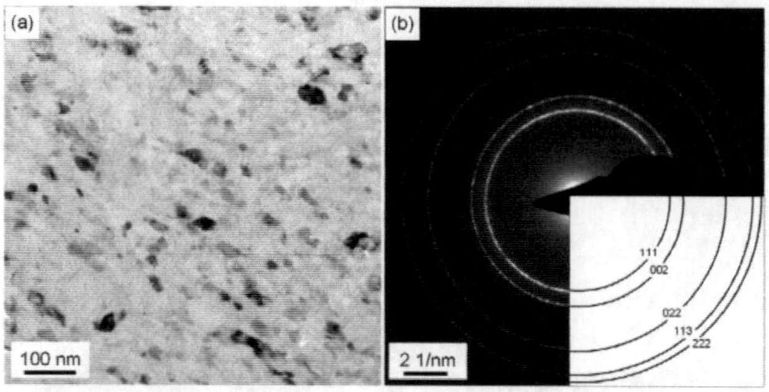

Figure 3. (a) Bright field TEM image of the Al-5 at.% Nb powder after 90 hours of milling and (b) corresponding SAED ring pattern, identifying nanocrystalline fcc structured Al.

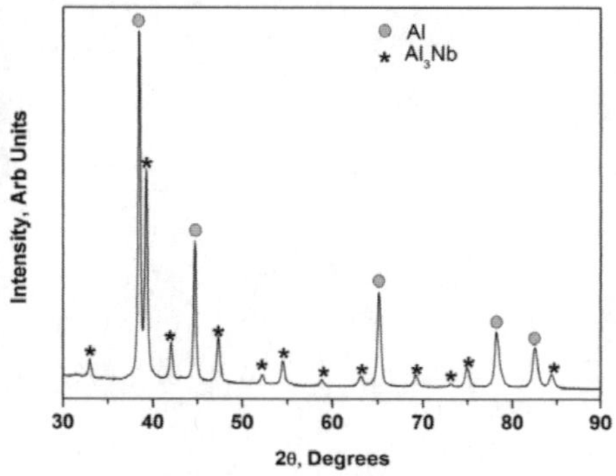

Figure 4. XRD pattern of the annealed Al-5 at.% Nb powder at 530°C for 1 hour.

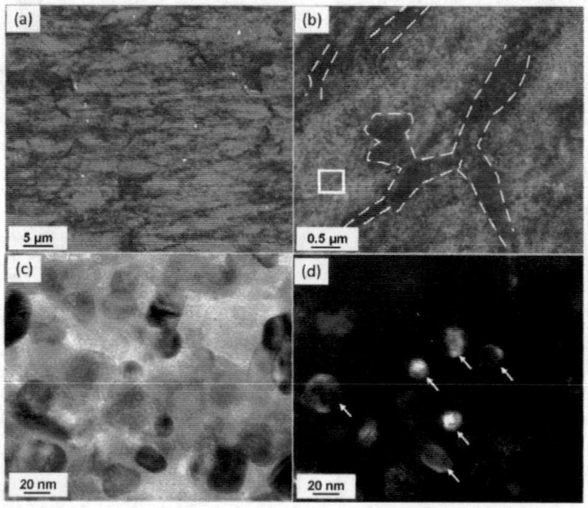

Figure 5. (a) and (b) Backscattered SEM microstructure of Al-Al_3Nb consolidated by ECAP at two magnifications, showing some pure Al areas delineated by dashed-lines. (c) Bright field TEM image from the boxed area in (b), and (d) corresponding dark field TEM image for Al_3Nb, showing uniformly distributed nano Al_3Nb precipitates (arrowed).

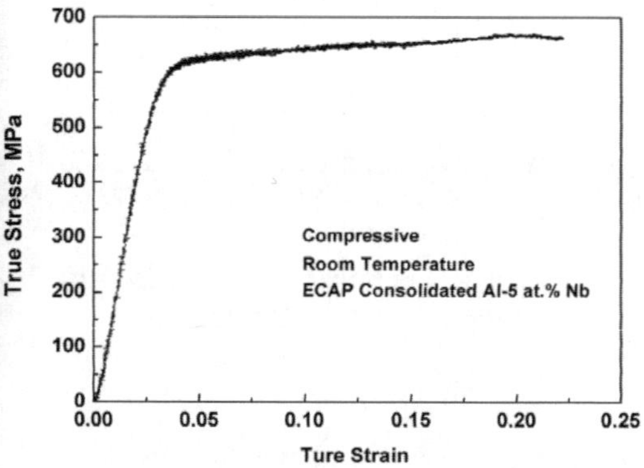

Figure 6. Compressive true stress-true strain curve for the ECAP consolidated Al-5 at.% Nb.

small number of precipitates. Image analysis indicates that the darker region accounts for an area fraction of ~12%.

Closer inspection of figures 5 c and d indicates that the Al_3Nb precipitates have an average size of ~30 nm and they are uniformly dispersed in the Al matrix. The identity of these round particles is also confirmed by EDS, giving the same composition as Al_3Nb.

No other intermetallic phases of Al and Nb are detected after the ECAP consolidation. Although the powder after milling and that after annealing appear to possess a uniform microstructure and composition, the ECAP consolidation has resulted in the formation of the thin band with relatively few precipitates. Further investigation is in progress to understand this phenomenon.

The compressive stress-strain curve of the consolidated material is shown in figure 6, with an average 0.2% proof stress of ~590 MPa, an ultimate compressive strength of ~670 MPa and a good plasticity of ~20%. The strength is comparable to or higher than conventional high strength Al alloys, and is attributed to several microstructural factors. The presence of nano precipitates provides dispersion strengthening. The nanocrystalline structure retained after the ECAP consolidation also results in grain boundary strengthening, as reported in other Al and Ti based alloys [12]. The unique intermixed microstructure and good bonding between the particles might be responsible for the good plasticity in compression. As reported in another SPD consolidated material [12], there is a small XRD peak shift towards lower angles after annealed and ECAP, indicating the internal strain energy relief [17].

4. Conclusions

(1) Supersaturated solid solution of Al-5 at.% Nb was achieved after 90 hours of ball milling at 400 rpm.
(2) All possible Al_3Nb was precipitated after annealing the milled particles at 530°C for one hour.
(3) The Al-Al_3Nb consolidated by ECAP was fully dense with regions of uniform dispersion of nano precipitates in Al, intertwined by nearly pure Al layers.
(4) The consolidated material showed excellent yield strength of 590 MPa with a good plasticity (~20%) in compression.

5. References

[1] Peng Z, Suryanarayana C and Froes F H 1996 *Metall. Mater. Trans. A* **27** 41-8
[2] Yoo D J, Hwang S M and Lee S M 2000 *J. Mater. Sci. Lett.* **19** 1327-9
[3] Suryanarayana C 2004 *Mechanical Alloying And Milling* (New York: CRC Press)
[4] Fan G J, Gao W N, Quan M X and Hu Z Q 1995 *Mater. Lett.* **23** 33-7
[5] Nayak S S, Pabi S K and Murty B S 2010 *J. Alloys Compd.* **492** 128-33
[6] Sasaki T T, Ohkubo T and Hono K 2009 *Acta Mater.* **57** 3529-38
[7] Peng Z, Suryanarayana C and Froes F H 1992 *Scr. Metall. et Materialia* **27** 475-80
[8] Nayak S S and Murty B S 2004 *Mater. Sci. Eng. A* **367** 218-24
[9] Wu X and Xia K *J. Mater. Process. Technol.* **192–193** 355-9
[10] Wu X, Xu W and Xia K 2008 *Mater. Sci. Eng. A* **493** 241-5
[11] Xia K 2010 *Adv. Eng. Mater.* **12** 724-9
[12] Lui E W, Xu W, Wu X and Xia K 2011 *Scr. Mater.* **65** 711-4
[13] Xia K and Wu X 2005 *Scr. Mater.* **53** 1225-9
[14] Nayak S S, Wollgarten M, Banhart J, Pabi S K and Murty B S 2010 *Mater. Sci. Eng. A* **527** 2370-8
[15] Suryanarayana C and Al-Aqeeli N 2013 *Prog. Mater. Sci.* **58** 383-502
[16] Xia K, Wu X, Honma T and Ringer S P 2007 *J. Mater. Sci.* **42** 1551-60
[17] Ungár T 2004 *Scri. Mater.* **51** 777-81

Acknowledgments

Prathap Chandran acknowledges the receipt of Melbourne India Post graduate Program (MIPP) scholarship. We also acknowledge the facilities and technical assistance of MCFP X-ray Diffraction Platform and the Bio21 Advanced Microscopy Unit at the University of Melbourne.

IOP Conf. Series: Materials Science and Engineering **194** (2017) 012009 doi:10.1088/1757-899X/194/1/012009

Microstructure and mechanical properties of Al-1050 during incremental ECAP

M Salamati[1], S Tamimi[1], S Moturu[1], G Sivaswamy[1], M Qarni[1], A Rosochowski[2]

[1] Advanced Forming Research Centre, University of Strathclyde,
85 Inchinnan Drive, Inchinnan, Renfrewshire, PA4 9LJ, United Kingdom
[2] Design, Manufacturing and Engineering Management, University of Strathclyde,
James Weir Building, 75 Montrose Street, Glasgow, G1 1XJ, United Kingdom

E-mail: mohammad.salamati@strath.ac.uk

Abstract. Incremental ECAP is a new method of ECAP process were the severe shear deformation is incrementally applied on the sample resulting in grain refining and new texture developing. The fundamental objective of the present work is an observation of effect of different passes of I-ECAP on microstructure and mechanical properties of AA1050 billet. To that end, 8 pass of I-ECAP have been carried out using Bc route and microstructure evolution and mechanical properties of the I-ECAPed samples have been studied. The EBSD and TEM analyses indicates that I-ECAP is as capable as conventional ECAP to grain refinements and a UFG structure is resulted after I-ECAP cycles. Tensile testing and hardness measurements indicates that mechanical properties of the Al-1050 billets increases dramatically by increasing the I-ECAP passes.

1. Introduction

Severe plastic deformation (SPD) is one of the metal forming processes used to obtain ultrafine grained (UFG) structure (below 1 micron) [1-3]. UFG metals exhibit enhanced physical and mechanical properties compared to that of coarse grained counterparts. Various SPD processes exist but the equal channel angular pressing (ECAP) is one of the few methods that can produce UFG billets for industrial scale applications [4-6]. Segal and his co-workers [7, 8] were the first researchers to study this method. This method induces large shear strain in the billet material by passing it repeatedly through the two channel intersect at an angle ϕ=90°, and have equal cross-sections. Typically at each pass an equivalent plastic strain of approximately 1.15 is produced in the billets. This imposes large cumulative strains eventually leads to UFG structure in the billet. To achieve varying microstructural characteristics, different routes such as billet rotation between consecutive passes can be used. One of the main drawbacks of ECAP is the length of the billets processed, due to the limitation of the stroke length of the actuator. The incremental equal channel angular pressing (I-ECAP) being an extension of the ECAP process, does not have this limitation therefore can process very long billets. This paper systematically examines the mechanical properties, microstructure of multiple passes I-ECAP samples, with the channel intersection angle ϕ=90°.

2. Materials and methods

The billet material comprised of 99.5% Aluminium with the following traces elements Cu, Mg, Si, Fe, Mn, Zn and Ti. The billets (10×10×60mm) samples were extracted from round bars in the extrusion direction using electrical discharge machining (EDM). The billet surface was prepared using sandblasting, conversion coating with calcium aluminate and finally applying a thin layer of Loctite 8009 (a graphite based anti-seize lubricant) from Henkel technologies. Room temperature I-ECAP was carried out with a cycle frequency of 0.5 Hz and with a punch amplitude of 1.6 mm. A feeding stroke of 0.2 mm per cycle was used. The details of I-ECAP have been reported in previous studies [10, 11]. The I-ECAP process was performed on a custom built 1000KN servo-hydraulic press with Cubus software. The punch is attached to the press actuator and follows a sinusoidal cyclic command during processing. The material feeding is carried out by a screw jack which is driven by a servo-motor with the help of a dedicated LabVIEW application controls and synchronises the material feeding by monitoring the punch oscillation.

3. Results and discussion

3.1. Mechanical properties

The Zwick ZHVμ micro hardness tester was used to produce the hardness maps on as-received, 1,2,3,4,6 and 8 pass I-ECAP samples as shown in Figure 1. The billets were cut in half, along the flow plane direction and the cut surface was ground and polished to 1μm. In total of 500 indents with 1mm spacing between each indents and 5kg load was employed on each samples. With the number of IECAP pass the hardness of the sample increased proportionally.

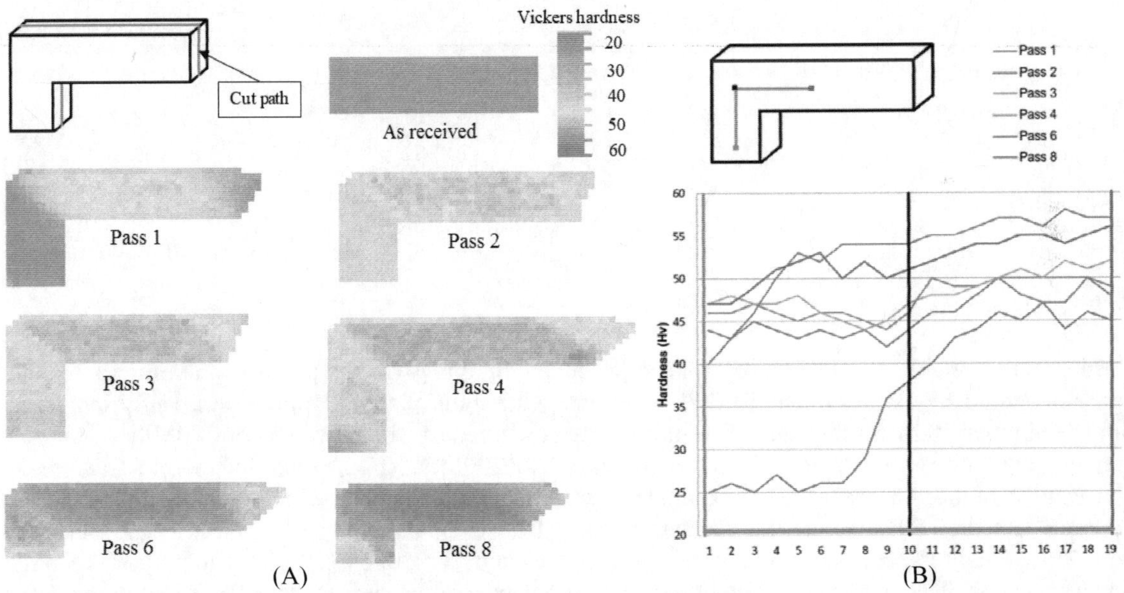

Figure 1: (A) Micro hardness measurement along the flow plain for partially processed billets after IECAP passes of ϕ=90° at 1, 2, 3, 4, 6 and 8. (B): Hardening results of billets after passing through the plastic shear zone for passes 1, 2, 3, 4, 6, 8. The green line represents the centre of the billet and the location of the hardness measurements. The blue dot (starting point of hardness results shown in graph), is located 10mm from the centre of the shear zone in the vertical axis and the red dot (ending point of hardness results shown in graph), is located 10mm from the centre of the shear zone in the horizontal axis.

In compression to other I-ECAP samples, the first pass sample experienced significant hardened, the hardness increased from 22HV to 45HV. The maximum hardiness measured after eight pass is 59HV. The subsequent passes also show a more uniform hardness being achieved within the billets. It is also possible to see that the head section of the billet is non-uniformly hardened from the rest of the billet. The first 10 to 15mm of the billet remains different from the rest of the billet even after 8 passes. It is also interesting to note that the hardness in the subsequent passes at the bottom of the tail section changes compared to the rest of the billet. The side towards the output channels appear to be hardening, whereas the side away from the output channel appears to be softening. It is clear to notice the changes in hardness measurement in all passes of the billets as they pass through the plastic shear zone. An area 10mm either side of the centre line of the shear zone was examined closer and the results can be seen in Fig. 1b.

Most of the hardness changes seem to occur 5mm before and after the centre of the plastic shear point shown in by the black dot in the schematic and the black line in the centre of the graph. Before and after this 5mm zone the hardness of the billet appears to be relatively consistent. This is true for all cases apart from the 8th pass where the partial processed billet was processed further than intended. Partially processing billets can be challenging as it is very difficult to see the billet during processing and determining the processing progress.

To investigate the effects of these changes, tensile testing was conducted on I-ECAP passed material. The processed I_ECAP samples and extracted tensile samples are stored in a freezer to minimize the dynamic recovery and recrystallization. Before tensile testing the specimens were brought to the room temperature. The results from the tensile testing are shown in Fig. 2A.

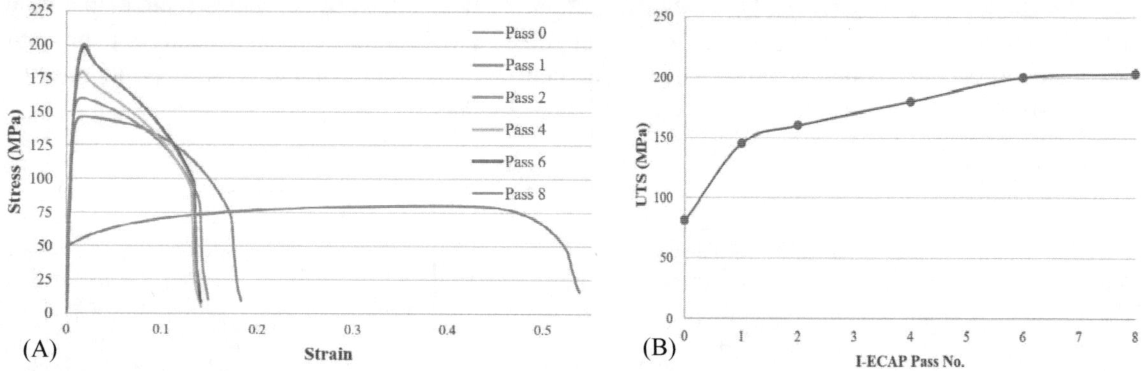

Figure 2:(A) Tensile testing results for I-ECAPed Al1050 specimens. Specimens were created from pass 0, 1, 2, 4, 6 and 8 billets. (B)The Ultimate Tensile Strength of I-ECAPed Al1050 specimens. The UTS of pass 0, 1, 2, 4, 6 and 8 have been plotted.

Due to the relative small size of billets produced in this I-ECAP process only micro tensile specimens could be extracted using EDM. The tensile strength of the Al1050 significantly increases with the number of I-ECAP passes. The ultimate tensile strength (UTS) was about 80MPa in the 'as received' condition, after 8 pass the UTS increased to 200MPa. The elongation before fracture is significantly higher in the as-received materials, After first pass the elongation of the material decreases less than 50%. While the ultimate tensile strength of material after first pass increases by about 75%. This trend continues with subsequent passes until the 6th pass but after that, the UTS only increase slightly. This pattern of stepped increase in mechanical properties with the initial passes was also seen in hardness measurements.

3.2. Microstructure analysis

In order to explain mechanical properties variation of the I-ECAP samples, the microstructure analysis of the samples after each pass was examined. The scanning electron microscope Quanta FEG 650, equipped with a field emission gun (FE-SEM) and with a NordlysF EBSD detector, was used. The AZtec acquisition software was utilized for data acquisition and Channel 5 software to analyse the data. All EBSD measurements were performed with a step size of 2μm, using the beam-scanning mode under dynamic focus conditions and with a SEM magnification of 200X. For EBSD analysis, samples were grinded and finally electro polished using Struers lectroPol 5 and electrolyte A2. Figure 3, shows the microstructure of the as-received and different stages of I-ECAP samples. The observed microstructures at different stages of the I-ECAP clearly indicates the shear deformation of grains. The density of the high angle grain boundaries (greater than 15°) and refinement of grain structure in I-ECAP samples increases significantly in comparison to the as-received sample. The average grain size of the material decreased significantly with increasing ECAP pass. The rate of grain refinement through the initial passes of I-ECAP is higher than the rate of grain refinement through the higher passes.

Figure 3: EBSD analyses of Al1050 billets: as received sample and after 1, 2 and 4 passes of I-ECAP.

For transmission electron microscopy (TEM) observations, the samples were sectioned to produce longitudinal sections (flow plane) using a slow speed abrasive saw to obtain slices of thickness less than 300µm. The slices were mechanically polished to reduce the thickness. Discs of 3mm diameter were spark eroded. The twin-jet electro-polishing with 25% HNO3 and 75% CH3OH (by volume) mixture as electrolyte was done as a final polishing. Bright field micrographs recorded from the flow plane of the I-ECAP process samples using TEM are presented and discussed in this section (Fig. 4). Evidence for the formation of banded structure in the flow plane of billet after first is shown in Fig. 4(a). This is primarily due to the intense shear experienced by the billet when it passes through the shear deformation existing between input and output channel. The orientation of these bands are at angle 25°.The dislocations got accumulated with in those deformation bands are clearly shown by the regions having better contrast in micrograph (Fig. 4(a)). Dislocations stored in the banded structure started to undergo recovery process called dynamic recovery, when the billet subjected to second pass. Micrograph of second pass samples given in Fig. 4(b) revealed the formation of dislocations sub-grain structure developed as a result of dynamic recovery process. Dislocations annihilating the sub-grain boundaries are noticed.

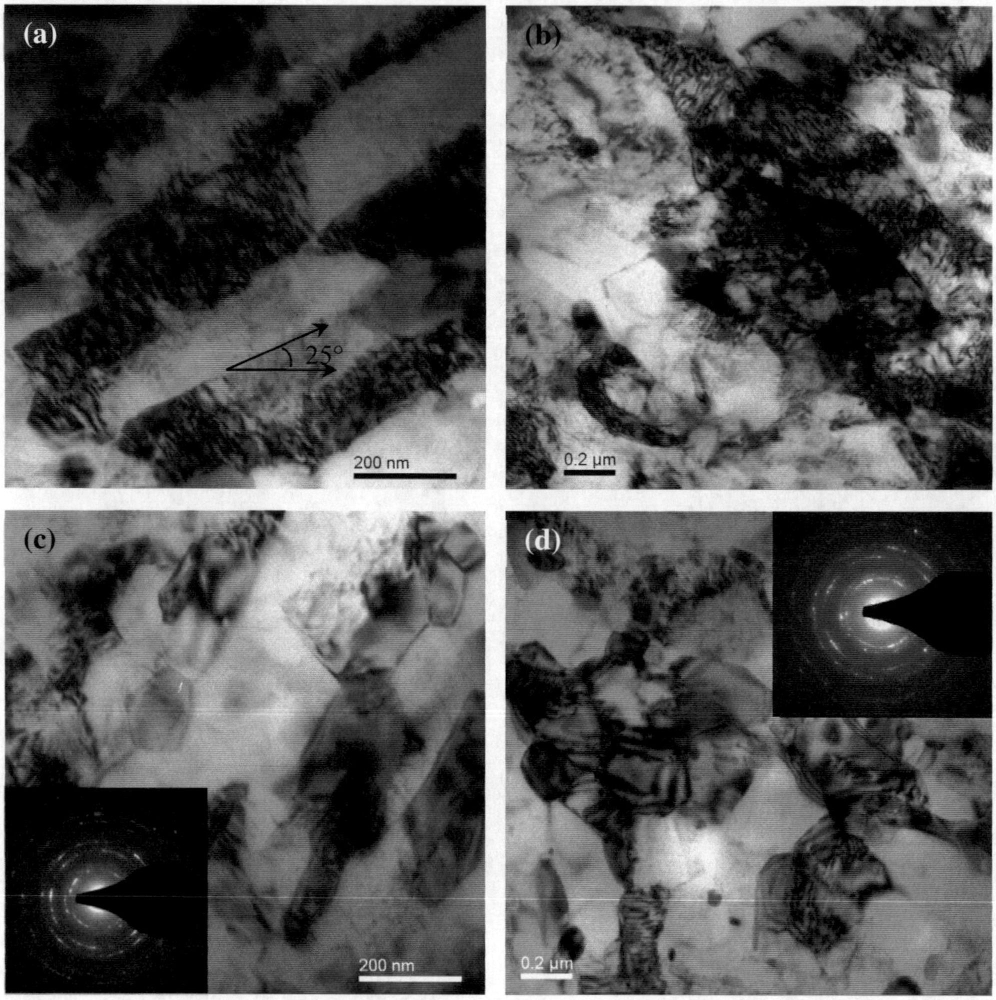

Figure 4: Bright field TEM micrographs showing the deformation characteristics of aluminium matrix in AA1050 alloy biller subjected I-ECAP process after (a) one pass, (b)two passes, (c) six passes and (d) eight passes. Corresponding SADP are given as inserts.

As amount of strain imposed on the billets during subsequent passes increases dynamic recovery assisted recrystallization started to dominate in the matrix leads to the formation of fine grains in the matrix (Fig. 4(c)-(d)). Temperature rise in the billet while crossing the shear deformation zone also influences the process of dynamic recrystallization to some extent. Dynamically recrystallized grains formed in the matrix after 6th and 8th passes are shown in Fig. 4(c) and Fig. 4(d), respectively. Dynamic recovery and dynamic recrystallisation at the 6-8 passes of I-ECAP can explain why the rate of increasing hardness and tensile strength of the samples at the final passes of I-ECAP reduces in compared to the initial stage of the process (see Figures 1 and 2).

Continuous ring patters shown by the selected area diffraction pattern (SADP) confirming the formations of fine grains in the regions he shown in bright field micrographs of 6th and 8th passes are provided as inserts.

IOP Conf. Series: Materials Science and Engineering **194** (2017) 012009 doi:10.1088/1757-899X/194/1/012009

4. Conclusions

In this study we examined the next natural evolution of the classical ECAP method for grain refinement and improved mechanical properties in billets. I-ECAP improves some of the limitations that are often associated with ECAP. The results of this study showed that the first pass leads to the greatest increase in mechanical properties and the following passes had a more gradual increase in these effect. The microstructure results indicate that grain size in the billet decreases with the increase in the number of passes of I-ECAP. In addition, dynamic recovery and recrystallization take place which influence the mechanical properties of the material at the later passes of I-ECAP.

References

1. R. Valiev, R. Mulyukov, V. Ovchinnikov, and V. Shabashov, 1991: "Mossbauer analysis of submicrometer grained iron", Scripta Metallurgica et Materialia (USA), 25, pp. 2717-2722.
2. N.A. Akhmadeev, R.Z. Valiev, V.I. Kopylov, and R.R. Mulyukov, 1992: "Formation of submicro grain structure in copper and nickel by extensive shear deformation", Russian Metallurgy, 2, pp. 96-101.
3. J. Wang, Z. Horita, M. Furukawa, M. Nemoto, N.K. Tsenev, R.Z. Valiev, Y. Ma, and T.G. Langdon, 1993: "An investigation of ductility and microstructural evolution in an Al– 3% Mg alloy with submicron grain size", Journal of Materials Research, 8, pp. 2810-2818.
4. R.Z. Valiev, 1997: "Structure and mechanical properties of ultrafine-grained metals", Materials Science and Engineering A, 234-236, pp. 59-66.
5. I.V. Alexandrov, Y.T. Zhu, T.C. Lowe, R.K. Islamgaliev, and R.Z. Valiev, 1998: "Microstructures and properties of nanocomposites obtained through SPTS consolidation of powders", Metallurgical and Materials Transactions A, 29A, pp. 2253 - 2260.
6. V.V. Stolyarov, Y.T. Zhu, T.C. Lowe, R.K. Islamgaliev, and R.Z. Valiev, 1999: "A two-step SPD processing of ultrafine-grained titanium", NanoStructured Materials, 11, pp. 947-954.
7. V.M. Segal, 1977: "The method of material preparation for subsequent working", Patent of the USSR No. 575892.
8. V.M. Segal, V.I. Reznikov, A.E. Drobyshevskiy, and V.I. Kopylov, 1981: "Plastic metal working by simple shear", Russian Metallurgy, 1, pp. 99-104.
9. M. Furukawa, Z. Horita, and T.G. Langdon, 2002: "Factors influencing the shearing patterns in equal-channel angular pressing", Materials Science and Engineering A, 332, pp. 97-109.
10. A. Rosochowski, and L. Olejnik, 2011: "Incremental Equal Channel Angular Pressing for Grain Refinement", Materials Science Forum, 674, pp. 19-28.
11. M.R. Salamati, M. J. Qarni, S. Tamimi, and A Rosochowski, 2016: "Effect of channel angle on the material flow, hardness distribution and process forces during incremental ECAP of Al-1050 billets", ESAFORM 2016 AIP Conf. Proc, 1769, 090004-1–090004-6

IOP Conf. Series: Materials Science and Engineering **194** (2017) 012010 doi:10.1088/1757-899X/194/1/012010

Equal Channel Angular Pressing (ECAP) of hollow profiles made of titanium

M Krystian[1], K Bryła[2], J Horky[1], B Mingler[1]

[1] Center for Health & Bioresources, Biomedical Systems, AIT Austrian Institute of Technology GmbH, 2700 Wr. Neustadt, Austria
[2] Institute of Technology, Faculty of Mathematics, Physics and Technical Science, Pedagogical University of Cracow, 30-084 Krakow, Poland

E-Mail: maciej.krystian@ait.ac.at

Abstract. Cylindrical tubes made of commercially-pure titanium (CP-Ti) filled with various mandrels (metallic as well as non-metallic materials) were processed by Equal Channel Angular Pressing (ECAP). Different temperatures (500°C down to room temperature), routes (Bc, B_{120}, C), tools and number of passes (1-6) were applied depending on the mandrel material. For reasons of comparison, solid bolts made of metallic mandrel materials were processed at the same ECAP conditions. The mechanical properties of as-received as well as ECAP tubes and core materials were characterised by hardness mappings (in order to reveal the homogeneity) and tensile tests. The results can be summarised as follows: i) It is feasible to process tubes by ECAP; ii) Using appropriate mandrels, tubes made of CP-Ti can be processed by ECAP at significantly lower temperatures, even at room temperature, as compared to solid bolts; iii) Mechanical properties of tubes and mandrel materials after ECAP are similar to or even better than those of their solid counterparts; iv) Tubes after ECAP at low temperatures show higher strength than what can be achieved in bulk material; v) Excellent homogeneity of microhardness is achieved at least when metallic mandrels and a sufficient number of ECAP passes are used.

1. Introduction

Among the different techniques of Severe Plastic Deformation (SPD) available to obtain bulk nanostructured materials with enhanced mechanical properties, Equal Channel Angular Pressing (ECAP) is the most prominent one, especially when larger sample dimensions are required [1, 2]. However, until now ECAP processing was almost exclusively performed on massive rods and bars, thus, samples with solid (full) cross-sectional area. The main aim of this comprehensive study was to expand the possible cross-sectional shapes of ECAP processed materials and semi-products to tubes (hollow profiles) and to shed more light into the effects of different mandrel materials, processing temperatures, numbers of passes and ECAP die angles on the mechanical properties of tubes and the core materials.

So far, investigations of ECAP of hollow profiles are limited. Moreover, in the great majority of these studies the tubes were not subject of investigations at all but either containers for powder consolidation [1, 3–6] or for encapsulating the target material for using an ECAP die with bigger dimensions [7] or used to protect the core material [8] or both [9]. Nonetheless, some papers deal with the properties of ECAP-processed tubes. Nagasekhar et al. [10] subjected tubes of commercially pure (CP) Ti grade 1 filled with sand to three ECAP passes. The tensile strength increased by approximately 28%, however, the die angle was rather large (150°) and grain refinement was limited. Another interesting study by Al-Mufadi et al. [11] compared polyurethane rubber, sand and grease as inserts into

IOP Conf. Series: Materials Science and Engineering **194** (2017) 012010 doi:10.1088/1757-899X/194/1/012010

copper tubes subjected to ECAP. It turned out that rubber and sand worked well as mandrel materials but grease failed. The hardness of the Cu tube increased with increasing number of ECAP passes, though, the minimum grain size achieved after 4 passes at room temperature (RT) was 19 µm. Djavanroodi et al. [12] also investigated copper tubes filled with flexible rubber and showed that ECAP did not change the wall thickness. Additionally, there are investigations on ECAP of Al tubes filled with hydraulic oil [13] and on ECAP of Al as core material in a Cu tube [14–16]. The latter study showed that the strength of Al encapsulated in a Cu tube during ECAP was lower than after conventional ECAP.

2. Experimental procedure

2.1. Materials and samples

Two types of samples were subjected to ECAP: 1) tubes made of CP-Ti grade 2 filled with different (metallic and non-metallic) mandrel materials and 2) solid, cylindrical bolts made of the same materials as the metallic mandrels. Both sample types had the same dimensions: outer diameter of 20 mm and length of 120 mm. The wall thickness of the tubes was 2 mm and thus the inner diameter was 16 mm. Mandrel materials consisted of the metals CP aluminium (hereafter abbreviated to CP-Al), CP-Ti grade 2 (CP-Ti) – nominal the same as the tube material, however, unfortunately with different mechanical properties in the as-received (AR) condition – and the Ti-6Al-4V alloy (Ti64) as well as of two non-metals: graphite (C) and oak wood (oak).

2.2. Equal Channel Angular Pressing (ECAP)

ECAP was performed at AIT Austrian Institute of Technology. Two ECAP tools with a channel diameter of 20 mm and with different angles of intersection (120° and 105°) were used. Routes Bc, B_{120} and C, in which between subsequent passes the billet is rotated in the same direction by 90°, 120° and 180°, respectively, were applied. Both, tools and samples were held at the same, constant temperature during the whole process, whereby, if needed, the samples were pre-heated prior to the first ECAP pass in an air furnace for 20 minutes. A graphite-based anti-seize paste Molykote® P-74 was used as lubricant.

The critical processing parameters, particularly temperatures and angles of intersection, were adjusted to the mandrel materials and are listed together with routes in table 1. Thus, for the Ti64 core 500°C had to be chosen regardless the fact, that the temperature is too high for processing pure titanium. In general, one of the aims of the study was to conduct ECAP on Ti tubes at the lowest possible temperature. In fact, for CP-Al and non-metallic cores, temperatures as low as 100°C and even RT were used. Further, it was intended to reveal the homogeneity of the microstructure particularly to clarify the question whether the simple shear deformation originating at the point of channels' intersection can spread through the mandrel, especially the non-metallic ones, and reach the outer section of the sample. Therefore, ECAP with one single pass as well as a higher (optimum) number of passes was performed.

Table 1. ECAP parameters for CP-Ti tubes with different mandrel materials.

	Mandrel material	Number of passes and route	Temperature [°C]	Angle of intersection [°]	Total uniform equivalent strain acc. [1]
Metallic	**CP-Al**	1x	100	105	0.9
		2xC			1.8
	CP-Ti	1x	400	105	0.9
		6xB_{120}			5.3
	Ti64	1x	500	120	0.6
		4xBc			2.5
Non-metallic	**C**	1x	RT 100	105	0.9
	oak	1x	100	105	0.9
		2xC	RT	120	1.3
		4xCBC*	100	120	2.5

* Rotation sequence: 180°, 90° and 180°

2.3. Mechanical properties

Vickers hardness measurements (HV1) were performed on cross sections perpendicular to the ECAP direction using a fully-automatic hardness tester (EMCO-TEST DuraScan 80) according to EN ISO 6507-1. Hardness mappings on the whole cross section of the tubes as well as of the cores were performed with a mesh of 0.5x0.5mm^2 to reveal the respective hardness distributions.

All tensile tests were carried out at RT according to ISO 6892-1 A222 procedure with an initial strain rate of 3x10^{-4} s^{-1}. A computerized universal Shimadzu Servopulser EHF-UV050K1 testing machine equipped with a high-precision clip-on extensometer was used. Two types of tensile samples were used. First, thread-end test pieces of round cross-section of type B6x30 according to DIN 50125 obtained by turning from mandrel materials – in as-received (AR) state, as cores of tubes as well as solid bolts after ECAP – were tested. This approach ensured the highest precision of the data. At least two samples per condition were tested in order to single out possible outliers. Second, properties of the tube material – both AR and after ECAP – were obtained on test pieces analogous to DIN 50125 - E2x5x20. Four samples per tube were fabricated to obtain an average value. The two transverse surfaces were machined by spark erosion to tight tolerances whereas the original thickness or the wall remained unmachined.

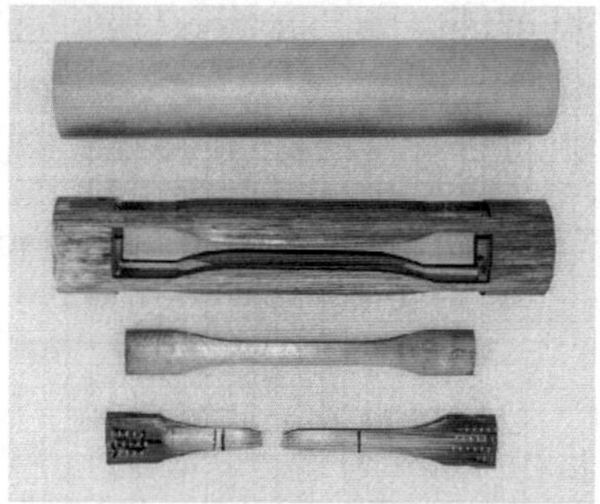

Figure 1. From top to bottom: a piece of tube AR, tube after ECAP and cutting of tensile samples, tensile sample prior to and after testing.

3. Results

3.1. Hardness and homogeneity

The results of hardness on tubes are shown as colour-coded mappings in table 2 and as numerical data (mean value) in table 5. Processing tubes with a graphite mandrel by ECAP is feasible, but only for one pass. Even so, a vast increase in hardness occurs due to the low deformation temperature (from 180 HV1 in AR condition to 233 HV1 and 217 HV1 after one ECAP pass at 100°C and RT, respectively).

Also in case of tubes with oak mandrel a significant increase in hardness (212 HV1) as well as an inhomogeneous hardness distribution (table 2) after one pass at 100°C were observed. Moreover, several ECAP passes could be performed and resulted in very high hardness (220 HV1 after 2xC at RT and 224 HV1 after 4xCBC at 100°C) as well as in fairly homogenous hardness distributions. However, oak was the only mandrel material which did not preserve the dimensions of the tube during ECAP processing. The tubes partially collapsed and a shortening as well as roughly a nonuniform doubling in wall thickness occurred due to the compressibility of wood.

As CP-Al is a quite soft metallic mandrel material the ECAP temperature can be as low as 100°C. This effectively increased the hardness of the CP-Ti tube, even after one pass, and resulted in a quite uniform hardness distribution on the cross-section of the tube after only 2 passes with route C.

CP-Ti as mandrel material leads to a homogenous hardness distribution as well, though, many passes (6) are needed for an increase in hardness compared to the AR material. This is due to the higher processing temperature of 400°C used for this rather strong mandrel material.

Using mandrels made of Ti64 the ECAP processing temperature had to be even higher (500°C) resulting in rather low hardness values, as a matter of fact, in case of only one ECAP pass even lower than in the AR condition. However, the tensile properties improved (see next chapter).

In case of non-metallic mandrels no investigations of mechanical properties were performed. Hardness results of metallic mandrels represented by colour-coded mappings are shown in table 3 for CP-Ti and in table 4 for Ti64. The mean values are specified in table 5. It can be seen that ECAP of metals as mandrels in CP-Ti tubes leads to higher hardness values and slightly more homogenous hardness distributions (for CP-Ti mandrels) than conventional ECAP processing of solid bolts.

Table 2. Hardness maps of CP-Ti tubes as received (AR) and after indicated number of ECAP passes and routes with different mandrel materials. The top side of the mappings corresponds to the channels' intersection in the ECAP tools. (For abbreviations see chapter 2.1.)

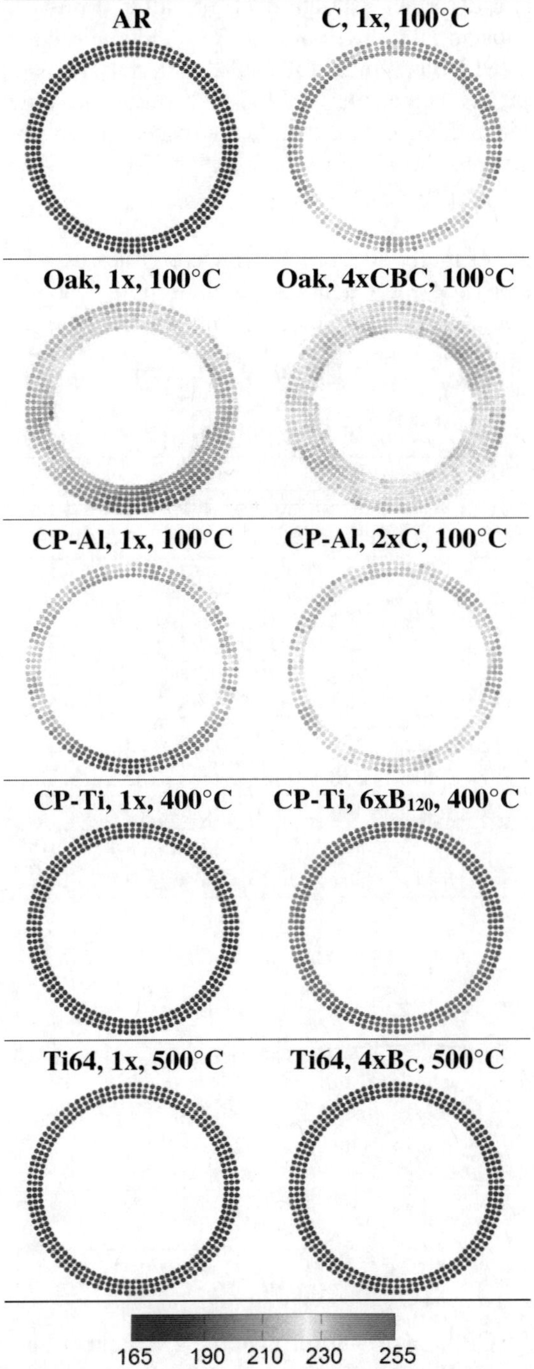

Table 3. Hardness maps of CP-Ti mandrel material AR and after one and six ECAP passes at 400°C as solid bolts and mandrels.

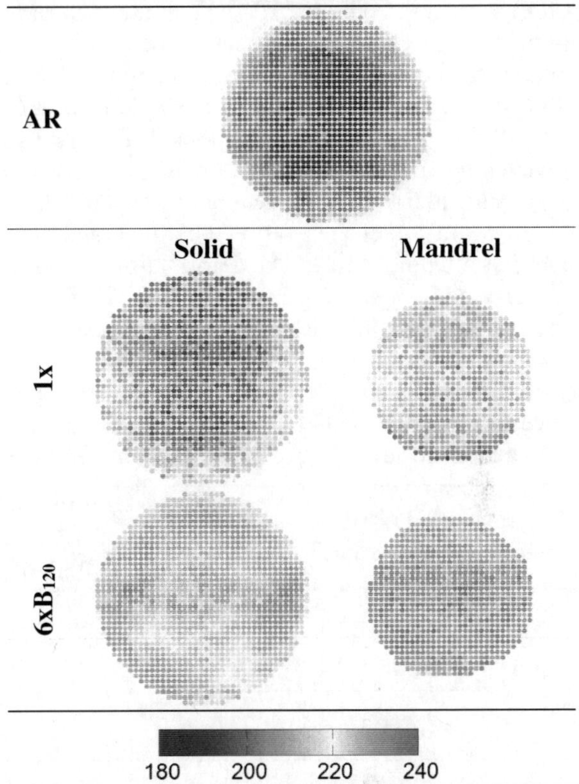

Table 4. Hardness maps of Ti64 in as-received state (AR) and after one ECAP pass at 500°C as solid bolt and mandrel.

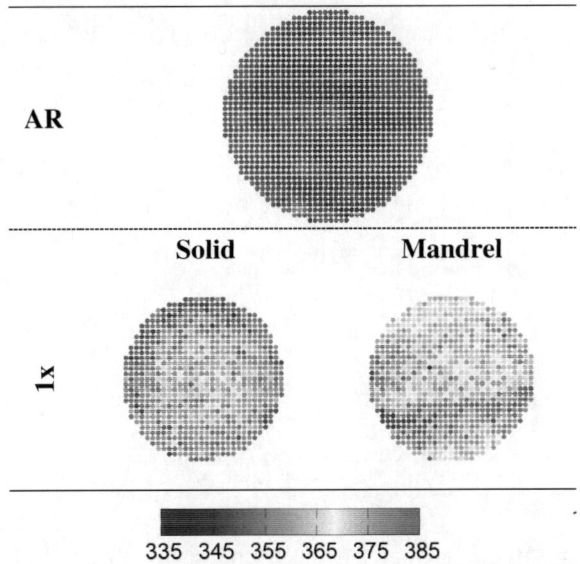

3.2. Tensile strength and ductility

The numerical results of tensile tests on CP-Ti tubes are summarised in table 5 while figure 2a shows stress-strain curves of selected samples. In all cases the yield strength and the ultimate tensile strength (UTS) are increased by ECAP, even at the highest processing temperature of 500°C, which is somehow contrary to the hardness measurements. This is due to a strong texture in the AR condition. The enhancement of the UTS is quite substantial; up to 50% increase is obtained after 6 passes at 400°C with CP-Ti as mandrel or even after one single pass at 100°C with CP-Al as mandrel. A moderate decrease in total elongation is mostly observed after one pass and a further reduction occurs after 6 passes, however, the ductility is still reasonable. Furthermore, a substantial reduction in uniform elongation due to a change in the strain hardening coefficient takes place, which is typical for ECAP materials.

The results of tensile tests on metallic mandrel materials as well as on solid ECAP bolts of the same materials after the same ECAP deformation are summarized in table 5 and exemplary stress-strain curves are depicted in figure 2b. In case of CP-Al the improvement of UTS is in the range of 20-25%. The highest increase in tensile strength is obtained for CP-Ti after 6 passes at 400°C (+37%).

Table 5. Mechanical properties of CP-Ti tubes and mandrel materials. The latter were either ECAP-processed as solid bolts (S) or as mandrels of CP-Ti tubes (M). YS denotes yield strength, UTS is ultimate tensile strength, Ag percentage uniform elongation, A percentage elongation after fracture.

Mandrel material	ECAP conditions	Tubes					Mandrel materials					
		HV1	YS [MPa]	UTS [MPa]	Ag [%]	A [%]	ECAP	HV1	YS [MPa]	UTS [MPa]	Ag [%]	A [%]
CP-Al	AR	180	289	427	15	27		79.5	198	223	7	18
	1x, 100°C, 105°	209	543	608	2.5	15	S	90.8	269	275	2.3	13
							M	92.6	266	276	2.8	13
	2xC, 100°C, 105°	227	–	–	–	–	S	92.3	263	267	2.0	15
							M	80.0	–	–	–	–
CP-Ti	AR	180	289	427	15	27		202	413	494	11	39
	1x, 400°C, 105°	178	432	491	5.5	27.5	S	208	527	576	4.2	28
							M	213	547	602	4.4	25
	$6xB_{120}$, 400°C, 105°	194	507	585	2.5	12.5	S	226	591	655	3.1	21
							M	230	609	676	2.2	22
Ti64	AR	180	289	427	15	27		346	977	1024	6	18
	1x, 500°C, 120°	174	422	494	7.5	20	S	356	1138	1152	1.1	10
							M	361	1125	1146	0.9	11

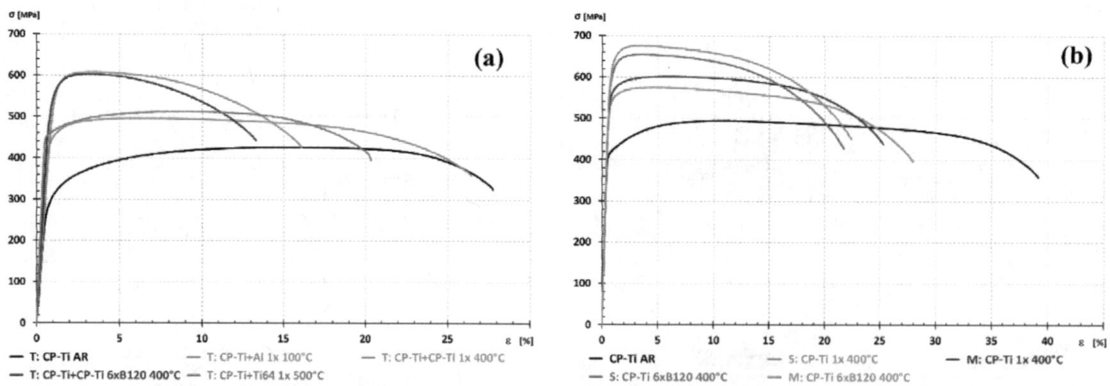

Figure 2. Tensile stress-strain curves. a) CP-Ti tube b) CP-Ti as mandrel material (M) and solid counterpart (S). Note the different mechanical properties in AR state.

4. Discussion and summary

The current study has proved the feasibility of processing tubular samples by ECAP. By selecting proper mandrel materials it is possible to considerably reduce the pressing force and therefore to lower the ECAP temperature of CP-Ti. This leads to a vast increase in hardness and strength of CP-Ti tubes, even higher than what can be obtained in solid titanium rods by numerous ECAP passes at elevated temperatures. E.g. the highest hardness was reached after only one pass at 100°C with C as a mandrel (233 HV1). However, close attention should be paid to the homogeneity of the microstructure especially when non-metallic mandrels are used in consequence of a non-uniform plastic deformation during processing: The sample area in the vicinity of the channels' intersection undergoes typical ECAP deformation by simple shear, whereas the opposite site of the tube is simply bent. The most uniform hardness distribution on tube cross-sections were achieved with CP-Ti as mandrel material after 6 passes at 400°C using route Bc. Using mandrel materials with higher strength than the tube material (e.g. Ti64) requires higher ECAP temperatures than necessary for pure titanium resulting in rather low mechanical properties after ECAP. Therefore, soft metallic materials, e.g. pure Al, are recommended for mandrels. CP-Al leads to the highest hardness of the CP-Ti tube compared to all other metallic mandrel materials tested, however, the hardness is lower than in case of C.

Mechanical properties of metals processed by ECAP as mandrels encapsulated in CP-Ti tubes are better (CP-Ti, Ti64) or at least the same (CP-Al) as properties of these materials after conventional ECAP deformation in the form of solid bolts under the same conditions.

Acknowledgments

The authors appreciate the financial support of the Austrian Research Promotion Agency (FFG), project number 854064, 12th funding programme 'Produktion der Zukunft 2015, transnational'.

References

[1] Valiev R Z and Langdon T G 2006 *Progress Mater. Sci.* **51** 881–981
[2] Estrin Y and Vinogradov A 2013 *Acta Mater.* **61** 782–817
[3] Nagasekhar A V, Tick-Hon Y, Guduru R K and Ramakanth K S 2007 *Physica C* **466** 174–180
[4] Derakhshandeh R H and Jahromi A J 2011 *Materials and Design* **32** 3377–3388
[5] Wang X-X, He M, Zhu Z, Xue K-M and Li P 2015 *Trans. Nonferrous Met. Soc. China* **25** 2122–2129
[6] Rezaei M R, Shabestari S G and Razavi S H 2017 *J. Mater. Sci. Tech.* in press (http://dx.doi.org/10.1016/j.jmst.2016.10.013)
[7] Jiang S-Y, Zhao Y-N, Zhang Y-Q, Tang M and Li C-F 2013 *Trans. Nonferrous Met. Soc. China* **23** 2021–2028
[8] Krystian M, Zehetbauer M J, Kropik H, Mingler B and Krexner G 2011 *J. Alloys Comp.* **509S** S449–S455
[9] Révész Á, Gajdics M, Varga L K, Krállics G and Spassov T 2014 *Int. J. Hydrogen Energy* **39** 9911–9917
[10] Nagasekhar A V, Chakkingal U and Venugopal P 2006 *J. Mater. Proc. Tech.* **173** 53–60
[11] Al-Mufadi F and Djavanroodi F 2015 *Arab. J. Sci. Eng.* **40** 2785–2794
[12] Djavanroodi F, Zolfaghari A A, Ebrahimi M and Nikbin K M 2013 *Acta Metall. Sin. (Engl. Lett.)* **26** 574–580
[13] Jafarlou D M, Zalnezhad E, Hassan M A, Ezazi M A, Mardi N A, Hamouda A M S, Hamdi M and Yoon G H 2016 *Mater. Design* **90** 1124–1135
[14] Djavanroodi F, Daneshtalab M and Ebrahimi M 2012 *Mater. Sci. Eng. A* **535** 115–121
[15] Shaeri M H, Shaeri M, Salehi M T, Seyyedein S H and Djavanroodi F 2015 *Trans. Nonferrous Met. Soc. China* **25** 1367–1375
[16] Shaeri M H, Djavanroodi F, Sedighi M, Ahmadi S, Salehi M T and Seyyedein S H 2013 *J. Strain Analysis Eng. Des.* **48** 512–521

IOP Conf. Series: Materials Science and Engineering **194** (2017) 012011 doi:10.1088/1757-899X/194/1/012011

Microstructural and superficial modification in a Cu-Al-Be shape memory alloy due to superficial severe plastic deformation under sliding wear conditions

C G Figueroa[1], F N Garcia-Castillo[2], V H Jacobo[2], J Cortés-Pérez[1] and R Schouwenaars[2]

[1] Centro Tecnológico Aragón, Facultad de Estudios Superiores Aragón, Universidad Nacional Autónoma de México. Av. Rancho Seco s/n, Col. Impulsora, Cd. Nezahualcóyotl, 57130, Estado de México, México.

[2] Departamento de Materiales y Manufactura, Facultad de Ingeniería, Edificio O, Universidad Nacional Autónoma de México. Avenida Universidad 3000, Coyoacán, 04510, Ciudad de México, México.

Email: carlosgabf@msn.com

Abstract. Stress induced martensitic transformation in copper-based shape memory alloys has been studied mainly in monocrystals. This limits the use of such results for practical applications as most engineering applications use polycristals. In the present work, a coaxial tribometer developed by the authors was used to characterise the tribological behaviour of polycrystalline Cu-11.5%Al-0.5%Be shape memory alloy in contact with AISI 9840 steel under sliding wear conditions. The surface and microstructure characterization of the worn material was conducted by conventional scanning electron microscopy and atomic force microscopy, while the mechanical properties along the transversal section were measured by means of micro-hardness testing. The tribological behaviour of Cu-Al-Be showed to be optimal under sliding wear conditions since the surface only presented a slight damage consisting in some elongated flakes produced by strong plastic deformation. The combination of the plastically modified surface and the effects of mechanically induced martensitic transformation is well-suited for sliding wear conditions since the modified surface provides the necessary strength to avoid superficial damage while superelasticity associated to martensitic transformation is an additional mechanism which allows absorbing mechanical energy associated to wear phenomena as opposed to conventional ductile alloys where severe plastic deformation affects several tens of micrometres below the surface.

1. Introduction

Shape memory alloys (SMA) exhibit particular mechanical properties represented by the stress-strain curve which for these materials shows a hysteresis loop with a residual deformation during the loading- unloading process performed at constant temperature. This deformation is not produced by dislocation movement but by detwinning of martensite. Therefore, if as effect of increasing temperature, the material transforms from martensite to austenite, the residual deformation could be recovered [1]. In the last decades, this behaviour known as shape memory effect (SME) has increased the attention in SMA because of the possibility of expanding their use in technological applications. In particular, copper-based SMA has received special attention because of the mechanical properties associated to martensitic transformation (MT) like SME and the super-elasticity effect (SE), understanding that the latter contributes to a damping capability and high fatigue resistance [2,3]. Given the above, both the temperature and external induced stress are factors which play an important role in the phase transformation mechanisms giving origin to SME and SE, this because of the driving force for the MT directly depends of them.

Although Cu-Al SMA has been studied mainly as monocrystals, in recent years several studies have shown important capabilities which make polycrystals of the mentioned alloys multifunctional materials with the possibility to be used in a wide range of applications due to its high thermal stability and a transformation temperature M_s which can be controlled by addition of Be [4]. It is known that MT in Cu-Al-Be SMA is a first order diffusionless phase transition induced by stress in which a high temperature face centred cubic (FCC) structure transforms into a low temperature orthorhombic unit cell [5].

Nowadays, some potential applications for SMA are focused in sensor-actuator devices applied for energy recovering [6-8], production of medical instrumental [9], or seismic isolation devices [10-12]. Nevertheless, in the case of Cu-Al-Be SMA, structural applications are limited because of its mechanical behavior is highly influenced by its high anisotropy and large grain size. This issue can be overcome by a grain size refinement produced by severe plastic deformation (SPD) in which a high dislocation density and deformation twins are induced [13].

As it is known, in SPD the increment of dislocation density and subsequent grain fragmentation produces an ultra-fine grain structure (UFGS) with predominantly high angle grain boundaries (HAGB), this refinement process known as continuum dynamic recrystallization (CDRX) increases the material resistance [14-16]. Moreover, in materials like low carbon steel, titanium, copper [17-19], inducing a gradient microstructure by SPD applied in the surface has shown to be a good processing method which improve the tribological performance due to the combination of high strength in the UFG zone and good ductility in coarse grained zone. In addition, a superficial SPD could generate the necessary stress induced martensitic transformation (SIMT) to obtain SE. This combination of SE and high strength in SPD zones could improve the performance of SMA when conducted under sliding contact conditions due to the SE contribute to the dissipation energy during cyclic loads, while high hardness reduces the superficial damage.

In previous works the authors have studied the superficial modification and microstructural layers induced by a gradual SPD process applied in Al-Sn, Cu-Mg-Sn and pure Cu (20, 21). In the present research, a coaxial tribometer developed by the authors (22) was used to characterise the tribological behaviour and microstructural modification of polycrystalline Cu-11.5%Al-0.5%Be SMA in contact with AISI 9840 steel under sliding wear conditions.

2. Experimental method

A cast slab of polycrystalline Cu-11.5%Al-0.5%Be SMA obtained in a high frequency furnace and betatized according to the method proposed in (23) was used for the present work. SMA specimens were thin square coupons with dimensions 20x20x2 mm^3 obtained by precision cutting and subsequently prepared by conventional mechanical polishing giving an initial roughness of 230nm±23nm. The superficial SPD was conducted at room temperature using a coaxial tribometer constructed by the authors and described in (22). The SPD process consists in putting a cylindrical pin made of AISI9840 steel into contact with the SMA specimen under a normal load of 400 N. Then, the pin rotates over its own axes during 300s at constant angular speed of 60 rpm. During the test, the normal load was constantly measured with a Burster 8524® load cell, while the torque with a Futek TRS300® sensor.

Worn surfaces were characterized using a Bruker Innova atomic force microscope (AFM) in contact mode and a Phillips XL 20 scanning electron microscope (SEM) equipped with tungsten filament, conventional Eberhart-Thornly secondary electrons detector and solid state backscatter detector. Transverse sections trough the centre of the wear tracks were prepared by conventional mechanical polishing. Microstructural observation of an etched specimen was conducted with SEM using the secondary electrons detector, while a unetched specimen was observed with a Zeiss Axio Imager.A2m reflected light microscope using crossed nicols. Microhardness measurements were made over the transverse section to worn surface using a mechanical Leitz Wetzlar 8068 Vickers indenter.

3. Results and discussion

The diameter of the wear tracks were 1.9mm±0.1mm, with a slight rougness increase in the worn zone of 400nm±20nm. These values suggest a very high wear resistance as compared to previously tested materials like Al-Sn (22) which is commonly used under sliding contact conditions. An important point to take in account for this comparison is that although the wear track diameters are similar and neither of the two materials presented weight loss; the applied load in Cu-Al-Be was four times higher than the applied in Al-Sn, according to the hardness of the former which is also four times higher

The tribological behaviour of the material can be related with the surface modification shown in figure 1 which corresponds to the SEM image obtained from the worn surface. The elongated flakes aligned to sliding direction are product of the plastic deformation induced during the tribological test and are sign of adhesive wear. A better detail of the zone into the red rectangle is presented in figure 1b, where the pile up of elongated flakes suggest the strain accumulation as described by Kapoor .

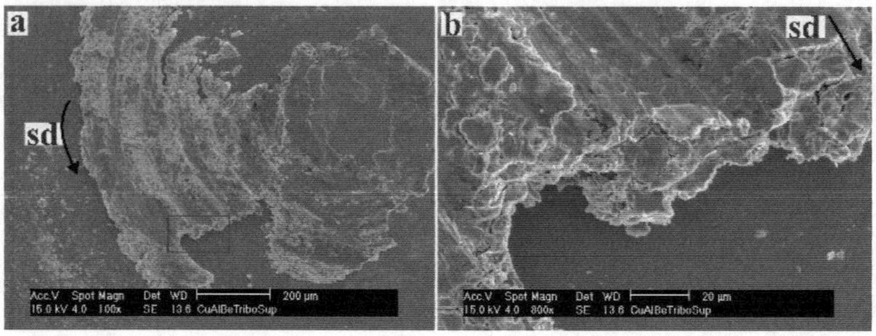

Figure 1. SEM images of the worn surface. (a) Elongated flakes aligned to sliding direction, (b) signs of adhesive wear and flakes pile up suggesting plastic ratchetting. (sd=sliding direction).

The surface AFM deflection signal maps inside of the wear track (figure 2) shows a portion of the modified surface in which is possible to distinguish the wear grooves in sliding direction. The shown zone doesn't present the elongated flakes associated to adhesive wear and plastic deformation. Nevertheless, at higher zoom (figure 2b) is possible to distinguish a regular pattern of cells aligned to sliding direction. The geometry and size of the "cells" are similar to sub-grains observed in transmission electron microscope images, this suggests that the "cells" were generated during the tribological process as effect of the accumulated plastic deformation producing a grain refinement in the closest zone to surface. The most visible patterns in figure 2b correspond to the highest zone of the grooves shown in figure 2a.

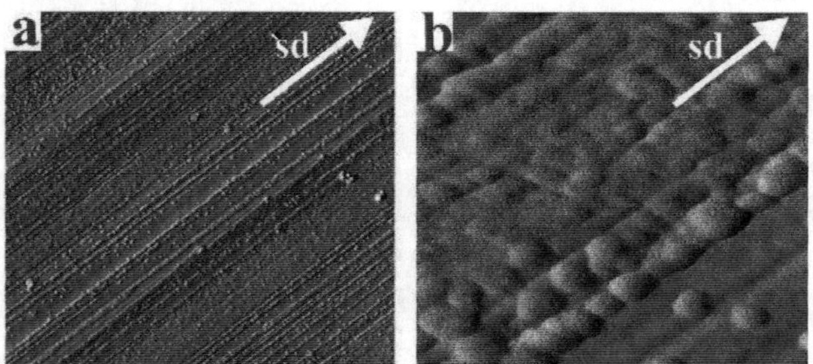

Figure 2. AFM maps of the surface inside the wear track. (a) Wear pattern following the sliding direction, (b) Detail over one wear line showing a cellular pattern produced during plastic deformation (notice the difference in scale).

Figure 3 shows the microstructure observed with optical microscope using a light polarizer. In figure 3a, the strong presence of martensite indicates that a SIMT took place during the tribological test, this modified the mechanical properties of the SMA due to the SE behaviour in the martensite, giving the material a high damping factor which allows to reduce the effect of the cyclic loads imposed during sliding wear conditions. In figure 3b an arrow points out the fragmented zone of a grain in the surface contact of the material. This fragmentation did not extend through grain; a reason for this could be that the high density of martensitic variants acts as a barrier for the dislocation mobility. The black ellipse indicates martensitic variants bended as effect of the plastic deformation induced at the surface.

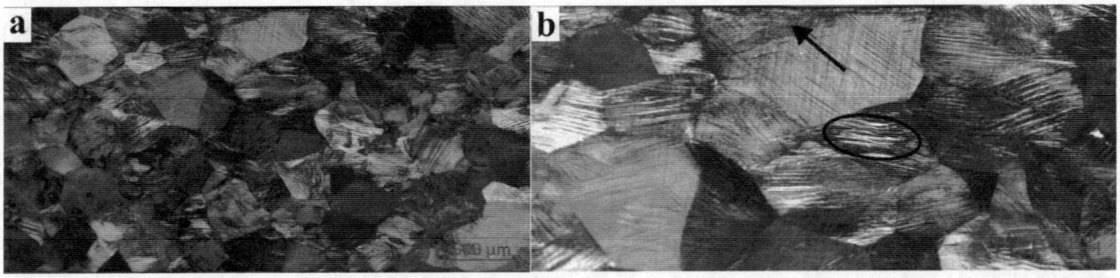

Figure 3. Microstructure observed with optical microscope under crossed nicols. (a) General view of the microstructure with coarse grains. (b) Detail of a zone close to surface where is possible to observe the high density of martensite. The arrow points out to the wear-induced structural modification whereas the ellipse shows bended martensitic variants. Sliding direction is perpendicular to image.

SEM images of the zone near the worn surface are shown in figure 4. The martensitic structure is observed in figure 4a; in the top layer (contact surface) is possible to distinguish a discontinuity in the martensitic variants and instead of them, flow lines are visible. These lines are product of the material flow during plastic deformation induced at the surface, In contrast to conventional ductile materials, this zone is confined to a very thin layer.

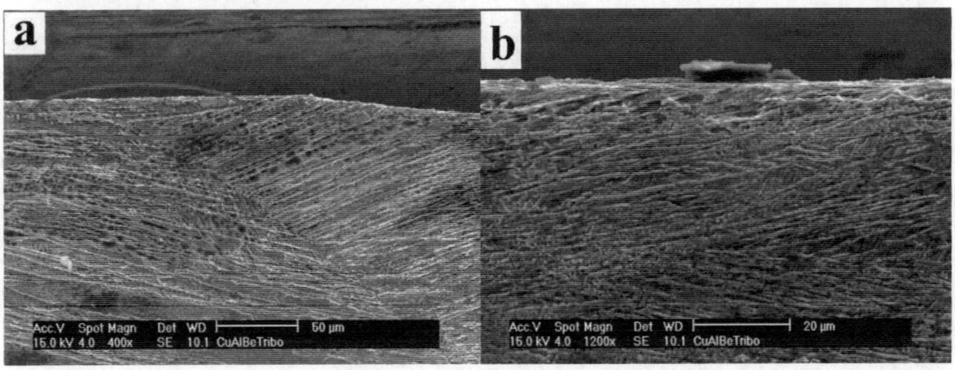

Figure 4. SEM images of the section transverse to the worn surface. (a) Strong presence of martensite in grains near to surface (b) zoom of the red ellipse zone in which is possible to see flow lines produced by plastic deformation in the top layer.

Hardness results are not shown as a chart because no substantial change was observed through the transverse section, while a slight hardness increment was measured after the surface modification regarding to initial condition. This could be attributed to the SIMT, also is important to notice that because of the MT was carried out homogeneously and the grain fragmentation was experienced only in the top layer, hardness keeps constant through the transverse section.

4. Conclusions

Cu-Al-Be SMA shows a high tribological compatibility when subject to sliding contact conditions against AISI9840 steel. The superficial damage was lower than observed in previous work using the classical tribological Al-Sn alloy, with loads applied which are proportional to the hardness of each material. The improvement of tribological behaviour associated to the high damping factor of SE materials was predicted in previous papers by Schouwenaars and co-workers [25,26].

Contrary to common ductile materials like Al or pure Cu, the grain refinement in Cu-Al-Be SMA was observed only in a very thin layer at the surface, while the expected microstructural gradient was not observed. Nevertheless, a future microstructural characterization of the ultra-fine tribolayer could show a gradual grain fragmentation.

Under the action of the shear deformation imposed during the coaxial tribometry test, a stress induced martensitic transformation was observed throughout the sample. One possible reason for the very limited grain refinement could be that the martensitic variants generated during the transformation act as a barrier for the dislocation mobility and because of this, the continuum dynamic recrystallization associated to obtaining an ultra-fine grain structure does not take place in zones away from the surface.

The combination of the ultra-fine grain structure at the top layer with the SE observed in SMA allows the material to support very high loads under sliding contact without presenting pronounced superficial damage.

Acknowledgments

C G Figueroa and F N García wish to thank DGAPA for their postdoctoral fellowship. This work was financially supported by DGAPA trough project IN115917.

References

[1] Akselsen O 2010 *INTECH Open Access Publisher*
[2] Otsuka K and Wayman C M 1999 *Shape Memory Materials* (Cambridge: Cambridge university press)
[3] Montecinos S, Cuniberti A and Sepúlveda A 2008 *Mater. Charact* **59** 117-23
[4] Belkahla S, Zuñiga H F and Guenin G 1993 *Mater Sci Eng* A **169** 119-24
[5] Kaouache B, Inal K, Berveiller S, Eberhardt A and Patoor E 2006 *Mater Sci Eng* A. **438** 773-778
[6] Lagoudas D 2008 *Shape Memory Alloys: Modeling and Engineering Applications* (Springer)
[7] Zhu J.J, Liang N G, Liew K M and Huang W M 2001 *J Intel Mat Syst Str* **12** 127-132
[8] Zhu J.J, Liang N G, Liew K M and Huang W M 2001 *J Intel Mat Syst Str* **12** 133-140
[9] Yamauchi K, Ohkata I, Tsuchiya K and Miyazaki S 2011 *Shape Memory and Superelastic Alloys: Applications and Technologies* (Amsterdam: Elsevier)
[10] Graesser E J and Cozzarelli F A 1991 *J Eng Mech* **117** 2590-608
[11] Casciati F and Faravelli L 2004 *J Phys IV (Proceedings)* **115** 299-306
[12] Torra V, Isalgue A, Lovey F C, Martorell F, Molina F J, Sade M and Tachoire H 2004 *J Phys IV (Proceedings)* **113** 85-90
[13] Zhang Y, Jiang S, Hu L and Liang Y 2013 *Mater Sci Eng* A **559** 607-14
[14] Rollett A, Humphreys F J, Rohrer G S and Hatherly M 2004 *Recrystallization and Related Annealing Phenomena* (Amsterdam: Elsevier)
[15] Sakai T, Belyakov A, Kaibyshev R, Miura H and Jonas J J 2014 *Prog Mater Sci* **60** 130-207
[16] Doherty R D et al 1997 *Mater Sci Eng* A. **238** 219-74
[17] Wang Z B, Tao N R, Li S, Wang W, Liu G, Lu J and Lu K 2003 *Mater Sci Eng* A **352** 144-9
[18] Stolyarov V V, Shuster L S, Migranov M S, Valiev R Z and Zhu Y T 2004 *Mater Sci Eng* A

371 313-7
[19] Moshkovich A, Lapsker I and Rapoport L 2013 *Wear* **305** 45-50
[20] Figueroa C G, Ortega I, Jacobo V H, Ortiz A, Bravo A and Schouwenaars R 2014 *IOP Conf Ser Mater Sci Eng* **63**
[21] Figueroa C G, Schouwenaars R, Jacobo V H, Ortiz A, Petrov R and Kestens L A I 2016 *Tribology Online* **11** 389-95
[22] Figueroa C G, Jacobo V H, Ortiz A and Schouwenaars R 2015 *Tribol Lett* **59** 1-10
[23] Zúñiga H F, Rios-Jara D, Lovey F and Guénin G 1995 *J Phys IV* **5** 171-74
[24] Kapoor A and Johnson K L 1994 *P Roy Soc Lond A Mat* **445** 367
[25] Schouwenaars R, Jacobo V H and Ortiz A 2007 *Wear* **263** 727-35
[26] Bravo A E, Durán H A, Jacobo V H, Ortiz A and Schouwenaars R 2013 *Wear* **302** 1528-35

Thermal stability and mechanical properties of HPT-processed CP-Ti

Y Huang[1], S Mortier[2], P H R Pereira[1,3], P Bazarnik[4], M Lewandowska[4] and T G Langdon[1]

[1] Materials Research Group, Faculty of Engineering and the Environment, University of Southampton, Southampton SO17 1BJ, UK
[2] Polytech Montpellier, 34095 Montpellier CEDEX 5, France
[3] CAPES Foundation, Ministry of Education of Brazil, Brasília 70040-020, DF, Brazil
[4] Faculty of Materials Science, Warsaw University of Technology, Woloska 141, 02-507 Warsaw, Poland

E-mail: y.huang@soton.ac.uk

Abstract. A grade 2 commercial pure titanium was processed by high-pressure torsion (HPT) at room temperature to 20 turns. X-ray analysis showed that an ω phase formed during HPT processing but disappeared immediately during 10 minutes post-HPT short-term annealing even at a very low temperature of 473 K. The thermal stability of HPT-processed microstructural evolution was studied by electron backscatter diffraction (EBSD). After short-term annealing at lower temperatures (473 and 673 K), the disc centre had relative higher hardness value than the edge area, and it was found that the centre retained the feature of deformed microstructure whereas the edge showed recovery / recrystallization of the microstructure. Short-term annealing at higher temperatures (873 and 973 K) led to almost uniformly distributed hardness and microstructures in the disc centre and edge areas. Grain structures and hardness measurements indicate complete recrystallization occurred at 873 K.

Keywords: Electron backscatter diffraction; Grain size; Hardness; High-pressure torsion; Titanium; x-ray.

1. Introduction

Commercial purity titanium (CP-Ti) is widely used for medical implants in trauma surgery, orthopedic and oral medicine due to its good biocompatibility [1]. Material strength is an important consideration for practical bio-implantation. However, the mechanical strength of CP-Ti is relatively low compared to other metals such as Ti-6Al-4V used in medical device. It is now well established that equal-channel angular pressing (ECAP) and high-pressure torsion (HPT) are effective techniques in generating ultrafine-grained structure / nanocrystalline structure, and can improve the mechanical strength of CP-Ti [2-12]. CP-Ti normally needs to be heated to high temperatures between 473 and 773 K for ECAP processing [8], whereas HPT can easily process CP-Ti at room temperature [9-12]. In practice, HPT is more effective in refining the grain structure and enhancing the strength.

To benefit from the high strength and the nanocrystalline microstructure developed during HPT processing for use in practical environments, it is essential the materials have the ability to maintain this nanocrystalline microstructure over a certain time. Therefore, the current research on the thermal stability and mechanical properties was initiated to investigate the microstructure and hardness evolution of CP-Ti during short-term annealing after HPT processing.

2. Experimental material and procedures

The experimental material was grade 2 CP-Ti in the form of a round bar with diameter of 12.7 mm, supplied by Titanium Industries UK Ltd (Birmingham, UK). The as-received material was first annealed at 973 K for 2 h in a vacuum furnace to obtain a microstructure with average grain size of ~65 μm, and this is henceforth designated the as-annealed state (N0). The rod was then machined to a diameter of 9.9 mm, sliced into discs with thicknesses of ~1.1 mm and ground with abrasive papers to final thicknesses of ~0.8 mm.

All discs were processed by HPT at room temperature through total numbers of turns, N, of 1, 5, 10 and 20 using an imposed pressure of 6.0 GPa and a rotational speed of 1 rpm. The HPT was conducted under quasi-constrained conditions where there is a small outflow of material around the periphery of the disc during the processing operation. To investigate the thermo-stability of the HPT-processed discs, samples processed through 10 turns were selected for 10 minutes short-term annealing at temperatures of 473, 673, 873 and 973 K.

The microhardness was measured on HPT-processed and post-HPT annealing samples using a FM300 hardness tester equipped with a Vickers indenter with a load of 500 gf and a dwell time of 15 s. A JEOL1200 transmission electron microscope (TEM) was used to characterize the deformation microstructure after 10 turns of HPT processing at higher magnifications. The centre of the TEM foil was 3 mm away from the centre of the HPT sample. The grain structures near the edge of disc after post-HPT annealing were examined by electron backscattered diffraction (EBSD) using a JSM6500F thermal field emission scanning electron microscope (SEM). The EBSD patterns were collected using a step size of 0.15, 0.21, 0.52, 0.7 μm for 473, 673, 873 and 973 K for post-HPT annealing samples respectively.

The microstructures of the HPT-processed and post-HPT annealing samples were also evaluated using a Bruker D2 Phaser X-ray diffractometer equipped with a Cu target using Cu Kα (wavelength λ = 0.15406 nm) radiation and a Ni monochromator with a 1D LYNXEYE detector. The whole disc surface was analysed by X-ray diffraction (XRD) and θ - 2θ scans were conducted from 2θ = 25° to 100° with scan steps of 0.02° to record the XRD patterns.

3. Experimental results

Figure 1 shows the variation of the microhardness along the diameter of the discs processed to 5, 10 and 20 turns. The as-annealed material had a microhardness of ~157 Hv. After 5 turns, the disc centre had a lower hardness than at the edge area, which is consistent with a rigid body assumption that strain imposed on the disc has a linear dependence on the distance from the disc centre and the strain is a maximum value at the outer edge of the disc [13]. After 10 turns, the hardness along the disc diameter tended to distribute more homogenously although there were some jumps in the local hardness values. The highest hardness values were achieved in the 20 turns sample, with reasonable homogeneity along the disc diameter.

The microstructure of CP-Ti after 10 turns HPT processing is shown in the TEM image of Fig. 2. At the half-radius of the disc, the microstructure consists of equiaxed grains with average grain size of ~70 nm. Therefore, HPT produced significant grain refinement in CP-Ti with an average grain size reduced from ~65 μm for as-annealed to ~70 nm after 10 turns of HPT processing. The ring-like selective area diffraction pattern (SADP) demonstrates that the microstructures have boundaries mainly with high angles of misorientation. The strengthening effect after HPT processing is obvious and significant.

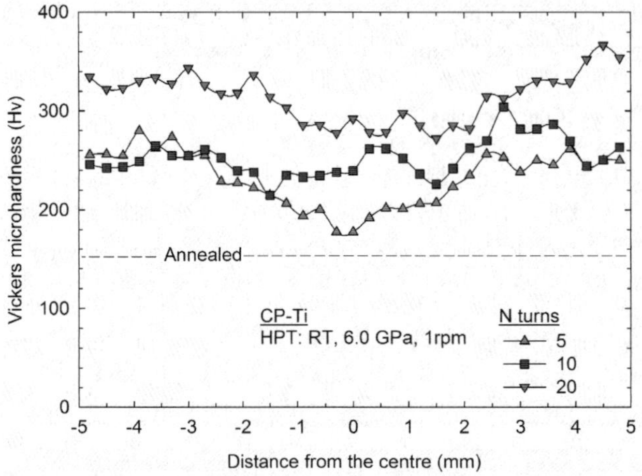

Fig. 1. Distribution of Vickers microhardness, Hv, along the diameter of disc processed by HPT through 5, 10, 20 turns.

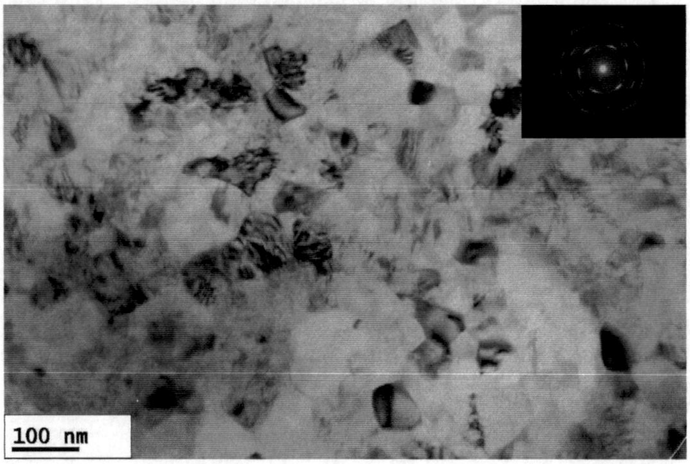

Fig. 2. Microstructure in CP-Ti after 10 turns in disc half-radius area.

Figure 3 shows the x-ray patterns of CP-Ti in the as-annealed (N0) condition and after HPT processing to different number of rotations. In the as-annealed condition, there are only α peaks and no ω phase was detected. After HPT processing to 1 turn, ω phase peaks become visible with main peak $(11\text{-}20)_\omega$ at $2\theta \approx 39°$ to replace the α peak $(0002)_\alpha$ at $2\theta \approx 38.42°$. As deformation proceeded to 5, 10 and 20 turns, three more ω peaks at $2\theta \approx 56.5°$, 84° and 97° appeared. This result confirms an α to ω phase transformation due to HPT deformation and indicates the intensity of the ω phase increases with increasing numbers of rotations.

The hardness variations along the disc diameter during 10 minutes post-HPT short-term annealing on 10 turns samples are shown in Fig. 4. After short-term annealing at lower temperatures of 473 and 673 K, the disc centre had relative higher hardness values than the edge area, and this indicates that the disc edge area had more rapid recovery / recrystallization than the centre area. Short-term annealing at higher temperatures of 873 and 973 K gave almost uniformly distributed hardness in disc centre and edge areas, and also the hardness at 873 and 973 K were almost at the same level. Overall, the hardness values in the disc edge area decreased with increasing temperature from 473 to 873 K. This indicates that HPT-processed CP-Ti is very sensitive to the applied temperatures.

IOP Conf. Series: Materials Science and Engineering **194** (2017) 012012 doi:10.1088/1757-899X/194/1/012012

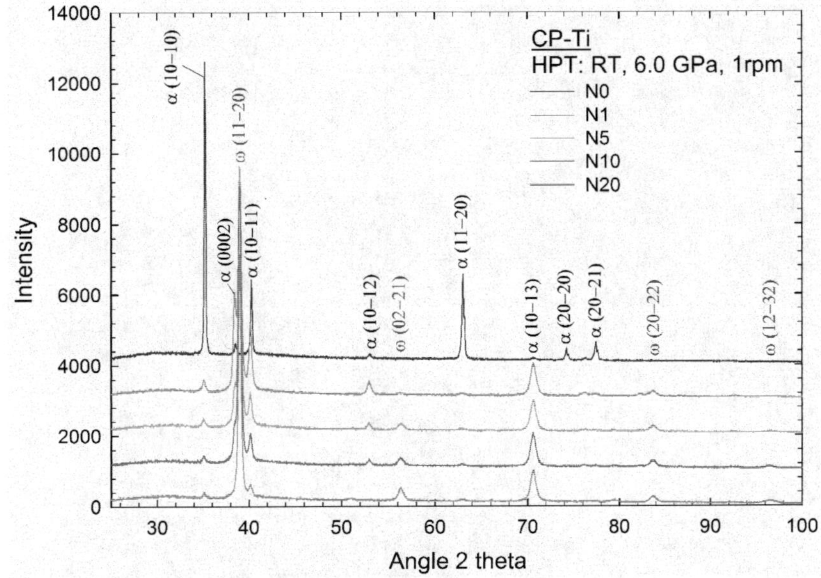

Fig. 3. X-ray patterns of CP-Ti in both the as-annealed (N0) and after HPT processing through different numbers of rotations.

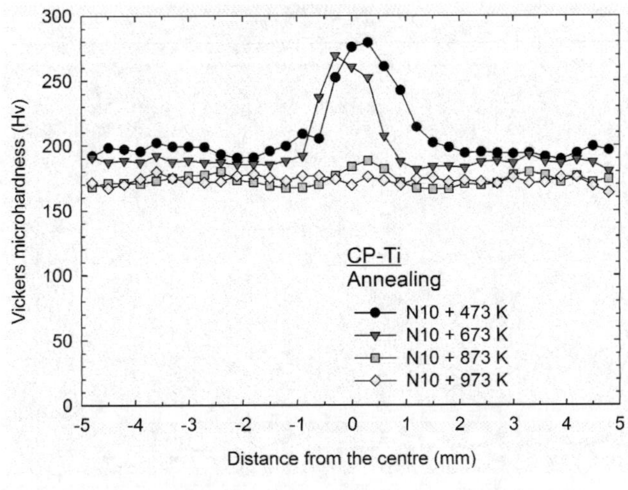

Fig. 4. Distribution of Vickers microhardness, Hv, along the diameter of disc during post-HPT short-term annealing at different temperatures.

Figure 5 shows the microstructure development in disc edge area during post-HPT short-term annealing at 473, 673, 873 and 973 K. At the lowest temperature of 473 K, the microstructure should have a high dislocation density and it is difficult to obtain a clear EBSD pattern, whereas at 673, 873 and 973 K the average grain sizes were 3.1, 3.9 and 9.7 μm, respectively. The grain structures coarsen rapidly from ~70 nm after 10 turns to the microns level after 10 minutes short-term annealing.

Figure 6 shows the x-ray patterns of CP-Ti in 10 turns samples during post-HPT short-term annealing. Upon heating to 473 K for 10 minutes, the ω phase peak $(11\text{-}20)_\omega$ is replaced by α phase peak $(0002)_\alpha$ at $2\theta \approx 38.42°$. Meanwhile the other three ω phase peaks disappeared at $2\theta \approx 56.5°$, 84° and 97°. At higher temperatures of 673, 873 and 973 K, no ω peak was detected. This confirms the reverse ω to α phase transformation due to heating.

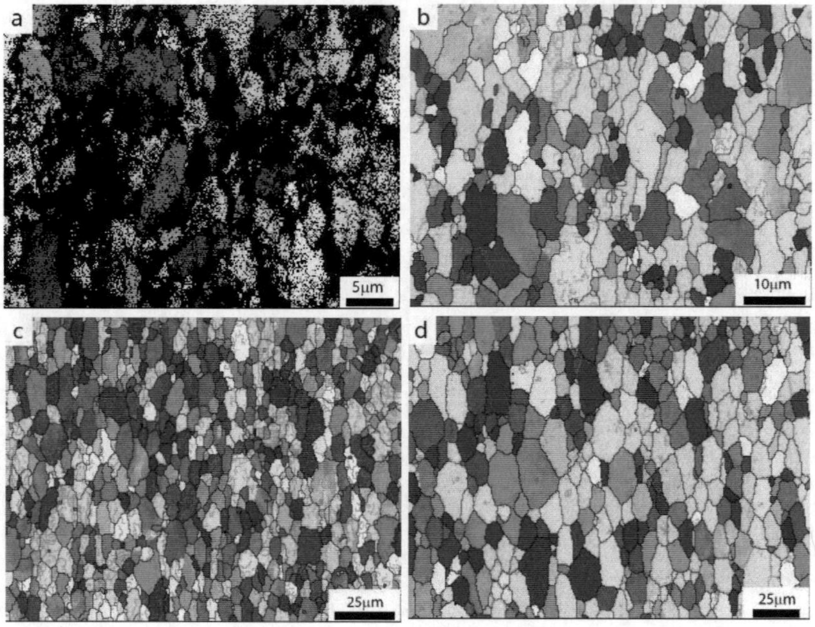

Fig. 5. Microstructure development during post-HPT short-term annealing at (a) 473, (b) 673, (c) 873, and (d) 973 K.

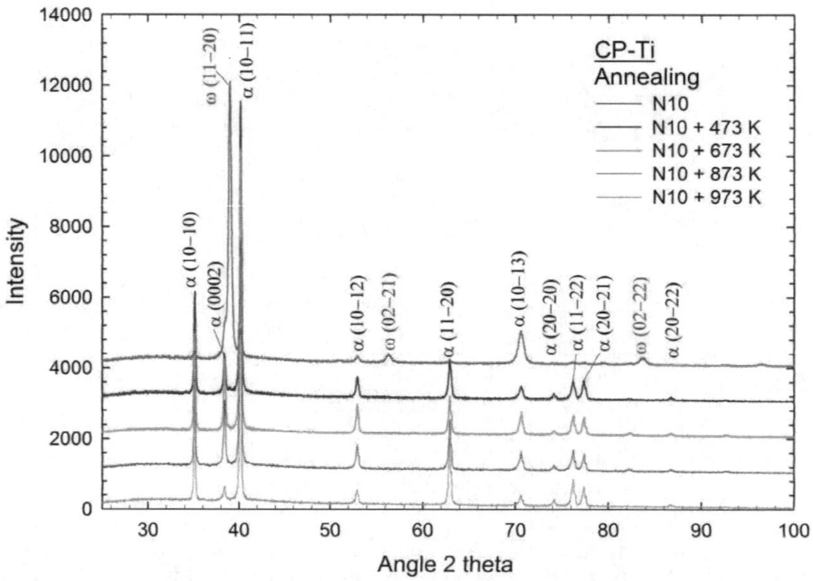

Fig. 6. X-ray patterns of CP-Ti in 10 turns samples during post-HPT short-term annealing.

4. Discussion

There are many reports on grain refinement and strength enhancement in HPT processed grade 2 CP-Ti [9-12]. Grain size and hardness measurement in the CP-Ti processed in this research confirmed the ability of HPT to achieve a nanostructure and to improve the material strength. HPT processing gave nanostructured grains with an average grain size of ~70 nm. However, after 10 minutes short-term annealing, the grain size changed to 3.1, 3.9 and 9.7 μm at 673, 873 and 973 K, respectively. The post-HPT short-term annealing behaviour indicates HPT-processed CP-Ti is temperature sensitive, with significant grain growth in short holding times. This may be related to the high dislocation density introduced by heavy shear deformation which places the material in a high stress and nonequilibrium state with therefore a high tendency for recovery and recrystallization.

Not all grade 2 CP-Ti processed by HPT showed the ω phase [11], but there are other reports of the ω phase in HPT samples [10, 12]. This may be due to the fact that different materials are from different suppliers with different composition control and this affects the α to ω phase transformation during HPT. It was found that the pressure and rotation speed during HPT processing affects the amount of ω phase transformation [12]. In this research, HPT was carried out at a constant pressure 6.0 GPa and using a constant rotation speed of 1 rpm, and Fig. 3 shows the amount of ω phase transformation increased with increasing numbers of rotations. This confirms that the α to ω phase transformation during HPT is deformation-induced. There is a report that the high-pressure ω phase exhibits a reverse phase transformation upon heating [14]. The x-ray patterns from short term annealing samples in Fig. 6 shows the reverse ω to α phase transformation at a very low temperature of 473 K for 10 minutes. This reverse ω to α phase transformation means that the short-term annealed material loses the strengthening effect from the ω phase.

5. Summary and conclusions

1. Significant grain refinement and strength enhancement in CP-Ti were achieved through HPT processing.
2. HPT processed CP-Ti showed temperature sensitivity during short-term annealing. The material loses the nanostructure feature derived from 10 turns HPT processing and turns into a micron coarse grain structure during the short-term annealing.
3. A deformation induced α to ω phase transformation and heating-induced reverse ω to α phase transformation occurred in CP-Ti during HPT processing and post-HPT short-term annealing.

Acknowledgements

This work was supported by the European Research Council under ERC Grant Agreement No. 267464-SPDMETALS and by CAPES in Brazil.

References

[1] Valiev R Z, Semenova I P, Latysh V V, Rack H, Lowe T C, Petruzelka J, Dluhos L, Hrusak D and Sochova J 2008 *Adv. Eng. Mater.* **10** B15.
[2] Stolyarov V V, Zhu Y T, Lowe T C, Islamgaliev R K and Valiev R Z 1999 *Nanostructured Mater.* **11** 947.
[3] Kim I, Jeong W S,Kim J, Park K T and Shin D H 2001 *Scr. Mater.* **45** 575.
[4] Stolyarov V V, Zhu Y T, Alexandrov I V, Lowe T C and Valiev R Z 2001 *Mater. Sci. Eng. A.* **299** 59.
[5] Stolyarov V V, Zhu Y T, Lowe T C and Valiev R Z 2001 *Mater. Sci. Eng. A* **303** 82.
[6] Valiev R Z 2004 *Nature Mater.* **3** 511.
[7] Valiev R Z, Sergueeva A V and Mukherjee A K 2003 *Scr. Mater.* **49** 669.
[8] Zhao X, Yang X, Liu X, Wang C T, Huang Y and Langdon T G 2014 *Mater. Sci. Eng. A* **607** 482.
[9] Edalati K, Masubara E and Horita Z 2009 *Metall. Mater. Trans. A* **40** 2079.
[10] Shirooyeh M, Xu J and Langdon T G 2014 *Mater. Sci. Eng. A* **614** 223.
[11] Wang C T, Fox A G and Langdon T G 2014 *Mater. Sci. Forum* **783-786** 2701.
[12] Shahmir H and Langdon T G 2016 *Mater. Sci. Eng. A* **667** 293.
[13] Valiev R Z, Ivanisenko Yu V, Rauch E F and Baudelet B 1996 *Acta Mater.* **44** 4705.
[14] Jamieson J C 1963 *Science* **140** 72.

IOP Conf. Series: Materials Science and Engineering **194** (2017) 012013 doi:10.1088/1757-899X/194/1/012013

Thermal stability and superplastic behaviour of an Al-Mg-Sc alloy processed by ECAP and HPT at different temperatures

P H R Pereira[1,2,a], Y Huang[1] and T G Langdon[1]

[1] Materials Research Group, Faculty of Engineering and the Environment, University of Southampton, Southampton SO17 1BJ, UK
[2] CAPES Foundation, Ministry of Education of Brazil, Brasília 70040-020, DF, Brazil

E-mail: [a] phrp1d13@soton.ac.uk

Abstract. An Al-3%Mg-0.2%Sc alloy was processed by ECAP and HPT at different temperatures. Afterwards, samples subjected to 10 turns of HPT at 300 and 450 K, 8 passes of ECAP at 300 K and 10 passes of ECAP at 600 K were annealed for 1 hour at 523 K and their mechanical properties and microstructure were examined using microhardness measurements and EBSD analysis. In addition, tensile specimens with similar dimensions were machined from the HPT and ECAP-processed materials and further tensile tested at 523 K. The results demonstrate that the Al alloy processed by HPT at 450 K exhibits higher microhardness values (~138 Hv) and a smaller average grain size (~0.28 μm) after annealing at 523 K among all SPD processing conditions. Accordingly, the material subjected to HPT at an elevated temperature displays superior superplastic properties such that an elongation of ~1020 % was attained after testing at 523 K at 1.0×10^{-3} s^{-1}. Furthermore, detailed EBSD analysis revealed a significant fraction of low-angle grain boundaries (LAGBs > 33 %) in the ECAP-processed material after annealing which may be responsible for the inferior superplastic behaviour by comparison with the HPT-processed samples (LAGBs < 13 %) tested at 523 K.

1. Introduction

There is a growing interest in the application of Al alloys in the transportation industry for replacing heavier structural components usually fabricated with steel and reduce the energy consumption and, thereby the pollutant emissions from motor vehicles [1]. Accordingly, in order to satisfy this demand, processing by severe plastic deformation (SPD) [2,3], as in high-pressure torsion (HPT) [4] and equal-channel angular pressing (ECAP) [5], has been successfully used to produce ultrafine-grained (UFG) metals having remarkably enhanced properties by comparison with materials subjected to conventional thermo-mechanical processing.

Among the various Al alloys already processed through SPD procedures, notable attention has been dedicated to Al-Mg-Sc alloys because of their excellent mechanical strength [6,7] and superplastic properties [8,9]. The addition of Mg solutes in aluminium delays its recovery kinetics and promotes further strain hardening and grain refinement during processing by ECAP [10,11] and HPT [12,13]. Furthermore, the Al_3Sc precipitates are effective in pinning the grain boundaries and retaining ultrafine grains even at elevated temperatures [14-16].

There are numerous studies showing the outstanding superplastic properties of Al-Mg-Sc alloys processed by ECAP [8,9,14,15]. The Al-3Mg-0.2Sc alloy subjected to 8 passes of ECAP at room temperature achieves an elongation of ~2580 % after tensile testing at 723 K at 3.3×10^{-3} s^{-1} [8].

Conversely, a record elongation of ~4100 % was attained in the Al-5Mg-0.2Sc-0.08Zr alloy processed through 10 passes of ECAP at 598 K and further tested in tension at 5.6×10^{-2} s^{-1} at 723 K [9]. It is worth mentioning that, although finer grain sizes are obtained after ECAP at ambient temperature [8], the Al alloy processed by ECAP at an elevated temperature displays an improved thermal stability and a larger proportion of high-angle grain boundaries [9,16].

Several investigations have demonstrated that HPT processing at ~300 K leads to additional grain refinement and improved strength in Al-Mg-Sc alloys by comparison with ECAP [17-23]. The Al-3Mg-0.2Sc alloy processed through 10 turns of HPT exhibits exceptional superplastic elongations at low homologous temperatures [19,20,24], although there are only few and isolated studies evaluating the microstructural changes in this metal when exposed to high temperatures [18,22,24]. Furthermore, it was revealed in recent studies that the microstructural stability in pure Ni [25] and the superplastic properties in Mg alloys [26,27] are significantly enhanced by conducting HPT processing at high temperatures and, to date, there have been no studies dedicated to assess the superplastic properties of Al-Mg-Sc alloys processed by HPT at an elevated temperature.

Therefore, the present study was initiated to evaluate the effect of the nature of the SPD procedure and the processing temperature on the thermal stability and low temperature superplasticity of an Al-3Mg-0.2Sc alloy subjected to ECAP and HPT at different temperatures.

2. Experimental material and procedures

The material used in this study was an Al-3% Mg-0.2% Sc (in wt. %) alloy supplied by China Rare Metal Material Corporation (Jiangxi Province, China) as forged bars with ~130 mm length and 10 mm diameter. These bars were solution treated at 880 ± 2 K for 1 h and quenched in water. Afterwards, billets with lengths of ~65 mm were cut from the bars and then processed through 8 passes of ECAP at room temperature (~300 K) or through 10 passes of ECAP at 600 ± 5 K using route B_c wherein each billet is rotated by 90° in the same sense between passes [28]. The processing was conducted using a pressing speed of ~2 mm s^{-1} and a solid die having an internal channel angle of 90° and an outer arc of curvature of ~20°. These angles impose an equivalent strain of ~1 on each ECAP pass [29].

Discs with ~1 mm thickness were cut from the solution treated bars and then ground to a thickness of ~0.8 mm. The discs were processed by HPT under quasi-constrained conditions [30] either at ambient temperature (~300 K) or 450 ± 5 K. The elevated processing temperatures were attained by using small heating elements around the anvils and the temperature was controlled by a thermocouple placed within the upper anvil at a position of ~10 mm from the HPT sample as described in earlier studies [26,27]. Initially, in the compression stage of HPT [31], the discs were compressed within the anvils under a nominal pressure of 6.0 GPa and then held at the HPT facility for ~1 and ~10 min for the procedures conducted at 300 and 450 K, respectively. Thereafter, the lower anvil was rotated at a constant rate of 1 rpm for 10 turns imposing high torsional straining within the HPT sample.

Following ECAP processing, discs with ~1 mm thickness were cut from the processed billets and then annealed for 1 h at 523 K and cooled in air. The same annealing procedure was performed in the HPT-processed material. Afterwards, both the SPD-processed metal and the annealed samples were ground and polished to obtain mirror-like surfaces and hardness measurements were taken at the middle-sections of the discs using the same method described as in earlier studies [23,32]. The microhardness was recorded using an FM300 microhardness tester equipped with a Vickers indenter under a load of 200 gf and a dwell time of 15 s. Since there is axial symmetry around the central axis in HPT processing, an area-weighted average microhardness was estimated in the HPT samples using the hardness values recorded along the diameters of the discs. Conversely, a normal average microhardness was calculated in the ECAP-processed samples.

The microstructures of the ECAP and HPT-processed materials were analysed using a JSM6500F thermal field emission scanning electron microscope. Samples were ground, final polished using 0.06 μm colloidal silica and then etched using an aqueous solution of 5 % HBF_4. The grain structures of the Al alloy were examined by scanning electron microscopy (SEM) and electron backscattered diffraction (EBSD) and the average grain size, d, was estimated using the linear intercept method.

IOP Conf. Series: Materials Science and Engineering **194** (2017) 012013 doi:10.1088/1757-899X/194/1/012013

EBSD patterns were collected for the samples annealed at 523 K using step sizes as small as 0.03 μm for the material processed by HPT at 450 K. A cleaning procedure, including grain dilatation, was performed after data collection such that the total number of modified points was <20 %. Low-angle grain boundaries (LAGBs) were defined as having misorientation differences between adjacent points from 2° to 15° and high-angle grain boundaries (HAGBs) had misorientations higher than 15°.

For tensile testing, miniature specimens with the same dimensions as in earlier investigations [7,24] were machined from the discs previously processed by ECAP and HPT at different temperatures. These specimens were tested in tension at 523 ± 2 K using a Zwick Z030 testing machine operating at a constant rate of cross-head displacement in order to provide an initial strain rate of 1.0×10^{-3} s^{-1}. All samples were held at the testing temperature for ~10 min prior to the application of the load.

3. Experimental results

Figure 1 shows the variation of the Vickers microhardness either along the diameter of the HPT discs or along the cross-section of the ECAP billets of the Al-3Mg-0.2Sc alloy processed by SPD at different temperatures and further annealed for 1 h at 523 K. It is clearly noted from these plots that the microhardness distribution along the diameter of discs is fairly uniform after post-SPD annealing, except for the metal processed by ECAP at 300 K which displays higher hardness values at the centre of the sample. In addition, the HPT-processed alloy exhibits superior microhardness compared with the material processed by ECAP and subsequently annealed at identical conditions, especially for the metal processed by HPT at 450 K in which the average microhardness is ~138 Hv after annealing.

Typical OIM images are shown in Fig. 2 for the Al-3Mg-0.2Sc alloy processed through 10 turns of HPT at (a) 300 and (b) 450 K or subjected to either (c) 8 passes of ECAP at 300 K or (d) 10 passes of ECAP at 600 K and further annealed for 1 h at 523 K. It follows from Fig. 2 that the microstructure of the Al alloy processed either by ECAP or HPT consists of a homogeneous array of ultrafine grains after post-SPD annealing, regardless of the processing temperature. The samples originally subjected to HPT at 300 and 450 K have nearly equiaxed grains mostly formed by HAGBs, as evident from Fig. 2(a) and (b). By contrast, inspection of Fig. 2 (c) and (d) reveals that the microstructure of the metals processed by ECAP displays a significant proportion of LAGBs after annealing at 523 K and average grain sizes of ~0.34 and ~0.94 μm for the Al alloy processed by ECAP at 300 and 600 K, respectively.

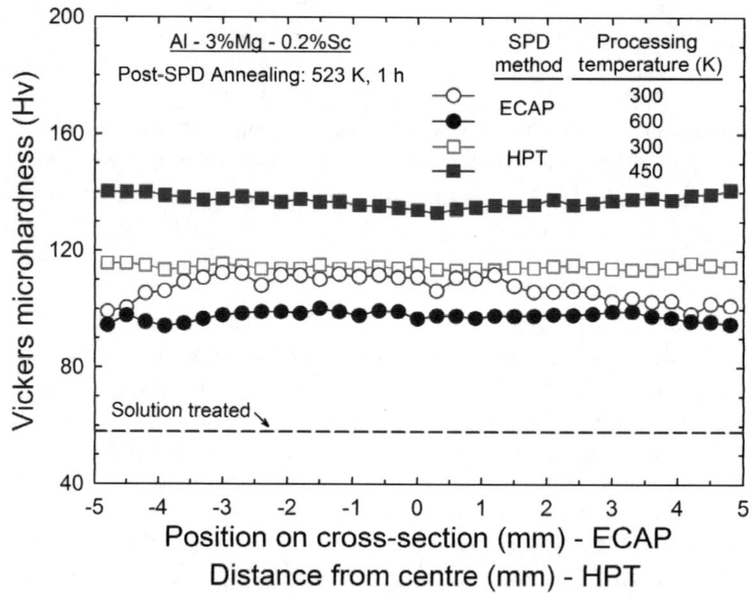

Fig. 1. Variation of the average microhardness recorded along the cross-section of the ECAP billets and the diameter of the HPT discs of the Al-3Mg-0.2Sc alloy subjected to SPD processing at different temperatures and further annealed for 1 h at 523 K.

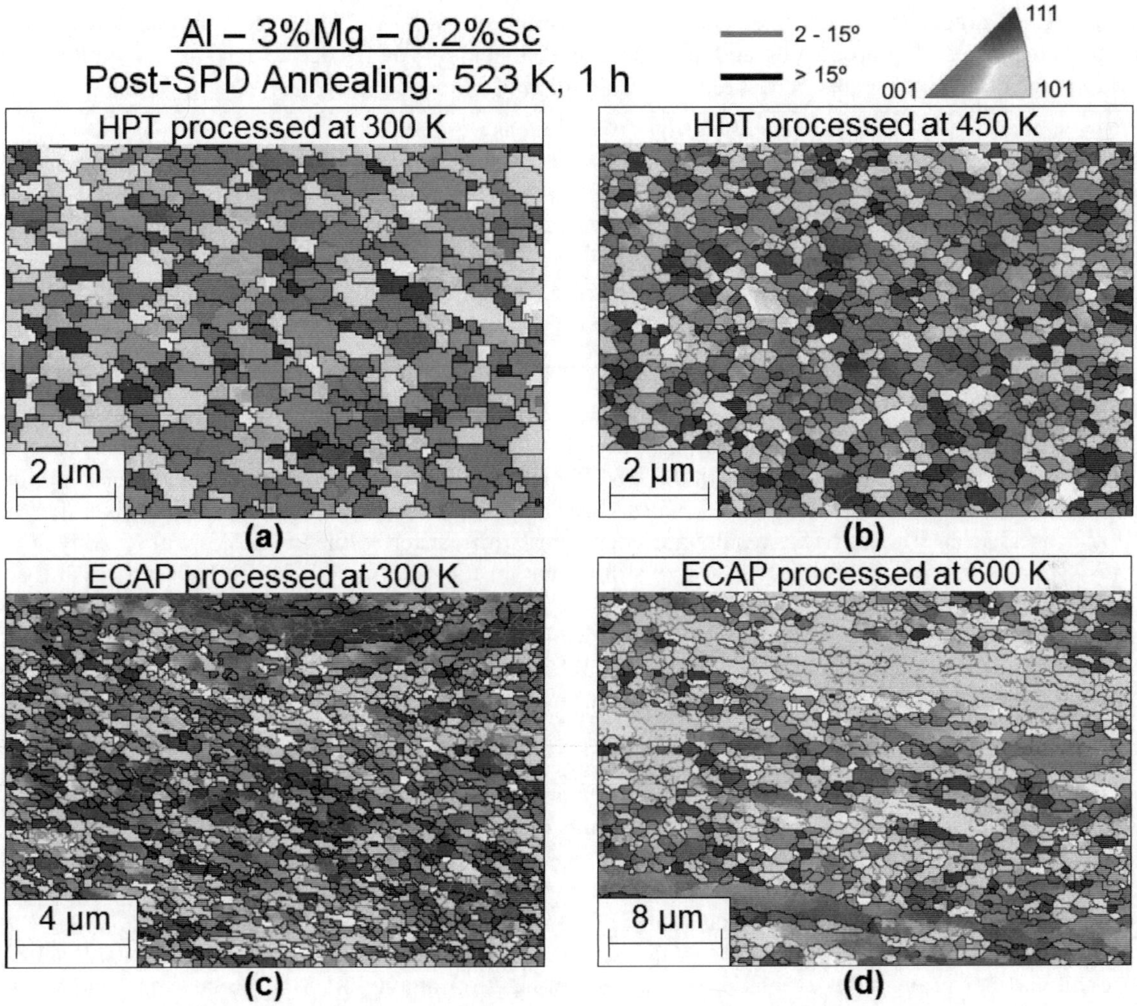

Fig. 2. OIM images of the Al-3Mg-0.2Sc alloy processed through 10 turns of HPT at (a) 300 and (b) 450 K or subjected to either (c) 8 passes of ECAP at 300 K or (d) 10 passes of ECAP at 600 K and further annealed at 523 K for 1 h.

The grain sizes of the Al-3Mg-0.2Sc alloy immediately after ECAP and HPT were estimated using SEM images obtained at positions located at distances of ~3 mm from the centres of the discs. Accordingly, in order to provide information on the coarsening kinetics and superplastic behaviour of this material, Table 1 shows the average microhardness and the average grain size in Al-3Mg-0.2Sc samples immediately after SPD processing and after annealing for 1 h at 523 K. The elongation and the average grain size are also listed for the Al alloy after tensile testing at 1×10^{-3} s^{-1} at 523 K.

It is clearly demonstrated in Table 1 that higher hardness values and notably smaller grain sizes are attained in the Al-Mg-Sc alloy immediately after HPT processing by comparison with ECAP. The thermal stability in this material is significantly enhanced by conducting HPT at 450 K as finer grains ($d \approx 0.28$ μm) are retained after annealing at 523 K. Furthermore, it is interesting to note that the samples processed by ECAP and HPT at 300 K have similar microhardness and grain sizes after post-SPD annealing, although a larger fraction of HAGBs was achieved through HPT.

It also follows from Table 1 that the Al-3Mg-0.2Sc alloy processed by HPT exhibits superior superplastic properties during tensile testing at 523 K at 1.0×10^{-3} s^{-1}. An exceptional elongation of ~1020 % was attained in the material processed by HPT at 450 K whilst the elongations in the Al alloy processed by ECAP at 300 and 600 K were ~540 and ~560 % after straining under identical conditions.

Table 1. Average microhardness and average grain size of Al-3Mg-0.2Sc samples measured immediately after SPD processing and after post-SPD annealing for 1 h at 523 K. The elongation and the average grain size are also measured for the Al alloy after testing at 1×10^{-3} s^{-1} at 523 K.

SPD method	*Processing temperature (K)*	*Average microhardness (Hv)*		*Average grain size (μm)*			*Elongation (%)*
		After SPD processing	*After post-SPD annealing for 1 h at 523 K*	*After SPD processing*	*After post-SPD annealing for 1 h at 523 K*	*After Tensile testing at 523 K at 1×10^{-3} s^{-1}*	*After Tensile testing at 523 K at 1×10^{-3} s^{-1}*
HPT	300	194	115	0.14	0.40	1.1	620
	450	179	138	0.15	0.28	1.0	1020
ECAP	300	138	106	0.25	0.37	1.1	540
	600	96	93	0.60	0.94	1.4	560

4. Discussion

This investigation reveals that processing through 10 turns of HPT leads to a superior average microhardness and smaller grain sizes in the Al-3Mg-0.2Sc alloy compared with ECAP processing. In addition, the material processed by ECAP at 600 K depicts significantly larger grain sizes than the Al alloy processed at 300 K. These results are consistent with earlier studies with Al-Mg-Sc alloys [7-9,17-23] and are associated with the higher strain rates in the UFG metal during HPT processing as well as with the increasing action of dynamic softening mechanisms during deformation at higher homologous temperatures, T/T_m, as reported for various fcc metals [25,33].

It is also clearly demonstrated in Fig. 2 and Table 1 that the microstructural stability in Al-Mg-Sc alloys is notably improved by conducting HPT at an elevated temperature. Although higher hardness values were attained in the material processed by HPT at 300 K, the grain sizes after HPT processing are essentially identical, regardless of the processing temperature. Nevertheless, finer grain structures are distinguished in the material processed by HPT at 450 K after subsequent annealing for 1 h at 523 K. This is attributed to the contribution of dynamic recovery on reducing the strain energy stored in the form of dislocations in Al-Mg-Sc alloys during HPT at elevated temperatures. It is worth noting that the temperature rise in the disc was estimated as ~30 K using an empirical relationship [34,35] and therefore there is no significant change in T/T_m during processing by HPT.

The Al-3Mg-0.2Sc alloy processed by HPT at 450 K and further tensile tested at 523 K at 1.0×10^{-3} s^{-1} exhibits finer grains and a remarkably superior elongation among all SPD processing conditions. This excellent superplastic behaviour is a direct consequence of the extremely small grain sizes and the larger proportion of grains with HAGBs in the HPT-processed metal. By contrast, although true superplastic elongations of >400 % [36] were also achieved in the material processed by ECAP at 300 K, the larger grain sizes and the presence of a significant fraction of substructures with LAGBs lead to an earlier failure in the specimens of the Al alloy pulled to failure in tension at 523 K at 1.0×10^{-3} s^{-1}.

It is therefore concluded the Al-3Mg-0.2Sc alloy processed by HPT at 450 K is a potential candidate for applications demanding materials with high strength to density ratio fabricated through superplastic forming at low temperatures.

5. Summary and conclusions

1. An Al-3% Mg-0.2% Sc alloy was processed through 10 turns of HPT at 300 and 450 K, 8 passes of ECAP at 300 K and 10 passes of ECAP at 600 K to produce grain sizes in the range of ~140-600 nm. Thereafter, samples were annealed for 1 h at 523 K or tensile tested at 523 K at 1.0×10^{-3} s^{-1}.

2. The Al alloy processed by HPT at 450 K displays finer grain structures (~0.28 μm) after annealing at 523 K among all SPD processing conditions. This is attributed to the increasing action of dynamic recovery during HPT at high temperatures which delays its grain coarsening kinetics.

3. Excellent superplastic properties were attained in the Al-3Mg-0.2Sc alloy processed by HPT. A maximum elongation of ~1020% was achieved in the material processed by HPT at 450 K and further tensile tested at 523 K at 1.0×10^{-3} s^{-1}. This is associated to the presence of a uniform array of equiaxed grains with a lower fraction of substructures by comparison with the ECAP-processed metal.

Acknowledgements
This work was supported by CAPES in Brazil and by the European Research Council under ERC Grant Agreement No. 267464-SPDMETALS.

References

[1] Miller W S, Zhuang L, Bottema J, Wittebrood A J, De Smet P, Haszler A and Vieregge A 2000 *Mater. Sci. Eng.* A **280** 37
[2] Valiev R Z, Islamgaliev R K and Alexandrov I V 2000 *Prog. Mater. Sci.* **45** 103
[3] Langdon T G 2013 *Acta Mater.* **61** 7035
[4] Zhilyaev A P and Langdon T G 2008 *Prog. Mater. Sci.* **53** 893
[5] Valiev R Z and Langdon T G 2006 *Prog. Mater. Sci.* **51** 881
[6] Zhemchuzhnikova D and Kaibyshev R 2015 *Adv Eng. Mater.* **17** 1804
[7] Pereira P H R, Wang Y C, Huang Y and Langdon T G. 2017 *Mater. Sci. Eng.* A **685** 367
[8] Komura S, Horita Z, Furukawa M, Nemoto M and Langdon T G 2001 *Metall. Mater. Trans.* A **32A** 707
[9] Avtokratova E, Sitdikov O, Markushev M and Mulyukov R 2012 *Mater. Sci. Eng.* A **538** 386
[10] Iwahashi Y, Horita Z, Nemoto M and Langdon T G 1998 *Metall. Mater. Trans.* A **29A** 2503
[11] Gubicza J, Chinh N Q, Horita Z and Langdon T G 2004 *Mater. Sci. Eng.* A **387-389** 55
[12] Andreau O, Gubicza J, Zhang N X, Huang Y, Jenei P and Langdon T G 2014 *Mater. Sci. Eng.* A **615** 231
[13] Lee H J, Han J K, Janakiraman S, Ahn B, Kawasaki M and Langdon T G 2016 *J. Alloys Compd.* **686** 998
[14] Lee S, Utsunomiya A, Akamatsu H, Neishi K, Furukawa M, Horita Z and Langdon T G 2002 *Acta Mater.* **50** 553
[15] Dám K, Lejček P and Michalcová A 2013 *Mater. Charact.* **76** 69
[16] Avtokratova E, Sitdikov O, Mukhametdinova O, Markushev M, Narayana Murty S V S, Prasad M J N V and Kashyap B P 2016 *J. Alloys. Compd.* **673** 182
[17] Sakai G, Horita Z and Langdon T G 2005 *Mater. Sci. Eng.* A **393** 344
[18] Dobatkin S V, Zakharov V V, Vinogradov A Yu, Kitagawa K, Krasil'nikov N A, Rostova T D, Bastarash E N 2006 *Russ. Metall. (Metally)* **6** 533
[19] Horita Z and Langdon T G 2008 *Scr. Mater.* **58** 1029
[20] Harai Y, Edalati K, Horita Z and Langdon T G 2009 *Acta Mater.* **57** 1147
[21] Valiev R Z, Enikeev N A, Murashkin M Y, Kazykhanov V U and Sauvage X 2010 *Scr. Mater.* **63** 949
[22] Sauvage X, Enikeev N, Valiev R, Nasedkina Y and Murashkin M 2014 *Acta Mater.* **72** 125
[23] Pereira P H R, Huang Y and Langdon T G 2017 *Mater. Sci. Forum* **879** 1471
[24] Pereira P H R, Huang Y and Langdon T G 2015 *Lett. Mater.* **5** 294
[25] Ghosh P, Renk O and Pippan R 2017 *Mater. Sci. Eng.* A **684** 101
[26] Harai Y, Kai M, Kaneko K, Horita Z and Langdon T G 2008 *Mater. Trans.* **49** 76
[27] Kai M, Horita Z and Langdon T G 2008 *Mater. Sci. Eng.* A **488** 117
[28] Furukawa M, Iwahashi Y, Horita Z, Nemoto M and Langdon T G 1998 *Mater. Sci. Eng.* A **257** 328
[29] Iwahashi Y, Wang J, Horita Z, Nemoto M and Langdon T G 1996 *Scr. Mater.* **35** 143
[30] Figueiredo R B, Cetlin P R and Langdon T G 2011 *Mater. Sci. Eng.* A **528** 8198
[31] Pereira P H R, Figueiredo R B, Cetlin P R and Langdon T G 2015 *Mater. Sci. Eng.* A **631** 201
[32] Kawasaki M and Langdon T G 2008 *Mater. Sci. Eng.* A **498** 341
[33] Edalati K and Horita Z 2011 *Mater. Sci. Eng.* A **528** 7514
[34] Figueiredo R B, Pereira P H R, Aguilar M T P, Cetlin P R and Langdon T G 2012 *Acta Mater.* **60** 3190
[35] Pereira P H R, Figueiredo R B, Huang Y, Cetlin P R and Langdon T G 2014 *Mater. Sci. Eng.* A **593** 185
[36] Kawasaki M and Langdon T G 2007 *J. Mater. Sci.* **42** 1782

IOP Conf. Series: Materials Science and Engineering **194** (2017) 012014 doi:10.1088/1757-899X/194/1/012014

Effect of severe plastic deformation on the structure and properties of Ni-Cu alloys

V V Popov, A V Stolbobsky and E N Popova

M.N. Miheev Institute of Metal Physics Ural Branch of Russian Academy of Sciences, 620990, 18 S. Kovalevskaya Str., Ekaterinburg, Russia

E-mail: vpopov@imp.uran.ru

Abstract. The structure of Ni-Cu alloys processed by high pressure torsion by 5 revolutions at room temperature (saturation stage) has been studied by the methods of transmission electron microscopy and microhardness measurements. It is demonstrated that moving from copper possessing lower melting temperature and lower stacking fault energy (SFE) to nickel the melting temperature and SFE of which are considerably higher, the submicrocrystalline structure of alloys is markedly refined (by about a factor of 2), whereas the microhardness increases by more than 1.5 times. The simultaneous effect of two factors (changing of homological temperature and SFE) results in the non-linear dependence of these parameters (crystallite sizes and microhardness) on alloy composition.

1. Introduction

Nowadays, great attention of materials science researchers is attracted by metal materials in which nanocrystalline (or submicrocrystalline) state is obtained by severe plastic deformation (SPD), because these materials demonstrate new unique properties, such as enhanced strength, plasticity, specific diffusion properties, etc. [1]. However, it is not always possible to obtain a true nanocrystalline state, i.e. the structure in which crystallites the sizes of less than 100 nm with high-angle boundaries are dominating. The main limitation in fabricating the nanocrystalline structure is the onset of the so-called saturation state under the SPD, when with increasing strain there is no more refinement and strengthening of the material [3],[4]. At this stage there is equilibrium between generation and annihilation of defects (vacancies, dislocations, high- and low-angle boundaries). The key process limiting structure refinement is migration of grain boundaries. The smallest grain sizes achievable under SPD are determined, first of all, by the type of a material, particularly, by its crystal lattice, melting temperature and stacking fault energy (SFE) [5]. Of not less importance are such factors as deformation temperatures, phase transformations and the possibility of two concurrent mechanisms of grain refinement, deformation twinning and dislocation activities, as demonstrated for steels in two recent publications [6,7].

Copper and nickel are quite plastic metals and can be deformed without fracture up to high strains at relatively low temperatures, which is necessary for the formation of ultrafine-grained (UFG) structure, and that is why their structure and properties after SPD were studied by numerous researchers [8]-[15]5 et al.]. These studies have shown an important role of deformation temperature and material purity. Thus, in [9]9-[11] it is demonstrated that when the deformation temperature is decreased from room to liquid nitrogen the average crystallite size in Ni and commercially pure Cu gets considerably smaller and the nanocrystalline structure is formed.

Another factor strongly affecting structure formation under the SPD is the stacking fault energy. There are experimental data on different effect of SPD on the materials with different SFE. In the materials with high SFE the dominating role is played by dislocation glide [[16]6-[17]. In such materials dislocation nets and low-angle boundaries develop into high-angle boundaries by dislocation redistribution and/or grain rotation. For materials with low SFE, in which twinning is dominating, the grain refinement was considered in less detail, but there are experimental data that at equal parameters of SPD the obtained grain sizes can be smaller in the materials with lower SFE [[5],[18]. The role of SFE in Cu, Cu-Al and Cu-Ni alloys subjected to ECAP and HPT is considered in [19-21], where it is demonstrated that homogeneous microstructures in materials with high or low SFE are much more readily achieved than in metals with medium SFE.

The Ni-Cu alloys considered in the present study are of interest because in this system composition and, consequently, SFE, can be changed continuously between constituent elements, which have relatively low SFE in Cu (78 mJ/m^2) [5] and high SFE in Ni (160 mJ/m^2) [22], which gives chance to analyze the effect of SFE on the structure formation under the SPD.

The goal of the present study is to analyze specific features of the structure of Ni-Cu alloys in its saturation state (in refinement and strengthening) after high pressure torsion at room temperature, and to reveal factors responsible for their capability for nanostructuring.

2. Experimental

The Ni-Cu alloys of three compositions (10, 34 and 90 at. % Cu) were studied in the present paper along with polycrystalline Ni 99.6 % pure and commercially pure Cu, subjected to severe plastic deformation. Initial specimens of 10 mm diameter and 0.5 mm thick were deformed at room temperature by 5 revolutions of high-pressure torsion in Bridgman anvils. The average pressure of axis compression of disks was 6 GPa which reliably excluded their slipping at torsion; the angular velocity of anvils rotation was 0.3 rev/min. According to preliminary studies, in all the specimens under study this treatment results in the saturation state.

The structure of as-deformed specimens was studied in transmission electron microscopes JEM-200CX and Philips CM30 with following treatment of the images with the computerized program SIAMS-600. Microhardness was measurement by a special unit adjusted to optical microscope Neophot-21 and calculated by formula $H=18192 \cdot P/C^2$, МПа, where P is load in grams, and C is indentation diagonal in microns. Every value of C was averaged over not less than 9 indentations. The microhardness measurement error was in average about 2-3 % and did not exceed 5 % with confidence probability of 0.95.

3. Results and discussion

According to numerous publications (see, e.g., [23]3-[24]), the structure fragmentation under deformation proceeds as follows. With the increasing strain dislocations are accumulated and localized in cell walls, and misorientation between cells is growing, which results in the formation of UFG structure, the basis of submicrocrystalline or nanocrystalline grain structure. Further deformation causes activation of rotation deformation modes throughout a specimen bulk, ensuring the achievement of the steady-state deformation (saturation state). The determining factor in structure fragmentation is dislocation mobility, which can be decreased by two ways, by decreasing of SFE (for example, by alloying) and by lowering of deformation temperature (or by decreasing the homological temperature if one deals with different materials).

The structure of Cu after HPT is shown in Fig. 1. It is seen that the HPT by 5 revolutions of anvils results in the formation of non-uniform structure with wide grain size scattering (from 110 to 740 nm) and average grain size of ~ 280 nm. Analysis of this structure (drastic contrast changing in dark-field images, point-wise reflections in the Debye rings in the electron diffraction patterns, intricate dislocation contrast in some grains and its absence in the others, wide grain size scattering, the presence of coarse grains with low dislocation density) allows to conclude that under the HPT not only relaxation processes, but even dynamic recrystallization are possible in copper. In some grains one can see a striated contrast which, in an opinion of some authors [25], indicates the grain boundary structure recovery. The results obtained are in agreement with the previous observations on high-pure and commercial-purity copper after room temperature HPT [14].

The structure of Ni-90%Cu alloy is shown in Fig. 2. It does not appreciably differ from that of copper. As in Cu, in some grain boundaries the striated contrast is observed, indicating relaxation processes and an approach of grain boundary state to equilibrium. There are areas with wide crystallite size scattering, and in small crystallites dislocation density is low and they have a polyhedron shape with high-angle boundaries. In neighboring coarser crystallites dislocation density is high, and smooth changing of contrast in dark-field images testifies the presence of low-angle boundaries. In the electron-diffraction patterns one can see a great number of point-wise reflections in the Debye rings and slight azimuth reflection blurring (insert in Fig. 2b).

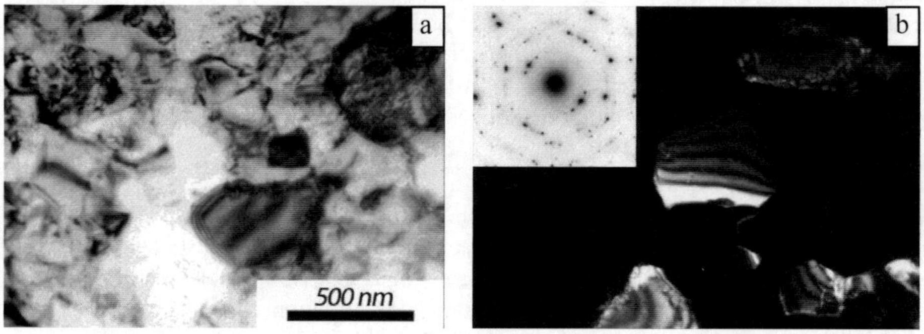

Fig. 1. The structure of copper after HPT by 5 revolutions at room temperature: a - bright-field image; b – dark-field image in $(111)_{Cu}$ reflection and electron diffraction pattern

In other areas coarse crystallites of polyhedral shape with low dislocation density are observed, and, in some of them, the dislocation contrast changes smoothly indicating the presence of low-angle boundaries. Thus, in general, it can be concluded that dynamic recrystallization is not suppressed in copper alloyed with 10 at. % of Ni. However, in the alloy the grain size scattering (130-470 nm) and the average grain size (250 nm) are lower compared to pure Cu.

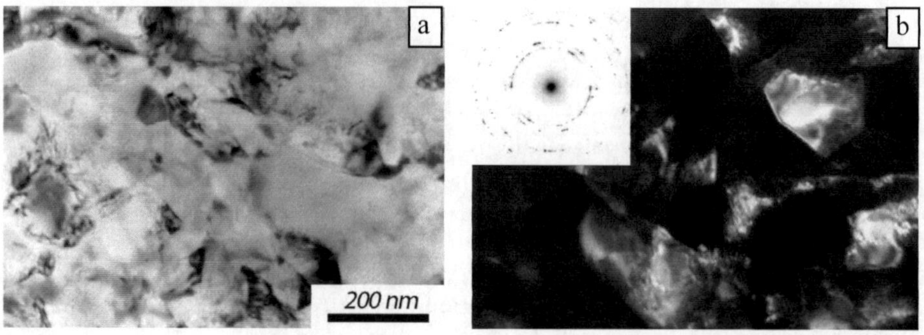

Fig. 2. The structure of Ni-90at.%Cu alloy after HPT by 5 revolutions at room temperature: a - bright-field image; b – dark-field image in $(111)_{Cu}$ reflection and electron diffraction pattern

Now let's consider the effect of Ni alloying with Cu on the structure obtained at the steady-state deformation after 5 revolutions of HPT at room temperature. The structure of Ni after such treatment is shown in Fig. 3. It is characterized with appreciably curved crystallite boundaries, not always clearly seen, curved contrast inside crystallites and high dislocation density. All these features indicate high level of internal stresses and indirectly testify the specific non-equilibrium state of grain boundaries. In dark-field images one can see drastic change of contrast between neighboring crystallites characteristic of high-angle boundaries, and smooth contrast changing inside crystallites indicating the presence of low-angle boundaries as well. The average grain size in Ni is 145 nm with grain-size scattering from 50 to 250 nm, i.e., as it was expected, the structure is much more uniform and dispersed compared to that of Cu and Cu-10at.%Ni alloy.

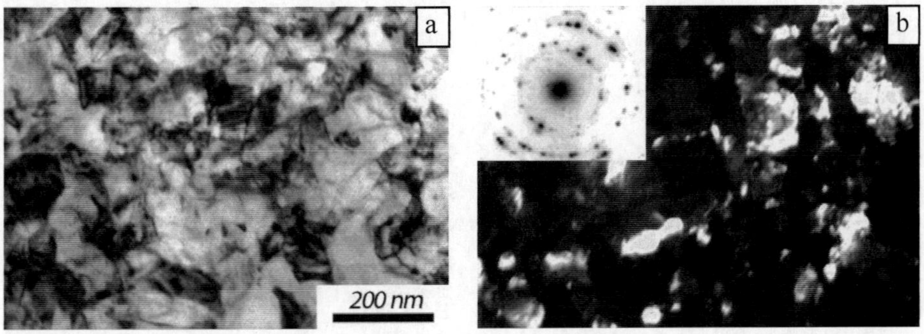

Fig. 3. The structure of Ni after HPT by 5 revolutions at room temperature: a - bright-field image; b – dark-field image in $(111)_{Cu}$ reflection and electron diffraction pattern

When Ni is alloyed with Cu, one could expect, on the one hand, the formation of uniform and less dispersed structure, as the homological temperature of such alloy is increased due to much lower melting temperature of copper. On the other hand, the SFE is decreased as the SFE of Cu is almost by a factor of 2 lower than that of Ni, and this factor could affect the structure oppositely. The analysis of TEM data has shown that combination of these two factors results in practically no changes of structure when Ni is alloyed with 10 at.% of Cu and only slight changes of structure in case of alloying it with 34 at.% of Cu (Fig. 4).

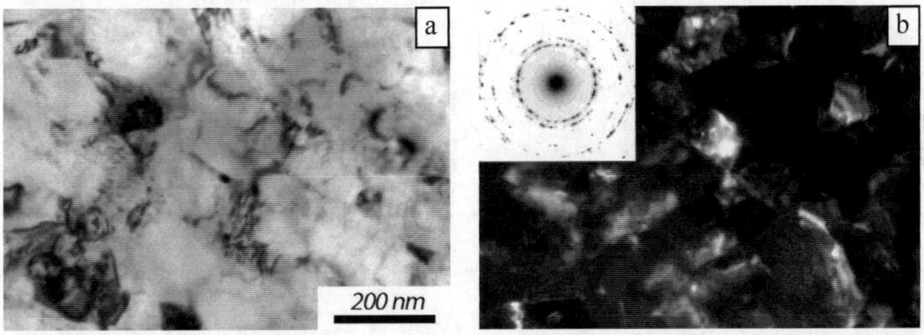

Fig. 4. The structure of Ni-34at.%Cu after HPT by 5 revolutions at room temperature: a - bright-field image; b – dark-field image in $(111)_{Cu}$ reflection and electron diffraction pattern

Figure 5a demonstrates dependence of average crystallite sizes on alloy composition. As mentioned above, when Cu is alloyed with 10 at % of Ni, the crystallite sizes and their scattering decrease, which could result mainly from the effect of two factors, decreasing of homological temperature (as Ni has much higher melting temperature) and changing of SFE (as the latter is 2 times higher in Ni). When the homological temperature is decreased, the structure is refined due to decreased dislocation mobility and suppression of relaxation processes. As seen from the structure obtained, these processes are really partly suppressed compared to those in pure copper, but not completely. On the other hand, alloying with Ni could increase the SFE and affect the structure oppositely. That is why some grain refinement in Ni-90at.%Cu alloy compared to Cu is most probably due to the decreasing homological temperature.

Absolutely another situation is observed when Ni is alloyed with Cu. When Cu concentration reaches 34 at %, the structure and crystallite average sizes change only slightly compared to pure Ni, but the crystallite size scattering somewhat increases. This slight influence of Ni alloying with Cu is probably due to the effect of two factors acting oppositely. On the one hand, when Cu content in an alloy increases, its homological temperature gets higher, and it should result in coarser crystallite sizes. On the other hand, the decrease of SFE in this case must affect oppositely. Obviously, the effect of these two factors in Ni-rich alloys is compensated, and up to 34 at % of Cu there is no appreciable change in crystallite sizes of alloys compared to those of pure Ni, and the average crystallite size obtained is considerably smaller than that predicted by the rule of mixtures.

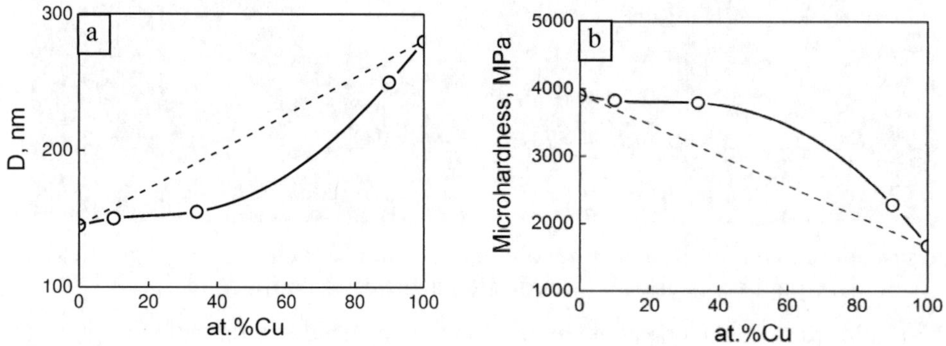

Fig. 5. Average crystallite size (a) and microhardness (b) of Ni-Cu alloys after 5 revolutions of room temperature HPT versus their composition. Dashed line – rule of mixtures.

The composition dependence of microhardness (Fig. 5b) correlates with the dependence of average crystallite size. In Ni-rich alloys the growth of Cu concentration almost does not affect the microhardness, whereas from the side of Cu-rich alloys the microhardness drastically decreases when approaching to pure Cu, but it should be stressed that in all the composition range the microhardness is higher than that predicted by the rule of mixtures.

If only one parameter affected the crystallite average size and, consequently, microhardness, dependently on composition, for example, homological temperature or SFE, then these parameters would change with composition linearly, i.e. in accordance with the rule of mixtures. The fact that it is not so, indicates that the structure and microhardness are affected by both factors. The results obtained allow to conclude that in Cu-rich alloys the determining role in structure formation is played by the homological temperature, whereas in Ni-rich alloys the changes of both homological temperature and SFE with composition to a great extent compensate each other acting in opposite directions.

4. Conclusion

The effect of composition of Ni-Cu alloys on their structure in the saturation state (steady-state deformation) obtained under room temperature HPT, has been studied and analyzed. It is demonstrated, that with composition changing average crystallite sizes and microhardness are changed non-linearly.

In Cu-rich alloys the increasing concentration of Ni results in considerable structure refinement and microhardness increase. The determining role in structural changes in this concentration range is obviously played by decreasing homological temperature with increasing Ni content.

A completely different situation is observed in Ni-rich alloys. When up to 34 at. % of Cu are added to Ni, the crystallite sizes and microhardness change only slightly, which obviously results from the effect of two factors acting oppositely and to a great extent compensating each other, namely, the increase of homological temperature and decrease of SFE.

The analysis of concentration dependences of average grain sizes and microhardness of Ni-Cu alloys has shown that the former are smaller and the latter is higher than those predicted by the rule of mixtures. It can be suggested that this results from a considerable effect of SFE changing dependently on alloy composition. Thus, the role of SFE in structure refinement and strengthening is an important factor, which must be taken into account in fabrication of alloys with a given complex of properties.

The electron microscopic studies were carried out on an equipment of the Collective Use Center of Nanotechnologies and Advanced Materials of Institute of Metal Physics UB of RAS. The work has been done in the framework of the state assignment of FASO of Russia (topic “Spin”, state registration number 01201463330), with partial support of the Program of Fundamental research of Ural Branch of RAS (project 15-9-2-44) and Russian Foundation on Basic Research (project 15-03-03103).

References

[1] Gleiter H 2000 Nanostructured materials: basic concepts and microstructure *Acta Mater.* **48** No. 1 pp 1-29.
[2] Pippan R, Scheriau S, Hohenwarter A and Hafok M 2008 *Mater. Sci. Forum* **584-586** pp 16-21.
[3] Korznikov A V, Tyumentsev A N and Ditenberg I A 2008 *Phys. Met. Metallogr.* **106** No. 4 pp 418-423.
[4] Pippan R, Scheriau S, Taylor A, Hafok M, Hohenwarter A and Bachmaier A. 2010 *Ann. Rev. Mater. Res.* **40** pp 319-343.
[5] Zhao YH, Liao XZ, Zhu YT, Horita Z and Langdon TG 2005 *Mater. Sci. Eng.* A. **410–411** pp 188–193.
[6] Cao Y, Wang YB, An XH, Liao XZ, Kawasaki M, Ringer SP, Langton TG and Zhu YT 2014 *Acta Mater.* **63** pp 16-29.
[7] An XH, Lin QY, Sha G, Huang MX, Ringer SP, Zhu YT and Liao XZ 2016 *Acta Mater.* **109** pp 300-313.
[8] Hebesberger T, Stuwe H P, Vorhauer A, Wetscher F and Pippan R 2005 *Acta Mater.* **53** pp 393-402.
[9] Stolbovsky A V, Popov V V, Popova E N and Pilyugin V P 2014 *Bul.RAS. Physics*. **78** No. 9 pp 908–916.
[10] Popov V V, Popova E N, Pilyugin V P, Kuznetsov D D and Stolbovsky A V 2014 *IOP Conference Series: Mater. Sci. Eng.* **63** 012096.
[11] Popov V V, Popova E N, Kuznetsov D D, Stolbovskii A V and Pilyugin V P 2014 *Phys. Met. Metallogr.* **115** No. 7 pp 682-691.
[12] Zhang HW, Huang X, Pippan R and Hansen N. 2010 *Acta Mater.* **58** pp 1698-1707.
[13] Zhang HW, Lu K, Pippan R, Huang X and Hansen N 2011*Scripta Mater.* **65** pp 481-484.
[14] Popov V V, Stolbovsky A V, Popova E N and Pilyugin V P 2010 *Def. Diff. Forum* **297-301** pp 1312-1321.
[15] Edalati K, Fujioka T and Horita Z 2008 *Mater. Sci. Eng.* A. **497** pp 168-173.
[16] Zhu YT, Huang J Y, Gubicza J, Ungar T, Wang YM, Ma E and Valiev RZ 2003 *J. Mater. Res.* **18** pp 1908-1917.
[17] Dobatkin SV, Bastarache E N, Sakai G, Fujita T, Horita Z and Langdon TG 2005 *Mater. Sci. Eng.* A **408** pp 141-146.
[18] Zhao YH, Zhu YT, Liao XZ, Horita Z and Langdon TG 2006 *Appl. Phys. Let.* **89** 121906.
[19] Qu S, An XH, Yang HJ, Huang CX, Yang G, Zang QS, Wang ZG, Wu SD and Zhang ZF 2009 *Acta Mater.* **57** pp 1586-1601.
[20] An XH, Lin QY, Wu SD, Zhang ZF, Figueiredo RB, Gao N and Langdon TG 2011 *Phil. Mag.* **91** (25) pp 3307-3326.
[21] An XH, Song M, Huang Y, Liao XZ, Ringer SP, Langdon TG and Zhu YT 2014 *Scripta Mater.* **72-73** pp 35-38.
[22] Gallagher P C J 1970 *Met. Trans* **1** pp 2429-2461.
[23] Horita Z, Smith DJ, Furukawa M, Nemoto M, Valiev RZ and Langdon TG 1996 *J. Mater. Res.* **13** pp 446-450.
[24] Zhilyaev AP, Ohishi K, Langdon TG, McNelley TR 2005 *Mater. Sci. Eng.* A. **410–411** pp 277–280.
[25] Mulyukov KY, Korznikova G F, Abdulov RZ, Valiev RZ 1990 *J. Magn. Magn. Mater.* **89** pp 207-213.

Dynamic strain aging precipitation of $Mg_{17}Al_{12}$ in AZ80 magnesium alloy during multi-directional forging process

Q F Zhu[1, 2*],G S Wang[1, 2], X J Wang[1, 2], F Z Liu[1, 2], C Y Ban[1, 2], J Z Cui[1, 2]

[1] Key Laboratory of Electromagnetic Processing of Materials, Ministry of Education, Northeastern University, Shenyang, P.R. China
[2] School of Materials Science and Engineering, Northeastern University, Shenyang, P.R. China

E-mail：zhuqingfeng@epm.neu.edu.cn

Abstract. Dynamic aging precipitation of $Mg_{17}Al_{12}$ phases in AZ80 magnesium alloy was studied by multi-directional forging (MDF) with decreasing temperatures from 410 to 300 ℃. The results show that the morphology of the dynamically precipitated β-$Mg_{17}Al_{12}$ phases (formed during forging process) exhibited granular shape. During the multi-directional forging process, the inhomogeneous dynamic precipitation of the β-$Mg_{17}Al_{12}$ phases result in the coexistence of the fine grains (with many granular $Mg_{17}Al_{12}$ phases) and coarse grains (without $Mg_{17}Al_{12}$ phases) in the samples. The fine grains (with many granular $Mg_{17}Al_{12}$ phases) area expands with the decreasing of final forging temperature. The inhomogenous Al content distribution in the Mg matrix leads to the non-uniform dynamic precipitation of the $Mg_{17}Al_{12}$ phase. These $Mg_{17}Al_{12}$ phase retards the growth of the DRX grains, which in turns results in the formation fine grains area during the during the MDF process with temperature decreasing.

1. Introduction

Dynamic strain aging (DSA) is a phenomenon in metals and alloys resulted from the interaction between the diffusing solute atoms and the moving dislocation. It has been found that DSA occurs in many alloys during uniaxial tension or compression process [1-5]. In traditional metal forming process, such as extrusion and rolling, it is difficult to control the DSA process because of the limited deformation time and deformation reduction. By contrast, the DSA process can be effectively controlled during the Sever Plastic Deformation (SPD) by constantly increasing accumulative strains. In recent years, the DSA behavior during SPD process has been widely studied [6-12]. Roven et. al [7] found that the globular β″ phase with size about 4 nm precipitated in 6063 aluminum alloy during Equal Channel Angular Pressing (ECAP) at room temperature and 175 ℃. Xia, Nie et. al [9] have studied the DSA process in magnesium alloys during the Multi-Directional Forging (MDF) process. The granular $Mg_{17}Al_{12}$ phases, which distributed along the grains boundaries, restricted the growth of dynamic recrystallized grains and therefore refined the grains. We have also found the dynamic precipitation behavior in AZ80 magnesium alloy during multi-directional forging process [12]. In this paper, dynamic strain aging precipitation of $Mg_{17}Al_{12}$ phases in the AZ80 magnesium alloy was studied by multi-directional forging (MDF) in the temperature range from 410 to 300 ℃.

2. Experimental

Rectangular samples with the size of 61 mm (z axis) × 52 mm (y axis) × 45 mm (x axis)) were cut from a commercial direct chill (DC) casting AZ80 magnesium alloy (annealed at 410 ℃ for 16 hours) with compositions of Al (7.85wt.%), Zn (0.43wt.%), Mn (0.21wt.%) and the balance Mg. The samples were heated at 410 ℃ for 1.5 hours, and then forged by a 300 t hydraulic press. Fig. 1 schematically shows the MDF process. First, the sample was forged down along the Z direction (the long side) to 35 mm (pass 1); then the sample was turned 90° around the X axis and forged down along the Y direction (newly formed long side) to 35 mm (pass 2); subsequently the sample was turned 90 ° around the Z axis and forged along the X direction to 35 mm (pass 3). The strain rate and the true strain of each pass were 0.13 s-1 and 0.5, respectively. The final forging temperature is controlled by time interval of each passes and the temperature of the forging hammers. The temperature of the samples were detected the thermocouple and thermometer. The detailed experimental parameters are shown in Table 1. Small specimens were cut from these samples. With a further grounding, polishing and oxidizing, these specimens were observed by a OM and SEM. The Al and Zn compositions in the primary Mg matrix were manually detected by the EPMA-1600/1610 Electron Probe Microanalysis (EPMA).

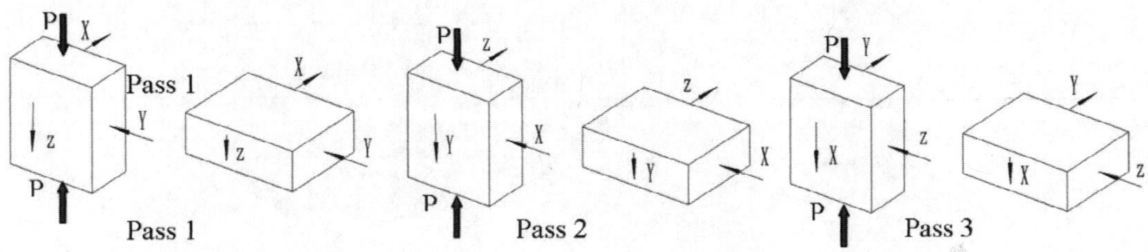

Fig. 1 Schematic diagram of the multi-directional forging

The differential scanning calorimetry (DSC) technique was employed to analysis the thermal behaviors of the AZ80 alloy. Disc-shaped sample of average weight of ~24 mg was placed in one of the pans and the aluminum reference was kept in other pan. The heating rate was 5 ℃/min. Scans were performed between room temperature and 400 ℃ in purified nitrogen flow at a rate of 50 ml min^{-1}.

Table 1. Experimental parameters

Sample number	Forging passes	Final forging temperature (℃)	Accumulative true strain (ε)
P1	4	335	2
P2	7	325	3.5
P3	14	300	7

3. Results and discussion

Fig. 2(a) shows the optical micrograph of the initial DC casting AZ80 alloy after homogenization heat treatment. It can be seen that the dendritic microstructure disappears to be replaced by the equiaxed grain. And the average grain size is about 210 μm. And some black contamination like substances are distributed along the grains boundaries. A further observation shows that these substances are composed of laminar second phase, as shown in SEM figure in Fig. 1(b). EDX analysis indicates that the phase is laminar β-$Mg_{17}Al_{12}$. A statistical measurement suggests that the mean area fraction of $Mg_{17}Al_{12}$ in the specimen is about 13.6%.

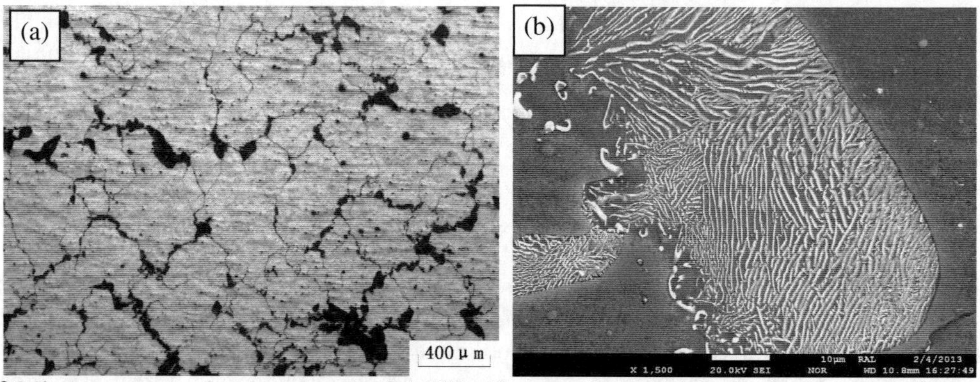

Fig.2 Microstructure of as-homogenized AZ80 alloy specimen, (a) Optical microscope images (b) magnified micrographs observed by SEM

Fig. 3 shows photography of the samples forged by different condition. It can be seen that the samples show similar outline dimensions and are generally consistent with the initial shape. There are no obvious crack and overlaps on the surface of these samples. This indicates that deformability of AZ80 sample can ensure the MDF process performed successfully.

Fig. 3 Photography of the samples forged by different condition, (a) P1-sample, (b) P2-sample and (c) P3-sample

The microstructures in the center of the forged samples are displayed in Fig. 4(a)-(c).It can be found that some mutual-mixed black and white structures distribute on the center of the samples. And the areas of the black structures increase with increasing the accumulative strains and decreasing temperature. The enlarged picture shows that the black structures are very fine grains with granular second-phase β-$Mg_{17}Al_{12}$ distributed along the grain boundaries. While, the white structures are relatively coarse-grains without second-phase β-$Mg_{17}Al_{12}$ formed.

In order to reveal the details of the coarse and fine grains areas, the EPMA were employed and the intersection area of coarse and fine grains were observed and analyzed. Fig.5 (a) shows the SEM picture in the intersection area of coarse and fine grain. The left side and right side of the picture present the typical fine grain area (with granular β-$Mg_{17}Al_{12}$) and the coarse grain area, respectively. It is shown that the grain size of the AZ80 alloy gradually gets bigger, and the amount of the granular β-$Mg_{17}Al_{12}$ gradually decreases from the left to the right on Fig. (a). And the granular β-$Mg_{17}Al_{12}$ phases mainly distribute along the grain boundary of the fine grain. The element distribution along the straight line in Fig (a) is shown in Fig 5(b). It is clearly show that the wave crest of the Al element correspond with the wave valley of the Mg element on the point of granular β-$Mg_{17}Al_{12}$ phases. The wave crest of the Al mainly appears on the left of the figure, and no wave crest appears on the right of the figure. The content of the Mg element increases gradually from the left to the right. On the contrary, the content of the Al element decreases gradually from the left to the right.

Fig. 6 presents the DSC scan obtained at a heating rate of 5 ℃/min for AZ80 alloy. The DSC curve shows that an endothermic peak forms between 300 ℃ and 350 ℃. This reaction can be ascribed to the dissolution of β-$Mg_{17}Al_{12}$ phases. The Al element will precipitate from the Mg matrix as the temperature lower than 330 ℃ for this alloy. So the granular β-$Mg_{17}Al_{12}$ phases began to dynamically precipitate from

IOP Conf. Series: Materials Science and Engineering **194** (2017) 012015 doi:10.1088/1757-899X/194/1/012015

the Mg matrix with the forging temperature constantly decrease from 410 ℃ to 335 ℃, as show in Fig.4 a. With the final forging temperature further decreasing to 300 ℃, more granular β-$Mg_{17}Al_{12}$ phases dynamically precipitated from the Mg matrix, as shown in Fig.4 b and c.

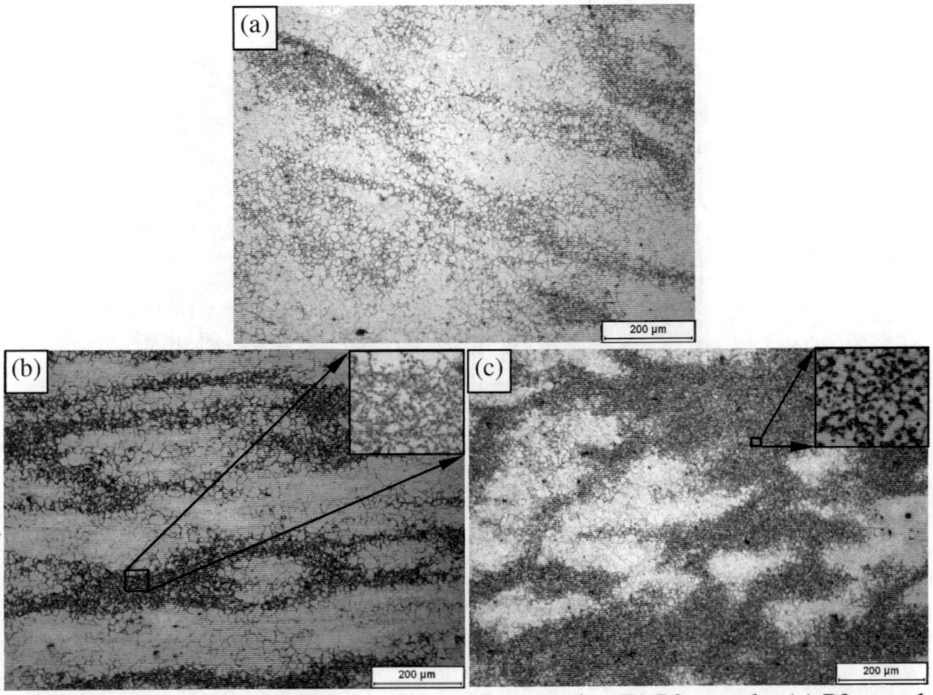

Fig. 4 Micro-structure of the MDF samples, (a) P1-sample, (B) P2-sample. (c) P3-sample

During the MDF process, the precipitation of $Mg_{17}Al_{12}$ phase should be related with the deforming temperature, accumulative strains and Al content in local position. The deforming temperature and the accumulative strains are probably consistent on the local micro region with or without $Mg_{17}Al_{12}$ phase. So the precipitation of $Mg_{17}Al_{12}$ phase should be mainly controlled by the Al content in the Mg matrix. As aforementioned, the solute distribution in the initial alloy is inhomogenous, which will naturally result in an ununiform precipitation of the $Mg_{17}Al_{12}$ phase. This has also been proved by the above-mentioned EPMA measurement. As a consequence, fine and coarse grains coexist in the microstructures.

Difference from the static aging precipitation, the precipitation of $Mg_{17}Al_{12}$ phase during the MDF process is be accompanied by constantly increasing the accumulative strain and dynamic recrystallization. Under the interaction among the strain, recrystallization and precipitation, the $Mg_{17}Al_{12}$ phase precipitate from the Mg matrix and formed granular shape. These granular $Mg_{17}Al_{12}$ phases can refine the grain of AZ80 alloys by pin up the grain boundaries. With the decreasing of forging temperature, more $Mg_{17}Al_{12}$ phase precipitate from the Mg matrix and the area of fine grains expanded, as shown in Fig. 4

4. Conclusions

(1) The β- $Mg_{17}Al_{12}$ phases with granular shape unevenly precipitated from the Mg matrix, as AZ80 Mg alloy multi-directional forged with decreasing temperature from 410 to 300 ℃. These $Mg_{17}Al_{12}$ phase retards the growth of the DRX grains, which in turns results in the formation fine grains area.

(2) With increasing the accumulative strains and decreasing temperature, the fine grains area with granular $Mg_{17}Al_{12}$ phase expanded.

(3) The EPMA analysis show that the inhomogenous Al content distribution in the Mg matrix leads to the non-uniform dynamic precipitation of the $Mg_{17}Al_{12}$ phase.

5. Acknowledgement
This work was supported by the National Natural Science Foundation of China (Nos. 51204053, 51674078 and 51374067), the Fundamental Research Fund for Central Universitis (Nos. N160913002, N130409005 and N130209001) and A Project supported by Scientific Research Fund of Liaoning Province (No. 2015022003).

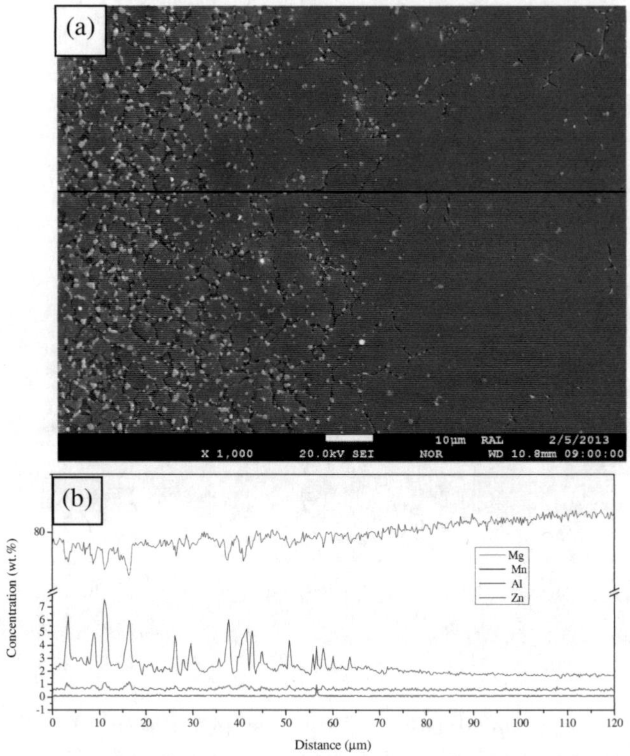

Fig.5 Microprobe measurements of Mg, Mn Zn and Al concentrations in the coarse and fine grain regions of P3-sample, (a) the SEM photo in the intersection area of coarse and fine grain (b) the distribution of the elements along the scanning line in (a).

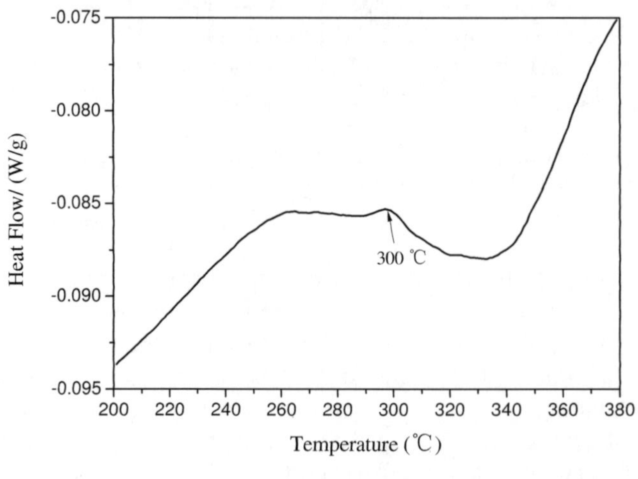

Fig. 6 Typical DSC scan at a heating rate of 5 ℃/min of the as casting AZ80

References

[1] Ha J.S. and Hong S I. 2016. *Materials Science and Engineering: A.,* **651** 805.
[2] Choudhary B.K. Samuel E.I. Sainath G, et al. 2013. *Metallurgical and Materials Transactions A.* **44** 4979.
[3] Nagesha A, Kannan R, Srinivasan V. S., et al. 2016. *Metallurgical and Materials Transactions A.* **47** 1110
[4] Araujo J., Gabriel S., Dille J. et al. 2015. *Journal of alloys and compounds.* **43(S1)** S256
[5] Aboulfadl H, Deges J, Choi P. et al. 2015 *Acta Materialia.* **86** 1110
[6] Zhilyaev A.P. Shakhova I. Morozova A. et al. 2016 *Materials Science and Engineering: A.,* **654** 131
[7] Roven H., Liu M., Werenskiold J. 2008. *Materials Science and Engineering: A.* **483** 54
[8] Zha M., Li Y., Mathiesen R. et al. 2015. *Acta Materialia,* **84** 42.
[9] Xia X., Chen Q., Zhao Z., et al. 2015. *Journal of Alloys and Compounds.* **623** 62.
[10] Nie K., Wang X., Ddeng. K., et al. 2015. *Journal of Alloys and Compounds,* **617** 979
[11] Jiang M., Yan H., Chen R. 2015. *Journal of Alloys and Compounds,* **650** 399
[12] Zhu Q.F., Li L., Zhang Z.Q. et. al. 2014. *Materials Transactions,* **55** 270

IOP Conf. Series: Materials Science and Engineering **194** (2017) 012016 doi:10.1088/1757-899X/194/1/012016

Change of deformation mechanisms in ultrafine grained Mg-Zn-Zr-Ca alloy

Ruixiao Zheng[1], Tilak Bhattacharjee[1,2], Akinobu Shibata[1,2], Taisuke Sasaki[3], Kazuhiro Hono[3], Nobuhiro Tsuji[1,2]

[1] Department of Materials Science and Engineering, Kyoto University, Yoshida Honmachi, Sakyo-ku, Kyoto 606-8501, Japan
[2] Elements Strategy Initiative for Structural Materials (ESISM), Kyoto University
[3] National Institute for Materials Science, 1-2-1 Sengen, Tsukuba 305-0047, Japan

E-mail: zheng.ruixiao.7m@kyoto-u.ac.jp

Abstract. In this study, a fully recrystallized ultrafine grained (UFG) Mg-Zn-Zr-Ca alloy was successfully fabricated by a process including high pressure torsion (HPT) and subsequent rapid annealing. Room temperature tensile test revealed that the UFG Mg alloy with a mean grain size of 0.98 μm exhibited simultaneously enhanced strength and ductility compared to those of the coarse grained counterpart (grain size 57 μm). Observation of deformation microstructures revealed that {10-12} deformation twinning and basal slip were the dominant deformation mechanisms in the coarse grained specimen, while deformation twinning was significantly inhibited but non-basal slip systems seemed activated in the UFG specimen. The reason for the enhanced mechanical properties in the UFG specimen was discussed based on the change of deformation mechanisms observed.

1. Introduction

In recent years, Magnesium (Mg) alloys have attracted extensive attention due to their low density and high specific strength. They are expected to be applied in transportation and aerospace industries as structural components, achieving significant weight reduction [1-3]. However, their hexagonal close-packed (HCP) crystal structure provides only a limited number of independent slip systems, resulting in relatively low tensile ductility as well as poor formability at room temperature [4]. Therefore, up to date, most Mg alloys are still used in their cast state, which cannot have taken their full advantage.

Grain refinement is considered as one of the most promising ways for optimizing strength without sacrificing ductility [5]. To achieving grain refinement, many strategies have been developed. Among them, severe plastic deformation (SPD) processes, such as equal channel angular pressing (ECAP), accumulative roll bonding (ARB) and high pressure torsion (HPT), have already been proven as effective approaches for realizing significant grain refinement [6]. A survey of available scientific literatures has demonstrated that for Mg alloys, ECAP is mostly applied at relatively high temperatures to avoid cracking [7]. However, due to the effect of dynamic recovery and recrystallization at elevated temperatures, minimum grain sizes obtained in the ECAP-processed Mg alloys are usually around several micrometers.

In contrast, high hydrostatic pressure is applied to disc-shaped specimens during torsion deformation in HPT, so that undesirable cracking can be avoided even at room temperature [8-10].

Very recently, Meng *et al.* [9] carried out HPT deformation on Mg-3.4Zn (at.%) alloy at ambient temperature and found that equiaxed grains with an average diameter of 140 nm could be obtained after HPT by 20 revolutions. However, due to the high density of retained lattice defects (such as dislocations), the HPT-processed NC/UFG Mg alloys did not show good ductility. Thus, their mechanical properties were characterized only by using simple hardness test. In order to overcome the above deficiency of HPT-processed Mg alloys, a process including room temperature HPT and subsequent annealing treatment is proposed in the present study. A Mg-Zn-Zr-Ca (ZKX600) alloy is selected for realizing optimal nanostructures. Microstructure observation and mechanical property test reveal that fully recrystallized UFG ZKX600 Mg alloy with simultaneously enhanced strength and ductility is successfully obtained. The deformation microstructures of several representative specimens are observed after tensile deformation to specified strains. The change of deformation mechanisms are discussed based on the deformation microstructures observed.

2. Experimental Procedures

The as-received ZKX600 Mg alloy ingot with a composition of Mg-6.2%Zn-0.5%Zr-0.2%Ca (mass%) was firstly homogenized at 350 °C for 48 h followed by a solid solution treatment (ST) at 400 °C for 1 h using a box-type resistance furnace. Disc-shaped samples with diameter of 10 mm and thickness of 0.80 mm sliced from the ingot were provided for HPT. HPT was conducted at room temperature under a compressive pressure of 6 GPa at a rotation speed of 0.2 rpm. The total rotation angle applied was 360°. After HPT, the specimen was provided for an annealing treatment at 400 °C for 1 min, using a salt bath furnace.

The areas at a radial distance of 3.0 mm from the center on transverse sections involving the rotation axis of the HPT-processed discs were observed by transmission electron microscopy (TEM) using JEOL 2010 operated at 200 kV. The areas at a distance of 3.0 mm from the center on sections perpendicular to the rotation axis of the annealed discs were characterized by field emission scanning electron microscopy (FE-SEM) using JSM 7100F equipped with an electron backscattering diffraction (EBSD) system. The grain size was measured by linear interception method on the obtained EBSD maps.

Mechanical properties of the samples were characterized by uniaxial tensile test. Small-sized tensile specimens with a gauge length of 2 mm and a cross section of 1 mm × 0.5 mm were cut from the discs at a radial distance of 2.5 mm from the center, so that the tensile direction is perpendicular to the radial direction. Tensile tests were carried out at a quasi-static strain rate of 8.3×10^{-4} s^{-1}, with a direct measurement of the displacement of the gauge section using a CCD video camera extensometer.

3. Results and Discussion

Figure 1 shows (a) an EBSD inverse pole figure (IPF) map indicating crystallographic orientations parallel to the normal direction of the as-ST disc specimen and (b) a bright field TEM micrograph of the 360° HPT processed specimen. The EBSD-IPF map (Figure 1 (a)) exhibits that the as-ST specimen shows a coarse grained structure with a mean grain size of about 57 μm and does not have strong crystallographic texture. Nearly equiaxed nanocrystlline structure was obtained by 360° HPT rotations, as shown in Figure 1 (b). The mean grain size is about 100 nm, which is significantly finer than that of pure Mg (grain size ~1 μm) subjected to the HPT process under the similar rotation angles [11]. The result indicates that the alloying elements in the present alloy play important roles in reducing the grain size obtained by SPD. In addition, the ring-like selected area electron diffraction (SAED) pattern inserted in Figure 1 (b) indicates that the nano-grains have various orientations.

IOP Conf. Series: Materials Science and Engineering **194** (2017) 012016 doi:10.1088/1757-899X/194/1/012016

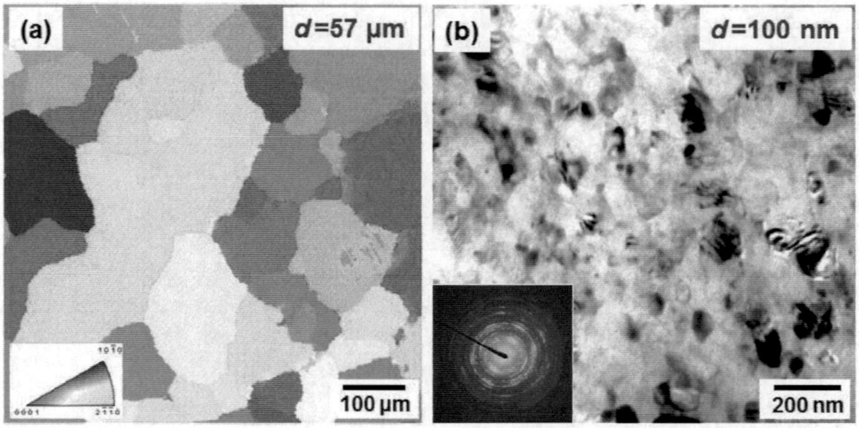

Figure 1. (a) EBSD-IPF map of the as-ST specimen showing coarse grained structure with a mean grain size of about 57 μm. (b) Bright field TEM image of the 360° HPT processed specimen showing nearly equiaxed nanocrystalline structure with a mean grain size of about 100 nm.

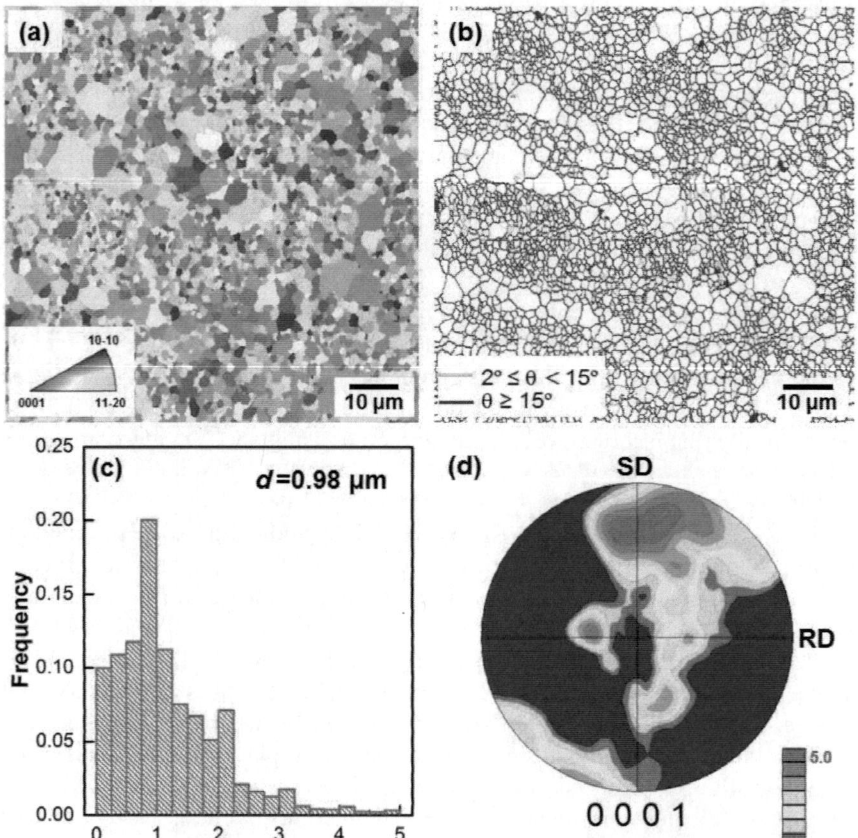

Figure 2. EBSD results of the specimen HPT processed by 360° rotation and then annealed at 400 °C for 1 min. (a) IPF map and (b) corresponding GB map showing fully recrystallized UFG structure. (c) Statistical histogram showing the grain size distribution. (d) (0001) pole figure showing weak texture. SD and RD represent shear direction and radius direction, respectively.

Figure 2 shows EBSD results of the specimen HPT processed by 360° and then annealed at 400 °C for 1 min. The colors in the IPF map (a) indicate crystallographic orientations parallel to the normal direction of the HPT disc. The blue and green lines in the grain boundary (GB) map (b) correspond to high angle grain boundaries (HAGBs) having misorientations larger than 15° and low angle grain boundaries (LAGBs) with misorientations between 2° and 15°, respectively. After the rapid annealing at 400 °C for 1 min, a fully recrystallized UFG structure with a mean grain size of 0.98 μm (Fig. 2 (c)) was successfully obtained. In addition, the peak intensity was low and the peak intensity regions located far from the center in the (0001) pole figure (Fig. 2 (d)), indicating that the basal texture usually observed in wrought Mg alloys is not seen in the present annealed specimen.

For understanding the mechanical properties of the alloy processed under different conditions, several representative specimens were subjected to the tensile test and obtained nominal stress-strain curves are presented in Figure 3 (a). For the as-ST specimen with a coarse mean grain size of 57 μm, the yield strength, ultimate tensile strength, uniform elongation and total elongation were 107 MPa, 256 MPa, 13.9% and 16.5%, respectively. After the HPT process by 360° rotation, the yield strength greatly increased to 311 MPa, which was almost three times higher than that of the as-ST specimen. However, the tensile ductility (total elongation) dropped to 1.2%. After annealing at 400 °C for 1 min, the yield strength and ultimate tensile strength of the specimen with the mean grain size of 0.98 μm (Fig.2) were 213 MPa and 317 MPa, respectively. The uniform elongation and total elongation significantly recovered to 19.3% and 21.8%, respectively. The results clearly mean that, by making the microstructure a fully recrystallized UFG, the tensile ductility could be remarkably regained without sacrificing of the tensile strength.

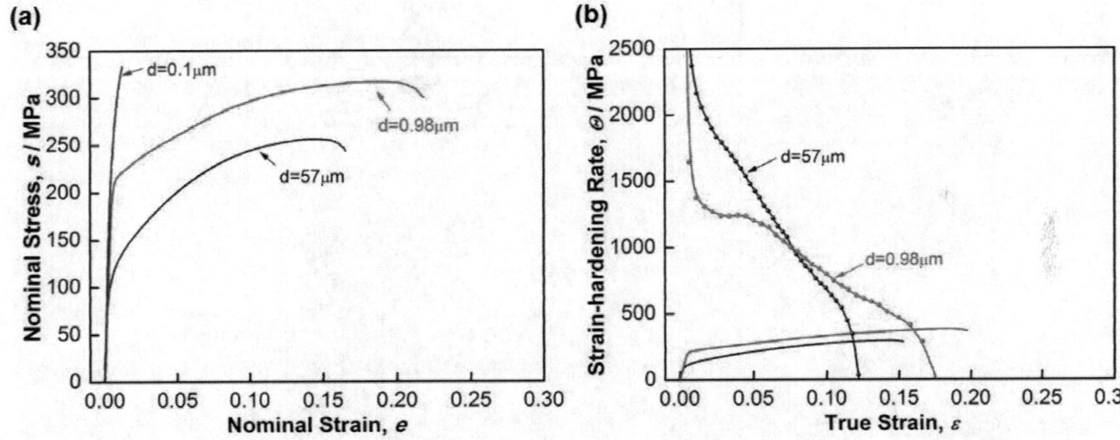

Figure 3. (a) Nominal stress-strain curves of the specimens having different grain sizes (*d*) after processing under different conditions. (b) The stain-hardening rate and true stress curves against the true strain of the specimens having different mean grain sizes of 0.98 μm and 57 μm.

The uniform elongation is an important parameter when considering ductility of materials. Figure 3 (b) shows strain-hardening rate and true stress against true strain of the as-ST (d=57μm) and UFG (d=0.98μm) specimens. According to the Considére criterion for plastic instability [12], the intersection between the strain-hardening rate curve and the true stress-strain curve indicates the point of plastic instability, i.e., the uniform elongation, after which necking progresses. For the as-ST specimen, the strain-hardening rate decreased quickly with the increase of the strain, resulting in an expected uniform tensile elongation of 12.4% (in nominal strain), which was not far from the experimental uniform elongation (13.9%) determined from the nominal stress-strain curve. In contrast, in the specimen having the fully recrystallized UFG structure (Fig.2: d=0.98μm) the drop of the strain-hardening rate was suppressed, compared with the as-ST specimen, so that the plastic instability was delayed. The plastic instability point obtained from Fig.3 (b) (16.8% nominal strain) was again not far from the uniform elongation (19.3%) obtained from the nominal stress-strain curve (Fig.3 (a)). Consequently, it can be concluded that the large tensile ductility in the present annealed specimen is attributed to the enhanced strain-hardening ability realized in the fully recrystallized UFG structure.

In order to reveal underlying deformation mechanisms realizing the enhanced strain-hardening behavior in the fully recrystallized UFG specimen, the deformation microstructures were observed after tensile deformation to specified strains. Figure 4 shows the EBSD maps of the as-ST and the UFG specimens at a tensile strain of 10%. Colors in the IPF maps (Fig.4 (a) and (c)) indicate crystallographic orientations parallel to the normal direction of the sheet-type tensile specimens. As shown in Figure 4 (a) and (b), high density of lenticular-shaped deformation twins with an area fraction of 12.1% were generated in the as-ST specimen. Crystallographic analysis indicated that the twins were exclusively {10-12} extension twins. It is known that {10-12} extension twinning has much lower critical resolved shear stress (CRSS) compared to other twinning systems [13]. In contrast, quite limited number of deformation twins with an area fraction of only 1.1% formed in the UFG specimen at the same tensile strain.

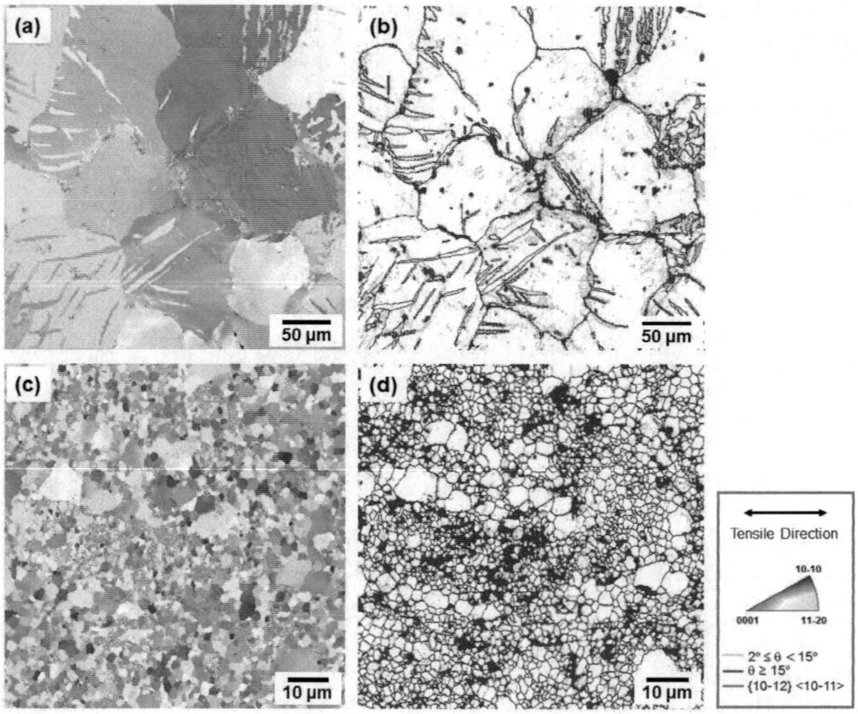

Figure 4. EBSD (a) IPF map and (b) corresponding GB map of the as-ST specimen (d=57μm) after 10% tensile deformation, showing large number of deformation twins. (c) IPF map and (d) corresponding GB map of the UFG specimen (d=0.98μm) after 10% tensile deformation, showing limited number of deformation twins.

Although deformation twinning was significantly inhibited in the UFG specimen, it exhibited enhanced strain-hardening and better ductility compared to the as-ST specimen. Thus, there must be some other deformation mechanisms responsible for the enhanced strain-hardening and ductility, for example, the activation of non-basal slip systems. As the major slip system, i.e., basal slip, of Mg and Mg alloys has only <***a***> component in its Burgers vector, any deformation mechanisms involving <***c***> deformation component is considered to be necessary for large plasticity [14, 15]. Koike *et al.* [14] reported a large tensile elongation up to 47% in a commercial AZ31 alloy having a mean grain size of

6.5 μm. They concluded that the superior tensile ductility was due to the activation of <***c+a***> non-basal slip. We are planning to make further investigations to clarify this issue in the present material.

4. Conclusion

In summary, a fully recrystallized UFG ZKX600 Mg alloy having a mean grain size of 0.98 μm was successfully fabricated by the combination of HPT and subsequent annealing. Simultaneously enhanced strength and ductility were realized in the UFG specimen, which were attributed to grain boundary strengthening and enough strain-hardening ability realized in the fully recrystallized UFG structure. Deformation microstructures indicated that deformation twinning was significantly inhibited in the UFG specimen. Activation of non-basal slip systems having <***c***> component was suggested in the UFG specimen, which will be confirmed in our future study.

Acknowledgments

This work was financially supported by the Elements Strategy Initiative for Structural Materials (ESISM) and the Grant-in-Aid for Scientific Research (S) (No. 15H05767), both through the Ministry of Education, Culture, Sports, Science and Technology (MEXT), Japan.

References

[1] Kim N 2014 *Mater. Sci. Technol.* **30** 1925.
[2] Suh B, Shim M, Shin K and Kim N 2014 *Scr. Mater.* **84** 1.
[3] Hono K, Mendis C, Sasaki T and Oh-Ishi K 2010 *Scr. Mater.* **63** 710.
[4] Agnew S and Nie J 2010 *Scr. Mater.* **63** 671.
[5] Kim W, Jeong H and Jeong H 2009 *Scr. Mater.* **61** 1040.
[6] Tsuji N, Saito Y, Lee Y and Minamino Y 2003 *Adv. Eng. Mater.* **5** 338.
[7] Kulyasova O, Islamgaliev R, Mingler B and Zehetbauer M 2009 *Mater. Sci. Eng. A* **503** 176.
[8] Zhilyaev A and Langdon T 2008 *Prog. Mater. Sci.* **53** 893.
[9] Meng F, Rosalie J, Singh A, Somekawa H and Tsuchiya K *Scr. Mater.* **78** 57.
[10] An X, Wu S, Zhang Z, Figueiredo R, Gao N and Longdon T 2012 *Scr. Mater.* **66** 227.
[11] Edalati K, Yamamoto A, Horita Z and Ishihara T 2011 *Scr. Mater.* **64** 880.
[12] Dieter G 1986 *Mechanical Metallurgy* New York, McGraw-Hill.
[13] Barnett M 2007 *Mater. Sci. Eng. A* **464** 1.
[14] Koike J, Kobayashi T, Mukai T, Watanabe H, Suzuki M, Maruyama K and Higashi K 2003 *Acta Mater.* **51** 2055.
[15] Sun J, Trimby P, Yan F, Liao X, Tao N and Wang J 2013 *Scr. Mater.* **69** 428.

IOP Conf. Series: Materials Science and Engineering **194** (2017) 012017 doi:10.1088/1757-899X/194/1/012017

Hardening and thermal stability of a nanocrystalline $CoCrFeNiMnTi_{0.1}$ high-entropy alloy processed by high-pressure torsion

H Shahmir[1,2,*], M Nili-Ahmadabadi[1,3], A Shafie[1], TG Langdon[2]
[1] School of Metallurgy and Materials, College of Engineering, University of Tehran, Tehran, Iran
[2] Materials Research Group, Faculty of Engineering and the Environment, University of Southampton, Southampton SO17 1BJ, UK
[3] Center of Excellence for High Performance Materials, School of Metallurgy and Materials, College of Engineering, University of Tehran, Tehran, Iran

* h.shahmir@ut.ac.ir

Abstract. A $CoCrFeNiMnTi_{0.1}$ high-entropy alloy (HEA) was subjected to high-pressure torsion (HPT) processing under 6.0 GPa pressure up to 10 turns. XRD results reveal that the initial and HPT-processed microstructures consist of a single *fcc* phase and there is no evidence for creating a new phase and the occurrence of a phase transformation during HPT processing. It is shown that there is a gradual evolution in hardness with increasing numbers of turns but full homogeneity is not achieved even after 10 turns. Microhardness measurements reveal that the material reaches a saturation hardness value of Hv ≈ 460 which is approximately three times higher than for the homogenized alloy. The nanostructured HEA was subjected to post-deformation annealing (PDA) at 473-1173 K and it is shown that the hardness increases slightly up to Hv ≈ 550 at 773 K due to a phase decomposition and the formation of new precipitates and then decreases to the hardness of the homogenized sample (Hv ≈ 140) at 1173 K due to a combination of recrystallization, grain growth and dissolution of the precipitates. The results reveal that an addition of only 2 at.% Ti will improve the hardness and thermal stability of the nanocrystalline CoCrFeNiMn HEA.

1. Introduction

High entropy alloys (HEAs) have attracted worldwide attention due to their potential beneficial mechanical and electrochemical characteristics, such as high strength, high thermal stability and oxidation resistance. An intriguing aspect of HEAs containing five or more elements with each elemental concentration between 5 at % and 35 at % is that their high configurational entropies may favor the formation of solid solutions over the precipitation of brittle intermetallic compounds [1,2]. These alloys surprisingly have relatively simple structure based on solid solution phases [1-3]. In fact, a potential combination of high solid solution strengthening and good ductility is achievable if the solid solution phase possesses a simple crystal structure, such as an *fcc* lattice with a large number of slip systems [4]. An example and one of the best studied such single-phase HEAs is the equiatomic

CoCrFeNiMn alloy [5]. Even though the elements in this five-element alloy possess different crystal structures, it crystallizes as a single-phase *fcc* solid solution [6]. CoCrFeNiMn HEAs exhibit very good ductility but their strength is relatively low, around ~300 MPa in the homogenized condition [7]. To increase the strength of this alloy without significantly sacrificing its ductility, additional strengthening methods may be induced. Grain refinement by severe plastic deformation processing may lead to superior properties in HEAs. High-pressure torsion (HPT) is a well established procedure for obtaining ultra-fine and even nanostructured grains in metals and alloys. In HPT processing, a disk-shaped specimen is deformed by simple shear between two anvils where it is constrained under a high pressure and subjected to concurrent torsional straining [8]. It is well known that processing by HPT will introduce a significant inhomogeneity into the material, but a gradual evolution towards a homogeneous structure has been reported in disks processed by HPT with this evolution occurring by increasing the applied pressure and/or the total numbers of revolutions [9,10]. There are only a few reports describing the influence of HPT processing on the HEAs [11-16]. Processing the CoCrFeNiMn high-entropy alloy by HPT leads to significant hardening and grain refinement and post deformation annealing (PDA) of such a nanocrystalline CoCrFeNiMn HEA provides a combination of high strength and good ductility including an ultimate tensile strength [15]. To further improve the strength without losing plastic stability, one must, therefore, rely on other mechanisms, such as solid solution or precipitation hardening, which would require a modification of the chemical composition of the alloy. It was well known that besides the principal elements, HEAs can also contain minor elements with each below 5 at.% [17]. Accordingly, the present research was initiated to provide detailed information on the adding of minor Ti element to CoCrFeNiMn HEA and processing it by HPT through different numbers of revolutions.

2. Experimental material and procedures

The experiments were conducted on an HEA with a nominal composition of $Co_{19.6}Cr_{19.6}Fe_{19.6}Ni_{19.6}Mn_{19.6}Ti_2$ (in at.%). The master alloy was prepared using a nonconsumable vacuum arc melting technique in a water-cooled copper crucible. After several remeltings for homogenization, the ingots were hot forged and then homogenized at 1000 °C for 720 minutes. The grain size after homogenizing was ~200 μm. Disks of 10 mm in diameter were prepared with electro-discharge machining to thicknesses of ~1 mm and they were then polished mechanically to final thicknesses of ~0.8 mm. These disks were processed by HPT at room temperature under quasi-constrained conditions [18] using an applied pressure, *P*, of 6.0 GPa and a rotation speed of 1 rpm through totals, *N*, of 1, 5 and 10 revolutions. The disks subjected to HPT through 5 rotations were used for PDA at 773-1373 K for 60 min. After processing, each HPT disk was polished to a mirror-like quality and hardness measurements were taken using a Vickers microhardness tester with a load of 500 gf and dwell times of 10 s. The average microhardness values, Hv, were measured along randomly selected diameters on each disk. These measurements were taken at intervals of ~0.5 mm and at every point the local value of Hv was obtained from the average of four separate hardness values. The phase constituents were determined using X-ray diffraction (XRD) employing Cu Kα radiation (wavelength λ = 0.154 nm) at 45 kV and a tube current of 200 mA with Rigaku SmartLab equipment. The XRD measurements were performed over 2θ ranging from 30° to 100° using a scanning step of 0.01° and a scanning speed of 2° min^{-1}. Grain sizes of samples after HPT and after PDA at 773 K were estimated from XRD results using the classic Williamson-Hall method [19] and the grain size of samples after PDA at >883 K were measured from scanning electron microscopy (SEM). For these measurements, samples were polished using a 40 nm colloidal silica suspension.

3. Experimental results and discussion

3.1. *Hardening after HPT Processing*

The results for the corresponding Vickers microhardness measurements are shown in Fig. 1(a) after processing through different numbers of turns with the average values of Hv plotted along each disk diameter and with the lower dashed line at Hv ≈ 140 corresponding to the initial hardness in the homogenized condition. Inspection of Fig. 1(a) shows that, with reference to the homogenized condition, the hardness at the edge of the disk increases significantly by a factor of ~3.2 to Hv ≈ 450 after only 1 turn whereas the hardness in the centre increases from Hv ≈ 140 to Hv ≈ 320. At the edges of the disks, the hardness values increase slightly with further straining to ~460 after 10 turns and in the centres the hardness values also gradually increase to Hv ≈ 380 after 10 turns. These results confirm the gradual evolution in hardness with increasing torsional straining. However, even after 10 turns it was not possible to produce a fully homogeneous hardness distribution and the results showed that after 5 and 10 turns the area within a radius of r < 1 mm at the centre of the disk continued to have a lower hardness. The results indicate that there is no difference between the hardness of samples processed through 5 and 10 turns at a radius of r > 1 mm. The variation in strain across each disk processed by HPT is estimated from the well-known equivalent strain [20] which indicates that the imposed strain is a maximum around the edge of the disk but it becomes equal to zero in the centre of the disk where r = 0. However, in practice there is generally a gradual evolution towards microstructural homogeneity as the numbers of turns increase and this has been demonstrated both experimentally [21] and theoretically [22]. For comparing the effect of Ti on the hardness of the HPT-processed CoCrFeNiMn HEA, the variation of Hv with equivalent strain is shown in Fig. 1(b) for CoCrFeNiMn and $CoCrFeNiMnTi_{0.1}$ HEA disks processed through a wide range of rotations. The plot shows a similar trend for both conditions and both alloys show significant hardening during HPT processing. The datum points tend to a saturated hardness value of ~450 and ~460 in CoCrFeNiMn [15] and $CoCrFeNiMnTi_{0.1}$ HEAs, respectively, after straining up to $\varepsilon_{eq} \approx 20$. It is important to note that the initial hardness of homogenized-CoCrFeNiMn HEA is ~120. Therefore, the hardening of both alloy are similar but the level of hardness in $CoCrFeNiMnTi_{0.1}$ HEA is higher.

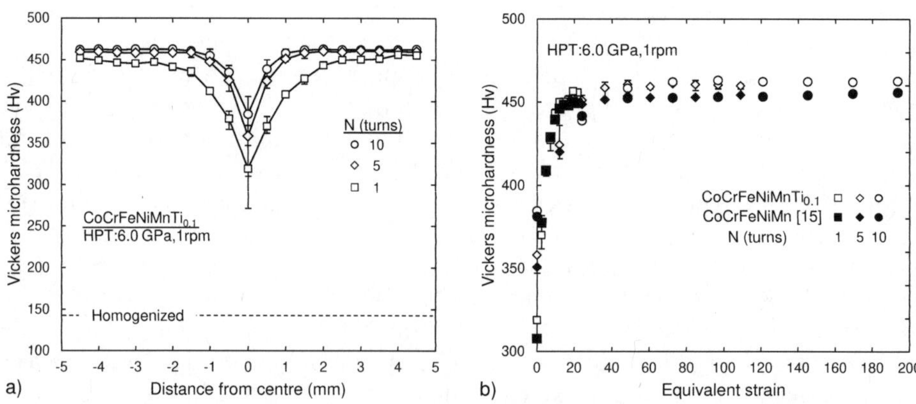

Figure 1 (a) Values of the Vickers microhardness measured across disks processed from 1 to 10 turns at a rotation speed of 1 rpm: the lower dashed line shows the homogenized initial condition. (b) Values of the Vickers microhardness plotted against the equivalent strain after processing by HPT through various numbers of turns for CoCrFeNiMnTi [15] and $CoCrFeNiMnTi_{0.1}$ HEAs.

3.2. Thermal stability of the microstructure

Microstructural evaluations of the nanostructured CoCrFeNiMn [15] and $CoCrFeNiMnTi_{0.1}$ HEAs after PDA were conducted by XRD and the results are presented in Fig. 2(a) and (b), respectively. It is apparent that the HPT-processed microstructures in both alloys consist of a single *fcc* phase. It is important to note that the initial microstructures in bot HEAs are single phase (not shown) and there is no evidence for creating a new phase and the occurrence of a phase transformation during HPT processing. The lattice parameters of the *fcc*-phase matrix are for CoCrFeNiMn and $CoCrFeNiMnTi_{0.1}$ HEAs $a \approx 3.599$ Å and $a \approx 3.610$ Å. Occurrence of significant peak broadening after HPT indicates energy storage during HPT processing and this is consistent with the hardness increment after HPT. These broadenings were examined for determinations of grain size in the HPT-processed alloy using the classic Williamson-Hall method [19]. Accordingly, the grain size of CoCrFeNiMn and $CoCrFeNiMnTi_{0.1}$HEAs after HPT processing were estimated as ~10 [15] and ~30 nm, respectively, which indicate significant grain refinement during HPT processing. It should be noted that the grain size estimated through the XRD analysis refers to the crystallite size which is considered equivalent to the grain size especially when the microstructure is refined to the nano-scale [23]. Inspection of Fig. 2 shows that several new peaks appear after annealing at temperatures from 773 to 1073 K and these peaks are of two different forms. First, the new peaks marked with inverted triangles with the main peak at $2\theta \approx 44.2°$ and 44.3° for CoCrFeNiMn and $CoCrFeNiMnTi_{0.1}$ HEAs are related to the formation of a new *bcc* phase which is precipitated within the *fcc*-phase matrix which is in good agreement with the Cr-rich phase reported earlier in this HEA [12,15]. Second, close inspection of Fig. 2 reveals the appearance of some additional peaks of very low intensities marked with open inverted triangles in the samples annealed at 873 and 973 K for CoCrFeNiMn HEA and at 873-1073 K for $CoCrFeNiMnTi_{0.1}$ HEA. The crystal structure of this new phase was identified as tetragonal where this is consistent with the σ-phase which is a hard Cr-rich phase reported earlier in HEAs [23-26]. It was suggested that Cr promotes the formation of a *bcc* phase and, based on thermodynamic calculations, solubility of Cr in the *fcc* phase is quite high at high temperatures but decreases with decreasing temperature giving rise to the formation of a Cr-rich *bcc* phase.

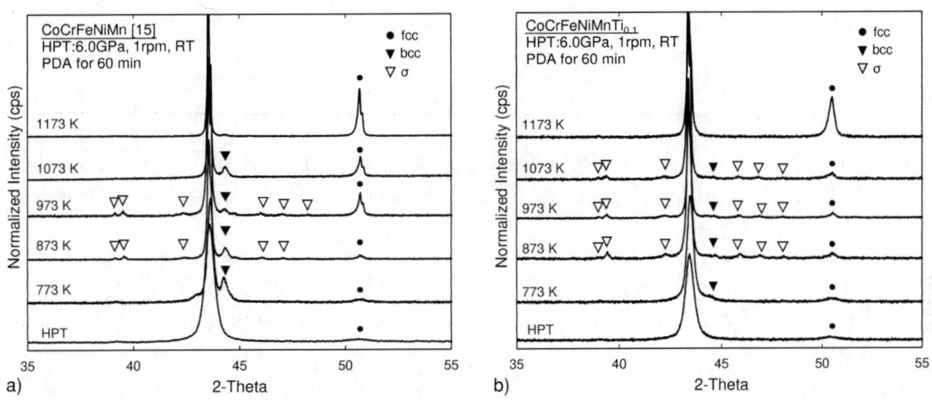

Figure 2 X-ray patterns near the edges of disks after HPT processing followed by PDA at 473 to 1173 K for 60 min for (a) CoCrFeNiMn [15] and (b) $CoCrFeNiMnTi_{0.1}$.

The nano-scale precipitates, consisting of a MnNi phase, were reported after HPT processing of the alloy followed by short-term annealing at 723 K and Cr- and Mn-rich grain boundary precipitates were observed after low strain rate creep at 1073 K [27]. In addition, grain boundary precipitates enriched in Cr, containing some Mn, Fe and Co but very little Ni (σ-phase), were reported recently after very long-term annealing of a coarse-grained CoCrFeNiMn HEA at 973 K [25,26]. It is important to note that, unlike the HPT condition, the σ phase was detected in the present sample after annealing the homogenized condition for a very long time (500 days) [25]. It appears that the nanocrystalline grain structure of the HPT-processed HEAs facilitates the phase decomposition owing to the large number of grain boundaries which serve both as fast diffusion pathways and also as preferential nucleation sites for the formation of new phases. A recent report demonstrated the occurrence of fast diffusion in a nanocrystalline HEA due to the presence of severe torsion stresses [13]. Thus, in the present study the heavy deformation leads more quickly to the formation of stable phases than in the absence of any mechanical treatment. Nevertheless, the XRD results in Fig. 2 indicate that the microstructure is again single phase after annealing at 1173 K. The results show that the volume fractions of precipitates decrease with increasing annealing temperature and finally approach zero at 1173 K in both HEAs.

IOP Conf. Series: Materials Science and Engineering **194** (2017) 012017 doi:10.1088/1757-899X/194/1/012017

Figure 3(a) shows the measured values of the microhardness for the HPT-processed samples of CoCrFeNiMn [15] and CoCrFeNiMnTi $_{0.1}$ HEAs after annealing at temperatures of 473 to 1173 K for a period of 60 min. The results demonstrate that the hardness of the nanocrystalline HEAs increases slightly up to Hv ≈ 530 and 550 for CoCrFeNiMn [15] and CoCrFeNiMnTi $_{0.1}$ HEAs, respectively, after annealing at 773 K and then decreases rapidly with increasing annealing temperatures up to 1173 K. At this latter temperature, the hardness is very close to the value for the homogenized condition. This behavior suggests that precipitates form at temperatures up to 773 K in both HEAs which is consistent with the XRD results. Close inspection of the hardness results shows that in both HEAs the hardness decreases significantly above 773 K due to dissolution of the precipitates and activation of the recrystallization mechanism up to 1173 K so that finally the hardness is equal to the homogenized condition due to grain growth.

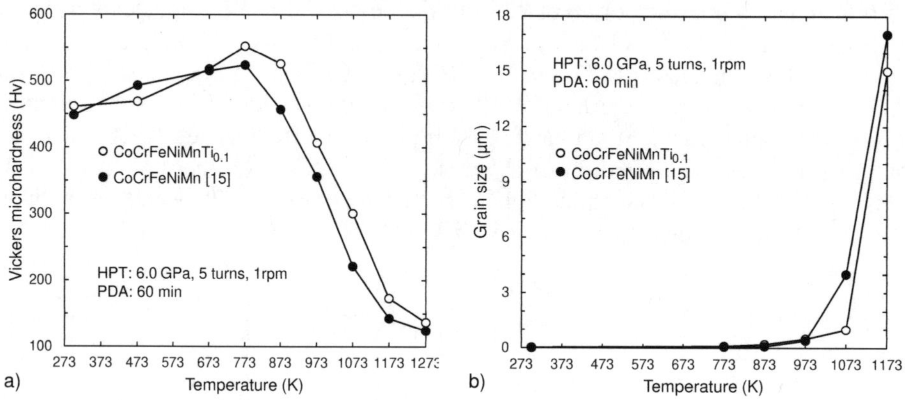

Figure 3 (a) Dependence of Vickers hardness and (b) grain size of the CoCrFeNiMn and CoCrFeNiMnTi$_{0.1}$ HEAs after HPT on the annealing time at different annealing temperatures.

The measured grain size is represented in Fig. 3(b) for the HPT-processed samples of CoCrFeNiMn [15] and CoCrFeNiMnTi $_{0.1}$ HEAs after annealing at temperatures of 773 to 1173 K for a period of 60 min. The present results reveal that the grain size increases significantly in nanocrystalline-CoCrFeNiMn and CoCrFeNiMnTi$_{0.1}$ HEAs after PDA at T>973 K and T>1073 K, respectively, which suggests that recrystallization occurs at these temperatures and this is consistent with the dissolution of the precipitates. This suggests that thermal stability in severely-deformed CoCrFeNiMnTi$_{0.1}$ HEA is higher than CoCrFeNiMn HEA. This behavior is consistent with XRD results which show stable σ phase after PDA of nanocrystalline CoCrFeNiMnTi $_{0.1}$ HEA at 1073 K. Generally, HEAs have a high recrystallization temperature and a strong resistance against grain coarsening during annealing due to a sluggish diffusion effect in these alloys [28]. A systematic study of grain growth in an *fcc* single-phase CoCrFeNiMn HEA after heavy plastic deformation suggested that the grains coarsen at a temperature above 1123 K and grain boundary motion is then controlled by a solute-drag mechanism [29]. The results show *fcc* single-phase grains with average grain sizes of ~17 and ~15 μm for CoCrFeNiMn and CoCrFeNiMnTi $_{0.1}$ HEAs after PDA at 1173 K.

4. Summary and conclusions

A CoCrFeNiMnTi$_{0.1}$ high-entropy alloy was processed by HPT under 6.0 GPa pressure up to 10 turns at room temperature. Following processing, the average grain size was reduced to ~30 nm and this was accompanied by an exceptional increase in the hardness. The hardening and thermal stability of this HEA was studied and compared with a CoCrFeNiMn HEA in a similar condition. The results suggest that hardening of two alloys is similar during HPT processing but the hardness of CoCrFeNiMnTi$_{0.1}$ HEA is higher. The hardness of the CoCrFeNiMnTi$_{0.1}$ HEA further increased after PDA at temperatures up to 773 K due to the formation of new precipitates including Cr-rich *bcc* and σ phases.

The hardness decreased after annealing at higher temperatures up to 1173 K due to recrystallization and precipitate dissolution. The results reveal that an addition of only 2 at.% Ti will improve the hardness and thermal stability of the CoCrFeNiMn HEA by stability of precipitates up to 1073 K.

Acknowledgements
This work was supported by the National Elites Foundation of Islamic Republic of Iran and in part by the European Research Council under Grant Agreement No. 267464-SPDMETALS (TGL).

References

[1] Tong CJ, Chen YL, Chen SK, Yeh JW, Shun TT, Tsau CH, Lin SJ and Chang SY 2005 *Metall. Mater. Trans. A* **36A** 881.
[2] Yeh JW, Chen SK, Lin SJ, Gan JY, Chin TS, Shun TT, Tsau CH and Chang SY 2004 *Adv. Eng. Mater.* **6** 299.
[3] Yeh JW, Chen SK, Lin SJ, Gan JY, Chin TS, Shun TT, Tsau CH and Chang SY 2004 *Adv. Eng. Mater.* **6** 299.
[4] Otto F, Dlouhý A, Somsen Ch, Bei H, Eggeler G and George EP 2013 *Acta Mater.* **61** 5743.
[5] Cantor B, Chang ITH, Knight P and Vincent AJB 2004 *Mater. Sci. Eng. A* **375-377** 213.
[6] Otto F, Yang Y, Bei H and George EP 2013 *Acta Mater.* **61** 2628.
[7] Wu Y, Liu WH, Wang XL, Ma D, Stoica AD, Nieh TG, He ZB and Lu ZP 2014 *Appl. Phys. Lett.* **104** 051910.
[8] Zhilyaev AP and Langdon TG 2008 *Prog. Mater. Sci.* **53** 893.
[9] Zhilyaev AP, Lee S, Nurislamova GV, Valiev RZ and Langdon TG 2011 *Scr. Mater.* **44** 2753.
[10] Zhilyaev AP, Nurislamova GV, Kim B-K, Baró MD, Szpunar JA and Langdon TG 2003 *Acta Mater.* **51** 753.
[11] Tang QH, Huang Y, Huang YY, Liao XZ, Langdon TG and Dai PQ 2015 *Mater. Lett.* **151** 126.
[12] Schuh B, Mendez-Martin F, Völker B, George EP, Clemens H, Pippan R and Hohenwarter A 2015 *Acta Mater.* **96** 258.
[13] Lee D-H, Choi I-C, Seok M-Y, He J, Lu Z, Suh J-Y, Kawasaki M, Langdon TG and Jang J-I 2015 *J. Mater. Res.* **30** 2804.
[14] Yu PF, Cheng H, Zhang LJ, Zhang H, Jing Q, Ma MZ, Liaw PK, Li G and Liu RP 2016 *Mater. Sci. Eng. A* **655** 283.
[15] Shahmir H, He JY, Lu ZP, Kawasaki M and Langdon TG 2016 *Mater. Sci. Eng. A* **676** 294.
[16] Shahmir H, He JY, Lu ZP, Kawasaki M and Langdon TG 2017 *Mater. Sci. Eng. A* **685** 342.
[17] Tsai MH and Yeh JW 2014 *Mater. Res. Lett.* **2** 107.
[18] Figueiredo RB, Cetlin PR and Langdon TG 2011 *Mater. Sci. Eng. A* **528** 8198.
[19] Williamson GK and Hall WH 1953 *Acta Metall.* **1** 22.
[20] Valiev RZ, Ivanisenko YV, Rauch EF and Baudelet B 1996 *Acta Mater.* **44** 4705.
[21] Xu C and Langdon TG 2009 *Mater. Sci. Eng. A* **503** 71.
[22] Estrin Y, Molotnikov A, Davies CHJ and Lapovok R 2008 *J. Mech. Phys. Solids* **56** 1186.
[23] Zhang Z, Zhou F and Lavernia EJ 2003 *Metall. Mater. Trans. A* **34A** 1349.
[24] Tsai MH, Yuan H, Cheng G, Xu W, Jian WW, Chuang MH, Juan CC, Yeh AC, Lin SJ and Zhu Y 2013 *Intermetallics* **33** 81.
[25] Otto F, Dlouhý A, Pradeep KG, Kuběnová M, Raabe D, Eggeler G and George EP 2016 *Acta Mater.* **112** 40.
[26] Pickering EJ, Muñoz-Moreno R, Stone HJ and Jones NG 2016 *Scr. Mater.* **113** 106.
[27] He JY, Zhu C, Zhou DQ, Liu WH, Nieh TG and Lu ZP 2014 *Intermetallics* **55** 9.
[28] Bhattacharjee PP, Sathiaraj GD, Zaid M, Gatti JR, Lee C, Tsai C-W and Yeh J-W 2014 *J. Alloy. Compd.* **587** 544.
[29] Liu WH, Wu Y, He JY, Nieh TG and Lu ZP 2013 *Scr. Mater.* **68** 526.

Effect of severe cold-rolling and annealing on microstructure and mechanical properties of $AlCoCrFeNi_{2.1}$ eutectic high entropy alloy

I S Wani[1], T Bhattacharjee[2,5], S Sheikh[3], Y Lu[4], S Chatterjee[1], S Guo[3], P P Bhattacharjee[1] and N Tsuji[2,5]

[1]Department of Materials Science and Metallurgical Engineering, IIT Hyderabad, India
[2]Department of Materials Science and Engineering, Kyoto University, Japan
[3]Materials and Manufacturing Technology, Chalmers University of Technology, SE-41296, Gothenburg, Sweden
[4]Key Laboratory of Solidification Control and Digital Preparation Technology (Liaoning Province), School of Materials Science and Engineering, Dalian University of Technology, Dalian 116024, PR China
[5]Elements Strategy Initiative for Structural Materials (ESISM), Kyoto, University, Japan
E-mail: (ms14resch11002@iith.ac.in, pinakib@iith.ac.in)

Abstract. The possibility of microstructural refinement and improvement of mechanical properties by severe cold-rolling was investigated in an $AlCoCrFeNi_{2.1}$ lamellar eutectic high entropy alloy (EHEA). The as-cast alloy revealed fine scale eutectic mixture of $L1_2$ (ordered FCC) and B2 (ordered BCC) phases. During severe cold-rolling up to 90% reduction in thickness the B2 phase maintained the ordered structure, while the $L1_2$ phase showed the evolution of a nanocrystalline structure and progressive disordering. Annealing of the severely cold-rolled material resulted in the formation of duplex microstructures composed of two different phases with equiaxed morphologies and significant resistance to grain growth up to 1200°C. Annealing at 1000°C resulted in an optimum strength-ductility balance with the tensile strength of 1175 MPa and the total elongation of 23%. The present results showed that severe cold-rolling and annealing can impart very attractive mechanical properties in complex EHEAs.

1. Introduction

High entropy alloys (HEAs) are originally proposed as a novel class of multicomponent alloys containing sufficiently large number of alloying elements (≥5) in equiatomic or near equiatomic proportions [1]. Despite the presence of large number of constituents, the HEAs may show simple phases due to their large configurational entropy. In spite of their complex chemistry, the HEAs show several attractive and intriguing mechanical properties which have generated tremendous research interest [2].

In order to further enhance the strength-ductility combination, multiphase HEAs possessing novel microstructures have been proposed. Development of Eutectic HEAs (EHEAs) is a noteworthy approach in this regard [3]. Although the as cast microstructure of the EHEAs show promising mechanical properties, it is envisaged that suitable thermo-mechanical processing route can tailor the microstructure and properties of these novel alloys significantly as observed in other HEAs [4] but needs to be clarified.

In this paper, the effect of severe cold-rolling and annealing on microstructure and mechanical properties of a $AlCoCrFeNi_{2.1}$ EHEA is investigated. It is believed that clarifying the mechanisms of microstructure development and enhancement of properties in the present EHEA by severe plastic deformation should be helpful in developing appropriate severe plastic deformation based processing strategies for other HEAs.

2. Experimental procedure

The $AlCoCrFeNi_{2.1}$ EHEA (15mm (width) x 90mm (length) x 3mm (thickness)) used in the present study was prepared by an optimized arc melting route starting from high purity elemental powders. The as-cast samples were cold-rolled up to 90% reduction in thickness (final thickness ~0.3 mm) using a laboratory scale two-high rolling machine (Fenn, USA). The cold-rolled samples were annealed at temperatures ranging between 800°C to 1200°C for 1 hour (h).

The microstructure of the different cold rolled and annealed samples were investigated using scanning electron microscope(FEG-SEM) (SUPRA 40, Carl-Zeiss, Germany) equipped with electron backscatter diffraction (EBSD) (Oxford Instruments, UK) and transmission electron microscope (TEM) (JEOL 2010) operated at 200 KV. The samples for EBSD and TEM were prepared by standard electro-polishing techniques using electrolyte content of 90% ethanol + 10% perchloric acid. The hardness of the constituent phases in the as-cast EHEA was analyzed by nanoindentation hardness mapping using a TI950 Triboindenter (Hysitron, USA) with the applied load linearly increasing from 1 μN to the maximum of 500 μN over a period of 0.25 s and unloaded to 1 μN over a period of 0.25 s before moving to the next indent position. Tensile tests were carried out at ambient temperature using a floor mounted testing machine (Shimadzu, Japan) using an initial strain rate of 8.3×10^{-4} s^{-1}.

3. Results and discussion

Figure 1(a) shows SEM micrograph of the as-cast EHEA revealing the fine eutectic mixture. The TEM micrograph (Fig.1 (b)) shows in detail the eutectic microstructure, while the selected area diffraction patterns (SADPs) obtained from the regions marked by green and red circles in (Fig.1(b)) show that the eutectic mixture consists of $L1_2$ and B2 phases having thickness of ~600 nm and ~200 nm, respectively.

Table 1. Detailed chemical composition of the phases present in the as-cast EHEA (at %)

Phases	Elements				
	Al	Co	Cr	Fe	Ni
Nominal	16.39	16.39	16.39	16.39	34.43
$L1_2$	8.66±0.2	18.47±0.25	20.53±0.05	19.48±0.25	32.85±0.36
B2	24.30±1.3	13.00±0.14	6.50 ±0.84	10.55±0.77	45.70±2.83

The chemical analysis of the as-cast alloy done by energy dispersive spectroscopy (EDS) shows the difference in elemental distribution in the two phase as summarized in Table 1. The B2 phase is enriched with Ni and Al but depleted in Cr. The B2 phase also show the presence of disordered Cr-rich BCC nano precipitates [5]. Figure 2 shows the nanoindentation mapping of an area of interest in the EHEA. It is clearly observed that the B2 is much harder than the $L1_2$ phase.

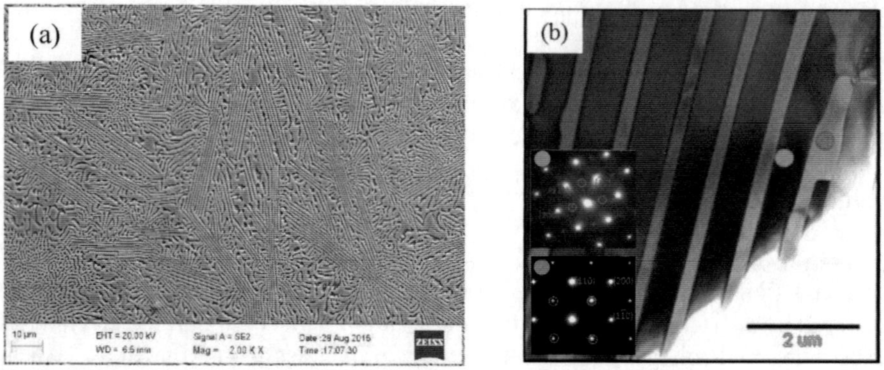

Figure 1. (a) SEM and (b) TEM micrographs of the as-cast EHEA showing the lamellar microstructure. The inset SADPs (marked by green and red spots) obtained from green and red spots in (b) correspond to the $L1_2$ with zone axis [011] and B2 phase with zone axis [001], respectively.

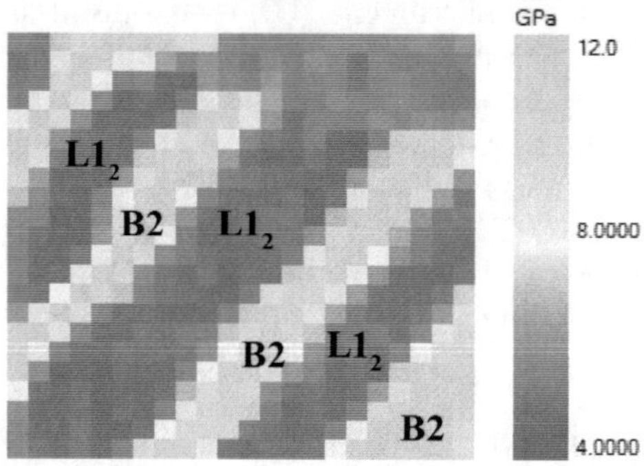

Figure 2. Nano-indentation map of the as-cast EHEA.

Figure 3 shows the TEM micrograph of the 90% cold-rolled material. The SADP in (Fig.3 (b)) (obtained from the region marked by green circle in Fig.3(a)) undergoes progressive disordering accompanied by the evolution of a nanocrystalline structure. The deformation induced disordering has been reported in other $L1_2$ alloys [6] and attributed to formation of deformation heterogeneities and the uncoupling of partial dislocations [7]. However, the B2 phase has retained its ordered structure after heavy deformation as confirmed by the SADP (obtained from the region shown in red circle in Fig.3(c)).

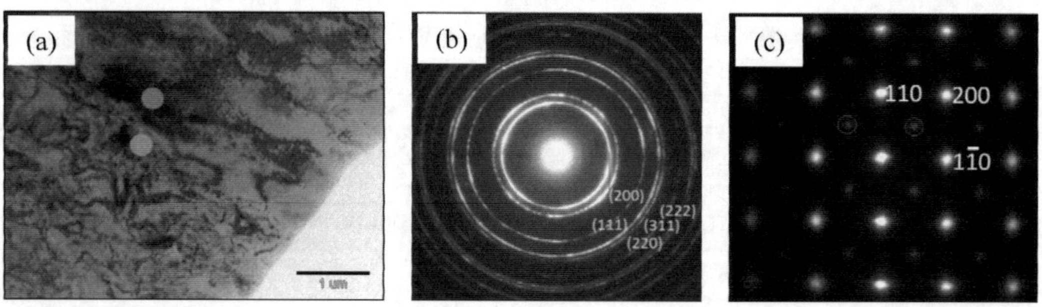

Figure 3. (a) TEM micrograph of the 90% cold-rolled EHEA; (b) and (c) are the SADPs of disordered FCC and B2 phase respectively.

IOP Conf. Series: Materials Science and Engineering **194** (2017) 012018 doi:10.1088/1757-899X/194/1/012018

Figure 4 shows the EBSD phase map of 90% cold-rolled material after annealing at 800°C (Fig.4(a)) and 1200°C (Fig.4(b)), respectively. The phase maps show the development of microduplex structure consisting of disordered FCC and B2 phases [8]. Table 2 shows the grain size of the two phases after different annealing temperature. The EHEA shows significant resistance to the grain growth with average grain size of ~2 μm at 1200°C.

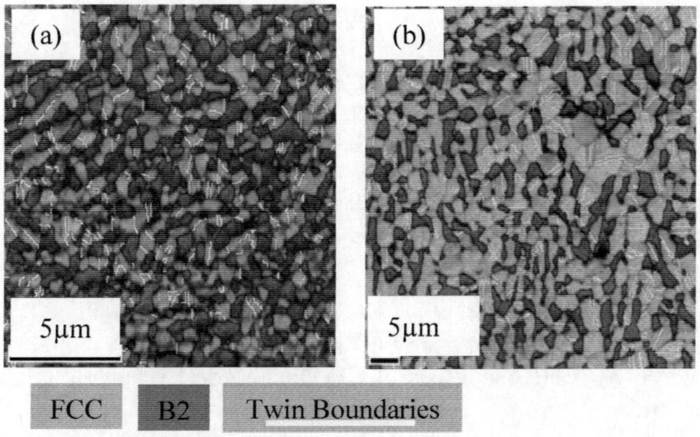

Figure 4. Phase maps of the 90% cold-rolled sample after annealing at (a) 800°C and (b) 1200°C respectively

Table 2: Grain size of the two phases in different annealed materials

90% Cold rolled	Average grain size (μm)
Annealed at 800°C	0.5±0.01
Annealed at 1000°C	0.7±0.05
Annealed at 1200°C	2.0±0.44

The tensile properties of the EHEA in different conditions are summarized in Table 3. The improvement in the mechanical properties of EHEA after severe cold-rolling and annealing is remarkable as compared to other two phase HEAs which undergo structural coarsening leading to significant decrease in strength [9]. It can be concluded that the severe cold-rolling and annealing can impart novel properties in EHEAs and render them suitable for a wide range of industrial applications.

Table 3: Summary of mechanical properties of the EHEA in various conditions

Material Condition	YS(MPa)	UTS(MPa)	e_f%
As-received	620	1050	17
90% Cold-rolled	1625	1800	6
Annealed at 800°C	1108	1200	12
Annealed at 1000°C	844	1175	23
Annealed at 1200°C	648	1075	27

4. Conclusion

The following conclusions may be drawn from the present work:

- The $AlCoCrFeNi_{2.1}$ EHEA consists of $L1_2$ (FCC ordered) and B2 (BCC ordered phase) in the as cast state.
- Severe cold-rolling results in progressive disordering of the $L1_2$ and evolution of a nanocrystalline structure.
- Development of duplex microstructure is observed after annealing at different temperatures.
- Severe cold-rolling and annealing can significantly enhance the strength-ductility combination in The EHEA. Annealing at 1000°C results in an optimum strength-ductility balance as evidenced by a high tensile strength of 1175 MPa and total elongation of 23%.

Acknowledgement

The authors would like to express appreciation for the support of DST, India under Grant SB/S3/ME-47/2013; MEXT, Japan under grant 15H05767; Chalmers University of Technology and Swedish Research Council (2015-04087); People Programme (Marie Curie Actions) of the European Union's Seventh Framework Programme (FP7/2007-2013) under REA grant agreement no 608743.

References

[1] Yeh J W, Chen S K, Lin S J, Gan J Y, Chin T S, Shun T T, Tsau C H and Chang S Y 2004 *Adv. Eng. Mater.* **6** 299-303.
[2] Tsai M H and Yeh J W 2014 *Mater. Res. Lett.* **2** 107-123.
[3] Lu Y, Dong Y, Guo S, Jiang L, Kang H, Wang T, Wen B, Wang Z, Jie J, Cao Z, Ruan H and Li T 2014 *Sci. Rep* **4** 1-5.
[4] Tang Q H, Huang Y, Huang Y Y, Liao X Z, Langdon T G and Dai P Q 2015 *Mater. Lett.* **151** 126-129.
[5] Wani I S, Bhattacharjee T, Sheikh S, Bhattacharjee P. P, Guo S and Tsuji N 2016 *Mater. Sci. Eng. A* **675** 99-109.
[6] Ball J and Gottstein G 1994 *Intermetallics* **2** 205-219.
[7] Ball J. and Gottstein G 1993 *Intermetallics* **1** 171-185.
[8] Wani I S, Bhattacharjee T, Sheikh S, Lu Y P, Chatterjee S, Bhattacharjee P P, Guo S and Tsuji N 2016 *Mater. Res. Lett.* **4** 174-179.
[9] Baker I, Meng F, Wu M and Brandenberg A 2016 *J. Alloys Compd.* **656** 458-464.

Temperature dependent structural evolution in nickel/carbon nanotube composites processed by high-pressure torsion

Andreas Katzensteiner[1], Katherine Aristizabal[2], Sebastian Suarez[2], Reinhard Pippan[1] and Andrea Bachmaier[1]

[1] Erich Schmid Institute, Jahnstr. 12, A-8700 Leoben, Austria
[2] Department of Materials Science, Campus D3.3, D-66123 Saarbrücken, Germany

E-mail: andreas.katzensteiner@oeaw.ac.at, katherine.aristizabal@uni-saarland.de, s.suarez@mx.uni-saarland.de, reinhard.pippan@oeaw.ac.at, andrea.bachmaier@oeaw.ac.at

Abstract. Nickel/Carbon nanotube (CNT) composites with varying amounts of CNTs were processed at different temperatures by high-pressure torsion (HPT) with the aim to optimize the process parameters to obtain a homogenous dispersion of CNTs in the metallic matrix. As the CNT distribution has an enormous influence on the composite properties, the structural evolution with increasing strain and the final microstucture of the composites are investigated by scanning and transmission electron microscopy. Microhardness measurements were additionally performed. Microhardness increases up to 800 Vickers (HV) and the mean grain size decreases to an equivalent radius smaller than 40 nm for HPT at room temperature (RT), while the CNTs form rather large agglomerates. HPT deformation at 200°C shows no significant change in hardness, grain size and CNT agglomerate size. For HPT deformation at 300°C and 400°C grain sizes increase to 60 nm respectively 90 nm, microhardness decreases to 500 HV respectively 400 HV and the size of the CNT agglomerates decreases from more than 5 times the grain size at RT to smaller than the grain size. It could be shown that the optimal HPT processing route to improve the CNT distribution is a combination of deformation at 400°C with subsequent deformation at RT.

1. Introduction

Nanocrystalline (NC) and ultrafine-grained (UFG) materials have been shown to surpass coarse grained materials both in mechanical and in functional utility [1]. During the last couple of decades, severe plastic deformation (SPD) has been used to obtain dense UFG or even NC bulk materials in reasonable volumes [2].

In the majority of the studies SPD processing by high-pressure torsion (HPT) deformation of metal matrix composites (MMCs) showed improved mechanical properties compared to the deformation of pure metals, due to combining matrix and reinforcement properties [3, 4]. These properties are controlled by the size of the reinforcement as well as by its distribution within the matrix [5].

In this study, MMCs with nickel as the metal phase and carbon nanotubes (CNTs) as reinforcement are deformed by HPT. CNTs are well suited for use as reinforcement phase because of their high thermal conductivity, high specific strength, low weight and large aspect ratio [6].

Preinvestigations of SPD deformed Ni and Cu/CNT composites with 1 and 3wt% CNTs and 3 to 4 vol% CNTs have been already performed [7–9]. The present study investigates Ni/CNT MMCs with a greater variety of CNT weight percentages and HPT deformation is additionally conducted at different temperatures to investigate the influence of temperature on the structural evolution. Since CNTs are known to form agglomerates in a metal matrix due to Van der Waals forces [10], a strong emphasis in this study is further laid on obtaining a uniform distribution of small CNT agglomerates. Hence, samples are deformed for different numbers of revolutions as well.

2. Experimental

The initial Ni/CNT MMC material was produced as described in [11]. The samples in the as-fabricated condition had a diameter of 8 mm, a thickness of about 1 mm and contained CNT weight percentages of 0.1, 0.25, 0.5, 1, 2 and 3. These samples were HPT deformed with 0.2 rpm at room temperature (RT) for up to 30 revolutions as described in [12]. Samples were additionally deformed for 30 revolutions at 200°C, 300°C and 400°C with 0.6 rpm, resulting in an equivalent strain of more than 700 at a radius r = 3 mm [13].
Microhardness was measured every 0.25 mm along the radii of the HPT discs (mean value of three indents at each measurement point). The hardness measurements were carried out with a BUEHLER Micromet 5100 using a load of 300 g ($HV_{0.3}$).
Microstructures were characterized in a scanning electron microscope (SEM) type LEO 1525 using back scattered electrons (BSE). Images were taken along the radii of the samples every 0.5 mm to determine the dependence of the microstructural evolution on the equivalent strain. Additionally, HPT deformed samples with 0.1wt% and 1wt% CNTs were prepared for transmission electron microscopy (TEM) investigations (Phillips CM12). HPT deformation parameters were the following: 30 revolutions at 200°C (0.1wt%); 30 revolutions at RT (1wt%); 30 revolutions at 400°C with additional 5 revolutions at RT (1wt%). TEM images were recorded in radial direction for the samples with 1wt% CNTs and in axial direction for the 0.1wt% sample. All TEM samples were obtained at r = 2mm.

3. Results and discussion

With increasing equivalent strain the microhardness increases until it reaches a saturation at high strains for all investigated Ni/CNT MMCs (except the sample containing 3wt% CNTs). This is exemplarily shown in figure 1a for the Ni MMC sample with 1wt% CNTs, which reaches a saturation hardness of 753 HV after 30 revolutions. Independent of the CNT weight percentage, all HPT deformed samples show saturation hardness values above 720 HV after HPT deformation at RT, as can be seen in table 1. The saturation onset for all samples is at a radius of about 3 mm after 30 revolutions, which corresponds to an equivalent strain of about 250 for the Ni/CNT MMC sample containing 1wt% CNTs (figure 1a) and up to an equivalent strain of 400 for other compositions. In pure metals and single phase alloys the saturation regime is reached at a much lower strain [14, 15]. Similar high strains to reach saturation in hardness were needed for HPT-consolidated samples of pure Ni powder [16]. In the Ni/CNT MMCs an additional effect due to the presence of the CNTs has also to be considered. For the sample containing 3wt% CNTs, HPT deformation at RT was limited to 5 revolutions due to the high hardness of the sample. Hence no steady state could be reached in this case.
To investigate the influence of the deformation temperature, HPT samples were deformed at 200°C, 300°C and 400°C. Figure 1b shows the hardness evolution with increasing strain at different deformation temperatures for Ni MMCs with 1wt% CNTs. The sample deformed at 200°C reaches almost the same saturation hardness as the sample deformed at RT, but higher strains are necessary to reach saturation. At 300°C, the saturation hardness decreases to 510 HV. At 400°C, the saturation hardness decreases further (451 HV), but almost the same amount

of strain is needed to reach saturation compared to the sample deformed at RT.

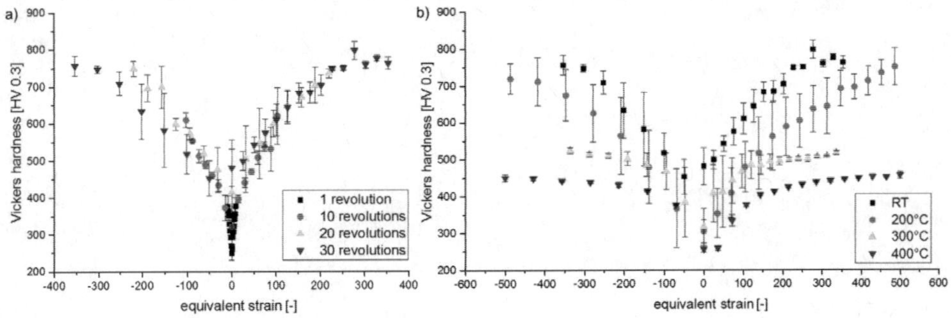

Figure 1. Vickers hardness as function of the applied equivalent strain of Ni MMCs with 1wt% CNTs after a) 1, 10, 20 and 30 revolutions at RT and after b) 30 revolutions at 200°C, 300°C and 400°C.

Table 1 summarizes the saturation hardness values for all compositions and deformation temperatures. For all Ni/CNT MMCs the saturation hardness decreases with increasing deformation temperature. Only for the samples containing 0.5wt% CNTs, nearly the same hardness is reached at RT and 200°C. The Ni/CNT MMC samples with 0.25wt% and 0.5wt% samples have the most distinct decrease in hardness with increasing HPT deformation temperature, the samples containing 2wt% CNTs, the smallest. As an example for the

Table 1. Hardness values for all compositions and deformation temperatures (r = 3 mm).

wt% CNTs	0.1	0.25	0.5	1	2	3
Saturation hardness RT	729±33	724±33	767±22	753±11	724±24	-
Saturation hardness 200°C	604±19	587±37	762±12	712±49	702±19	776±27
Saturation hardness 300°C	501±5	478±6	520±5	510±6	497±4	600±1
Saturation hardness 400°C	401±3	395±3	438±5	451±3	458±8	465±3

microstructural evolution of the Ni/CNT MMCs with increasing strain, figure 2 shows SEM images of the Ni MMCs with 1wt% CNTs deformed at RT at different amounts of applied strain. The size and the distribution of the CNT agglomerates does not significantly change with increasing number of revolutions, while the grain sizes of the Ni matrix decreases. Surprising is that Ni MMCs with different CNT weight percentages show a similar microstructural evolution of the Ni matrix and CNT distribution independent of the CNT content.

Figure 3 displays the saturation microstructure of the Ni MMCs with 1wt% CNTs at different deformation temperatures recorded at r = 3 mm. With increasing HPT deformation temperature, the saturation grain size increases, while CNT agglomerates become smaller and more globular. After deformation at 300°C the CNTs still show some clustering, although these clusters consist of smaller agglomerates (figure 3c). After deformation at 400°C a homogeneous distribution of small CNT agglomerates is achieved (figure 3d). Ni MMC samples containing higher weight percentages of CNTs show a higher density of agglomerates, but there is no significant influence of CNT content on agglomerate size and shape.

Processing of the Ni MMC samples by HPT leads to strong grain refinement. SEM investigations show a decrease of the grain size with increasing number of revolutions and saturation values

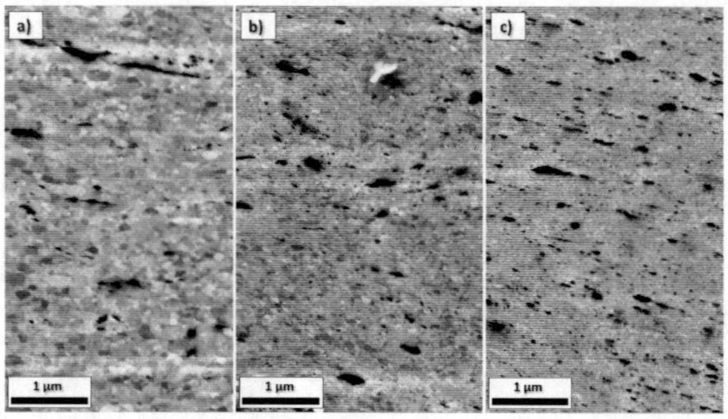

Figure 2. SEM images of Ni MMCs with 1wt% CNTs at r = 3 mm after a) 1, b) 10 and c) 20 revolutions deformed at RT.

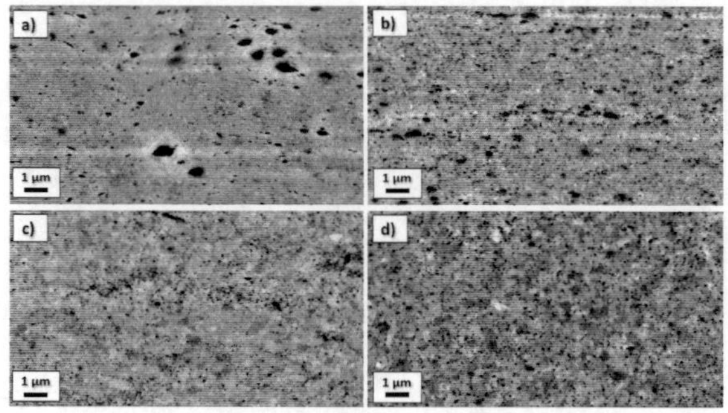

Figure 3. SEM images of Ni MMCs with 1wt% CNTs after 30 revolutions at r = 3 mm at a) RT, b) 200°C, c) 300°C and d) 400°C.

smaller than 40 nm for all compositions after 30 revolutions. However, a slight decrease of the grain size with increasing amount of CNTs is observed.
At a deformation temperature of 200°C, grain sizes increase slightly, but are below 100 nm. For deformation temperatures of 300°C and 400°C grain sizes increase further for nearly all compositions.
Since grains are hardly resolvable in SEM in the saturation region, the microstructure in the steady state of selected samples is additionally investigated in detail by TEM.
Figure 4 shows bright and dark field images with corresponding selected area diffraction (SAD) pattern of the Ni MMC sample with 1wt% CNTs deformed at RT, which was recorded at an equivalent strain of 450. Both images illustrate a nanocrystalline microstructure with grains having a size of about 100 nm and below, showing various defects displayed as contrast variations inside the grains. The grains are clearly elongated along the shear direction (as indicated in figure 4a). In the SAD pattern shown in figure 4a fcc Ni and traces of NiO Debye-Scherrer rings are observed. CNT agglomerates could not be discerned.
The generated microstructures of the Ni MMC sample with the lowest CNT content (0.1wt%) in the saturation regime are shown in bright and dark field images in figure 5. A comparable nanocrystalline structure with few larger but as well very small grains is revealed, where the different grains can hardly be distinguished from each other. The different shape of the grains (equiaxed in this case) is due to the sample being observed in axial direction. In the SAD pattern only fcc Ni rings are visible.

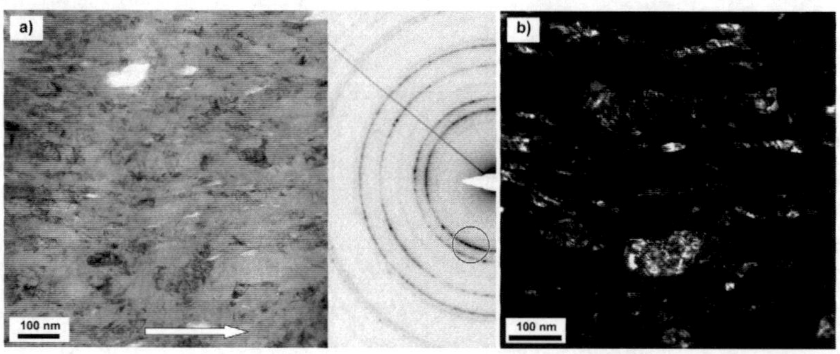

Figure 4. a) Bright field TEM image with corresponding SAD pattern and b) dark field TEM image of Ni MMCs with 1wt% CNTs after 30 revolutions at RT (r = 2 mm). The arrow marks the shear direction.

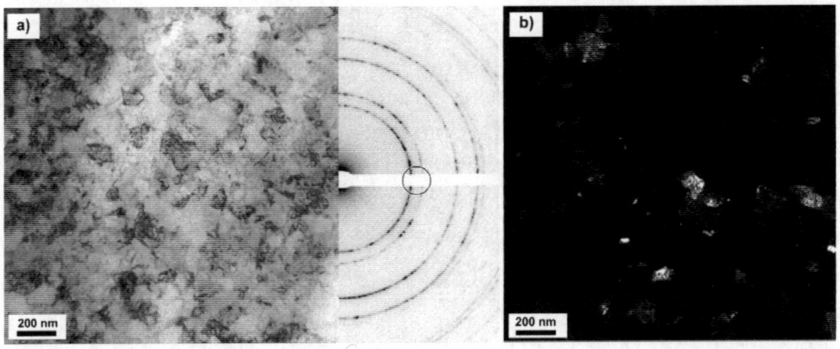

Figure 5. a) Bright field TEM image with corresponding SAD pattern and b) dark field TEM image of Ni MMCs with 0.1wt% CNTs after 30 revolutions at 200°C (r = 2 mm).

Unfortunately, HPT samples deformed at RT show strong crack formation. At deformation temperatures of 300°C and 400°C, no cracks are observed. Furthermore, a better homogeneity of CNT distribution is obtained at 400°C (see figure 3d). A drawback is, however, the decrease of the saturation hardness at this deformation temperature. By contrast, HPT deformation at 200°C prohibits crack formation without significant hardness decrease compared to RT and reduces the grain size compared to 400°C HPT (compare figure 4 and 5).
To achieve a microstructure consisting of a Ni matrix with small grain sizes and small, evenly distributed CNT agglomerates, HPT samples were initially deformed for 30 revolutions at 400°C and subsequently deformed for 5 revolutions at RT. Figure 6 shows the microstructure of a Ni MMC sample with 1wt% CNTs deformed with the aforementioned parameters. In the SEM image (figure 6a) a homogeneous distribution of small, globular CNT agglomerates is visible, while the grains of the Ni matrix are very small. The processed microstructure is also investigated by TEM (figure 6b and c). A nanocrystalline Ni matrix can be seen, in which the grains are below 50 nm in size. The grains are somewhat elongated due to the sample being observed in radial direction (as indicated in figure 6b). Again, no CNT agglomerates could be identified in TEM. The SAD pattern (inset in figure 6b with annotated dark field spot) shows a slight broadening of the Ni diffraction rings, due to the smaller grain size compared to the sample deformed at RT and 200°C (figure 4 and 5). Additionally, weak diffraction rings of NiO are visible.

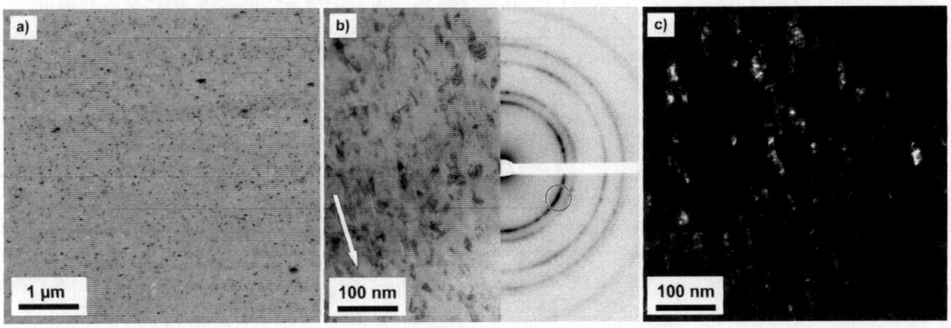

Figure 6. a) SEM image, b) bright field TEM image with corresponding SAD pattern and c) dark field TEM image of Ni MMC with 1wt% CNTs after 30 revolutions at 400°C and 5 additional revolutions at RT (r = 2 mm). The arrow marks the shear direction.

Ni MMC samples deformed for 30 revolutions at 400°C with additional 5 revolutions at RT exhibit the same crack formation tendency as samples deformed at RT. Hence, HPT parameters were adjusted to 30 revolutions at 400°C with additional 10 revolutions at 200°C. First microstructural investigations show that comparable microstructures can be achieved using this deformation parameters with the advantage of no crack formation after HPT deformation. Both samples also exhibit the same high hardness of 769 HV. Further adjustments of the HPT parameters and their influence on microstructural evolution are certainly of interest and are currently conducted. For the analysis of the structural defects present in the CNTs after SPD, Raman spectroscopy investigations are underway.

4. Conclusion

Ni MMCs with differing amounts of CNTs were processed by HPT. It could be shown that grain size and microhardness depend strongly on the number of revolutions, while CNT agglomerate size, shape and distribution change only slightly. The applied equivalent strain has a more pronounced effect on the grain size and hardness than the differing weight percentages of CNTs. Increasing the deformation temperature leads to larger grain sizes and a decrease of the hardness in the steady state. CNT agglomerates become smaller, more globular and more evenly distributed with increasing deformation temperature.

Samples, which were deformed for 30 revolutions at 400°C followed by additional 5 revolutions at RT, exhibit the most homogenous microstructure. Crack formation could be prevented by changing the deformation parameters to additional 10 revolutions at 200°C.

Acknowledgments

A. Katzensteiner and A. Bachmaier gratefully acknowledge the financial support by the Austrian Science Fund (FWF): I2294-N36. S. Suarez and K. Aristizabal gratefully acknowledge the financial support from DFG (Grant: SU911/1-1). Additionally, K. Aristizabal would like to thank the German Academic Exchange Service (DAAD) for financial support.

References

[1] Gleiter H 1989 *Prog. Mater. Sci.* **33** 223–315
[2] Zhilyaev A P and Langdon T G 2008 *Prog. Mater. Sci.* **53** 893–979
[3] Islamgaliev R K, Buchgraber W, Kolobov Y R, Amirkhanov N M, Sergueeva A V, Ivanov K V and Grabovetskaya G P 2001 *Mater. Sci. Eng.* A **319–21** 872–6
[4] Bachmaier A and Pippan R 2011 *Mater. Sci. Eng.* A **528** 7589–95
[5] Bachmaier A and Pippan R 2013 *Int. Mater. Rev.* **58** 41–62
[6] Jorio A, Dresselhaus G and Gresselhaus M S 2008 *Carbon Nanotubes: Advanced Topics in the Synthesis, Structure, Properties and Applications* (Topics in Applied Physics vol 111) ed C E Ascheron and A H Duhm (Berlin: Springer) chapter 5 pp 165–94
[7] Jenei P, Yoon E Y, Gubicza J, Kim H S, Labar J L and Ungar T 2011 *Mater. Sci. Eng.* A **528** 4690–95
[8] Akbarpour M R, Farvizi M, Lee D J, Rezaei H and Kim H S 2015 *Mater. Sci. Eng.* A **638** 289–95
[9] Suarez S, Lasserre F, Soldera F, Pippan R and Mücklich F 2015 *Mater. Sci. Eng.* A **626** 122–7
[10] Reinert L, Zeiger M, Suarez S, Presser V and Mücklich F 2015 *RSC Adv.* **5** 95149–59
[11] Suarez S, Lasserre F, Prat O and Mücklich F 2014 *Phys. Stat. Sol.* A **211** 1555–61
[12] Hohenwarter A, Bachmaier A, Gludovatz B, Scheriau S and Pippan R 2009 *Int. J. Mater. Res.* **100** 1653–61
[13] Stüwe H P 2003 *Adv. Eng. Mater.* **5**, 291–5
[14] An X H, Lin Q Y, Wu S D, Zhang Z F, Figueiredo R B, Gao N and Langdon T G 2011 *Phil. Mag.* **91:25**, 3307–26
[15] Pippan R, Scheriau S, Taylor A, Hafok M, Hohenwarter A and Bachmaier A 2010 *Annu. Rev. Mater. Res.* **40**, 319–43
[16] Bachmaier A, Hohenwarter A and Pippan R 2009 *Scripta Mater.* **61**, 1016–9

IOP Conf. Series: Materials Science and Engineering **194** (2017) 012020 doi:10.1088/1757-899X/194/1/012020

Characterization of microstructure and mechanical property of pure titanium with different Fe addition processed by severe plastic deformation and subsequent annealing

Guanyu Deng[1,*], Tilak Bhattacharjee[1,2], Yan Chong[1], Ruixiao Zheng[1], Yu Bai[1,2], Akinobu Shibata[1,2], Nobuhiro Tsuji[1,2]

[1]Department of Materials Science and Engineering, Kyoto University, Japan
[2]Elements Strategy Initiative for Structural Materials (ESISM), Kyoto University, Japan
E-mail: deng.guanyu.46s@st.kyoto-u.ac.jp, gd577@uowmail.edu.au

Abstract. Titanium and Ti-alloys are widely used in the marine, aerospace, and biomedical industries due to their high strength to weight ratio, excellent corrosion resistance, and biocompatibility. Increasing the strength level of pure Ti via grain refinement without a considerable decrease in ductility is an attractive approach. Fabrication of nanostructured or ultrafine grained (UFG) metals using severe plastic deformation (SPD) techniques has attracted a lot of interest in the last two decades. The main purpose of the present study is to explore the influence of Fe addition in grain refinement in pure Ti. As-cast pure Ti with two different levels of Fe was firstly deformed by high pressure torsion (HPT) up to 10 rotations at room temperature, and then annealed in a vacuum at a temperature of 500 °C for half an hour. Detailed microstructures were characterized by transmission electron microscopy (TEM) and electron backscatter diffraction (EBSD), and mechanical property was examined by microhardness test.

1. Introduction

Severe plastic deformation (SPD) techniques including equal channel angular pressing (ECAP), accumulative roll bonding (ARB) and high pressure torsion (HPT) have drawn significant attention in the last two decades due to their capacity in fabricating bulk metallic materials with nanostructure or ultrafine grained (UFG) structures without changing the dimension of samples [1-3]. Among various SPD techniques, HPT is advantageous for applying very high plastic strains to materials without fracture. During the HPT process, a thin disc sample is placed between two anvils under a high hydrostatic pressure and intense shear strain is introduced by rotating the two anvils with respect to each other. The shear strain γ at a certain radial positionin the disc can be calculated as $\gamma = {}^{2\pi Nr}/_{h}$, where N is the number of HPT rotation, r is the radial distance from the disc center and h is the disc thickness [3].

Ti and Ti-alloys are widely used in the marine, aerospace, and biomedical industries because they have high strength to weight ratio, excellent corrosion resistance and good biocompatibility. Several efforts have been done to fabricate pure Ti with nanostructure or UFG microstructures using HPT process. The first report was conducted by Popov et al. [4] on HPT processing of a commercial pure (CP)Ti (99.53% Ti), and microstructure with anaverage grain size of about 100 nm was obtained. Then, low temperature annealing of those HPT processed pure Ti was further conducted in other studies [5,6] to understand its influence on the mechanical properties. Recently, deformation heterogeneity in a Grade-2 CP Ti during HPT process has been extensively investigated in literatures [7-9] in terms of examining the microstructure and microhardness evolutions. Edalati et al. [10] has studied the influence of applied hydrostatic pressure varying from 1.2 to 40 GPa on the microstructural change, microhardness distribution and phase transformation in HPT processed CP Ti. Their results revealed that the phase transformation from α to ω phase occurred when the pure Ti was deformed under pressure larger than 4 GPa and a larger pressure led to a finer microstructure. Edalati et al. [11] has also studied the effect of grain size and HPT processing temperature on the phase transformation in pure Ti based on experimental observations and first-principles calculations. In addition, phase transformation in HPT deformed pure Ti has also been investigated by Todaka et al. [12], considering the influence of impurity oxygen (O). In another study by Todaka et al. [13], effect of the strain pathwas studied and it has been found that deformations with different strain paths in HPT led to larger Vickers microhardness in pure Ti than monotonic and cyclic HPT processes. Even a number of reports on HPT processing of pure Ti are available, however, influence of impurity in pure Ti on the grain refinement, phase transformation and mechanical properties during HPT process has not been well studied except for the effect of O by Todaka et al. [12]. In the present study, pure Ti with two different levels of impurity Fe, which is the element always included in pure Ti as well as O, is investigated. The pure Ti was firstly deformed by HPT process, and some deformed specimens were further annealed. Microstructure and mechanical property of both deformed and subsequent annealed samples were characterized.

2. Materials and experimental details

The experiments in this study were carried out on as-cast pure Ti samples with two different levels of Fe but the same amount of O, namely Ti-0.04%Fe (Ti-0.04O-0.04Fe in mass%) and Ti-0.25%Fe (Ti-0.04O-0.25Fe in mass%). The average colony size of the lamellar α structures in the as-cast Ti-0.04%Fe and Ti-0.25%Fe was 2.6 mm and 1.5 mm, respectively. Disc samples with a diameter of 10 mm and a thickness of 0.85 mm were cut out from the ingots and provided for the HPT process, which was conducted at room temperature at a speed of 0.5 revolution per minute (rpm) under a compressive pressure of 6 GPa. The disc samples were deformed to 1/4, 1, 5 or 10 HPT rotations (N). The samples processed by 10 HPT rotations (N=10) were selected for further annealing in a vacuum at a temperature of 500 °C for half an hour.

Both the HPT processed and annealed disc samples were firstly polished to obtain mirror-like surfaces. Then the Vickers microhardness was measured with an applied load of 200 g and a dwell time of 15 s across the diameter of each discon the section perpendicular to the normal direction of the disc. Hereafter, RD, SD and ND indicates the radial direction, shear direction and normal direction of the disc, respectively. Deformed microstructures of the pure Ti specimens were characterized by transmission electron microscopy (TEM) at the position of 3 mm away from the disc center using thin foils parallel to the SD-ND plane, and the annealed microstructures were characterized by electron backscatter diffraction (EBSD) at the position of 3 mm away from the disc center on the RD-SD plane.

3. Results and discussion

3.1. Microhardness heterogeneity after HPT process

Figure 1 shows Vickers microhardness distributions in the pure Ti specimens deformed by HPT process up to N=10 rotations. As can be seen in Fig. 1a, the average hardness of the as-cast Ti-0.04% is about 123 HV. The HPT deformation leads to a rapid rise of the micrhardness. When N is less than 5, the microhardness is lowest at the center of the disc and the hardness increases quickly with increasing the radial distance from the center. The heterogeneity in the microhardness distribution in Ti-0.04%Fe decreases gradually with increasing the number of HPT rotation and relatively uniform microhardness distribution is observed after 5 HPT rotations. With further deformation up to 10 rotations, the microhardness increases slightly to about 355 HV. In Fig. 1b, the microhardness of Ti-

0.25%Fe is about 132 HV before HPT deformation and increases to about 370 HV after 10 HPT rotations. It has been found that the microhardness difference along the diameter after 5 HPT rotations is larger in Ti-0.25%Fe than that in Ti-0.04%Fe. The heterogeneous distribution in hardness remains till 5 rotations and the hardness distribution becomes homogeneous only after 10 rotations in Ti-0.25%Fe.

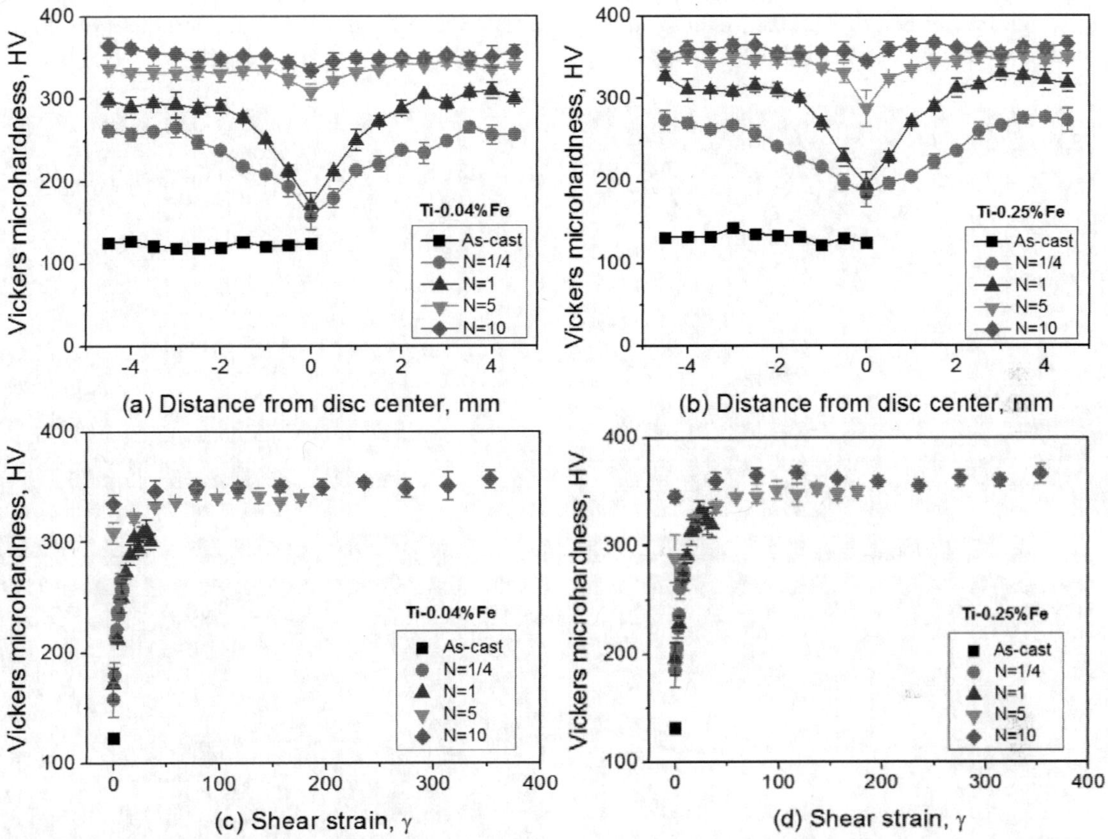

Fig. 1 Vickers microhardness distributions against radial distance from the disc center in the HPT processed (a) Ti-0.04%Fe and (b) Ti-0.25%Fe, and microhardness plotted as a function of shear strain in the HPT processed (c) Ti-0.04%Fe and (d) Ti-0.25%Fe.

The microhardness evolution in the HPT process has been plotted as a function of shear strain (γ) in Fig. 1c and d for Ti-0.04%Fe and Ti-0.25%Fe, respectively. It is evident that the microhardness in Ti-0.04%Fe increases significantly with increasing shear strain up to γ=39.25, and then increases slightly above γ=40. On the other hand, the microhardness in Ti-0.25%Fe increases up to γ=78.5, and then saturates. Therefore, it can be concluded that higher Fe contents in pure Ti leads to larger critical shear strain for the microhardness saturation during HPT process. The final hardness values achieved by γ=353. 25 are 357 HV and 366. 5 HV for Ti-0.04%Fe and Ti-0.25%Fe, respectively. In the previous study by Edalati et al. [10], the microhardness in a commercial pure Ti reached to only about 320 HV, which is much lower than the present value of Ti-0.25%Fe even though the amount of Fe in pure Ti and the HPT processing parameters are the same. The different microhardnesses values might be attributed to the different level of impurity O, which is about 0.2% in Ref. [10] but 0.04% in Ti-0.25%Fe in the present study. It has been reported by Todaka et al. [12] that α to ω phase transformation during HPT processing of pure Ti is suppressed by the increase in O content while the extra-hardening of pure Ti is brought by the formation of ω.

IOP Conf. Series: Materials Science and Engineering **194** (2017) 012020 doi:10.1088/1757-899X/194/1/012020

3.2. Deformed microstructures

Figure 2a and b show TEM microstructures of Ti-0.04%Fe HPT processed to N=1/4 and N=10 rotations, respectively. The shear strains at the positions of observations are γ=5.89 and 235.5, respectively. After N=1/4 rotation, ultrafine lamellar microstructures elongated along the shear direction indicated by white arrows in Fig. 2a are observed. When the Ti-0.04%Fe sample is further deformed up to 10 HPT rotations, more equiaxed nanostructure with unclear boundaries develops. The average grain size of the equiaxed nano-grains is estimated as 125 nm. Both nanostructures in Fig.2a and 2d involve high density of dislocations.

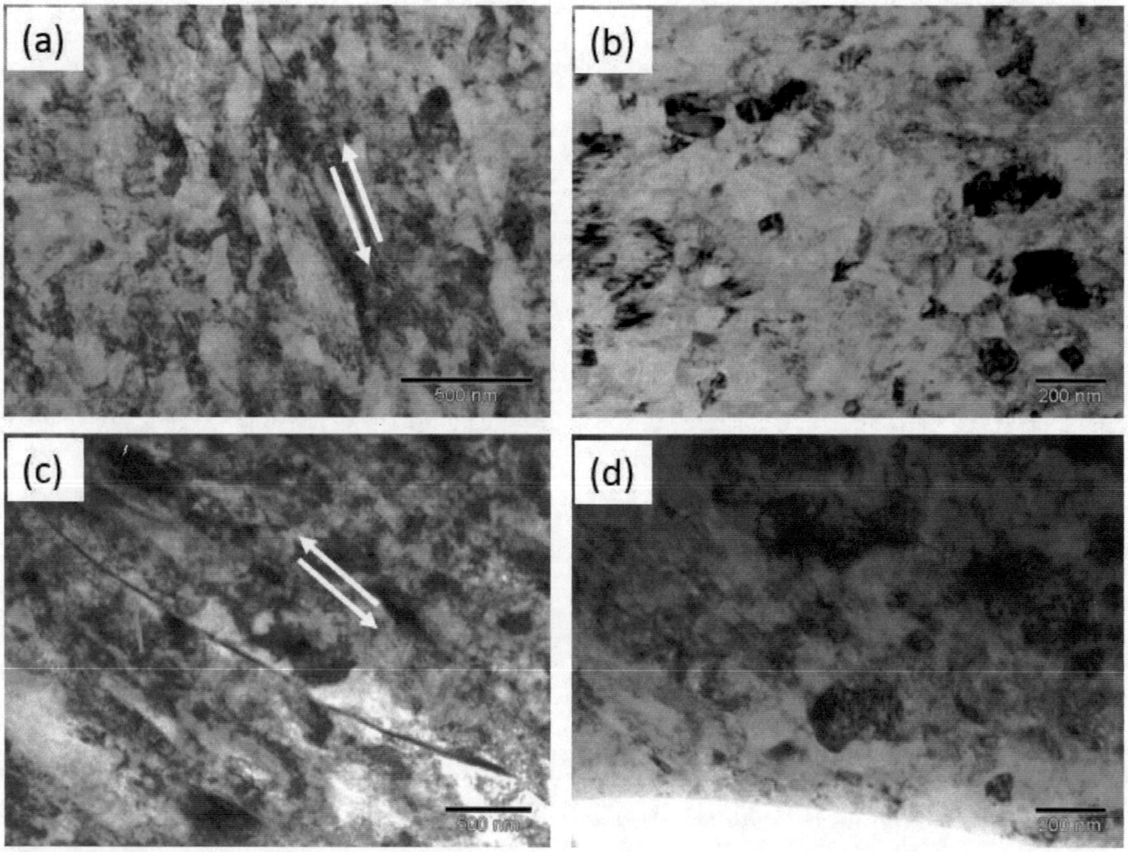

Fig. 2 TEM microstructures of Ti-0.04%Fe after (a) N=1/4 and (b) N=10 HPT rotations, and of Ti-0.25%Fe after (c) N=1/4 and (d) N=10 HPT rotations. The microstructures were observed at 3 mm radial position from the center on the SD-ND planes of the HPT discs. The white arrows in (a) and (c) indicate the shear direction in HPT, and the blue arrow in (c) points out the original β phase.

TEM microstructures of Ti-0.25%Fe HPT deformed by N=1/4 and N=10 rotations are shown in Fig.2c and 2d, respectively. Similar microstructure evolution to that in Ti-0.04%Fe is observed. Ultrafine lamellar microstructures develop first (Fig.2c) and they are broken into equiaxed nano-grains when the shear strain becomes very large (Fig.2d). The nanostructure contains large densities of dislocations and the grain boundaries are ill-defined and relatively diffuse. The average grain size of the equiaxed nano-grains in Ti-0.25%Fe is about 100 nm. The microstructures in Fig. 2b and d are similar to the nanostructures shown in early reports [1-13] on HPT processing of commercial pure Ti, where the average grain size varies from ~54 nm to 300 nm depending on the chemical compositions, disc dimension, applied hydrostatic pressure, number of HPT rotations and the rotating speed.

Comparison between Fig. 2b and Fig. 2d indicates that the Fe addition in pure Ti is effective in grain refinement in SPD.

3.3. Annealing after HPT process
As observed in Fig. 2, the deformed microstructures in pure Ti after the HPT process contain high densities of dislocations, which are typical nanostructures obtained by SPD. However, such deformed characteristics is one of the main reasons for limited tensile ductilities in SPD processed materials. On the other hand, fully recrystallized nanostructures managing both high strength and good ductility have been reported in some materials [14-16]. For example, Zheng et al. [17] have recently reported fully recrystallized UFG structures exhibiting excellent strength and tensile ductility in a Mg-Zn-Zr-Ca alloy through HPT and subsequent annealing under adequate conditions. In order to obtain fully recrystallized UFG structures in the present materials, the samples after 10 HPT rotations were further annealed at 500 °C for 30 minutes.

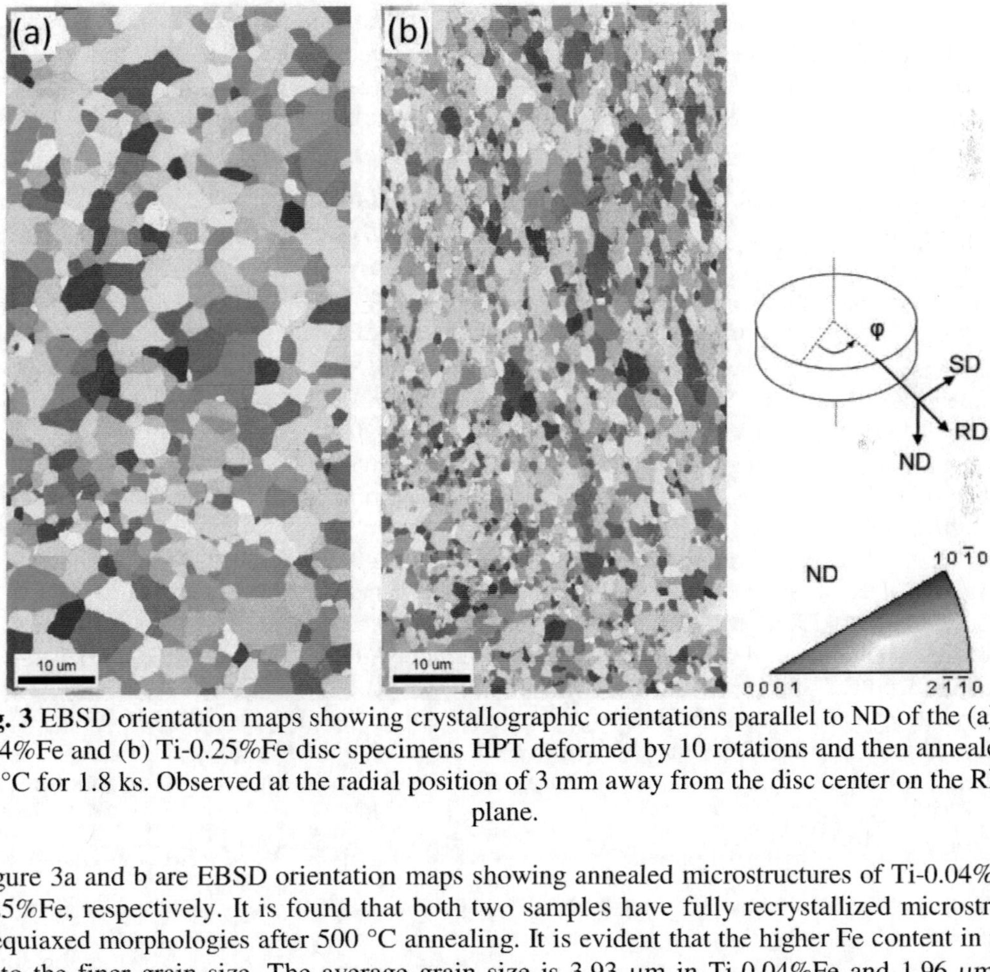

Fig. 3 EBSD orientation maps showing crystallographic orientations parallel to ND of the (a) Ti-0.04%Fe and (b) Ti-0.25%Fe disc specimens HPT deformed by 10 rotations and then annealed at 500 °C for 1.8 ks. Observed at the radial position of 3 mm away from the disc center on the RD-SD plane.

Figure 3a and b are EBSD orientation maps showing annealed microstructures of Ti-0.04%Fe and Ti-0.25%Fe, respectively. It is found that both two samples have fully recrystallized microstructures with equiaxed morphologies after 500 °C annealing. It is evident that the higher Fe content in pure Ti leads to the finer grain size. The average grain size is 3.93 μm in Ti-0.04%Fe and 1.96 μm in Ti-0.25%Fe, respectively.

The Vickers microhardness distribution in the annealed pure Ti discs is quite uniform and the average value is 153 HV in Ti-0.04%Fe and 166 HV in Ti-0.25%Fe, respectively. Significant decrease in the microhardness compared to that in the as-deformed samples is mainly attributed to recrystallization and grain growth as well as phase transformation during subsequent annealing after HPT deformation. It has been reported that ω phase formed in HPT deformation in pure Ti transforms to α phase when the annealing temperature is higher than 150 °C [10,12]. However, the influence of Fe addition on the ω to α phase transformation temperature needs to be systematically studied in future.

4. Conclusions

Pure Ti with two different Fe contents were deformed at room temperatureby HPT process up to 10 rotations. It was found that the higher Fe content led to the higher Vickers microhardness in the HPT processed discs and also the larger critical shear strain required for the microhardness saturation. Ultrafine lamellar microstructures were observed after 1/4 HPT rotation and equiaxed nano-grains were obtained after 10 HPT rotations in both materials. The average grain size in pure Ti having the larger amount of impurity Fe was smaller at both as-HPT deformed and subsequently annealed (fully recrystallized) states.

Acknowledgements

The first author would like to acknowledge the Japan Society for the Promotion of Science (JSPS) for awarding him a fellowship. This work was financially supported by the Elements Strategy Initiative for Structural Materials (ESISM), and the Grant-in-Aid for Scientific Research (S) (No.15H05767), both through the Ministry of Education, Culture, Sports, Science and Technology (MEXT), Japan.

References

[1] Valiev RZ, Langdon TG. Prog Matet Sci 51 (2006) 881-981.
[2] Saito Y, Ulsunomiya H, Tsuji N, Sakai T. Acta Mater 47 (1999) 579-583.
[3] Zhilyaev A, Langdon TG. Prog Matet Sci 53 (2008) 893-979.
[4] Popov AA, Pyshmintsev IY, Demakov SL, Illarionov AG, Lowe TC, Sergeyeva AV, Valiev RZ. Scripta Mater 37 (1997) 1089-1094.
[5] Sergueeva AV, Stolyarov VV, Valiev RZ, Mukherjee AK. Scripta Mater 45 (2001) 747-752.
[6] Valiev RZ, Sergueeva AV, Mukherjee AK. Scripta Mater 49 (2003) 669-674.
[7] Shirooyeh M, Xu J, Langdon TG. Mater Sci Eng A 614 (2014) 223-231.
[8] Wang CT, Fox AG, Langdon TG. J Mater Sci 49 (2014) 6558-6564.
[9] Chen YJ, Li YJ, Walmsley JC, Gao N, Roven HJ, Starink MJ, Langdon TG. J Mater Sci 47 (2012) 4838-4844.
[10] Edalati K, Matsubara E, Horita Z. Metall Mater Trans A 40 (2009) 2079-2086.
[11] Edalati K, Daio T, Arita M, Lee S, Horita Z, Togo A, Tanaka I. Acta Mater 68 (2014) 207-213.
[12] Todaka Y, Sasaki J, Moto T, Umemoto M. Scripta Mater 59 (2008) 615-618.
[13] Todaka Y, Umemoto M, Yamazaki A, Sasaki J, Tsuchiya K. Mater Trans 49 (2008) 47-53.
[14] Saha R, Ueji R, Tsuji N. Scripta Mater. 68 (2013) 813-816.
[15] Tian YZ, Zhao LJ, Chen S, Shibata A, Zhang ZF, Tsuji N. Sci. Rep. 5 (2015) 16707.
[16] Zhao L, Park N, Tian Y, Shibata A, Tsuji N. Sci. Rep. 6 (2016) 39127.
[17] Zheng R, Bhattacharjee T, Shibata A, Sasaki T, Hono K, Joshi M, Tsuji N. Scripta Mater. 131 (2017) 1-5.

IOP Conf. Series: Materials Science and Engineering **194** (2017) 012021 doi:10.1088/1757-899X/194/1/012021

Ultra-fine grained microstructure of metastable beta Ti-15Mo alloy and its effects on the phase transformations

K Václavová[1], J Stráský[1], P Zháňal[1], J Veselý[1], V Polyakova[2], I Semenova[2], M Janeček[1]

[1] Department of Physics of Materials, Charles University in Prague, Ke Karlovu 5, 121 16 Prague, Czech Republic
[2] UFA State Aviation Technical University, Institute of Physics of Advanced Materials, K. Marx Street 12, Ufa 450 000, Russia

E-mail: kristina.vaclavova@gmail.com

Abstract. Processing of metastable titanium alloys by severe plastic deformation provides an opportunity to achieve exceptional grain refinement, to enhance the strength and to affect phase transformations occurring during thermal treatment. The main aim of this study is to investigate the microstructure of ultra-fine grained (UFG) material and effect of microstructural changes on phase transformations in metastable β-Ti alloy Ti-15Mo.
Metastable β-Ti alloys are currently the most studied Ti-based materials with prospective use in medicine. Ti-15Mo alloy after solution treatment contains metastable β-phase. Metastable ω-phase and stable α-phase particles are formed upon annealing,.
Solution treated Ti-15Mo alloy was deformed by high pressure torsion (HPT) at room temperature. Severely deformed structure after HPT with grain size of ~200 nm was studied by transmission electron microscopy. In-situ electrical resistance measurements showed significant changes in undergoing phase transformations when compared to coarse-grained (CG) material. Scanning electron microscopy revealed heterogeneous precipitation of α-particles at grain boundaries (GB). Due to the high density of GBs in UFG structure, these precipitates are very fine and equiaxed.
The study demonstrates that SPD is capable of enhancing mechanical properties due to grain refinement and via affecting precipitation processes in metastable β-Ti alloys.

1. Introduction

Ultra-fine grained (UFG) metallic materials processed by severe plastic deformation (SPD) have attracted significant attention due to the enhancement of mechanical properties as compared to coarse-grained (CG) counterparts [1], which make UFG materials favorable for industrial and medical use [2, 3]. High pressure torsion (HPT) and equal channel angular pressing (ECAP) are the most widely used SPD techniques, producing extreme grain refinement (grain size below 100 nm) [4, 5].

Metastable β-titanium alloys gained significant interest thanks to their excellent mechanical properties [6, 7]. Ti-15Mo (titanium alloy with 15 wt% of molybdenum) is a thoroughly investigated simple binary alloy and a promising candidate for biomedical use [8]. Upon thermal treatment it undergoes several phase transformations. The solution treated (ST) material consists of β-phase with body-centered cubic structure and metastable ω-phase with hexagonal symmetry [9]. The ω-phase is formed already during quenching [10] and is referred to as ω_{ath} (ω athermal). ω_{ath}-phase particles are coherent with the β matrix and their diameter is only few nanometers [11]. Upon heating ω-particles become stabilized by diffusion process by expelling Mo atoms (as β-stabilizing element) – this phase being referred to as ω_{iso} (ω isothermal). The ω_{iso} – phase particles grow during annealing and simultaneously become more chemically stabilized [12]. Further annealing of the material leads to precipitation of α-phase particles. In some studies, it is claimed that ω_{iso}-particles serve as nucleation sites for precipitation of the α-phase [13]. Alternative preferential precipitation sites of α-phase particles are β/β grain boundaries.

The main aim of this study is to investigate the effect of the UFG structure in the Ti-15Mo alloy on the ongoing phase transformations in the material upon heating.

2. Material and experimental

Ti-15Mo alloy supplied by Carpenter Technology Corp. in a form of a rod with the diameter of 10.5 mm was used for this study. The as-received rod was solution treated (810 °C, 20 min) in a protective Ar atmosphere followed by a water quench. It was further cut to cylinders (diameter 10.5 mm, height approx. 5 mm) and pressed in HPT machine at room temperature to obtain the disk-shaped specimen of the diameter of 20 mm. The principle of HPT method is described in detail in [1]. Samples with the thickness of 1 mm were prepared by HPT at Ufa State Aviation Technical University (USATU) Ufa, Russia at room temperature and the pressure of 2 GPa. Sample processed by a single HPT rotation was used for all investigations presented in this study.

Scanning electron images using back-scattered electrons were performed using Zeiss Auriga Compact SEM equipped with the field emission gun (FEG) operated at 10 kV. Samples for SEM observations were prepared by mechanical grinding and polishing followed by a three-step vibratory polishing.

Transmission electron microscope (TEM) Jeol JEM 2200 operated at 200 kV was used for detail microstructure observation. Thin foils for TEM observations were prepared by electrochemical thinning using Tenupol-5 jet polishing unit at -20°C followed by ion polishing using Leica EM RES102 ion polisher.

The four-point method was employed for electrical resistance measurements of samples during heating [14]. This experimental set-up allows simultaneous measurement of the voltage and the electrical current. The samples of the thickness of approximately 0.7 mm and the size of 6.5x15 mm were cut to a special 'U' shape [15] and mechanically grinded. The electrical resistance of the samples was measured in-situ during linear heating up to 750 °C and 800 °C with the heating rate of 5 °C/min.

3. Results and discussion

3.1. Microstructure of HPT-deformed Ti-15Mo alloy

TEM images in Figure 1 show the periphery part of the sample of the HPT-processed (N = 1) Ti-15Mo alloy. Figure 1(a) shows the bright field (BF) image of the severely deformed microstructure. The corresponding dark field (DF) image from the β(110) reflection as well as selected area electron diffraction (SAED) pattern is shown in the Figure 1(b). The non-uniform contrast in the bright regions in Figure 1(b) and a distortion of a diffraction spots in SAED pattern reveal the lattice distortion associated with the high internal stress [16]. The individual grains in the UFG microstructure cannot be distinguished due to their overlapping through the thickness of the sample and the bright areas in the DF image are parts of grains rather than individual grains [17]. No ω_{ath}-particles were observed by TEM, which is in contrast to statements in previous report [16]. Wang et al. explained the absence of the ω-phase in the UFG material by occurrence of reverse ω-β transformation due to ultra-fine grains and high dislocation density [18].

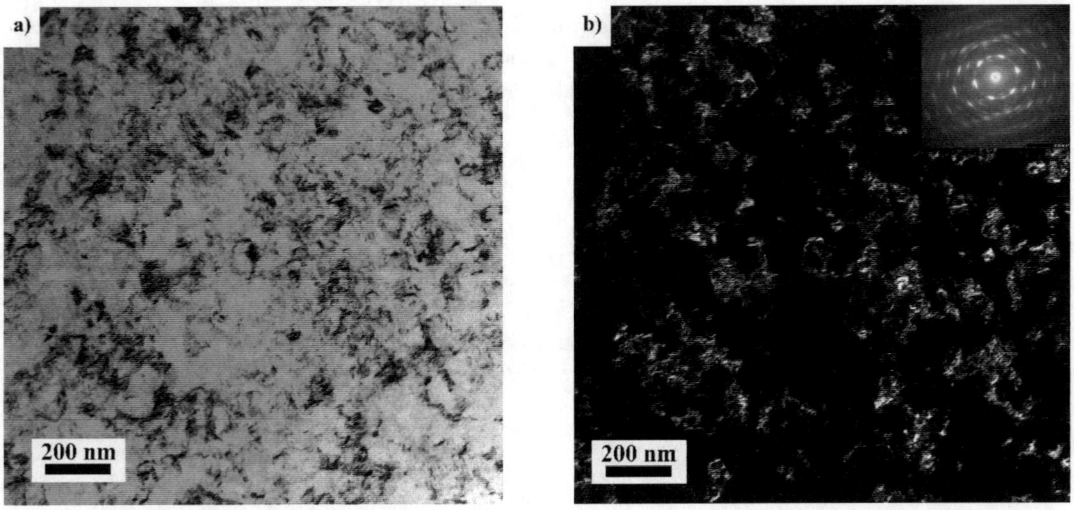

Figure 1. Deformed microstructure of HPT deformed Ti-15Mo alloy – periphery part of the sample (a) bright field image, (b) corresponding dark field image from β(110) reflections.

3.2. In-situ electrical resistance measurement

The relative electrical resistance was measured during two heating cycles for each studied specimen. Each run consisted of heating up to 750 °C and cooling back to room temperature. The evolution of relative resistance ($R(T)/R_0$), where $R(T)$ is the electrical resistance measured at the temperature T and R_0 is the resistance at room temperature after the second cycle, was studied. The selection of 'normalization constant' R_0 is based on assumption that after two heating cycles runs, the material is fully recovered and recrystallized and therefore both original HPT processed and solution treated conditions are comparable. Figure 2 shows the evolution of relative resistance of solution treated (solid line) and HPT processed (dashed line) Ti-15Mo alloy during linear heating up to 750 °C (first cycle) with the constant rate of 5 °C/min. The resistance evolution of ST specimen may be divided into four temperature zones (regions). Initially, the relative resistance of ST material decreases to the temperature of 255 °C – this stage is referred to as zone I. The decline in R/R_0 can be attributed to the dissolution of the ω_{ath} [9, 19]. Near the coherent β/ω interfaces elastic strain field is formed which acts as scattering zone for conducting electrons. In the absence of these interfaces (after dissolution of ω_{ath}) the elastic strain field is released which allows smooth electron drift associated with the decrease of the electrical resistance [20]. The electrical resistance of the HPT material also decreases in zone I. However, the decrease is less pronounced than in the ST specimen. It suggests that the HPT material contains less ω_{ath} particles due to high deformation of the parent β-phase. On the other hand, some ω_{ath} must be present in the material after HPT to cause initial resistance decrease.

In the zone II (255°C – 360°C) the relative resistance of ST material increases with increasing temperature. In this range the ω_{iso}-particles grow due to diffusion-driven chemical stabilization [11]. In contrast to the previous case, the amount of β/ω interfaces increases raising the electrical resistance [21]. In the HPT deformed material, the resistance increase is again much less pronounced when compared to the ST condition. The lack of ω_{ath} nuclei reduce the formation of ω_{iso}-particles and furthermore the outset of the recovery processes may reduce the resistance increase.

The decrease of relative resistance in zone III (360°C - 560°C) in ST material is attributed to the coarsening and continuous dissolution of ω_{iso}-particles in the ST material [22]. The decrease is even more significant in the HPT deformed material. Considering that the HPT deformed material contains lower fraction of ω_{iso}-phase, the resistance decrease is therefore related mainly to the recovery and recrystallization.

The abrupt increase of the relative resistance of ST material above 560 °C (zone IV) was observed. Distinct change of slope at 560°C is associated with the sudden dissolution of ω_{iso}-particles [9]. On the other hand in HPT material, the resistance decrease becomes slower between 450°C and 500°C, which suggests that the recovery and recrystallization processes become exhausted. In the zone IV, the resistance increase in the ST material is linear, which corresponds to the common electron phonon interaction in the absence of any phase changes. It is therefore assumed that α-phase particles do not homogeneously precipitate in the coarse-grained material during heating with the rate of 5 °C/min. On the other hand, small peak in the zone IV observed in resistivity evolution of HPT deformed material suggest increasing volume fraction of α-phase up to the temperature of 650°C. This behavior is typical also for other β-Ti alloys, e.g. TIMETAL LCB [23].

The effect of the ultra-fine grained microstructure prepared by HPT on the phase transformations upon thermal treatment was also studied in other materials [24].

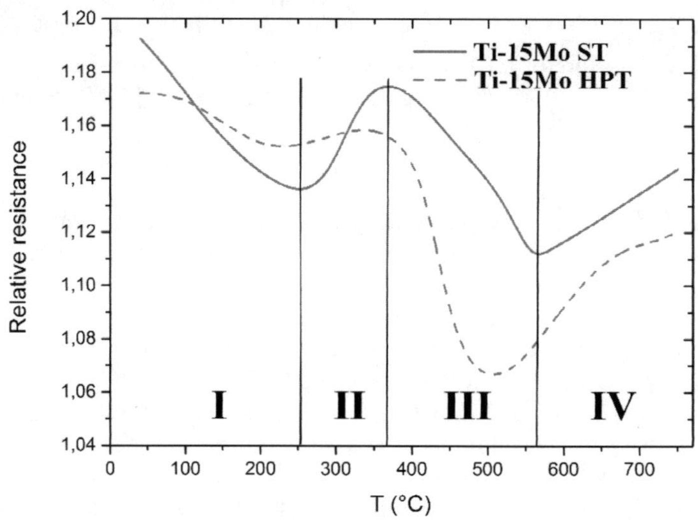

Figure 2. The dependence of the electrical resistance on the temperature of heating for ST and HPT-processed one.

3.3. Microstructure of annealed Ti-15Mo alloy after HPT deformation

Based on the measurement of the dependence of the electrical resistance on the heating temperature, the annealing temperatures of 400°C and 500 °C and annealing periods of 1 and 16 h were chosen for the investigation of the microstructure by SEM. Samples from the periphery part of the HPT material after N=1 HPT rotation were annealed to chosen temperatures in Ar atmosphere and subsequently water quenched. The back-scattered electron images in the Figure 3 clearly show the precipitation of the α-phase in the HPT deformed and annealed samples. It can be observed that the α-phase precipitates form already after annealing at 400 °C for 1h (Figure 3(a)). The α-particles are ultra-fine and equiaxed. Coarsening and further growing of α-particles can be observed after further annealing, see the Figure 3(b). After annealing at 500°C for 1 and 16 hours (Figure 3(c) and (d)) the α-phase particles are larger; moreover, enhanced contrast suggests that α-phase is more chemically stabilized (i.e. Mo depleted). Associated Mo diffusion to β-phase causes its stabilization and may cause reverse ω to β transformation.

The size of α-phase particles significantly differ in different regions. The origin of such more-refined and less-refined areas remains unclear. The size of the α-phase is still in sub-micrometer range. Note,

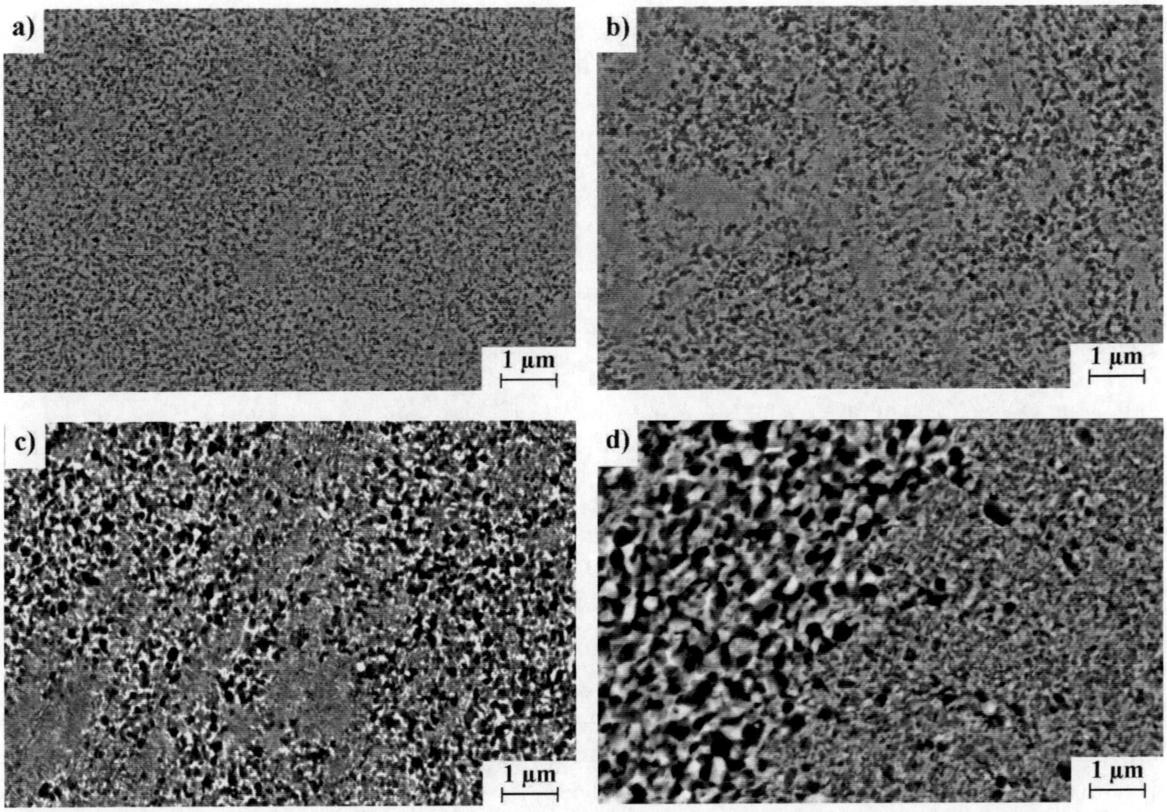

Figure 3. SEM backscattered images of Ti-15Mo alloy on the periphery part of the sample after HPT deformation and annealing (a) at 400 °C for 1 h, (b) at 400 °C for 16h, (c) at 500 °C for 1 h, (d) at 500 °C for 16h.

that no α-phase form in some areas (for instance in Figure 3(b)). Jiang et al. [16] claim that α-phase precipitates in so-called shear bands which are formed during HPT straining. These shear bands are surrounded by alpha-free zones. For the detail explanation of this phenomenon further investigation is needed by in-situ methods.

The shape of α-precipitates is rather polygonal than round or plate-like. It is well known that α-phase precipitate heterogeneously at GBs [23, 25]. Therefore, we assume that all α-phase particles formed during annealing of HPT processed material precipitated as grain boundary α-phase.

4. Conclusions

The microstructure of ultra-fine grained metastable beta Ti-15Mo alloy and its effect on the ongoing phase transformations in the material were investigated. The following conclusions can be drawn from this experimental study:

- Transmission electron microscopy proved severely deformed nanocrystalline microstructure.
- In-situ electrical resistance measurement indicated that ultra-fine grained structure has a significant effect on the phase transformations in metastable β-Ti alloys. High deformation and easy precipitation of GB α-phase suppress the formation of ω_{ath}-phase and subsequent evolution of ω_{iso}-phase, respectively. More exact explanation of ongoing processes requires further investigations, in particular in-situ scanning and transmission electron microscopy.
- Annealed samples at 400°C and 500 °C contain the high volume fraction of homogeneously distributed, sub-micrometer sized equiaxed α-phase particles which precipitate as GB α-phase.

This study shows that SPD processing is capable of complete alteration of phase transformations and precipitation sequence in metastable β-Ti alloys. The resulting microstructural condition is promising for achieving high strength.

Acknowledgement
This work was financially supported by Czech Science Foundation under the grant 17-20700Y. K.V. also acknowledges for the financial support by the Grant Agency of Charles University GAUK project No. 1106216.

References

[1] Valiev R Z, Korznikov A V and Mulyukov R R 1993 *Mater. Sci. Eng. A* **168** 141-48
[2] Geetha M, Singh A K, Askomanani R and Gugia A K 2009 *Prog. Mat. Sci.* **54** 397-425
[3] Long M and Rack H J 1998 *Biomaterials* **19** 1621-39
[4] Zhilyaev A P and Langdon T G 2008 *Prog. Mater. Sci.* **53** 893-979
[5] Valiev R Z, Islamgaliev R K and Alexandrov I V 2000 *Prog. Mater. Sci.* **45** 103-89
[6] Suresh K S, Gurao N P, Satyaveer Singh D, Suwas S, Chattopadhyay K, Zhrebtsov S V and Salishchev G A 2013 *Mater. Char.* **82** 73-85
[7] Zhang Z B, Hao Y L, Li S J and Yang R 2013 *Mat. Sci. Eng. A* **577** 225-33
[8] Williams D F 2001 *Titanium for medical applications* (Berlin: Springer-Verlag)
[9] Zháňal P, Harcuba P, Šmilauerová J, Stráský J, Janeček M, Smola B and Hájek M 2015 *Acta Phys. Pol. A* **128** 779
[10] de Fontaine D, Paton N E and Williams J C 1971 *Acta Metall.* **19** 1153-62
[11] Ng H P, Devaraj A, Nag S, Bettles C J, Gibson M, Fraser H L, Muddle B C and Banerjee R 2011 *Acta Mater.* **59** 2981-91
[12] Devaraj A, Nag S, Srinivasan R, Williams R E A, Banerjee S, Banerjee R and Fraser H L 2012 *Acta Mater.* **60** 596-609
[13] Prima F, Vermaut P, Texier G, Ansel D and Gloriant T 2006 *Scr. Mater.* **54** 645-48
[14] Hájek M, Veselý J and Cieslar M 2007 *Mater. Sci. Eng. A.* **462** 339-42
[15] Václavová K 2013 *Experimental characterization of ultrafine - grained Ti alloys for biomedical use* (Charles University, Prague: Bachelor's thesis)
[16] Jiang B, Tsuchiya K, Emura S and Min X 2014 *Mater. Trans.* **55** 877-84
[17] Popov A A, Pyshmintsev I Y, Demakov S L, Illarionov A G, Lowe T C, Sergeyeva A V and Valiev R Z 1997 *Scr. Mater.* **37** 1089-94
[18] Wang Y B, Zhao Y H, Lian Q, Liao X Z, Valiev R Z, Ringer S P, Zhu Y T and Lavernia E J 2010 *Scripta Mater.* **63** 613-16
[19] Williams J C and Blackburn M J 1980 *Met. Soc AIME* **2** 1309
[20] Rossiter P L 1991 *The electrical resistivity of metals and alloys* (Cambridge Univ. Press)
[21] Sun F, Prima F and Gloriant T 2010 *Mater. Sci. Eng. A* **527** 4262-69
[22] Prima F, Vermaut P, Thibon I, Ansel D, Debuigne J and Gloriant T 2002 *J. Metastable Nanocryst. Mater.* **13** 307
[23] Šmilauerová J 2012 *Phase transformations and microstructure changes in TIMET LCB alloy* (Charles University, Prague: Master's thesis)
[24] An X H, Lin Q Y, Sha G, Huang M X, Ringer S P, Zhu Y T and Liao X Z 2016 *Acta Mater.* **109** 300-13
[25] Kokuoz B 2008 *The role of crystallographic relationships between alpha and beta phases on the elevated temperature isothermal phase transformation kinetics in TIMETAL LCB (Ti-6.5Mo-4.5Fe-1.5Al)* (Clemson University: Master's thesis)

Structural stability of ultra-fine grained magnesium alloys processed by equal channel angular pressing

M. Janeček, T. Krajňák, P. Minárik, J. Čížek, J. Stráská, J. Stráský

Charles University, Faculty of Mathematics and Physics, Ke Karlovu 3, 121 16 Praha 2, Czech Republic

E-mail: janecek@met.mff.cuni.cz

Abstract. Structural stability of two magnesium alloys AZ31 (MgAlZn) and AX41 (MgAlCa) processed by equal channel angular pressing is investigated. The mechanical properties, microstructure evolution and dislocation density were studied by microhardness, electron back scatter diffraction and positron annihilation spectroscopy, respectively. The loss of microstructure stability at high temperatures and the coarsening of the ultrafine-grained structure as a result of isochronal annealing is accompanied by the sharp decrease of microhardness and the decrease of dislocation density. The differences in thermal stability of both alloys are related to different conditions of ECAP pressing and the phase stability, namely the presence of stable Al_2Ca precipitates in AX41 alloy. Microscopic mechanisms controlling the structure stability of both alloys are discussed.

1. Introduction

Magnesium and its alloys are light-weight metallic materials which belong to prospective materials that have extensive application potential as structural components in many industrial branches, e.g. in automobile and aircraft industry. Due to the low density and relatively good mechanical properties magnesium-based materials have been increasingly demanded in weight-critical applications where they replace aluminum, steel and other traditional materials. The interest in magnesium-based materials have recently been revived primarily due to their gradually decreasing costs and the effort of scientists, researchers and engineers to reduce energy consumption and greenhouse gas emissions [1].

The mechanical and other physical properties determining the application of magnesium alloys may be significantly improved by refining the grain size down to submicrometer ($d<1\mu m$) or even nanometer ($d<100$ nm) level by severe plastic deformation (SPD) [2].

A wide variety of special techniques have been proposed and successfully employed for the production of bulk ultra-fine grained (UFG) materials, e.g. equal channel angular pressing (ECAP) [3], high pressure torsion (HPT) [4], accumulative roll-bonding (ARB) [5], twist extrusion [6], multidirectional forging [7], etc.

Among these techniques, which introduce the severe plastic deformation in the material and result in exceptional grain refinement, ECAP, for the first time reported by Segal [8] already more than 40 years ago, became a very popular and extensively used technique due to its versatility, scalability and namely the effectiveness in producing UFG structure in a wide variety of materials with different crystal structures [9,10].

However, the application of UFG materials is limited due to the structure stability at elevated temperatures and concomitant processes of recovery and recrystallization which significantly limit the application potential of these materials. Thermal stability depends on many microstructural parameters, e.g. the phase composition, the stacking fault energy, conditions of material preparation, properties of grain boundaries, etc. Investigation of thermal stability is therefore a challenge to obtain a full portrait of properties of materials after SPD. Unfortunately, this has been done only rarely [11]. Poor thermal stability was found in pure FCC metals Cu, Al [12, 13]. Enhanced thermal stability was reported in Cu- and Al-alloys and attributed to the grain boundary stabilization role of alloying elements [14,15]. However, detail studies describing the positive effect of alloying elements on grain growth suppression in magnesium alloys are only sporadic [16, 17].

The objective of this work is the investigation of the structure stability of two commercial magnesium alloys by a detail characterization of the microstructure, dislocation density development and the loss of stability of UFG structure as a result of isochronal annealing and to correlate these properties with microhardness evolution.

2. Experimental procedures

Two commercial cast alloys were used in this investigation, namely the AX41 (Mg - 4 wt.% Al - 1 wt.% Ca) and the AZ31(Mg - 3 wt.% Al - 1 wt.% Zn). The alloy AZ31 was subsequently extruded with the extrusion ratio of 22:1 at 350 °C and a speed of 60 mm/min. Homogeneous ultrafine-grained structure of both materials consisting of fine equiaxed grains was produced by ECAP following route B_C [18] at the temperature of 180 °C and the velocity of 50 mm/min. by 4 passes, and 220 °C, 20 mm/min. by 8 passes in AZ31 and AX41 alloy, respectively. These different processing procedures were employed purposely to obtain approximately the same grain sizes in both alloys (d=1.4±0.1 μm).

Thermal stability of the ultrafine-grained microstructure was investigated by isochronal annealing of the respective samples in the temperature range of 150-500 °C for 1 h followed by a water-quench. Characterization of the microstructure prior and after annealing was performed by scanning electron microscopy, electron diffraction and positron annihilation spectroscopy (PAS) [19]. The evolution of mechanical properties with the temperature of annealing was determined by Vickers microhardness. Both the microstructure investigations and the microhardness measurements were performed on the cross section of the billets lying perpendicular to the pressing direction (plane X) [20]. The samples for SEM observations and microhardness measurements were first mechanically grinded and polished using diamond suspensions (particle size down to 250 nm) to the mirror like surface quality. Specimens for PAS measurements were additionally etched for 30 s in the mixture of nitric acid, water and ethylene glycol. For EBSD investigations, the mechanically polished surface was subsequently ion milled using a Leica RES ion mill at 5 kV and an incidence angle of 5°.

The microstructure was investigated by a Zeiss Auriga Compact scanning electron microscope operated at 10 kV, equipped with the BSE detector and EDAX EBSD camera. Measured EBSD data were cleaned and evaluated by software TSL OIM Analysis 7.

The density of dislocations was evaluated from PAS data using a simple two-state trapping model [19] and employing the specific trapping rate of positrons to dislocations in Mg $\nu_D = 10^{-5}\ s^{-1}m^2$. The details of PAS data evaluation and the experimental device are described in detail elsewhere [21-23]. Vickers microhardness measurements were carried out employing an applied load of 100 g for 10 s, using automatic microhardness tester Qness Q10a. The hardness values were determined as the average of 36 separate measurements.

3. Experimental results and discussion

3.1. Microstructure evolution

EBSD inverse pole figures maps, captured the evolution of the microstructure of AX41 alloy at selected temperatures of isochronal annealing, are shown in Figure 1. The microstructure of the original ECAP processed AX41 alloy (in non-annealed condition) is homogenous, formed by equiaxed grains with the average grain size of 1.6 μm, see Figure 1 (RT). After isochronal annealing in the temperature range of 160-240 °C the microstructure remained almost unchanged, only a slight grain coarsening was observed.

IOP Conf. Series: Materials Science and Engineering **194** (2017) 012022 doi:10.1088/1757-899X/194/1/012022

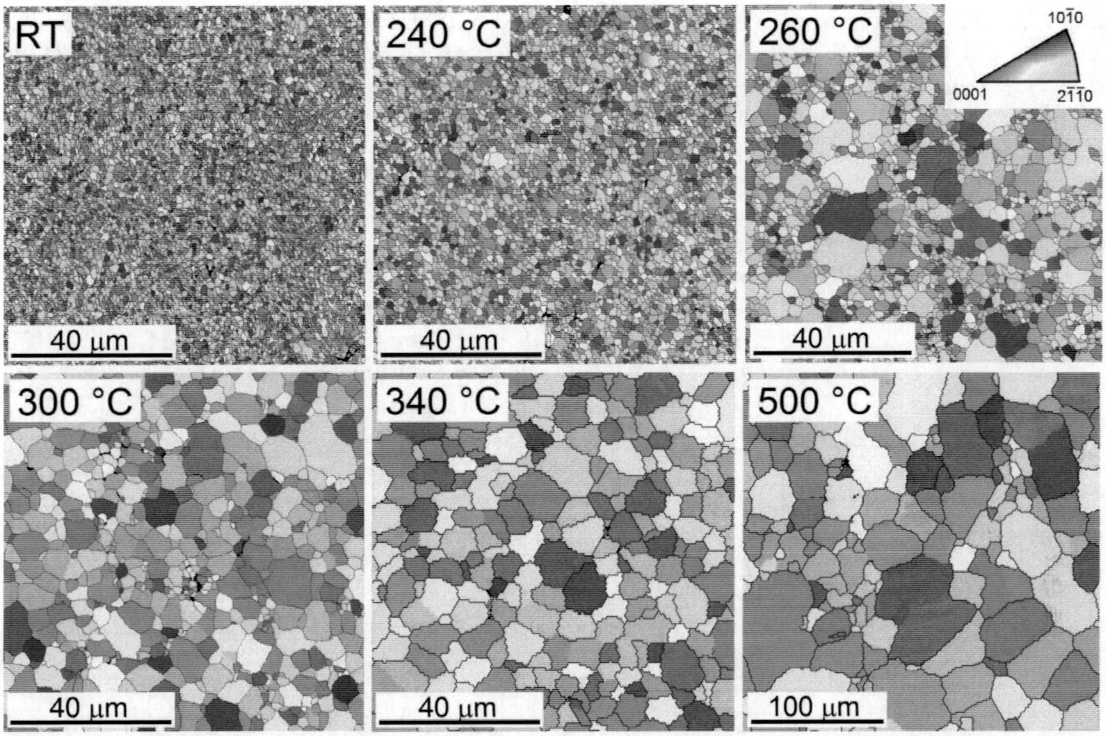

Figure 1 - EBSD inverse pole figure maps of the non-annealed (labeled as RT) and annealed AX41 magnesium alloy at various temperatures for 1h

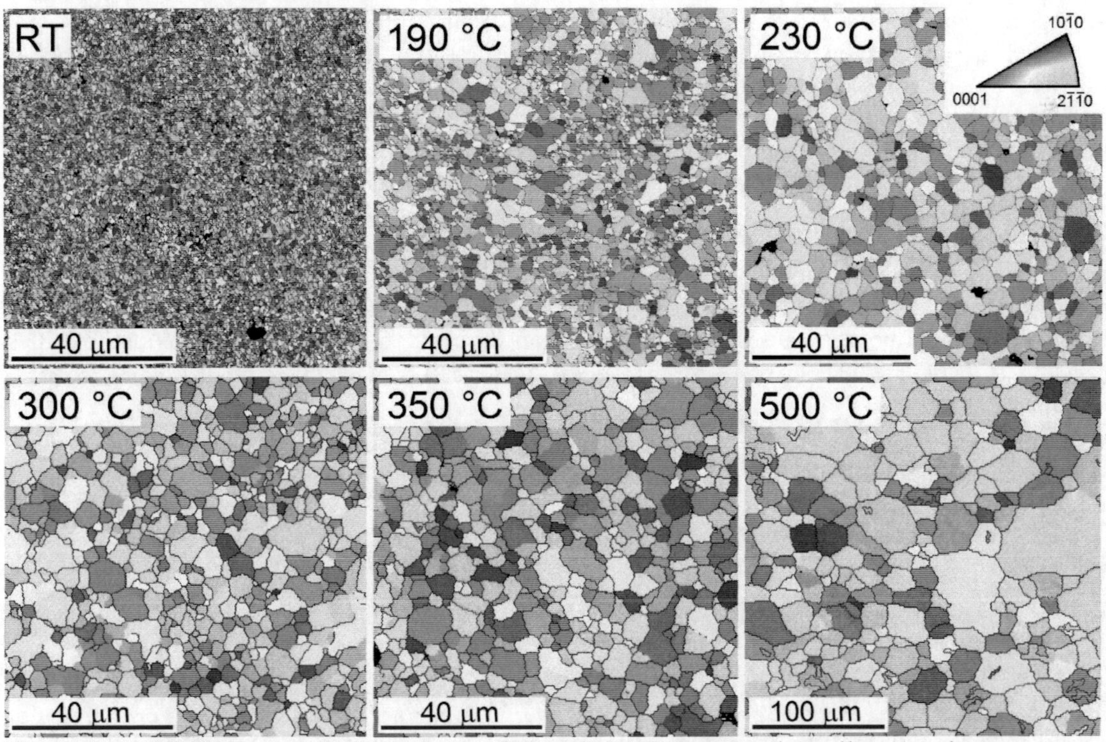

Figure 2 - EBSD inverse pole figure maps of annealed AZ31 magnesium alloy at various temperatures for 1h

Starting from the temperature of 260 °C, inhomogeneous grain growth, preferentially in the areas of microstructure without Al_2Ca phase (see below) was observed. During further annealing the fraction of coarse grains gradually increases and original small grains disappeared after annealing at 340 °C and homogeneous microstructure coarsening up to the final annealing temperature of 500 °C occurred.

The microstructural evolution in AZ31 magnesium alloy during isochronal annealing is shown in Figure 2. The original microstructure with the average grains size of 1.3 μm starts to grow already around the temperature of 190 °C. A bimodal microstructure was observed at this temperature consisting of coarse grains and the original ultrafine grains. Further annealing results in the gradual and homogeneous coarsening of the microstructure in the whole volume. Above the temperature of 230 °C no original ultrafine grains were observed.

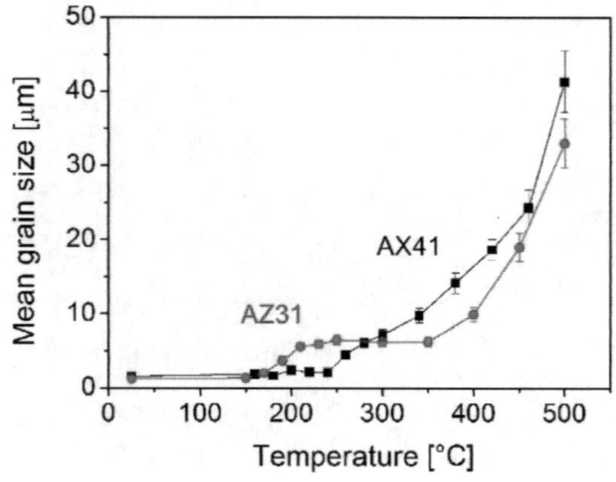

Figure 3 - The mean grain size of AZ31 and AX41 magnesium alloys after 1h of isochronal annealing

The mean grain size calculated from EBSD maps by software TSL OIM Analysis 7 using the area fraction statistics is plotted in Figure 3 as a function of the temperature of annealing for both AX41 and AZ31 alloys. The area fraction statistics was selected instead of the number fraction statistics, as the former is more sensitive to the inhomogeneous grain growth. Thus, the coarsening of even a small number of grains will affect the average grain size. Whereas in AZ31 alloy the grain size increases already from the temperature of 190 °C, in AX41 alloy the grain size starts to increase only above 240 °C. Nevertheless, the rate of the coarsening is much higher in AX41 alloy than in AZ31 resulting in higher average grain size in AX41 than in AZ31 above the temperature of 300 °C.

Improved thermal stability of AX41 alloy in the range of 180 – 260°C can be attributed to higher temperature and lower strain rate of ECAP processing of AX41 alloy as compared to AZ31 alloy. The achieved dynamic equilibrium grain size during ECAP at 220°C in AX41 alloy remains constant upon static annealing up to 240°C. Moreover, it is important to note that despite AX41 alloy was processed at higher temperature and lower strain rate, the achieved grain size is comparable to that of AZ31 alloy. This can be attributed to thermally stable secondary Al_2Ca particles (melting point 1079 °C, unsolvable in Mg matrix [24]), that are present during ECAP processing of the AX41 alloy, which contrasts to thermally instable $Mg_{17}Al_{12}$ particles in AZ31 alloy as shown in SEM micrographs in Figure 4. Increased grain growth of AX41 in the temperature range of 300 – 400°C is not fully resolved yet. It might be also caused by the absence of dissolved zinc in AX41 in contrast to AZ31 alloy.

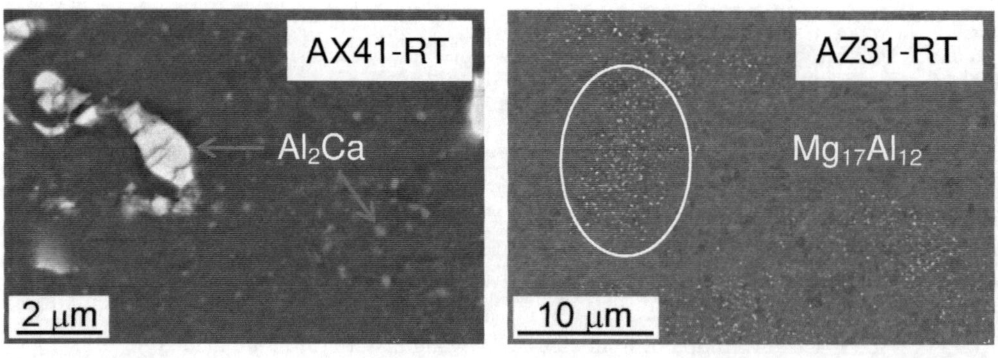

Figure 4 - SEM micrographs (BSE) of secondary phases in AX41 and AZ31 magnesium alloys at RT

IOP Conf. Series: Materials Science and Engineering **194** (2017) 012022 doi:10.1088/1757-899X/194/1/012022

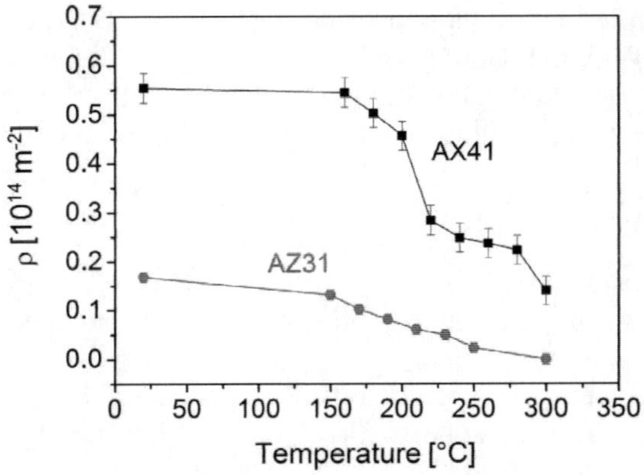

Figure 5 - The evolution of dislocation density with temperature of isochronal annealing in AX41 and AZ31 alloy

The values of dislocation density (ρ), obtained from PAS measurements are plotted in Figure 5 as a function of the annealing temperature. The dislocation density in both alloys starts to decrease at the temperature of ≈ 150 °C. After annealing at 300 °C, ρ in AZ31 drops below the detectable limit of PAS technique ($\rho < 5 \times 10^{12}$ m^{-2}). Note, that AX41 in the initial condition exhibits almost 4 times higher ρ than AZ31. The reason of this difference is rather complex and is influenced by many parameters, e.g. different processing conditions (temperature and number of passes) and the microstructure differences (thermally stable Al_2Ca particles which are present already in the cast AX41 alloy vs. rather instable $Mg_{17}Al_{12}$ which were probably formed in AZ31 only during cooling after ECAP), etc. The detail discussion of these effects is beyond the scope of this paper. On the other hand, the thermal stability and mechanical properties are determined by the temperature dependence of dislocation density rather than by its absolute value.

3.2. Microhardness evolution

Microhardness values of the ECAP processed samples and samples after 1 h of isochronal annealing are plotted in Figure 6. The first data points in both curves correspond to the initial UFG conditions after ECAP. A slightly higher microhardness (HV) was found in AZ31 alloy as compared to the AX41 alloy which corresponds to slightly smaller average grain size in AZ31 than in AX41, cf. Figures 3 and 6. HV values of both alloys do not differ significantly up to the temperature of 190 °C. For higher temperatures (T>190 °C) HV values of AZ31 decline abruptly up to 250 °C and continue to decrease up to 500 °C. On the other hand, AX41 alloy exhibits apparently higher thermal stability. HV values of AX41 decrease only slightly up to 240 °C, while a sharp drop in the HV can be observed only in the temperature range of 250-330 °C, which continues up to the temperature of 500 °C. After the annealing at 500 °C both alloys exhibit approximately the same values of HV. The evolution of HV is consistent with the evolution of grain size, cf. Figures 3 and 6.

In order to assess the role of grain size and the dislocation density in strengthening of both alloys the Hall-Petch (H-P) plot relating the microhardness with the grain size ($d^{-1/2}$) as determined from EBSD

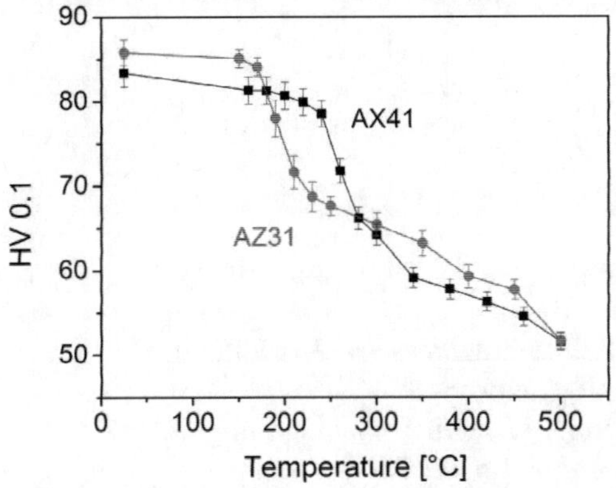

Figure 6 - The evolution of Vickers microhardness the with temperature of isochronal annealing

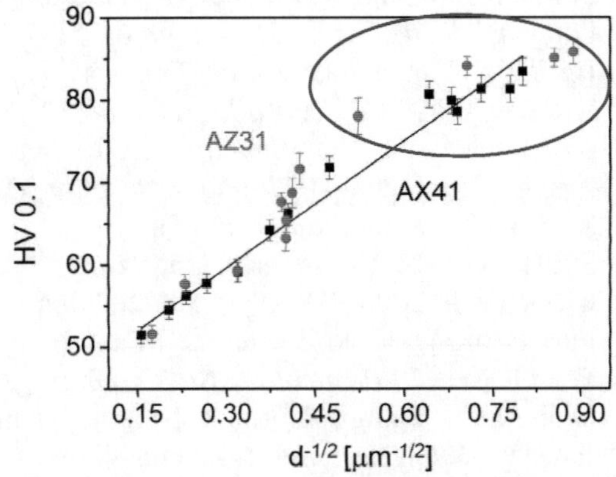

Figure 7 - The Hall-Petch plot for both alloys

analysis is displayed in Figure 7. It is clearly seen that the data of AX41 alloy obey the H-P law in the whole temperature range confirming the controlling role of the grain size on mechanical properties while the role of dislocations is minor. On the other hand, the data of AZ31 fit well the H-P relation only for high annealing temperatures (T>250 °C). At low temperatures (T< 250 °C) the data deviate from the linear relationship indicating that both the grain size and the dislocation density contribute to the strength of the material.

Conclusions

The following conclusions may be drawn from this investigation:

- the thermal stability of UFG AX41 alloy is superior to the thermal stability of AZ31 of the same initial grain size,
- enhanced thermal stability of the UFG structure of AX41 in the range of 180-260°C is caused by the conditions of ECAP processing and the presence of thermally stable Al_2Ca precipitates which stabilize the microstructure during ECAP,
- mechanical properties of AX41 are predominantly controlled by the grain size in the whole range of annealing temperatures. On the other hand, in AZ31 alloy the grain size affects the materials strength at high temperatures while at low temperatures (T< 250 °C) both the grain size and the dislocation density contribute to the strengthening.

Acknowledgement

This work was financially supported by the Czech Science Foundation under the project GB 14-36566G and by ERDF under the project "Nanomaterials centre for advanced applications", project No. CZ.02.1.01/0.0/0.0/15_003/0000485. J.C acknowledges financial support by the Czech Science Foundation project no. 16-12828S.

References

[1] Gupta M and Sharon N M L 2011 *Magnesium, Magnesium Alloy, and Magnesium Composites*, (New Jersey, John Wiley &Sons, Inc.)
[2] Koch C C 2009 *Bulk nanostructured materials* (Wiley, Weinheim) 9
[3] Valiev R Z and Langdon T G 2006 *Progress in Materials Science* **51** 881
[4] Zhilyaev A P and Langdon T G 2008 *Progress in Materials Science* **53** 893
[5] Saito Y et al 1999 *Acta Materialia* **47** 579
[6] Varyutkhin N V et al 2006 *Mater. Sci. Forum* **503-504** 335
[7] Yang X et al 2012 *J. Mater. Sci.* **47** 2823
[8] Segal et al 1981 *Russ. Metall.* **1** 99
[9] Estrin Y and Vinogradov A 2013 *Acta Materialia* **61** 782
[10] Langon T G 2013 *Acta Materialia* **61** 1237
[11] Hellmig R J et al 2008 *Mater. Trans.* **49** *31*
[12] Kužel et al 2008 *Z. Krystallogr. Suppl.* **27** 73
[13] Azushima et al 2008 *Manufacturing Technology* **57** 716
[14] Vinogradov A et al 2005 *Acta Materialia* **53** 2181
[15] Sitdikov 2008 *Acta Materialia* **56** *821*
[16] Stráská J et al 2014 *Mater. Char.* **94** 69
[17] Čížek J et al 2006 *Phys. Stat. Sol.* **203** 466
[18] Furukawa M, Horita Z, Nemoto M and Langdon T G 2001 *J. Mater. Sci.* **36** 2835
[19] West R N 1979 *Positrons in Solids*. (Springer Berlin, Heidelberg) 89
[20] Iwahashi Y, Wang J, Horita Z, Nemoto M and Langdon T G 1998 *Mater. Sci. Eng.* **A 257** 328
[21] Janeček M, Čížek J, Gubicza J and Vrátná J 2012 *J. Mater. Sci.* **47** 7860
[22] Bečvář F, Čížek J, Procházka I and Janotová J 2005 *Nucl. Instrum. Methods* A **539** 372
[23] Procházka I, Novotný I and Bečvář F 1997 *Mater. Sci. Forum* **255-257** 772
[24] Massalski T B, Okamoto H *Binary alloy phase diagram* (Mater. Information Society, Ohio)

IOP Conf. Series: Materials Science and Engineering **194** (2017) 012023 doi:10.1088/1757-899X/194/1/012023

Cryomilled and spark plasma sintered titanium: the evolution of microstructure

Jiří Kozlík[1], **Hanka Becker**[2], **Petr Harcuba**[1], **Josef Stráský**[1], **Miloš Janeček**[1]

[1] Department of Physics of Materials, Charles University, Prague, Czech Republic
[2] Institute of Materials Science, TU Bergakademie Freiberg, Freiberg, Germany

E-mail: `Jiri.Kozlik@seznam.cz`

Abstract. Bulk ultra-fine grained (UFG) commercially pure Ti was prepared by cryogenic milling in liquid argon and subsequent spark plasma sintering (SPS). During cryogenic milling, individual powder particles are repetitively severely deformed by attrition forces. Powder particles were not significantly refined, but due to severe repetitive plastic deformation, ultra-fine grained microstructure emerges within each powder particle. Cryogenic milling can be therefore considered as a specific severe plastic deformation (SPD) method.

Compactization of cryomilled powder by SPS technique (also referred to as field assisted sintering technique – FAST) requires significantly lower sintering temperatures and shorter sintering times for successful compaction when compared to any other sintering technique. This is crucial for maintaining the UFG microstructure due to its limited thermal stability.

Several specimens were prepared by varying processing parameters, in particular the sintering temperature. The microstructure of powders and compacted samples was observed by scanning electron microscopy (SEM). Increased sintering temperature results in recrystallization and grain growth. A trade-off relationship between the density of compacted material and grain size was identified. Microhardness of the material was found to depend on residual porosity rather than grain size.

This contribution presents cryogenic milling and spark plasma sintering as a viable alternative for achieving UFG microstructure in commercially pure Ti.

1. Introduction

Titanium belongs to modern materials widely used in aerospace and biomedical applications. It has been known for a few decades that processing by severe plastic deformation, e.g. equal-channel angular pressing (ECAP) [1] or high pressure torsion (HPT) [2], can lead to superior material properties [3]. Powder metallurgy methods are bottom-up alternatives to these bulk top-down SPD methods. Subsequent consolidation of processed powder is a crucial step, because of the limited stability of the milled nanoscaled microstructure [4, 5].

Cryogenic milling is a form of conventional attritor milling carried out under cryogenic temperatures, typically in liquid nitrogen or liquid argon. During milling, not only powder particles are refined, but also and more importantly in this case, due to the repeated plastic deformation, grain refinement within individual particles occurs [6]. At low temperatures dynamic recrystallization and grain growth are suppressed and grain refinement is enhanced, which was reported for Ti in several previous studies [4, 7].

The spark plasma sintering method (SPS) [8] used for powder consolidation allows high heating/cooling rate and short sintering time, which is beneficial for conserving the UFG microstructure [9]. The principle of this method is the heating of the powder by pulse direct current passing through the sample under concurrent application of uniaxial pressure.

2. Experimental methods

Initial Ti Grade 2 powder (supplied by TLS Technik GmbH), prepared by electrode induction-melting gas atomization, was in the from of spherical particles with diameter from 20 to 60 µm. The powder was milled in Union Process 01-HD attritor using 6.35 mm stainless steel balls with ball to powder ratio 16:1. 8 wt.% of stearic acid was used as a process control agent to prevent cold welding [6]. The milling was performed at 700 rpm for 4 h in liquid argon (LAr) at (-185 ± 5)°C (since milling in liquid nitrogen is known to cause the embrittlement of Ti and its alloys [10]). The powder was cleaned of stearic acid after milling to reduce C and H contamination by repeated washing in ethanol. The powder was handled in air.

HDP 25 machine (FCT Systeme GmbH) was used for SPS. The sintering started by heating up to 600 °C at a constant rate of 100 °C/min, followed by 1 min dwell accompanied by an increase of uniaxial pressure to 80 MPa. The heating continued subsequently up to the target temperature, where isothermal sintering was performed for 3 min. The following sintering temperatures were employed: 650, 750 and 850 °C. Cooling rate was not controlled, the initial cooling rate was approx. 200 °C/min and decreased at lower temperatures. 9 g of powder was used to produce one sample with a diameter of 20 mm and an approximate height of 6 mm.

H, N and O content was determined using carrier gas hot extraction (CGHE) method, energy-dispersive X-ray spectroscopy (EDS) was used for the identification of other elements.

Seifert XRD 7 diffractometer in Bragg-Bretano geometry using Cu K_α line was employed for X-ray diffraction (XRD) measurement.

Microstuctural investigations were performed by scanning electron microscope Zeiss Auriga equipped with a field emission gun (FEG), focused ion beam (FIB) and camera for electron back-scattered diffraction (EBSD). Samples were prepared by diamond wheel abrasive cutting, SiC paper grinding and vibratory polishing (0.3 and 0.05 µm alumina and 0.05 µm colloidal silica suspension). Residual porosity was determined by SEM micrograph analysis.

Microhardness measurements were performed on Qness Q10A automatic microhardness testing machine using Vickers method employing the load of 0.5 kgf and 10 s dwell time.

3. Results and discussion

Particles of as-milled CP Ti powder exhibited a plate-like morphology (Fig. 1b). No cold-welding was observed. The mean crystallite size of 48 nm was determined by XRD measurement.

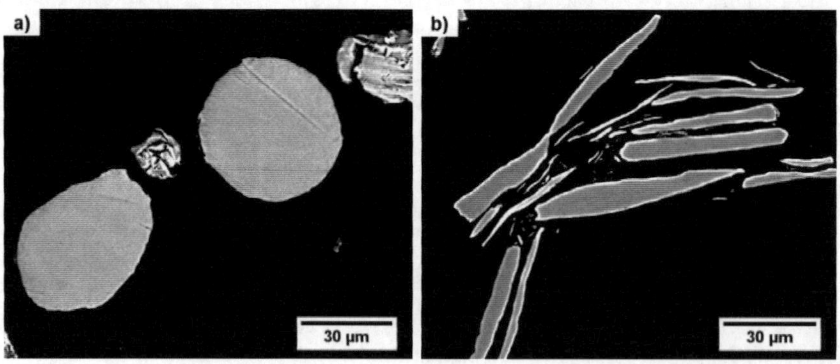

Figure 1. Cross-sectional SEM image of powder particles in a) initial and b) as-milled state.

IOP Conf. Series: Materials Science and Engineering **194** (2017) 012023 doi:10.1088/1757-899X/194/1/012023

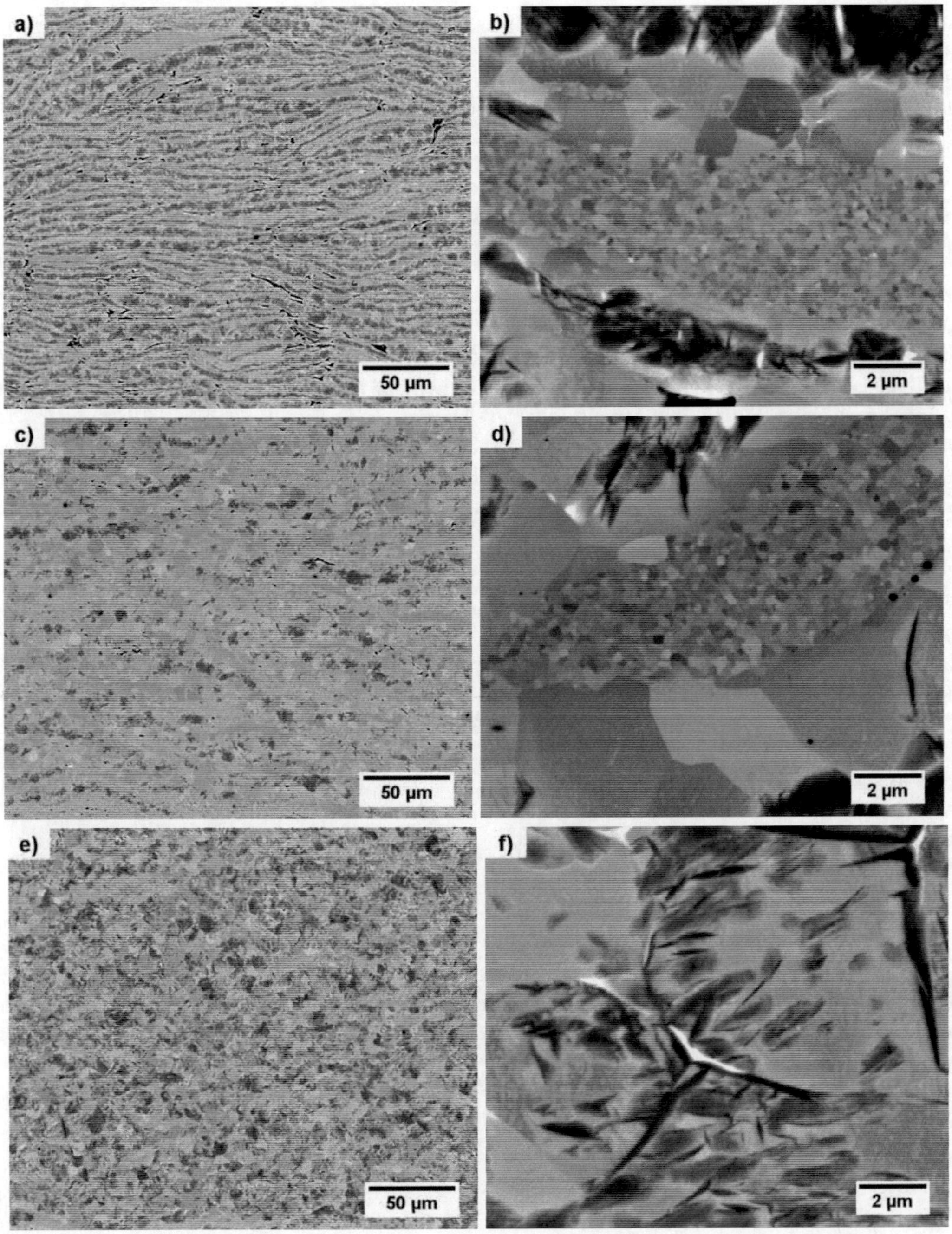

Figure 2. The microstructure (overview and detail) of samples sintered at different temperatures: a) and b) 650 °C; c) and d) 750 °C; e) and f) 850 °C obtained by back-scattered electrons (BSE). FCC phase appears dark in BSE contrast. The direction of a load applied during SPS corresponds to the vertical direction in the images.

3.1. Microstructural investigations on bulk material

Fig. 2 shows the microstructure of samples sintered at different temperatures. The dark areas, which can be seen in all micrographs, were identified as an FCC phase from EBSD Kikuchi diffraction patterns with high average confidence index and image quality (the mechanical damage, e.g. surface scratches or cracks can be ruled out). Kikuchi lines belonging to HCP or BCC phase were not observed in such regions.

The microstructure of a sample sintered at 650 °C (Fig. 2a) shows layered structure formed by stacking of plate-like particles in a die during sintering. Two different regions of HCP grains were observed (Fig. 2b – recrystallized regions of a typical grain size of 2 µm (containing the FCC phase within) and UFG regions with a grain size well below 500 nm. These regions are parallel with particles' packing plane. Residual porosity after sintering is 1.62 vol.%.

The sample sintered at 750 °C (Fig. 2c and 2d) exhibited lower porosity (0.34 vol.%). Similarly to the sample sintered at 650 °C, two regions – the recrystallized coarse-grained (∼ 5 µm) and the UFG (< 500 nm) – were observed (Fig. 2d). It was reported by [11] that enhanced mechanical properties can be achieved in material with such bimodal microstructure.

After sintering at 850 °C the microstructure is almost completely recrystallized (Fig. 2e and 2f) with grain size about 10 µm. UFG regions were observed only rarely. Almost fully dense sample was produced, with residual porosity of 0.11 vol.%. Small amount of Fe was found segregated at grain boundaries (see the bright spots in the Fig. 2). Increased fraction of FCC phase in the form of thin needles in the interior of recrystallized grains was also observed (Fig. 2f).

It was reported earlier in [4, 12] that recrystallization and grain growth starts already at approx. 450 °C. These phenomena are therefore inevitable at sintering temperatures sufficient to achieve full density.

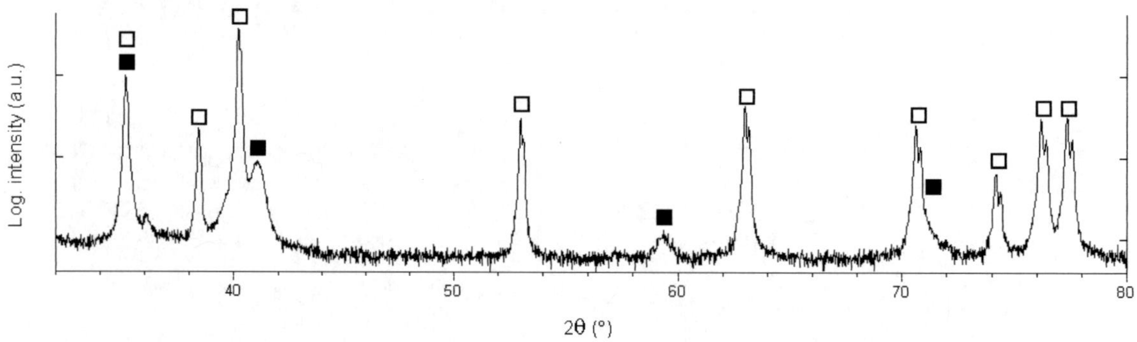

Figure 3. XRD diffraction pattern of the sample sintered at 850 °C. Peaks corresponding to hcp α-Ti (□) and fcc phase (■) are marked. We could not identify the peak at 36°.

XRD analysis (Fig. 3) and an area EBSD mapping (Fig 4a) was performed to investigate the FCC phase using the sample sintered at 850 °C. In the XRD pattern, peaks corresponding to FCC phase with lattice parameter of 0.440 nm can be observed. The phase fraction determined by XRD was significantly lower than that by EBSD. We can therefore conclude, that the observed FCC phase is present only at the sample surface. This was confirmed by a FIB cross-section (Fig 4b). The FCC phase is undoubtedly related to the microstructural condition. In particular, it emerges only in recrystallized regions (see Fig. 2). It is however unclear, whether creation of this phase is directly caused by sample preparation (and possible contamination) [13] or as a result of a newly formed free surface. The very similar HCP to FCC transformation was recently observed by Zhao et al. [14] in cold-rolled Hf. The detailed investigation of composition and structure of this unusual phase requires further experiments.

Ball milling is well known to increase the amount of contaminating elements, typically O, N, H, C and Fe originating from the milling environment, balls, milling tank and process control agent (stearic acid in this case). Analysis of C content was done by EDS point measurement in

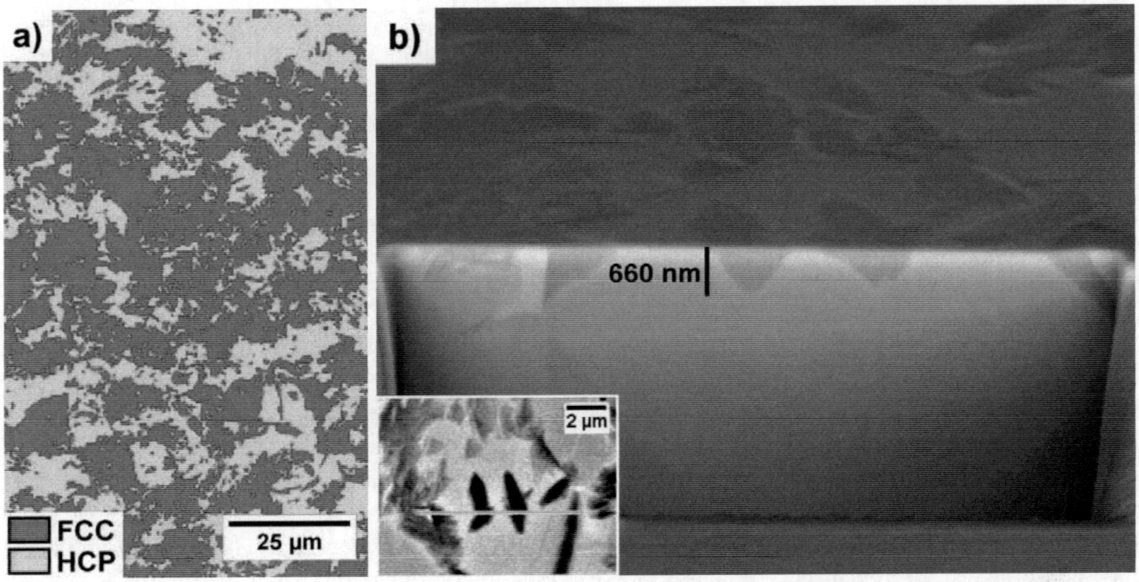

Figure 4. a) EBSD phase map of the 850 °C SPSed sample (60 % FCC, 40 % HCP). b) Cross-section (along the green line in the inset image) of the grain containing FCC lamellae showing their depth. BSE contrast, FCC phase appears darker.

identified HCP and FCC phase and no significant difference was observed. Only trace amount of Fe (< 0.5 wt.%) was detected by EDS (area scan).

The content of H, N and O in powders and bulk material analyzed by CGHE method is shown in Tab. 1. The contamination by nitrogen, oxygen and hydrogen occurs mainly during milling. Subsequent handling in air has only a negligible negative effect. The final nitrogen and oxygen content is slightly above the technical standard for conventionally prepared CP Ti Grade 2 (0.05 wt.% N, 0.25 wt.% O), while it meets technical standard of Grade 4 (0.07 wt.% N, 0.4 wt.% O) [15]. The content of hydrogen is lower after sintering because of the thermal decomposition and evaporation of stearic acid during the SPS process.

Table 1. The concentration of contaminating elements H, N and O. The as-milled condition is after cleaning the powder in ethanol. No significant difference between samples sintered at different temperatures was observed.

	H (wt.%)	N (wt.%)	O (wt.%)
Initial	0.001	0.017	0.16
As-milled	0.051	0.059	0.36
As-sintered	0.017	0.065	0.37

3.2. Microhardness

The microhardness measurements (Fig. 5) exhibit a strong dependence on the relative density of the sample. The inter-particle slip was observed around indentations in the sample sintered at 650 °C (not shown here), which indicates that the sintering is not completed yet, resulting in a porous and softer material. The microhardness of 750 °C and 850 °C sample is the same within the experimental error. The measured values are consistent with those given by [16].

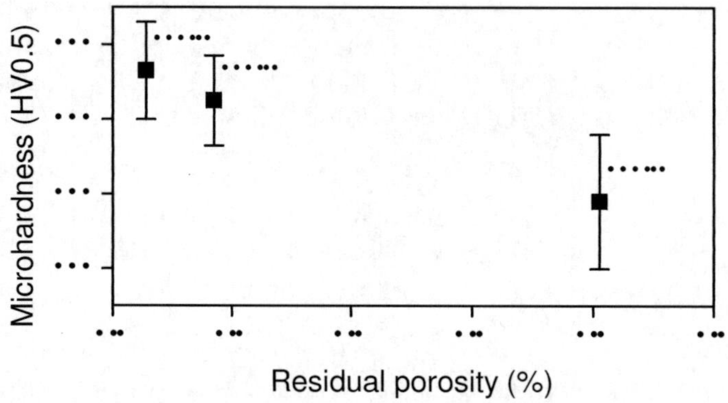

Figure 5. Dependence of microhardness on residual porosity.

4. Conclusions

The microstructure and microhardness of samples prepared by cryogenic milling and spark plasma sintering was investigated. The following conclusions may be drawn form this study:

- Bulk samples with bimodal and partially UFG structure can be produced, the fraction of UFG regions decreases with increasing sintering temperature as a result of recrystallization and grain growth.
- Sintering at minimum 750 °C is required to produce fully dense sample with relative density $> 99.5\,\%$.
- Microhardness depends on a residual porosity rather than on the microstructure/grain size.
- FCC phase was observed by in recrystallized grains on the sample surface. The amount and morphology of FCC phase is related to the sample microstructure. Further experiments are needed to analyze this phase in detail.

5. Acknowledgments

Financial support by the Czech Science Foundation under the project GB 14-36566G is gratefully acknowledged.

References

[1] Zhao X, Yang X, Liu X, Wang X and Langdon T G 2010 *Mat. Sci. Eng.* A **527** 6335–9
[2] Zhilyaev A P and Langdon T G 2008 *Prog. Mater. Sci.* **53** 893–979
[3] Langdon T G 2013 *Acta Mater.* **61** 7035–59
[4] Sun F, Zúñiga A, Rojas P and Lavernia E J 2006 *Metall. Mat. Trans.* A **37** 2069–78
[5] Witkin D B and Lavernia E J 2006 *Prog. Mat. Sci.* **51** 1–60
[6] Suryanarayana C 2001 *Prog. Mat. Sci.* **46** 1–184
[7] Sun F, Rojas P, Zúñiga A and Lavernia E J 2006 *Mat. Sci. Eng.* A **430** 90–7
[8] Guillon O, Gonzalez-Julian J, Dargatz B, Kessel T, Schierning G, Räthel J and Herrmann M 2014 *Adv. Eng. Mater.* **16** 830–49
[9] Munir Z A, Anselmi-Tamburini U and Ohyanagi M 2006 *J. Mater. Sci.* **41** 763–77
[10] Ertorer O, Zúñiga A, Topping T, Moss W and Lavernia E J 2008 *Metall. Mat. Trans.* A **40** 91–103
[11] Ertorer O, Topping T, Li Y, Moss W and Lavernia E J 2009 *Scripta Mater.* **60** 586–9
[12] Strasky J, Harcuba P, Hajek M, Vaclavova K, Zhanal P, Janecek M, Polyakova V and Semenova I 2014 *IOP Conf. Ser.: Mat. Sci. Eng.* **63** 012072
[13] Vullum P E, Pitt M, Walmsley J, Hauback B and Holmestad R 2009 *Appl. Phys. A* **94** 787–93
[14] Zhao H, Song M, Ni S, Shao S, Wang J and Liao X 2017 *Acta Mat.*
[15] Welsch G, Boyer R and Collings E 1993 *Materials Properties Handbook: Titanium Alloys* Materials properties handbook (ASM International)
[16] Ertorer O, Topping T D, Li Y, Moss W and Lavernia E J 2011 *Metall. Mater. Trans.* A **42** 964–73

IOP Conf. Series: Materials Science and Engineering **194** (2017) 012024 doi:10.1088/1757-899X/194/1/012024

Investigation of graded strengthened hyper-deformed surfaces by impact treatment: micro-percussion testing

David Tumbajoy-Spinel[1,2], Sylvie Descartes[1], Jean-Michel Bergheau[3], Halim Al-Baida[4], Cécile Langlade[4], Guillaume Kermouche[2]

[1] Université de Lyon, INSA-Lyon, LaMCoS, CNRS UMR5259, Villeurbanne F-69621, France
[2] Université de Lyon, École des Mines de Saint-Étienne, LGF, CNRS UMR5307, Saint-Étienne F-42023, France
[3] Université de Lyon, ENISE, LTDS, CNRS UMR5513, Saint-Étienne F-42023, France
[4] Université Technologique de Belfort-Montbéliard, IRTES-LERMPS EA 7274, 90010 Belfort, France

E-mail: david.tumbajoy-spinel@insa-lyon.fr

Abstract. In the industry, mechanical surface treatments could improve the mechanical behaviour of materials by the means of local hyper-deformation and graded strengthening. Micro-percussion test represents an interesting case scenario to emulate these kinds of conventional treatments (shot-peening, SMAT, roller-burnishing, etc) and go further on microstructural and mechanical characterization at local and global scales. For this technique, every impact is made at the same position by a rigid conical indenter, controlling the number, angle and velocity of impacts. The main issue of this work is to establish a complete description of the transformed microstructures; to understand the mechanisms involved on the formation and growth of refined structures; to make a parametric sensitivity analysis of different impact conditions.

1. Introduction
The mechanical surface treatments confer better local mechanical properties against wear or fatigue service conditions. In the case of conventional impact treatments (shot-peening [1]-[2], SMAT [3]), the material is exposed to repeated mechanical loadings, producing a severe plastic deformation (SPD) in the near-surface. It leads to a local progressive refinement of the microstructure over a few tens of microns [2]-[3], commonly known as Tribologically Transformed Structure (TTS) [4]. In this work, a model surface treatment (micro-percussion [5]-[6]) is used to explore several impact conditions (angle and amount of impacts) and its effects on the microstructural transformation. In this work, three main goals are considered: (i) characterize the transformed surfaces and evaluate the possibility to emulate industrial techniques (ii) make a parametric analysis, (iii) describe the microstructural transformation process and the mechanisms involved on it. This investigation is carried out in pure α-iron.

2. Process and Methods: micro-percussion tests in pure α-iron samples

2.1. Material: α-iron
The micro-percussion tests are done in a high-purity α-iron with less than 15 ppm of carbon. This material is produced by the cold crucible melting method. The resulting metallic bar is thermo-mechanically treated by forging and annealing at 800 °C during 60 minutes. This treatment produces a

IOP Conf. Series: Materials Science and Engineering **194** (2017) 012024 doi:10.1088/1757-899X/194/1/012024

homogeneous microstructure with equi-axed grains (average size in the order of 1 mm). No inclusions have been observed in this model material.

2.2. Micro-percussion tests

The principle of micro-percussion tests [5]-[6] is to carry out several impacts in the same position of the sample with a rigid conical punch (carbide tungsten). For metallic materials, the SPD produces a remaining print of millimetric diameter. The indenter tip has a spherical shape with a radius of 0.5 mm and an angle of 60º (**Figure 1**-a). A schematic representation of the test is shown in **Figure 1**-b. The impact set-up permits to control the number and velocity of impacts [7]. The impact angle is introduced directly on the sample geometry (tilted plane). Three angles were explored for the impact conditions: 0º, 15º and 30º. At each angle, the prints are done for different amounts of impacts: 10, 30, 100, 300, 1000, 3000 and 10000. Prints are arranged in a rectangular matrix (2x4), spaced of 2 mm for each line and 3 mm between both lines (**Figure 1**-b). Only the 10000 impacts print is spaced of 4 mm because of its high diameter. Two main conditions were considered to make a parameter sensitivity analysis: (i) the angle and (ii) the number of impacts. Concerning the velocity of impacts, some preliminary tests pointed out that the set-up working range for α-iron is enclosed between 80 mm/s and 200 mm/s [7]. Then, the velocity used for all the tests was in the order of 150 mm/s.

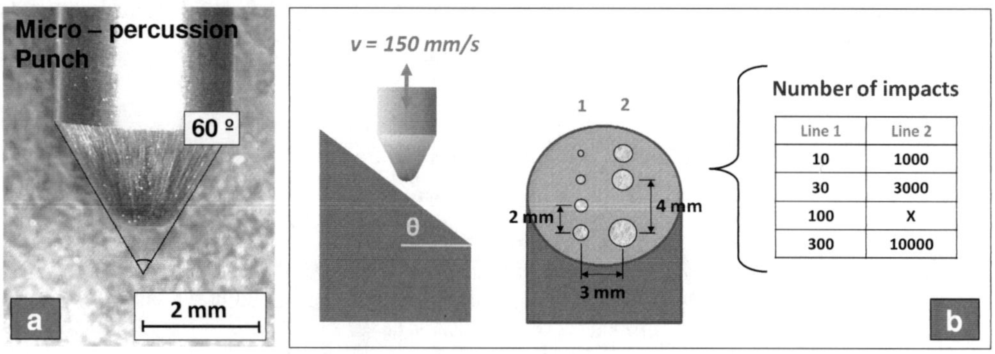

Figure 1. (a) Micro-percussion punch. (b) Schematic representation of the micro-percussion tests for three angles (0º, 15º, 30º) and seven impact amounts (10, 30, 100, 1k, 3k, 10k) at 150 mm/s.

2.3. Near-surface microstructural characterization

The characterisation of the resulting prints was done using scanning electron microscopy (SEM) and electron back-scattered diffraction (EBSD) mapping. Both techniques were carried out on the print cross-section in order to observe the near-surface microstructural transformation. For both methods, the samples were cutted with a metallic wire saw and then mechanically polished up to the middle plane of each print. The sample surfaces were polished with several abrasive papers (P240 to P1200), two diamond suspensions (3 μm and 1 μm) and finished with colloidal silica. The SEM images were done in a Zeiss Supra 55 VP at 20 kV using a back-scattered detector (BSE). The EBSD maps were carried out in a Tescan Lyra3/XMU (FEG/FIB) using an indexation step of 2 μm.

3. Results: mechanical surface treatment by micro-percussion

3.1. Characteristics of a print: different deformed regions

An example of a characteristic print is shown in Figure 2-a (10000 impacts at 15º). This print has a diameter of 1750 μm and a non-axisytmetric shape due to the tilt of the impact direction. Different regions are identified in the cross-section of the print. The entire mechanically affected zone (MAZ) [8] has a thickness of hundreds of microns (430 μm). The virgin bulk microstructure (initial milimetric grains) is observed in the outside of this region (MAZ). Moreover, two main layers are observed in the mechanically affected zone (MAZ). On the one hand, the mechanically attrited structure (MAS) [6] is essentially characterized by the presence of high local misorientations and new well-defined grain boundaries [9]. On the other hand, the low angle boundaries (LAB's) zone corresponds to a non-microstructured region in spite of the plastic deformation (Figure 2-a) [10]. Indeed, some previous works [11]-[12] revealed that this kind of region has some slight crystal orientation nuances and it could be considered as a single crystal zone due to the initial grain sizes in the bulk material.

Furthermore, the MAS region (total thickness of 150 μm) is defined by the TTS and the transition zones (Figure 2-b). The tribologically transformed structure (TTS) corresponds to the nearest surface layer with a progressive grain size refinement in depth [4]. The transformed microstructure in this case is considerably similar to the TTS regions obtained by other surface treatements (shot peening [2],[12],[14], SMAT [3],[15], roller-burnishing [16]-[17]). Beyond the TTS layer, the transition zone is characterized by the presence of shear bands (white arrow) [13] and sub-grain boundaries (not well-formed grains) [9]. This zone is an intermediary region between the well-refined microstructure (TTS) and the "single crystal" region deformed at low strain rates (LAB's zone).

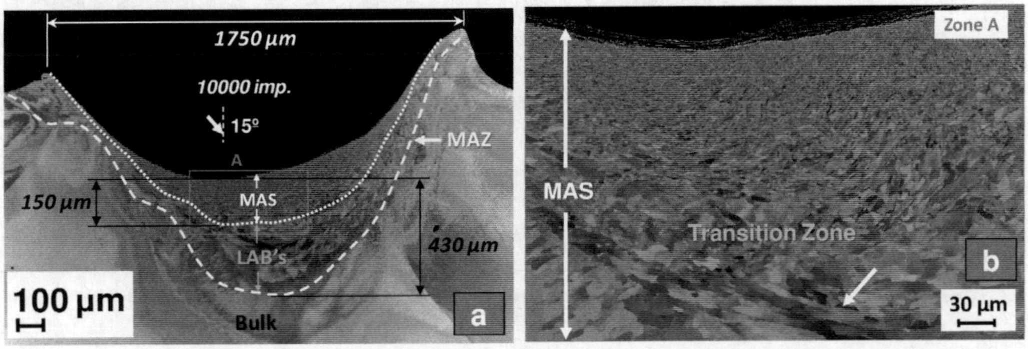

Figure 2. (a) Description of the different regions observed in the cross section of the print formed at 10000 impacts and 15°. (b) Detail of the *mechanically attrited structure* (MAS).

3.2. *Kinetics of the TTS layer transformation*

The wide thickness and homogeneity of the TTS layer observed in Figure 2-a,b is definitively an interesting case scenario to emulate and investigate the transformed surfaces by the means of industrial impact treatments. However, the impact conditions will be decisive to create a significant TTS layer for this purpose. For example, as shown in Figure 3, three different prints were formed at 10000 impacts for different impact angles: (a) 0°, (b) 15° and (c) 30°. The normal impacts (0°) produce a shallow transformed region with less than 50 μm in-depth. On the contrary, the tilted impacts (15° and 30°) can produce TTS layers three times wider (more than 120 μm in-depth). Moreover, the TTS layer at 0° presents more surface heterogeneities, crack formations (white arrows) and material release [18].

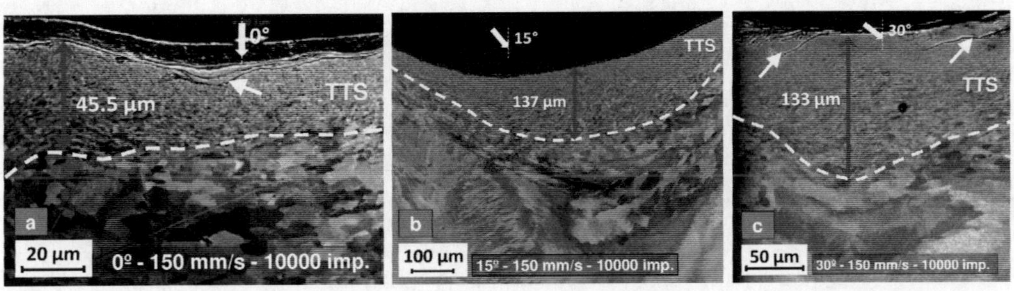

Figure 3. Cross-section of the TTS layer for 10000 impacts at (a) 0°, (b) 15° and (c) 30°.

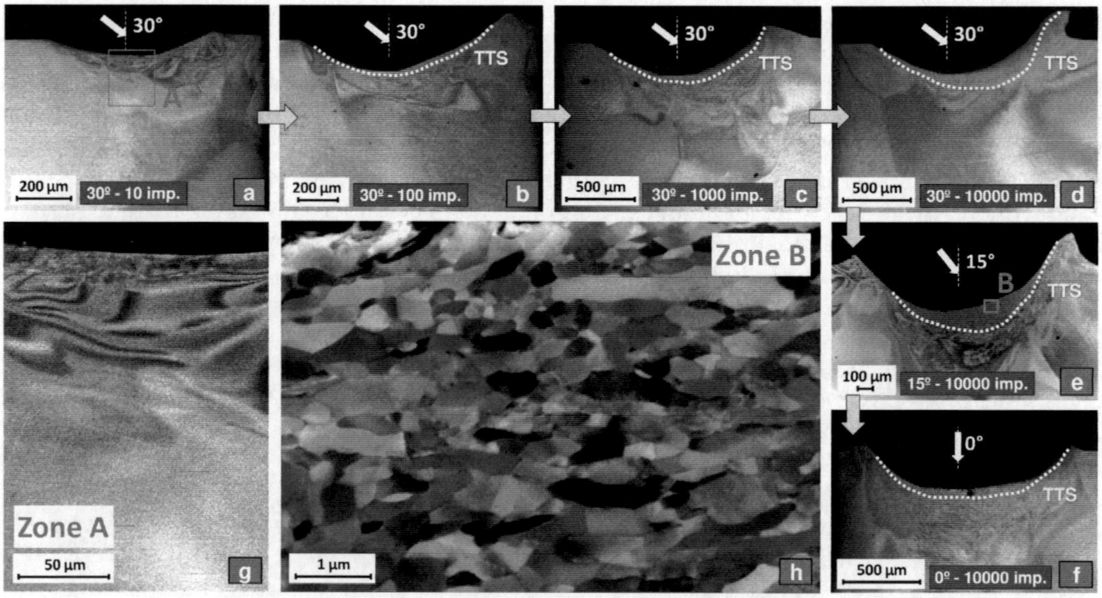

Figure 4. TTS layer growth at different amounts of impacts (30°): (a) 10, (b) 100, (c) 1000 and (d) 10k. Global views of 10k impact prints are compared for three different angles: (d) 30°, (e) 15° and (f) 0°. Local views in the near-surface are presented for the prints formed at (g) 10 and (h) 10k impacts.

The evolution of the TTS layer is shown in Figure 4 as a function of the impact amounts for 30°: (a) 10, (b) 100, (c) 1000 and (d) 10000 impacts. The TTS layer and the print diameter grow progressively with the number of impacts. The microstructural refinement leads sub-micrometric grain sizes (d < 1 µm) in the TTS layer for the highest amounts of impacts (Figure 4-h). On the contrary, the TTS layer is almost invisible beneath 100 impacts and the formation of well-defined grains is essentially nonexistent (Figure 4-g). Indeed, the mechanical deformation of few impacts is not enough to refine the near-surface microstructure [10],[13]. Likewise, considering the same number of impacts (10000) at different angles (0°, 15°, 30°), the refined microstructure and the TTS layer evolves differently (Figure 4-d,e,f). As shown in Figure 3, the tilted impacts (15°, 30°) are more propitious to have wider transformed regions.

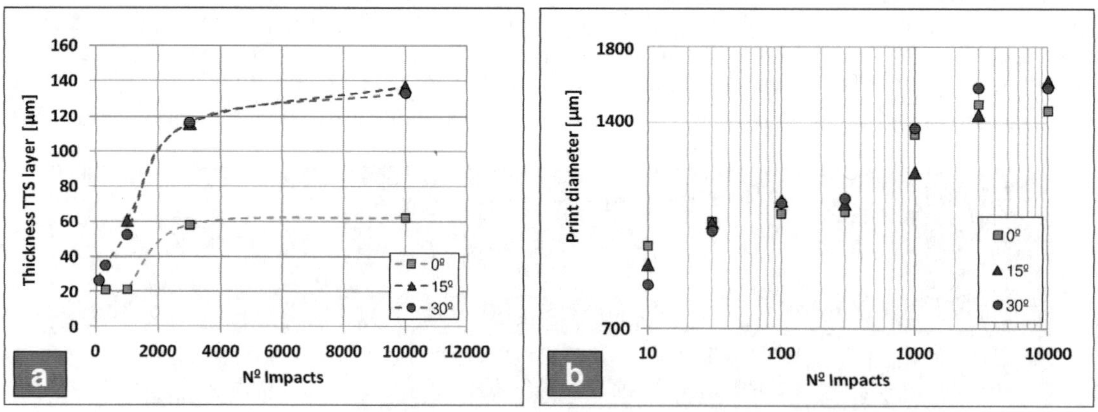

Figure 5. Parametric analysis of (a) the TTS layer thickness and (b) the print diameter as a function of the number of impacts at 0°, 15° and 30°.

Two parametric graphics are presented in Figure 5-a,b in order to quantify the impact condition sensitivity on the TTS layer formation as a function of the number of impact [6]: (a) the thickness of the TTS layer and (b) the mean diameter of the print. The first graphic (Figure 5-a) shows that the thickness of the TTS layer evolves following a power law behaviour. Indeed, the transformed region grows progressively up to 3000 impacts and then it stabilizes in a plateau. As observed in the SEM images presented below, the higher widths of the refined layer are measured for the cases of tilted impacts (15° and 30°). As shown in this graphic (Figure 5a), the thickness of these TTS layers could be more than twice the transformed region with normal impacts (0°). Likewise, it is particularly interesting that the curves for 15° and 30° are juxtaposed. The second graphic (Figure 5-b) shows the growth of the mean diameter in a log-log scale. The print average diameter evolves in the same manner for all the impact angles (0°, 15°, 30°). It would mean that the impact tilt only affects the width of the refined region and not the global size of the print, in spite of the bulge formed on the side.

4. Discussion

The parametric analysis pointed out that the impact angle has a relevant local effect on the microstructural refinement without changing the global dimensions of the print. This could mean that the impact tilt is strongly related with the microstructural transformation mechanisms in the material due to the SPD [9],[13]. To disclose this question, two sequences of the TTS layer growth at 15° are presented in Figure 6: (a, b, c) SEM images and (d, e, f) EBSD maps. In both cases, the transformed layer grows from the external side of the print and evolves progressively further below to the center, just underneath the normal impact axis. It suggests that the microstructural transformation starts from a shearing contact process (white arrows) and spreads in-depth (red and black arrows) with a combined phenomenon of shear and normal impact. Considering that shear bands and dislocation cells are quite significant for the refinement process in high stacking fault energy materials (as pure α-iron [3]), it is not surprising that shear contact and normal impact take a major role on the near-surface transformation [9]-[10]. Furthermore, the EBSD maps show that well-defined grains are not formed anymore outside the TTS region. Indeed, the remaining bulk grains beneath the refined layer perceive a low crystal misorientation in spite of the SPD induced in the mechanically affected zone (MAZ).

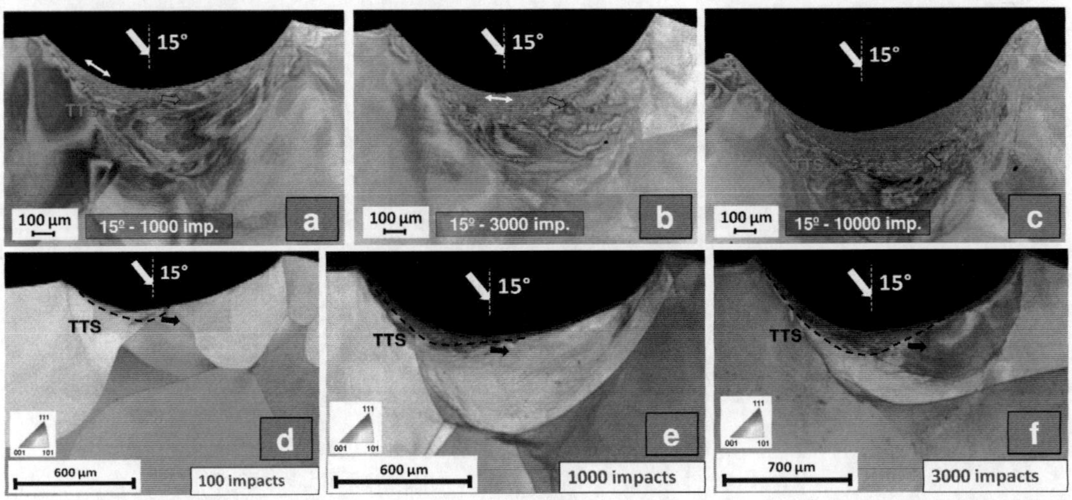

Figure 6. Sequence of the TTS layer formation and growth observed by the means of (a, b, c) SEM imagery (BSE detector) and (d, e, f) coarse EBSD mapping (2 µm indexation step).

5. Conclusions

A parametric investigation had been presented in this work in order to understand the effects of different impact conditions (micro-percussion) on the microstructural refinement of hyper-deformed surfaces. The main results had shown that tilted impacts (θ >> 0º) could produce TTS layers much wider (at least twice) than those formed by normal impacts (θ = 0º). Furthermore, it had been pointed out that the refined layer (sub-micrometric grains) grows progressively with the number of impacts. However, the microstructural transformation is done by a combined effect of shearing contact and normal impact, probably related with the microstructural refinement mechanisms in high stacking fault energy materials (pure α-iron): (i) the shear bands and (ii) the dislocation mobility [10].

From this work, micro-percussion tests could be an interesting method to emulate the transformed surfaces obtained by the means of conventional techniques (as shot-peening or SMAT). Firstly, the refined microstructure in the near surface (for α-iron) is absolutely comparable with the one obtained using industrial procedures [11],[12],[14]. Secondly, the size of the transformed layer is wide enough to be compared with conventional treatments and to analyse the microstructural and the mechanical gradients with several characterization techniques (as SEM, EBSD, nano-indentation, micro-pillar compression, etc.) [11]-[12]. Furthermore, this work highlights that α-iron is an appropriate model material to obtain well-defined TTS and it is favourable for experimental characterization methods. For a future work, it would be interesting to go further on the local mechanical characterization (nano-indentation [19] and micro-pillar compression [15]) and match it with the microstructural characterization presented above. This coupled experimental analysis would reveal the influence of different strengthening effects in surfaces as the Hall-Petch effect or the work hardening [20]-[21].

6. Acknowledgment

This work is supported by the LABEX MANUTECH-SISE (ANR-10-LABX-0075) of Université de Lyon, within the program "Investissements d'Avenir" (ANR-11-IDEX-0007) operated by the French National Research Agency (ANR). The authors would like to thank Dr. Xavier Maeder and Dr. Johann Michler from EMPA-Thun (Switzerland) for their expertise and help on the EBSD mapping characterization. Finally, the authors thank the laboratory IRTES-LERMPS of the UTBM in France for their contribution on the micro-percussion testing.

References

[1] Nordin E and Alfredsson B 2016 *Journal of Materials Processing Technology* **235** 143–148
[2] Marteau J and Bouvier S 2016 *Surface & Coatings Technology* **296** 136–148
[3] Lu K and Lu J 2004 *Materials Science and Engineering A* **375–377** 38–45
[4] Descartes S, Busquet M and Berthier Y 2011 *Wear* **271** 1833-1841
[5] Al-Baida H, Langlade C, Kermouche G and Ambriz R-R 2015 *J. Mater. Res.* **30** 2222-2230
[6] Kermouche G, Pacquaut G, Langlade C and Bergheau J-M 2011 *C. R. Mecanique* **339** 552–562
[7] Al-Baida H, Langlade C, Kermouche G and Ambriz R-R 2014 *Matériaux & Techniques* **102** 604
[8] Juran P, Liotier P-J, Maurice C, Valiorgue F and Kermouche G 2015 *C. R.Mecanique* **343** 344–353
[9] Tao N-R, Wang Z-B, Tong W-P, Sui M-L, Lu J and Lu K 2002 *Acta Materialia* **50** 4603–4616
[10] Sakai T, Belyakov A, Kaibyshev R, Miura H and Jonas J 2014 *Progress in Materials Science* **60** 130–207
[11] Tumbajoy-Spinel D, Kermouche G, Descartes S, Bergheau J.-M, Lacaille V, Guillonneau G and Michler J 2015 Matériaux & Techniques **103** 303
[12] Tumbajoy-Spinel D, Descartes D, Bergheau J-M, Lacaille V, Guillonneau G, Michler J and Kermouche G 2016 *Materials Science and Engineering A* **667** 189–198
[13] Segal V-M 2005 *Materials Science and Engineering A* **406** 205–216
[14] Lacaille V, Kermouche G, Tumbajoy-Spinel D, Feulvarch E, Morel C and Bergheau J-M 2014 *IOP Conf. Series: Materials Science and Engineering* **63** 012124
[15] Sun Z, Retraint D, Guelorget B and Waltz L 2015 Matériaux & Techniques **103**, 304
[16] Okada M, Suenobu S, Watanabe K, Yamashita Y and Asakawa N 2015 *Mechatronics* **29** 110–118
[17] Jacquet G, Kermouche G, Courbon C, Tumbajoy-Spinel D and Rech J 2014 *IOP Conf. Series: Materials Science and Engineering* **63** 012039
[18] Descartes S, Desrayaud C, Niccolinic E and Berthier Y 2005 *Wear* **258** 1081–1090
[19] Deng S-Q, Godfrey A, Liua W and Hansen N 2016 *Scripta Materialia* **117** 41–45
[20] Frutos E, Multigner M and González-Carrasco J-L 2010 *Acta Materialia* **58** 4191–4198
[21] Zou Y, Maiti S, Steurer W and Spolenak R 2014 *Acta Materialia* **65** 85–97

Effect of grain refinement on mechanical properties of martensitic steel

R K Islamgaliev[1], M A Nikitina, A V Ganeev, M V Karavaeva

Ufa State Aviation Technical University, Institute of Physics of Advanced Materials 12 K. Marx str., Ufa 450008 Russia

E-mail: [1]rinatis@mail.ru

Abstract

Microstructure and strength of commercial martensitic steel processed by equal channel angular pressing (ECAP) has been investigated using transmission electron microscopy and tensile tests. Application of ECAP led to significant grain refinement to a grain size of 0.8 μm and precipitation of particles with a size of 90 nm. Impact of annealing on thermal stability of ultrafine-grained structure was investigated. The contribution of various strengthening mechanisms to strength of steel was discussed.

1 Introduction

High strength in martensitic steels at room and elevated temperatures is traditionally ensured by two-stage thermal treatment. At the first stage strengthening takes place due to martensitic transformation as a result of quenching. At the second stage strengthening is provided by dispersion hardening at the expense of precipitation during tempering of the quenched steel [1-4].

It has been recently shown that the strength of steels can be enhanced by grain refinement using severe plastic deformation (SPD) [5-8]. In this case an ultrafine-grained microstructure with grain sizes less than 1 μm can be fabricated allowing additional strengthening in accordance with the Hall-Petch relationship [9,10]:

$$\sigma_y = \sigma_0 + K_y d^{-1/2} \tag{1}$$

where σ_0 – the friction stress of a crystal lattice, K_y – the coefficient of strengthening at the expense of grain boundaries, d – the mean grain size.

Usually a grain size depends on the parameters of deformation and thermal treatment, for example, when a deformation temperature increases, grain sizes grow [11]. Therefore, in order to achieve the maximum strengthening effect due to grain refinement, one should reduce the deformation temperature. However, several publications have appeared recently [7,8,12], which demonstrate that the dependence of strength on the deformation temperature is not always decreasing for steels. For example, the maximum strength for carbon steels 10 and 45 was observed after severe plastic deformation at 350 °C [7, 8]. The observed strength in the austenite steel after deformation at 300 °C is higher than that after deformation at room temperature [12]. This is explained by a complex microstructure observed in steels as a result of severe plastic deformation. The deformation temperature affects not only grain sizes, but also such structural features as sizes and distribution of secondary particles, formation of segregations, phase composition, formation of twins [7,8,11,12].

Therefore the influence of temperature on the microstructure and properties of complex-alloyed steels, for which dispersed particles are one of the main factors for strengthening, is especially ambiguous.

This work deals with the possibility of strengthening of martensitic steel as a result of grain refinement by equal channel angular pressing.

2 Experimental procedure

Martensitic steel in the shape of a hot-rolled rod were taken as initial material for investigations. Table 1 displays the chemical composition of the steel under study according to optical emission analysis using Q4 Tasman spectrometer.

Table 1. Composition of the steel, wt.%

C	Si	Mn	Ni	S	P	Cr	Mo	W	V
0.14	0.31	0.32	1.63	0.009	0.02	11.53	0.45	1.66	0.18

Processing by equal-channel angular pressing (ECAP) was performed using 6 passes on the route Bc with the angle of intersection of the channels equal to 120 degrees at the temperatures of 550°C on samples of 20 mm diameter and 100 mm long.

The microstructure of the alloy was characterized using JEM-2100 transmission electron microscope. Thin foils for TEM investigations were punched from slices of the ECAP-processed samples. Then they were mechanically ground to a thickness of 0.15mm, and finally double-jet electropolished to perforation using an electrolyte based on n-butyl alcohol.

The microhardness was measured on a Micromet-5101 device at a load of 100 g for 10 seconds.

3 Results

The microstructure of the steel in the initial state is a ferrite matrix with non-uniformly distributed particles of strengthening phases, which can be clearly seen on the image taken with the SEM (Fig.1a). The average grain size of the matrix phase is 1400 nm. The majority of particles is localized on grain boundaries of the ferrite matrix. The average size of particles is 170 nm, the volume fraction is 5 %. Clear grain boundaries can be observed in the fine structure (Fig. 1b). The majority of grains are free of dislocations. Pile-ups of dislocations can be seen in separate grains.

Non-uniform distribution of particles in the volume, as a rule, decreases the deformability of the material. Therefore, prior to deformation the samples are quenched in oil after heating and holding at 1050°C, for 1 hour, then they are tempered at 800°C for 1 hour.

After quenching martensitic microstructure can be observed (Fig.1 c, d). The average thickness of martensite plates is 518 nm. The martensite is characterized by high dislocation density within a plate. Separate dislocations as well as wide dislocation walls are observed within martensite plates. There are no secondary phase particles, which are typical of the initial state, in the martensitic structure.

After tempering uniform distribution of secondary phase particles in the volume is observed (Fig. 1 e, f). The particles precipitate in the boundaries of prior austenite grains and along the boundaries of martensite plates. Only about 25 % of particles are located in martensite plate bodies. The average size of particles is 160 nm, the volume fraction achieves 8.3 %. The plate structure of the martensite retains. The average width of plates is 550 nm.

As a result of ECAP an ultrafine-grained structure with an average grain size of about 0.8 μm has been fabricated (Fig. 2 a). The diffraction pattern with numerous spots located around the circle testifies to high-angle misorientations. The dark-field images display the particles of carbides in the body of grains with a volume fraction of 1.5% and a size of 90 nm (Fig. 2b, Table 2).

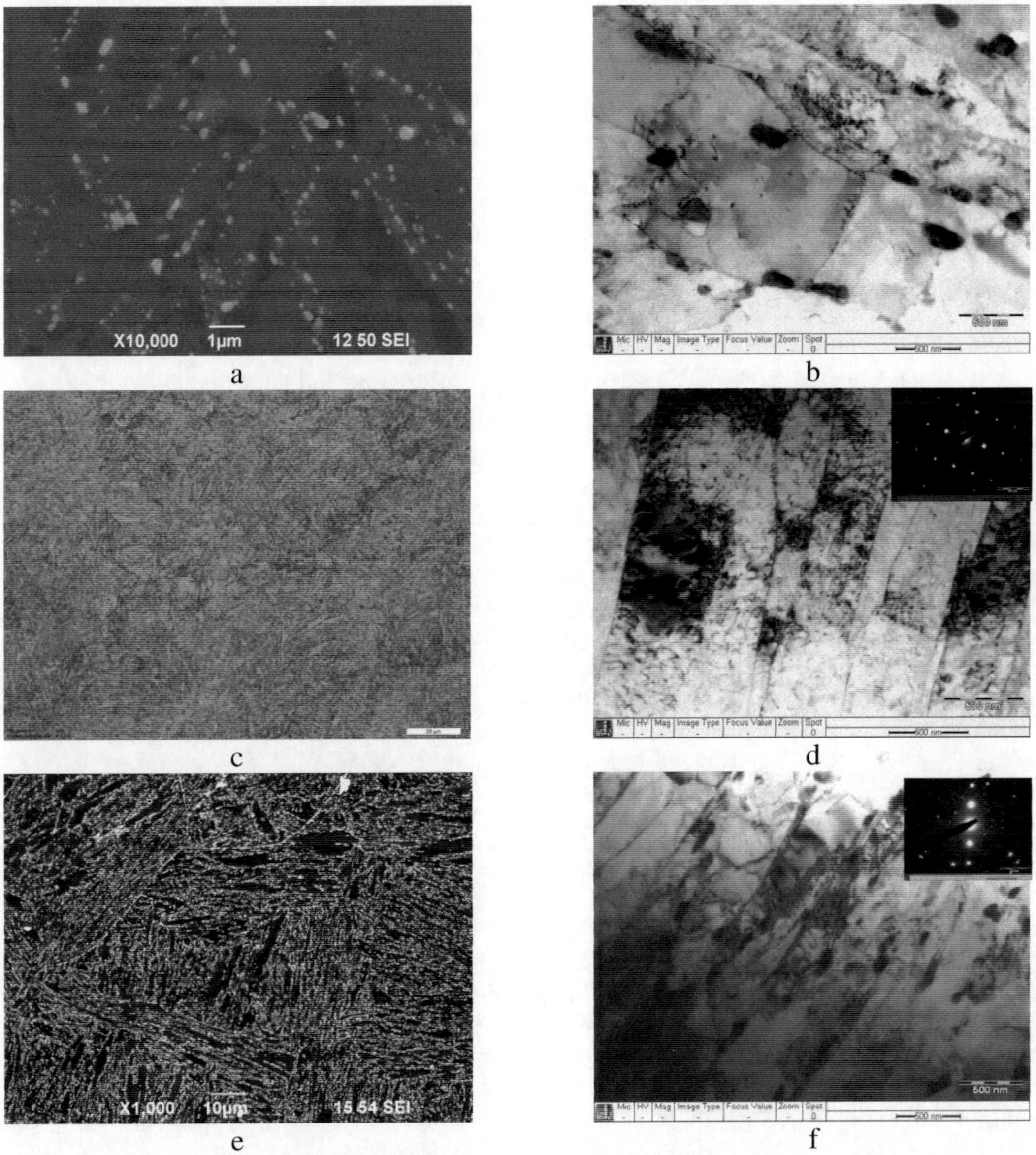

Figure 1. Microstructure of the initial (a, b); quenched (c, d) and tempered (e, f) samples

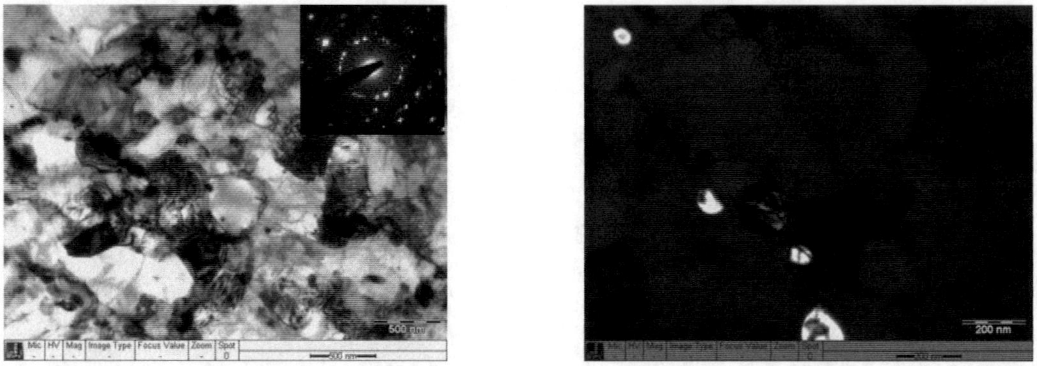

a b

Figure 2. Microstructure of steel after ECAP: a– bright-field image and diffraction pattern, b – dark-field image of carbides

Table 2. Characteristics of structure after various treatments

Treatment	Mean grain size, μm	Size of particles, nm	Volume fraction of particles, %
Tempered	1.4	160	8.3
ECAP	0.8	90	1.5

Transformation of microstructure after ECAP affects considerably the mechanical properties (Table 3). The microhardness in the initial state is 3140 MPa, after quenching and tempering the microhardness increases up to 4820 MPa, and after ECAP it demonstrates the value of 5230 MPa. After ECAP in comparison with tempered sample one can observe the increase of ultimate tensile strength and yield stress up to 1100 MPa and 1040 MPa without visible changes in ductility (Table 3).

Table 3. Mechanical properties of steel

Treatment	HV, MPa	UTS, MPa	YS, MPa	δ, %
Tempered	3500	950	790	8,0
ECAP	5230	1100	1040	8,1

4 Discussion

Enhancement of strength in martensitic steel as a result of severe plastic deformation is ensured by formation of a unique microstructure, in which a number of strengthening factors is combined. First of all, it is an ultrafine-grained matrix that ensures grain boundary hardening in accordance with the Hall-Petch relationship [9, 10]. Besides, solid solution hardening of an alloyed ferrite and dispersion hardening at the expense of high-dispersion particles of the strengthening phases take place for the complex-alloyed steel.

Solid solution hardening ($\Delta\sigma_{SS}$) is determined by the content of the alloying element and its strengthening effect [13]:

$$\Delta\sigma_{SS} = k_i c_i \tag{2}$$

where k_i – the strengthening coefficient for ferrite, which is an increment of the yield stress when 1 % (wt) of the *i*-th alloying element dissolves in it; (for carbon k = 4670 MPa/ % [14]) c_i – the concentration, % (wt), *i*-th alloying element.

Hardening at the expense of dispersion particles can be determined in accordance with the Orowan's ratio [15]:

$$\Delta\sigma_{or} = 1.2Gbf^{1/3}d^{-1} \tag{3}$$

where G=78 GPa – the shear modulus of the ferrite, b=0.25 nm – the Burgers vector for dislocations ½<111> of the ferrite, f, d – the volume fraction and size of particles.

The dislocation hardening can be calculated via the formula:

$$\Delta\sigma_d = \alpha M b G \rho^{1/2} \quad (4)$$

where $\Delta\sigma_d$ – the dislocation hardening, where α is the coefficient depending on the character of dislocation interaction in the course of work hardening; M is the orientation multiplier: for α -iron M = 2.75, and the product αM = 0.5, b – the Burgers vector, ρ - the dislocation density [14].

In general at first approximation the contribution of various mechanisms to the hardening is additive:

$$\sigma_y = \sigma_0 + \Delta\sigma_{ss} + \Delta\sigma_{or} + \Delta\sigma_{gb} + \Delta\sigma_d \quad (5)$$

where σ_0 - the friction stress of the α-iron lattice; $\Delta\sigma_{ss}$ - solid solution hardening; $\Delta\sigma_{or}$ - dispersion hardening; $\Delta\sigma_{gb}$ - grain boundary hardening; $\Delta\sigma_d$ - dislocation hardening.

The friction stress of the α-iron lattice (σ_0) is determined by the Peierls–Nabarro stress [16]:

$$\sigma_o = 2\times10^{-4} G \quad (6)$$

Table 4 lists the results of calculation of the contribution of various strengthening mechanisms in the yield stress of steel in accordance with equations (1-6).

The following assumptions were accepted during calculation of solid solution hardening. One-phase structure, namely solid solution of all the alloying elements, is taken for the quenched structure. The carbon content, which most significantly hardens α-Fe, in the tempered structure is taken as equilibrium (0.006 % according to diagram "iron-cementite").

Table 4. Contribution of various strengthening mechanisms to the yield stress of steel

Treatment	Calculated data					Experimental values	
	$\Delta\sigma_{gb}$, MPa	$\Delta\sigma_d$, MPa	$\Delta\sigma_{SS}$, MPa	$\Delta\sigma_{or}$, MPa	σ_y, MPa	UTS, MPa	YS, MPa
tempered	507	20	28	295	867	950	790
ECAP	670	33	28	297	1028	1100	1040

The results listed in the table show that the estimated values are very close to the experimental ones, which proves the possibility of considering the principle of linear superposition when evaluating the contribution of various mechanisms in the yield stress in accordance with (5). Thus, the strength of the tempered steel is ensured mainly by grain boundaries and dispersion hardening. After ECAP the dispersion hardening reduces and contribution of grain boundaries increases whereas solid solution strenghening is not changed.

An interesting fact is that calculated data of the yield stress for the ECAP samples is not coincided with experimental value. One can suppose there are some changes in contribution of strengthening mechanism, for example, at the expense of segregations at grain boundaries [18, 19].

Conclusions

1. As a result of ECAP the ultrafine-grained structure has been produced in martensitic steel. Significant reduction of the volume fraction of secondary phases from 8.3 % in the tempered state to 1.5 % in the deformed state is observed.
2. Evaluation of the contribution of various strengthening mechanisms showed that the strength of ECAP samples ensures mainly by grain boundary hardening, and the combination of various strengthening mechanisms makes it possible producing steels with enhanced mechanical properties.

Acknowledgements
RK is grateful to the Russian Science Foundation for the support of this work through project No. 15-19-00144. MA and AV are grateful to the Russian Ministry of Education within the project part of the program for universities through project No. 16.2061.2017/PH.

References
[1] Song M, Zhu R, Foley D C, Sun C, Chen Y, Hartwig K T and Zhang X 2013 *J Mater Sci* **48** 7360
[2] Song M, Wub Y D, Chen D, Wang X M, Sun C, Yu K Y, Chen Y, Shao L, Yang Y, Hartwig K T and Zhang X 2014 *Acta Materialia* **74** 285
[3] Fan Z Q, Hao T, Zhao S X, Luo G N, Liu C S and Fang Q F 2013 *Journal of Nuclear Materials* **434** 417
[4] Yang G, Huang C X, Wang C, Zhang L Y, Hu C, Zhang Z F and Wu S D 2009 *Materials Science and Engineering* A **515** 199
[5] Valiev R Z, Islamgaliev R K and Alexandrov I V 2000 *Prog Mater Sci* **45** 103
[6] Zhilyaev A P and Langdon T G 2008 *Prog Mater Sci* **53** 893
[7] Ganeev A V, Karavaeva M V, Sauvage X, Courtois-Manara E, Ivanisenko Y and Valiev R Z 2014 *IOP Conf. Series. Mater. Sci. Eng.* **63** 012128
[8] Karavaeva M V, Kiseleva S K, Ganeev A V, Protasova E O, Valiev R Z, Ganiev M M and Simonova L A 2015 *J Mater Sci* **50** 6730
[9] Hall E O 1951 *Proc Phys Soc* **B64** 747
[10] Petch N J 1953 *J Iron Steel Inst.* **174** 25
[11] Huang X, Yangb G, Gao Y L, Wu S D and Zhang Z F 2008 *Mat Sci Eng* A. **485** 643
[12] Abramova M M, Enikeev N A, Valiev R Z, Etienne A, Radiguet B, Ivanisenko Y and Sauvage X 2014 *Materials Letters* **136** 349
[13] Schmauder S and Kohler C 2011 Atomistic simulations of solid solution strengthening of a-iron. *Computational Materials Science* **50, 4** 1238
[14] Goldstein MI, Litvinov VS, Bronfin BM (1986) Metal physics of the high-strength state. Metallurgiya, Moscow (in Russian)
[15] Ning J l, Courtois-Manara E, Kurmanaeva L, Ganeev AV, Valiev R Z , Kübel C and Ivanisenko Y 2013 *Mat Sci Eng* A **581** 8
[16] Nabarro F R N 1997 *Mat Sci Eng* A **234-236** 67
[17] Pavlina E J and Van Tyne C J 2008 *J Mater Eng Perf* **17, 6** 888
[18] Sabirov I, Murashkin M Yu and Valiev R Z 2013 *Mat Sc. Eng* A **560** 1
[19] Nurislamova G, Sauvage X, Murashkin M, Islamgaliev R and Valiev R 2008 *Phil Mag Lett* **88, 6** 459

IOP Conf. Series: Materials Science and Engineering **194** (2017) 012026 doi:10.1088/1757-899X/194/1/012026

Realizing Ultrafine Grained Steel by Simple Hot Deformation Using Dynamic Transformation and Subsequent Dynamic Recrystallization Mechanisms

L Zhao[1,2], N Park[1,3], Y Tian[1,4], A Shibata[1] and N Tsuji[1]

[1] Kyoto University, Kyoto, Japan
[2] Colorado School of Mines, Golden CO, U.S.
[3] Yeungnam University, Gyeongsan Gyeongsangbuk-do, Korea
[4] Institute of Metal Research, CAS, Shengyang, China
Email: zhao@mines.edu, nobuhiro-tsuji@mtl.kyoto-u.ac.jp

Abstract. We have found a new strategy for ultra grain refinement without high strain deformation by combining dynamic transformation (DT) and dynamic recrystallization (DRX) mechanisms. Through simple thermomechanical processes using a total plastic strain of 0.92 at elevated temperatures, ultrafine grained microstructures having mean grain sizes down to 0.35 microns could be obtained in a 10Ni-0.1C steel. DRX phenomenon occurring in the dynamically transformed ferrite significantly reduced the strain necessary for the formation of ultrafine grains. The DRX of DT ferrite showed an unconventional temperature dependence, which suggested an optimal condition for grain refinement. The obtained UFG steel exhibited superior mechanical properties, for example, the tensile strength of 970 MPa and the total elongation over 20% at the same time.

1. Introduction

Ultrafine-grained steels having simple chemical compositions (e.g. plain carbon steels) and mean grain sizes smaller than 1 μm have large potential and benefits to replace some high-cost alloy steels with similar mechanical properties. Thermomechanically controlled processes (TMCP) can realize ultrafine ferrite (UFF) through various kinds of solid-solid reactions, i.e., recrystallization and phase transformation, occurring during the high-temperature deformation processes [1-3]. Among the metallurgical mechanisms in TMCP, dynamic transformation (DT), which is the transformation from austenite to ferrite occurring during deformation of austenite, can achieve UFF of 1~3 μm [4-6]. However, the high strain required for fabricating UFF is the main obstacle to the practical application of the TMCP including DT. To obtain a homogeneous UFF structure throughout the thickness of materials, a large accumulated strain over 2 is usually required [4]. As a result, so far, UFF has been only achieved in the surface layers of plates [5]. Recently, some reports have revealed that in the TMCP where DT occurs, DT may not be the only mechanism for ferrite grain refinement [7,8]. Dynamic recrystallization (DRX) of DT ferrite may occur, which suggests a possibility of further development in ferrite grain refinement.

In this work, we investigated refinement mechanisms of ferrite grains formed during dynamic austenite to ferrite transformation using a 10Ni-0.1C steel [8,9]. Occurrence of DRX was confirmed and it was suggested that DRX was the essentially important mechanism for the grain refinement of ferrite. We found an unconventional temperature dependence of the DRX (of DT ferrite) behavior, which indicated an optimal condition for grain refinement, leading to a new strategy proposed in this work which could make it possible to produce UFG steels without high-strain deformation in industrial processes.

2. Experiment and Methods

The material used in this study is a 10Ni-0.1C steel (C: 0.111, Ni: 10.08, Mn: 0.01, P: 0.001, Si: 0.006, Al: 0.33, S: 0.0017, Fe: bal. (wt.%)). Cylindrical specimens with a height of 12 mm and a diameter of 8 mm were machined from the homogenized plate heat treated at 1100 °C for 172,800 s in vacuum and then installed in a thermomechanical processing simulator (Thermecmastor-Z). The specimens were austenitized at 1000 °C for 300 s, cooled to a temperature of 520 °C (austenite + ferrite region) at a rate of 30 °C s^{-1}, held for 60 s, and then the specimens with single austenite phase were uniaxially compressed at 520 °C to various strains at strain rates ranging from 10^{-1} s^{-1} to 10^{-3} s^{-1} followed by water-quenching.

Microstructures at the center of the specimens on the sections parallel to the compression axis were characterized by a field-emission type scanning electron microscope (FEG-SEM) equipped with an electron back-scattering diffraction (EBSD) and also by transmission electron microscopy (TEM, Philips CM200FEG). The specimens for the EBSD analysis were mechanically polished and then electrically polished in a solution of a 10% $HClO_4$ and 90% CH_3COOH at 20 °C. The thin-foil TEM specimens were prepared by twin-jet electropolishing using the same solution as that for EBSD. Mechanical properties after the TMCP were examined by an uniaxial tensile test at room temperature at an initial strain rate of 8.3×10^{-4} s^{-1} [10,11].

3. Results and Discussion

3.1. Occurrence of DRX

Figure 1a-1d shows microstructure evolution of ferrite during deformation. Martensite transformed from austenite during quenching could be distinguished from ferrite due to higher misorientations [8-11], and is painted in black in Fig.1. At a strain of 0.11 (Fig. 1a), very fine and nearly equiaxed ferrite grains are formed along a prior austenite grain boundary (PAGB). With increasing the strain to 0.36 (Fig. 1b), the ferrite grains grow into prior austenite grains and exhibit an irregular morphology. Some low-angle boundaries (LABs) are observed within the coarse ferrite grains, indicating that a deformed structure is introduced during the growth. The fraction of LABs increases with increasing the strain to 0.60 (Fig. 1c), at which equiaxed UFF grains surrounded by high-angle boundaries (HABs) form along the grain boundaries of coarse ferrite (as marked by blue arrows in Fig. 1c). The fraction of the equiaxed UFF increases with increasing the strain to 1.39 (Fig. 1d). A typical microstructure of the equiaxed UFF grain being formed is shown in Fig. 1e by TEM image with misorientation map. There are mainly two coarse DT ferrite grains divided by HAB (blue solid lines), and several subgrains. Dislocations are generated in ferrite grains due to straining and gradients of dislocation densities seem to be developed near the grain boundaries of the DT ferrite, possibly because of deformation incompatibilities between the grains. This is accompanied by subgrain formation (i.e., formation of LABs), leading eventually to the development of boundary corrugations or serrations. With straining, higher concentration of deformation would occur near high-angle grain boundaries of ferrite [19], which promotes the accumulation of dislocations and the development of misorientation near the grain boundary area of DT ferrite. The features of UFF formation in this process accord well with those of DRX of ferrite [12].

The mechanism of DRX of DT ferrite is interpreted by Fig. 2. Basically there are two regions in Fig. 2a: a region prior to the onset of DT (No-DT region) and a region after the onset of DT (DT-region). In the DT-region, there is another stage during which DRX (UFF formation) occurs. At the beginning of DT, ferrite grains are transformed from austenite during deformation. The grain size of DT ferrite firstly increases, indicating a grain growth process at a relatively early stage of transformation, but then decreases after a certain strain. The decrease in the ferrite grain size accords well with the development of DRX grains (Fig. 2b) in subsequent deformation.

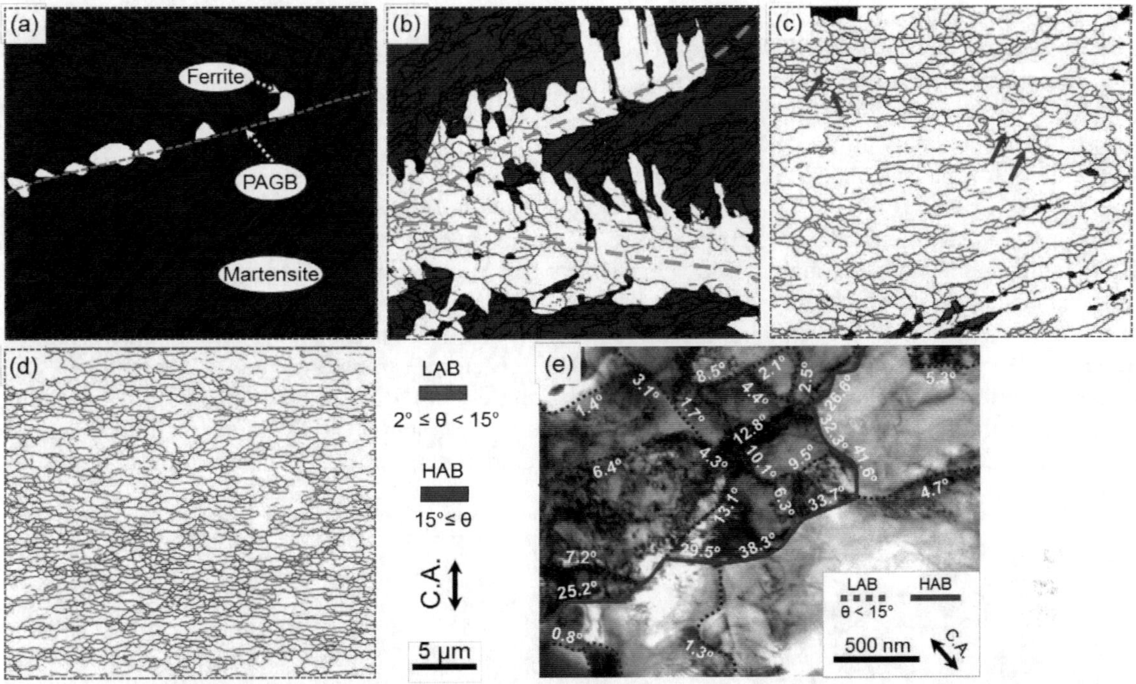

Fig. 1 (a-d) EBSD grain boundary maps of the 10Ni-0.1C steel deformed to different strains of (a) 0.11, (b) 0.36, (c) 0.60, and (d) 1.39 at a strain rate of 10^{-2} s^{-1} at 520 °C. (e) TEM image corresponding to (d). C.A. indicates the compression axis.

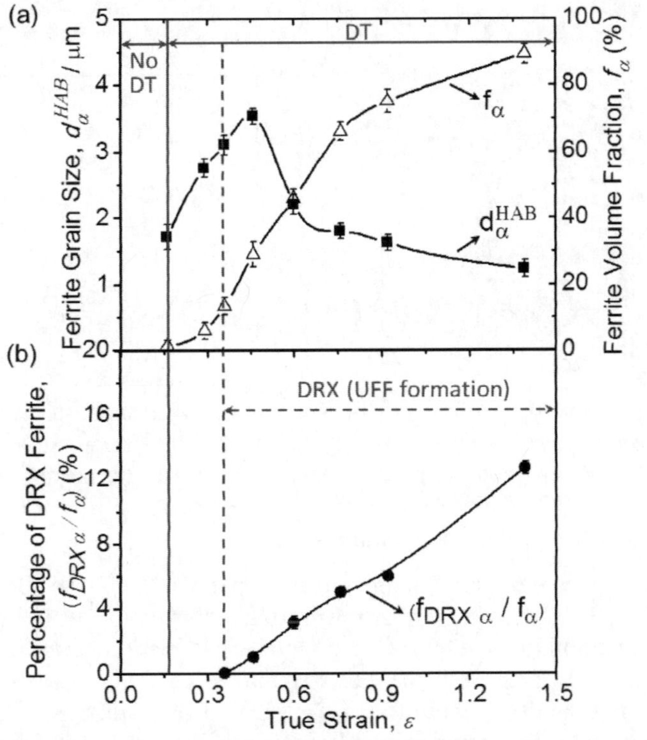

Fig. 2 Variations of (a) ferrite grain size (counting only HABs), volume fraction of the dynamically transformed ferrite and (b) volume fraction of the DRX ferrite as a function of hot-compression strain.

3.2. Unconventional Temperature Dependence of DRX of DT Ferrite.

The temperature dependence of DRX of DT ferrite is further investigated. Figure 3a-3c shows EBSD grain boundary maps of specimens deformed to a strain of 0.92 at different temperatures. At 560 °C (Fig. 3a), the ferrite mostly shows coarse and elongated morphologies. The volume fraction of DT ferrite is 52%, and only a few DRX grains having fine and equiaxed morphologies can be observed, as marked by blue arrows. At 520 °C (Fig. 3b), the fraction of DRX grains increases compared to that at 560 °C. At 440 °C (Fig. 3c), on the other hand, the fraction of DRX grains decreases. Most of the grains are elongated and contain a large amount of LABs, exhibiting more deformed structures. The HABs surrounding ferrite grains at 440 °C are more like geometrically necessary boundaries (GNB) [13] rather than being formed by DRX mechanism. Figure 3d summarizes the changes in the grain size, volume fraction of DT ferrite and volume fraction of DRX ferrite as a function of deformation temperature. At temperatures from 440 °C to 520 °C, the ferrite transformation (solid rectangles) is completed after a strain of 0.92. The grain size (solid circles) of DRX ferrite increases with increasing temperature from 440 °C to 560 °C. It is interesting that the fraction of DRX ferrite (solid triangles) exhibits a peak value at 520 °C, indicating an optimal temperature for the formation of DRX grains.

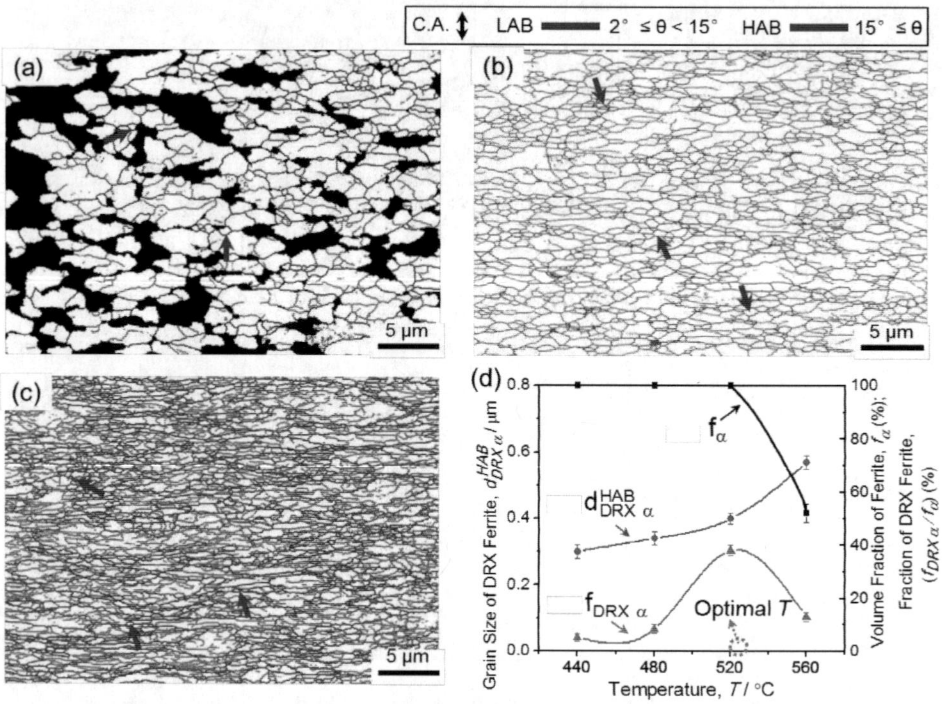

Fig. 3 (a-c) EBSD grain boundary maps of the specimens with an austenite grain size of 10 μm deformed to a strain of 0.92 at a strain rate of 10^{-2} s^{-1} at different temperatures. Non-ferrite phases were painted in black. C.A. indicates the compression axis. (d) Variations of the mean grain size ($d_{DRX\alpha}$), volume fraction of DRX grains ($f_{DRX\alpha}$) and volume fraction of dynamically transformed (DT) ferrite (f_α), as a function of the deformation temperature.

The reasons for the unconventional temperature dependence of the DRX of DT ferrite are explained as follows. It has been known that lowering deformation temperature (above the nose temperature in transformation kinetic C-curve) shortens the incubation period for the onset of DT and accelerates its kinetics [14]. At 560 °C, the driving force and nucleation density for ferrite transformation are low, but the diffusivity of atoms is high. Therefore, the grain growth of DT ferrite is rather dominant than nucleation, resulting in a large grain size. The combining effect of large grain size and temperature retards the onset of DRX. At 520 °C, both the driving force and the nucleation density for ferrite transformation increase, which leads to faster nucleation kinetics in DT. The fast nucleation results in finer ferrite grain size due to the sufficient impingement effect between ferrite grains. The fine ferrite grain size enhances the initiation of DRX and accelerates its kinetics [7,15]. At 440 °C, the diffusivity of atoms is so low that DRX is suppressed. The DRX behavior of DT ferrite exhibits a C-curve according to the deformation temperature, which implies an optimized temperature for the progress of DRX, to be more specific, for the formation of UFF grains.

3.3. New Strategy for Fabricating UFG Steel without High-Strain Deformation.

The novelty of the DRX of DT ferrite is that the initial grain size of ferrite formed by DT (prior to DRX) could be very small. This means that the required strain for the initiation of DRX and fabrication of fully ultrafine microstructure can be greatly reduced [7,15]. The present study has proved two important points: (1) DT can greatly affect the subsequent DRX behavior; (2) There exists an optimal temperature at which the kinetics of DRX of DT ferrite is the fastest. These two points give a new grain refinement strategy by controlling DT and DRX. That is, at the optimal temperature, DRX should be enhanced by accelerating DT kinetics and consequently leads to finer grains.

Figure 4a schematically illustrates a new TMCP route for grain refinement. The novelty of the new TMCP route is that at a relatively low temperature (T_{d2}), a small pre-deformation is applied for accelerating DT as well as subsequent DRX at a high temperature (T_{d1}). As a result, the final ferrite structure can be significantly refined to 0.46 μm with 75% HABs by applying only a true strain of 0.92. Figure 4b shows a possibility of further grain refinement through the new TMCP route with a true strain of 1.39. The fraction of HABs is 70% and the mean grain size is 0.35 μm.

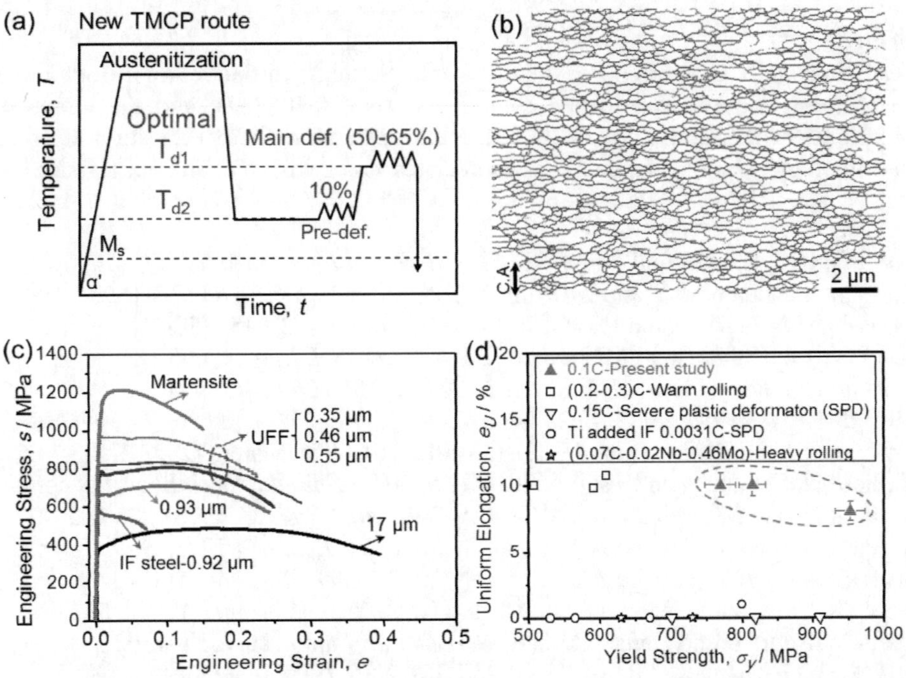

Fig. 4 (a) New TMCP route using DT and DRX for grain refinement; (b) EBSD grain boundary map of ferrite structures with a mean grain size of 0.35 μm obtained by the new TMCP route; (c) Mechanical properties of the UFF specimens processed in the new TMCP route, the specimen with coarse ferrite grains (17 μm), the specimen with full martensite, and an UFG IF steel (0.92 μm) obtained by accumulative roll-bonding and subsequent annealing [16]; (d) Comparison of the mechanical properties between the present UFF structures and those in previous works [17-20].

Figure 4c shows engineering stress-strain curves of the specimens processed in the new TMCP route (marked as UFF, mean grain sizes of 0.55-0.35 μm), the specimen with a statically transformed coarse ferrite grains (17 μm), the specimen with full martensite of the present 10Ni-0.1C steel, and an

UFG interstitial free (IF) steel (0.92 μm) obtained by accumulative roll-bonding and subsequent annealing. Compared to the specimen with statically transformed coarse ferrite, the UFF specimens showed significantly higher yield strength of 770-953 MPa and tensile strength of 810-973 MPa, which are much superior to the UFG IF steel and even close to the yield strength of martensite (1100 MPa). It has been known that UFF ferrite usually shows limited uniform elongation due to early plastic instability [21]. In such a context, it is rather surprising that the present UFF maintains large tensile ductility, i.e., uniform elongation of 8-10% and total elongation of 23-29%. Compared to the results in literatures [17-20], the plots of uniform elongation and yield strength of the present UFF structures are out of the trend in UFF steels obtained through processes including heavy deformations (Fig. 4d). And importantly, the present TMCP route combining DT and DRX mechanisms requires so far the smallest strain to achieve UFF structures.

Conclusions

We could obtain homogeneous UFF structures having a mean grain size down to 0.35 μm through a new TMCP route combining DT and DRX mechanisms, which required a true strain of only 0.92-1.39 in hot deformation. The DRX of DT ferrite showed an unconventional temperature dependence, which suggested the existence of an optimal condition for grain refinement. The obtained UFG steel exhibited superior mechanical properties with a tensile strength of 970 MPa and the total elongation over 20%. The present study could bring new excitement to the production of UFG steels without high-strain deformation in industries.

Acknowledgement

This study was financially supported by the Elements Strategy Initiative for Structural Materials (ESISM), the Grant-in-Aid for Scientific Research (S) (No.15H05767), and the Grant-in-Aid for Challenging Exploratory Research (No.26630365), all through Ministry of Education, Culture, Sports, Science and Technology (MEXT), Japan. L.Z. appreciates support from China Scholarship Council.

References

[1] Hodgson PD, Hickson MR and Gibbs RK 1999 *Scripta Mater.* **40** 1179–1184.
[2] Ueji R, Tsuji N, Minamino Y and Koizumi Y 2002 *Acta Mater.* **50** 4177–4189.
[3] Song R, Ponge D, Raabe D and Kaspar R 2005 *Acta Mater.* **53** 845–858.
[4] Hu J, Du LX, Xie H, Yu P and Misra RDK 2014 *Mater. Sci. Eng. A.* **605** 186–191.
[5] Hurley PJ and Hodgson PD 2001 *Mater. Sci. Eng. A.* **302** 206–214.
[6] Eghbali B 2010 *Mater. Sci. Eng. A* **527** 3407–3410.
[7] Galindo-Nava EI and Rivera-Díaz-del-Castillo PEJ 2014 Scr. Mater. 72–73 1–4.
[8] Zhao L, Park N, Shibata A and Tsuji N 2014 *TMS 2014 Annual Proceedings* 919–926.
[9] Zhao L, Park N, Tian Y, Shibata A and Tsuji N 2016 *Adv. Eng. Mater.* **19 (3)** 1600778.
[10] Zhao L, Park N, Tian Y, Shibata A and Tsuji N 2016 *Sci. Rep.* **6** 39127.
[11] Zhao L, Park N, Tian Y, Shibata A and Tsuji N 2017 *Mater. Res. Lett.* **5(1)** 61-68.
[12] Murty SVSN, Torizuka S, Nagai K, Kitai T and Kogo Y 2005 *Scripta Mater.* **53** 763-768.
[13] Kuhlmann-Wilsdorf D and Hansen N 1991 *Scr. Metall. Mater.* **25** 1557–1562.
[14] Shin DH, Park JJ, Chang SY, Lee YK and Park KT 2002 *ISIJ Int.* **42** 1490–1496.
[15] Humphreys FJ and Hatherly M 2004 *Recrystallization and related annealing phenomena*, Elsevier, Amsterdam.
[16] Gao S, Chen M, Chen S, Kamikawa N, Shibata A and Tsuji N 2014 *Mater. Trans.* **55** 73–77.
[17] Song R 2004 *Microstructure and mechanical properties of ultrafine grained C-Mn steels*, PhD thesis, Max-Planck-Institut für Eisenforschung.
[18] Shin DH, Park JJ, Chang SY, Lee YK and Park KT 2002 *ISIJ Int.* **42** 1490–1496.
[19] Liu MY, Shi B, Wang C, Ji SK, Cai X and Song HW 2003 *Mater. Lett.* **57** 2798–2802.
[20] Song R, Ponge D, Raabe D, Speer JG and Matlock DK 2006 *Mater. Sci. Eng. A* **441** 1–17.
[21] Tsuji N, Ito Y, Saito Y and Minamino Y 2002 *Scr. Mater.* **47** 893–899.

IOP Conf. Series: Materials Science and Engineering **194** (2017) 012027 doi:10.1088/1757-899X/194/1/012027

Microstructure, accumulated strain, and mechanical behavior of AA6061 Al alloy severely deformed at cryogenic temperatures

D C Magalhães, A M Kliauga, M Ferrante, V L Sordi

Universidade Federal de São Carlos, Department of Materials Engineering, 13565-9058 São Carlos, Brazil

kliauga@ufscar.br

Abstract. The combination of Severe Plastic Deformation (SPD) and cryogenic temperatures can be an efficient way to obtain metals and alloys with very refined microstructure and thus optimize the strength-ductility pair. However, there is still a lack of studies on cryogenic SPD process and their effects on microstructure and mechanical properties, especially in precipitation-hardenable aluminum alloys. This study describes the effect of low temperature processing on microstructure, aging kinetic and tensile properties of AA6061 Al alloy after cryo-SPD. Samples of AA6061 Al alloy in the solutionized state was processed by Equal-channel angular pressing (ECAP) at 77 K and 298 K, up to accumulate true strains up to 4.2. Results indicated that the aging kinetic is accelerated when deformation is performed at cryogenic temperature, dislocation density measurement by x-ray and diffraction analysis at TEM achieved a saturation level of 2x 10^{15} m^{-2} by ECAP at 298K and 5x10^{15} m^{-2} after cryogenic ECAP plus precipitation hardening. The same level of yield strength was observed in both deformation procedures but an improvement in uniform elongation was achieved by cryogenic ECAP followed by a T6 treatment

Introduction

Among the Al alloys the AA6xxx series is usually employed in structural applications due to excellent mechanical strength, high corrosion resistance, good weldability and machinability[1, 2]. In the recent years, different alternative materials have been used to cryogenic conditions, such as aluminum alloys for building ships, liquid natural gas carriers, icebreakers and off shores plants that operated at low temperature in polar and subsea regions [3]. Thus, it is important to understand the mechanical behavior of the AA6xxx group particularly for the determination of processing parameters under different conditions.

This alloy is normally thermomechanically processed above the precipitation hardening range, which yields a coarse grained structure, which will be further age hardened in order to achieve higher yield strength. The reported severe plastic deformation (SPD) of this alloys is done in the overaged state [4,5], or shortly after solid solution treatment [6,7] , that is during natural ageing, because their ductility at the peak hardened state its very low. There is few information about the interaction of

IOP Conf. Series: Materials Science and Engineering **194** (2017) 012027 doi:10.1088/1757-899X/194/1/012027

precipitation *after* SPD for this group of alloys. One of the purposes of this work is thus to study the deformation behavior of the AA6061 alloy at cryogenic temperatures and the influence of post precipitation heat treatments on the mechanical properties.

1. Materials and Methods

Initially, samples of AA6061 Al alloy were submitted to solution heat treatment at 530 °C for 2 hours to obtain a supersaturate solid solution (SSSS). Some specimens were submitted to aging heat treatments at 298 K (natural aging – T4) , 173 K (in salt bath – 173K T6), and 443 K (in salt bath – 443K T6).

The tensile tests were performed at 298, 173 and 77 K. The test at 173 K was performed in a closed cryogenic chamber INSTRON model 3119-610; the temperature was measured and controlled throughout the test making use of a type T thermocouple.The 77 K test was conducted with the specimen completely immersed in liquid nitrogen.

Prior to ECAP the samples were solution heat treated at 803K followed by quenching in water; for the room temperature route or in liquid nitrogen for the cryogenic route (see sketches in Figure 1a). The ECAP deformation was performed in a die with a corner angle of 120° and an external fitting angle of 22°, route Bc was applied. The deformation was performed either at room temperature or in the same cryogenic chamber used for the tensile tests, as sketched in Figure 1b . For the low temperature process the samples were maintained in liquid nitrogen between passes and reprocessing took only few minutes. The samples were deformed up to a equivalent deformation of 4.2. After deformation precipitation heat treatments were performed at 298 K (T4) or at 373 K (T6).

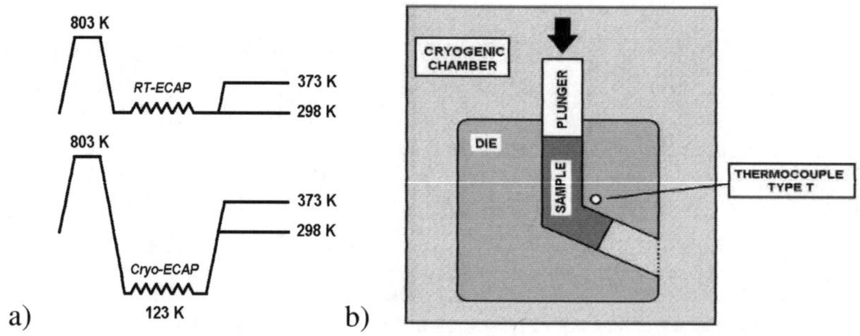

Figure 1. Schematic representation of the thermomechanical procedure (a) and of the cryogenic experiment (b).

Vickers microhardness measurements were carried out after different aging times. Miniature samples with a gauge length of 7 mm and 3 mm x 2 mm cross section area were employed for the tensile tests in the ECAP plus heat treated samples, tested at room temperature and at a nominal strain rate of $1 \times 10^{-3}\ s^{-1}$, with the elongation monitored by an optical extensometer.

The dislocation density was measured by x-ray diffraction in a SIEMENS D5000 equipment using Cu Kα radiation. Background subtraction was performed by using, alternatively, Si powder standard. The instrumental contribution was subtracted from the peak breadths, according to Cagliotti's equation [8]. On those diffractograms, peak broadening was analyzed by the Williamson-Hall (W-H) method [9], modified by Warren [10] Ungár [11-12] by using the following equation:

$$\frac{Breathcos\theta}{\lambda} - \beta W_g = \frac{1}{d} + \left(\frac{\pi M^2 b^2}{2}\right)\rho^{1/2}K^2C \qquad \text{(Equation 1)}$$

where d is the diffraction domain size, λ is the wavelength of the radiation, β is the twin density, W_g are the Warren constants related to the stacking faults, ρ is the dislocation density, b is the Burgers vector, M is a constant related to the cut-off radius of dislocations (which is smaller for more compact arrays), C are the average contrast factors of dislocations and $K=2\mathrm{Sin}(\theta)/\lambda$.

Samples for transmission electron microscopy (TEM) were prepared by electrolytic polishing (20% HNO_3 in methanol, 20V - 30°C) and observed in a CM120 FEI microscope.

2. Results and Discussion

Figure 2 shows the hardening behavior of the 6061 alloy at room temperature (T4) and at 373 and 443 K (T6). The temperature is an important factor to aging kinetics for AA6061 Al alloy. Detailed studies of precipitation sequence of AA6061 alloy can be founded in [13-16]. The simplified precipitation sequence of AA6061 [13] can be described as super-saturated solid solution (SSSS) $\rightarrow$ clustering stage $\rightarrow$ Mg, Si co-clusters $\rightarrow$ Guiner–Preston (GP)-I zones, pre-β" $\rightarrow$ β" $\rightarrow$ B', β' $\rightarrow$ β-Mg_2Si. Various other precipitates [14] have also been reported. The early stages of aging resulting in the formation of Mg, Si co-clusters [15].The natural aging, or T4 treatment, is very accentuated for this alloy, and after three hours it is already effective. In this condition, the microstructure is formed mainly cluster of Si and Mg, also called co-clusters. Other important feature is about the T6-443K treatment which reach a peak around 18 hours. This maximum is associated with formation of β" needles precipitates in the aluminum matrix. For the T6 treatment at 443K, shown in Figure 3, the major part or the precipitates are the pre-β" and β" needles are in the early stage of formation.

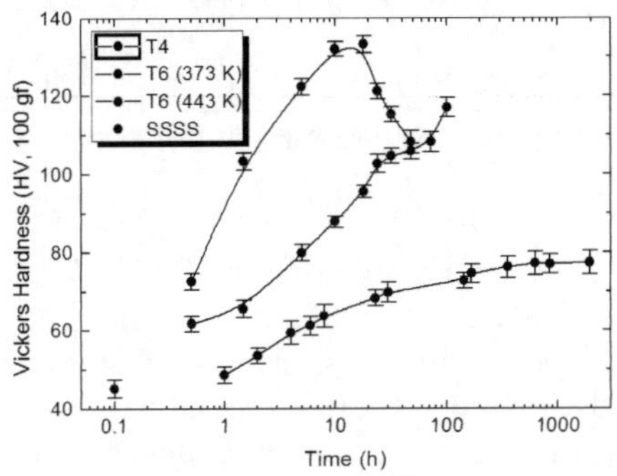

Figure 2. Hardening curves of the AA6061 ally at room temperature (T4) and at 373K (T6) .

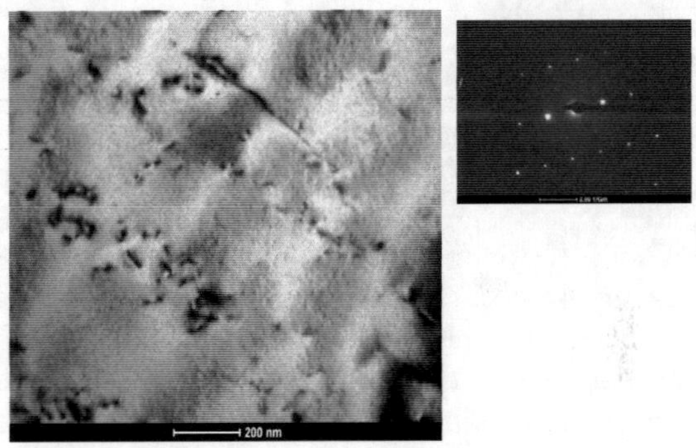

Figure 3. Transmission electron microscopy of the 6061 ally at the 443 T6 peak hardening condition

Xu, Roven and Jia [17], using an AA6060 alloy in T6 condition, determined significant changes in fracture surfaces after tensile tests at 77 K in comparison to room-temperature. Furthermore, the mechanical behavior was strongly affected by testing temperature, and the deformation mechanism by dislocation slip is more homogeneous at low temperature deformation. Esmaeili *et al.* [18] examined the AA6111 alloy subjected to tensile test between 298 K and 4.2 K under different precipitation treatments, indicating a complex behavior depending on the nature of the obstacles to the movement of dislocations, which implies strong changes in strain rate sensitivity and work-hardening response. Halim and co-authors [19] carried out tensile tests at 298 K and 223 K with AA5754 Al alloy, and their results indicated that the decrease of the test temperature inhibits the Portevin-Le Chatelier effect (PLC), leading to a more homogeneous distribution of solute atoms and dislocations than at room-temperature. All the above investigations concluded that both mechanical strength and ductility increase at low temperatures [17-21]. Fig. 4 shows the true stress true strain curves obtained at 298 K, 173 K and 77 K, with the alloy in the solid solution condition. The stress- strain curve at room temperature is characterized by the presence of the Portevin-Le Chatelier effect, which was suppressed at 173K and 77 K. A higher uniform elongation was obtained at 77 K as well as higher yield stress and strain hardening. this results are summarized in Table 1

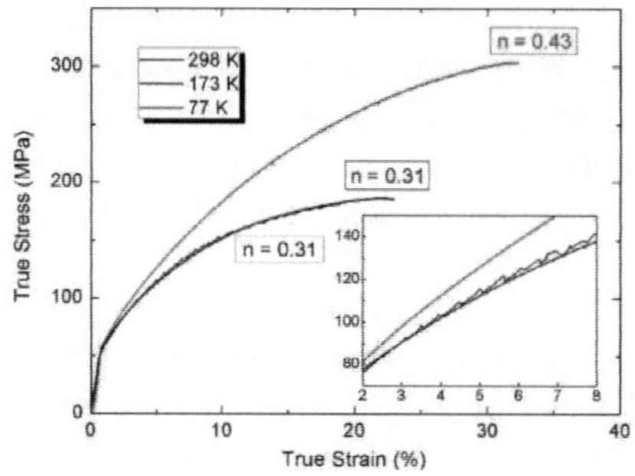

Figure 4. Tensile curves for solution-treated samples of AA6061, at 298 K, 173 K and 77 K, with respective work-hardening exponent. Magnification of the initial region of the curves (insert) shows the PLC effect at room-temperature. Initial strain-rate equal to 10^{-3} s^{-1}.

Table 1. Sumary of the tensile properties of AA6061 Al alloy in the solution heat-treated condition, at different testing temperatures. Strain rate = 10^{-3} s^{-1}.

Test temperature	σ_o (MPa)	σ_u (MPa)	ε_t (%)	ε_u (%)	RA (%)
298 K	54.5	150.2	30.1	20.2	76.9
173 K	60.1	158.0	33.2	23.0	72.0
77 K	72.3	233.6	46.6	35.1	61.6

Note: σ_o= yield stress; σ_u= ultimate stress; ε_t= total elongation; ε_u= uniform elongation; RA = reduction of area.

Figure 5 shows the samples deformed after 6 passes at room temperature and at 123 K. The room temperature sample presented cracks after the fourth pass and fracture was very evident after the sixth pass, whereas no fracture was observed at 123 K. Observation of the deformed samples in optical microscope showed that the strain was more homogeneous distributed at room temperature than at 123 K where deformation in macroscopic was more concentrated in macroscopic shear bands, as shown in Figure 6. Complementary to this information the hardness measurements just after deformation shows that the saturation level was reached faster at room temperature than for the cryogenic route. (see Figure 7). This indicates that the higher ductility at low temperature may be correlated to the strain distribution as well as to the absence of precipitation hardening.

After 6 passes the hardness for the two sets of samples was very similar, but the degree of grain subdivision at room temperature was more accentuated than at 123 K (see the TEM micrographs in Figures 8a - and 8c with the respective diffraction patterns in Figures 8b and 8d)

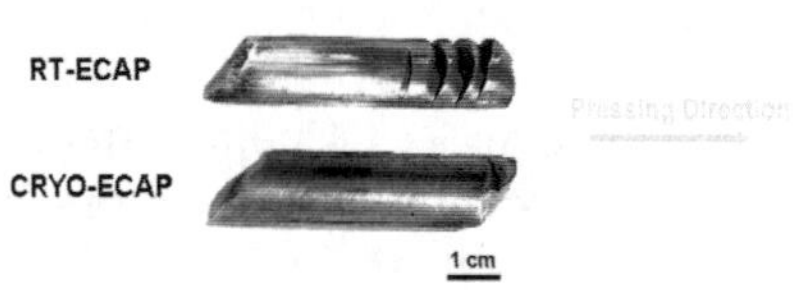

Figure 5. ECAP deformed samples after 6 passes

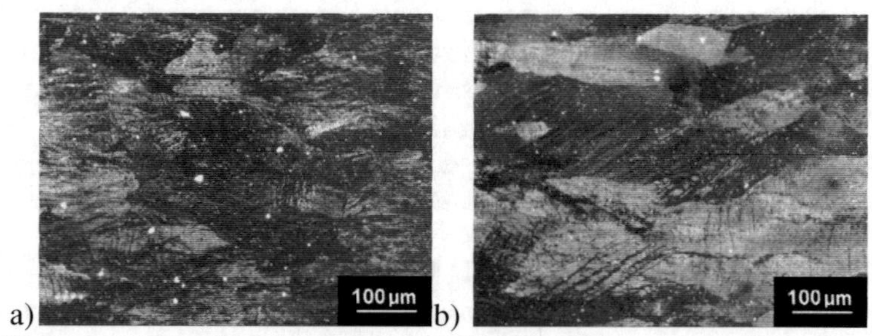

Figure 6. Optical microscopy of ECAP smples after 2 passes. a) room temperature; b) 123K

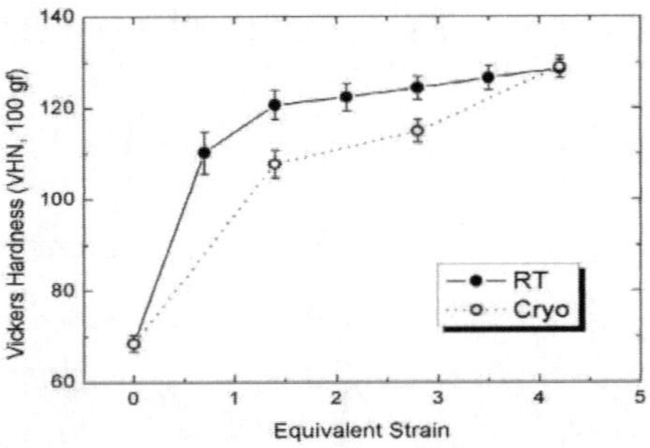

Figure 7. Hardness evolution after ECAP at room temperature and at 123K. Measurements performed just after deformation for both temperatures.

a) b) c) d))
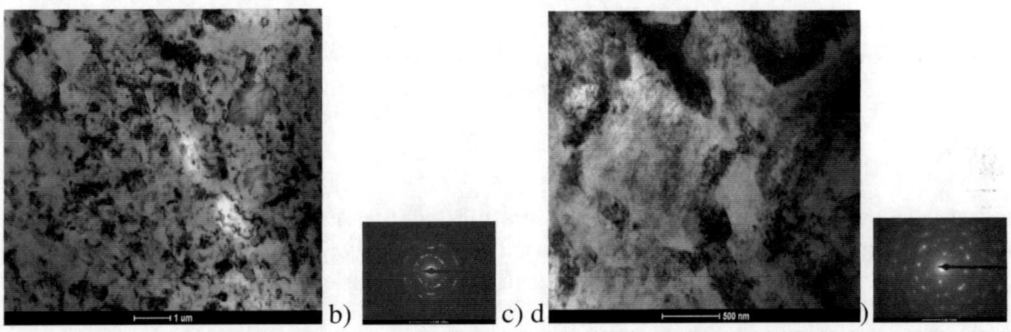

Figure 8. TEM micrographs of the AA6061 alloy after ECAP at room temperature and at 123K plus (T4) .

There was no accentuated difference in the dislocation density after 6 passes at room temperature and at 123K in the natural aged condition. The recovery process at 373K (T6) decreased the dislocation density at the same rate for both samples, as shown in Figure 9a. Figure 9b shows the hardness evolution after six ECAP passes for the T4 and the 373 K - T6 treatments The hardness evolution at T4 condition was very similar for both sets of samples. Whereas for the T6 condition the cryogenic samples presented a faster precipitation kinetics , which demonstrates the effectiveness of the cryogenic process in suppressing the early stages of clustering.

a) b)
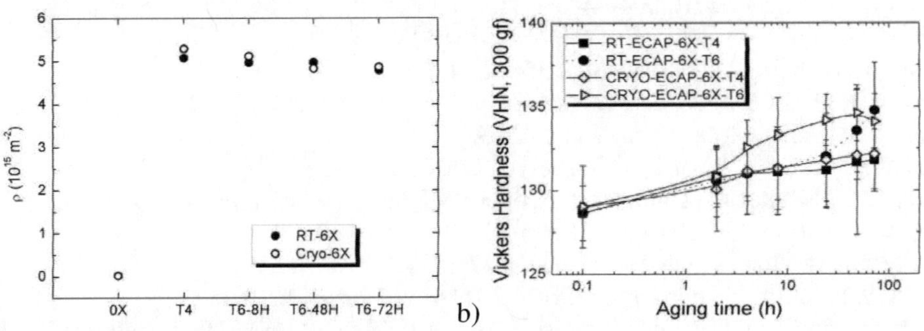

Figure 9. a) Dislocation density evolution after 6 ECAP passes at room temperature and 123 K plus heat treatments in T4 and T6 conditions. b) Hardness evolution

As for hardness, the yield strength for the room temperature and cryogenic routes were very similar after 80 aging hours. Figure 10 shows the results of uniform elongation after ECAP plus T4 and T6 heat treatments for both processing routes. The cryogenic deformation showed an improvement of this property in comparison with the room temperature route.

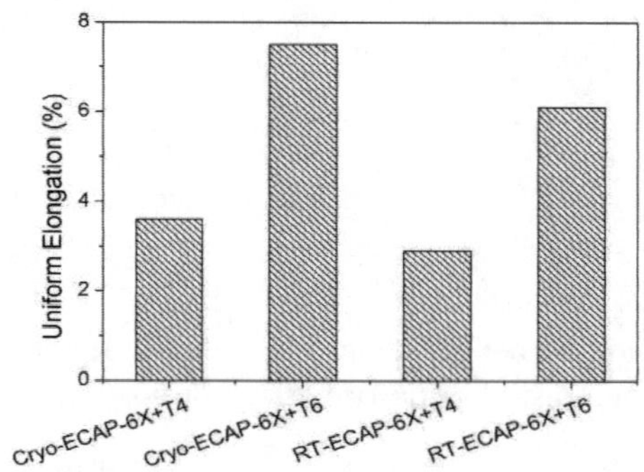

Figure 10. Uniform elongation obtained in tensile tests after 6 ECAP passes at room temperature and 123 K plus heat treatments in T4 and T6 conditions.

Conclusions

Deformation at cryogenic temperatures successfully suppressed the formation of β" and made possible the SPD process without the superimposed effect of precipitation hardening of the AA6061 alloy. The subsequent T6 treatment yielded higher uniform elongation for the deformed material.

Acknowledgments
This work was funded by FAPESP grants 2014/15091-4 and 2016/10997-0

References

[1] Starke JR , E. A., Staley J T. 1996 *Progress in Aerospace Sci* **32** 131.
[2] Roven HJ, Nesboe H, Werenkiold J C Seibert T 2005 *Mat Sci Eng A* **410-411** 426.
[3] Park D H, Choi S, W, Kim J, Lee J 2015 *Cryogenics* **68** 44.
[4] Rezaei M R , Toroghinejad M R, Ashrafizadeh F *Mat Sci Eng A* 2011 **529** 442.
[5] Kima JH, Hwanga SK, Ima YT, Sonb IH, Bae CM 2012 *Mat Sci and Eng A* **552** 316
[6] Farshidi MH, Kazeminezhad M, Miyamoto H 2014 *Mat Sci Eng A* **615** 139.
[7] Fujda M, Milos Matvija M, Tibor Kvackaj T, Ondrej Milkovic O, Pavol ZubkoP ,Katarína Nagyová 2012 *Mat Techn* **46** 465.
[8] Caglioti G, Paoletti A, Ricci F. 1958 *Nucl Inst Methods* **3** 223.
[9] Williamson GK Hall WH 1953 *Acta Metall.* **1** 22.
[10] Warren BE 1959 *Prog Met Phys* **8** 147.
[11] Ungár T, Borbély A1996 *Appl Phys Lett* **69** 3173.
[12] Ungár T, Gubicza J, Ribárik G, Borbély A 2001 *J App. Cryst* **34** 298.
[13] Edwards GA, Stiller K,. Dunlop GL, Couper MJ 1998, *Acta Mater.* **46** 3893
[14] Ravi C, Wolverton C, 2004 *Acta Mater.* **52** 4213.
[15] Murayama M, Hono K 1999 *Acta Mater* **47** 1537.
[16] Manes A, Peroni L, Scapin M, Giglio M 2011 *Procedia Eng.* **10** 3477.
[17] Picu R C, Vincze G, Ozturk F, Gracio JJ, Barlat F, Maniatty AM 2005 *Mat Sci Eng A* **390** 334.
[18] Hadianfard M J., Smerd R., Winkler S., Worswick M. 2008 *Mat Sci Eng A* **492** 283.
[19] Chen Y., Clausen AH, Hopperstad OS, Langseth M *Int J Solids Struct* **46** 3825.
[20] Xu Z, Roven H, Jia Z 2015*Mat Sci Eng A* **648** 350.
[21] Halim H, Wilkison DS,Niewczas M. 2007 *Acta Mat* **55** 4151.

IOP Conf. Series: Materials Science and Engineering **194** (2017) 012028 doi:10.1088/1757-899X/194/1/012028

Microstructure and texture evolution during severe plastic deformation of CrMnFeCoNi high-entropy alloy

W Skrotzki[1], A Pukenas[1], B Joni[2], E Odor[2], T Ungar[2,3], A Hohenwarter[4], R Pippan[4], E P George[5,6]

[1] Institut für Strukturphysik, Technische Universität Dresden, D-01062 Dresden, Germany
[2] Department of Materials Physics, Eötvös University, H-1117 Budapest, Hungary
[3] Materials Performance Centre, School of Materials, The University of Manchester, Manchester M13 9PL, UK
[4] Department of Materials Physics, Montanuniversität Leoben, A-8700 Leoben, Austria
[5] Materials Science and Technology Division, Oak Ridge National Laboratory, Oak Ridge, TN 37831, USA
[6] Department of Materials Science and Engineering, University of Tennessee, Knoxville, TN 37996, USA

werner.skrotzki@tu-dresden.de

Abstract. An equiatomic high-entropy alloy CrMnFeCoNi was severely deformed at room temperature by high pressure torsion up to shear strains of about 170. Its microstructure and texture were analyzed by X-ray diffraction (X-ray line profile analysis and X-ray microdiffraction, respectively). It is shown that at a shear strain of about 20 a steady state domain/grain size of 24 nm and a dislocation density of 3×10^{16} m^{-2} is reached, while the twin density goes over a maximum of 2% at this strain. The texture developed is typical for sheared face-centred cubic metals, but it is extremely weak. The results are discussed in terms of the mechanisms of deformation, including dislocation slip, twinning and grain boundary sliding.

1. Introduction

High entropy alloys (HEAs) represent a new class of single-phase multi-element (≥ 5) solid solution alloys with near-equiatomic concentrations of the individual elements [1]. In some cases, due to the large number of constituent elements the contribution of configurational entropy to the Gibbs free energy is high enough to suppress compound formation and phase separation. Among the wide variety of reported HEAs the most thoroughly investigated alloy is the quinary equiatomic face-centred cubic (FCC) HEA CrMnFeCoNi [2]. This alloy is stable as a single-phase solid solution at high temperatures [3], but decomposes into several different metallic and intermetallic phases at intermediate temperatures [4 - 6]. In the solid-solution state it exhibits certain noteworthy mechanical properties, including simultaneous strength and ductility increase with decreasing temperature [7] leading to outstanding fracture toughness at cryogenic temperatures [8].

To unravel the deformation mechanisms of these advanced alloys numerous investigations of microstructure and texture have been carried out on polycrystalline samples deformed in tension [9 - 11], and after swaging [12], rolling [13] and high pressure torsion (HPT) [5]. It is found that at room temperature and below, above a certain stress (equivalently strain) in addition to dislocation slip mechanical twinning contributes to deformation. Twinning is more pronounced at cryogenic temperatures. There is texture formation during deformation, but its intensity is quite low. The observed dislocation dissociation and texture are typical of medium stacking fault energy (SFE) metals and alloys.

The present paper extends recent work initiated by three of the present authors on CrMnFeCoNi HEA processed by HPT [5]. Here a detailed study is conducted of microstructure and texture evolution during severe plastic deformation.

2. Experimental

The CrMnFeCoNi HEA was synthesized from high-purity elements (> 99.9 wt.%) by arc melting and drop casting under pure argon atmosphere into cylindrical molds (diameter: 25.4 mm, length: 127 mm). The drop-cast ingots were encapsulated in evacuated quartz ampules and homogenized for 48 h at 1200°C. Discs with a radius $r = 4$ mm and an initial thickness $t_i \approx 0.8$ mm were cut from the cast and homogenized ingots and deformed by HPT [14]. During HPT the shear strain along the radius is approximately given by $\gamma = 2\pi rn/\langle t\rangle$, where n is the number of rotations (Rot n) and $\langle t\rangle = (t_i + t_f)/2$ with t_f = final thickness. With this approximation the maximum error in shear strain is less than 15%. HPT was conducted at room temperature (RT) at a nominal speed of 0.2 rotations/min yielding a maximum initial shear strain rate of $10^{-1}s^{-1}$ at the outer radius. The initial grain size was several hundred micrometers, while the saturation microstructure after HPT as analysed by transmission electron microscopy was estimated to consist of grains with a size of about 50 nm. During HPT the alloy did not decompose as checked by 3D atom probe tomography [5].

To further investigate the microstructure, X-ray diffraction measurements were carried out in a special high-resolution diffractometer dedicated to line-profile-analysis with a plane Ge (220) primary monochromator operated at the Cu $K\alpha$ fine focus rotating copper anode (Rigaku, RA-MultiMax9) at 40 kV and 100 mA [15]. The distance between the source and the monochromator is 240 mm and a slit of about 160 μm is placed before the monochromator, 200 mm from the X-ray source. In this configuration the $K\alpha_1$ and $K\alpha_2$ components arrive at the Ge (220) crystal with a large enough separation to allow the $K\alpha_2$ component to be cut off by the 160 μm wide slit. The Cu $K\alpha_1$ beam has a size of about 0.2 × 1.5 mm on the specimen surface. The scattered radiation is registered by two curved imaging plates (IP) with a linear spatial resolution of 50 μm. The IPs are placed at distances of 200 mm from the specimen covering the angular range $30° \leq 2\theta \leq 153°$. The distances between the specimen and detector are selected such that the instrumental effect is always less than 10% of the physical broadening. The diffraction geometry is of parallel-beam type; therefore the specimen does not have to be moved and sufficient angular resolution is achieved over the entire angular range of measurement. The diffraction patterns are obtained by integrating the intensity distributions along the corresponding Debye-Scherrer arcs on the IPs. Only the central parts of the arcs are used for the integration where the curvature does not affect line broadening. The X-ray beam is positioned on the specimen surface by using a low depth-resolution microscope coupled to a TV-screen. The measurements are done at about the centre, half radius and close to the edge of the samples, i.e. at positions where the shear strains are different.

The line profiles are evaluated by using the convolutional multiple whole profile (CMWP) procedure [16]. The measured diffraction pattern is matched by the theoretically calculated and convoluted profile functions accounting for the effects of size, distortion, planar defects and instrumental effects, while the background is determined separately. Because of the double-crystal high resolution diffractometer, described above, the instrumental effect is neglected. The parameter values that characterize the substructure, namely the area average crystallite (subgrain) size $\langle x\rangle_{area}$, dislocation density ρ, dislocation character q (edge versus screw), dislocation arrangement parameter $M = R_e\sqrt{\rho}$ (R_e = effective outer cut-off radius of dislocations), twin density β (number of twin boundary planes within hundred lattice planes parallel to the twin boundary) and average distance between adjacent twin boundaries $d_{Tw} = 100\ d_{\{111\}}/\beta$ are obtained by using the CMWP method.

In order to resolve the local texture of the deformed discs along the radial direction, i.e. as a function of increasing shear strain from the middle to the edge of the disc, two-dimensional X-ray micro diffraction was performed using the system D8 Discover (BRUKER AXS GmbH) equipped with an Euler cradle including x-y-z-stage, a laser-video adjustment system, a low-power micro-focus X-ray tube IµS (spot size about 100 µm) and a two-dimensional detector VÅNTEC 2000. The intensities of the measured Debye-Scherrer rings were integrated along their curvature to calculate the pole figures (200, 220, 111) on a 5°×5° grid. For calculation of the orientation distribution function (ODF) with these pole figures Multex 3 [17] and LaboTex software [18] were used. The Euler angles given are in the Bunge notation [19] with crystal and sample reference systems defined as $x \parallel$ shear direction, $y \parallel$ shear plane normal and $z \parallel$ transverse direction yielding an ODF representation appropriate for simple shear [20]. The textures are represented by φ_2=45° ODF sections, which for FCC metals contain all major shear components.

3. Results and Discussion

Typical diffraction patterns of HPT deformed CrMnFeCoNi HEA specimens are shown in Fig. 1. The quality of the fitting procedure by using the CMWP method can be assessed by the measured (open circles) and the fitted (red line) diffraction patterns.

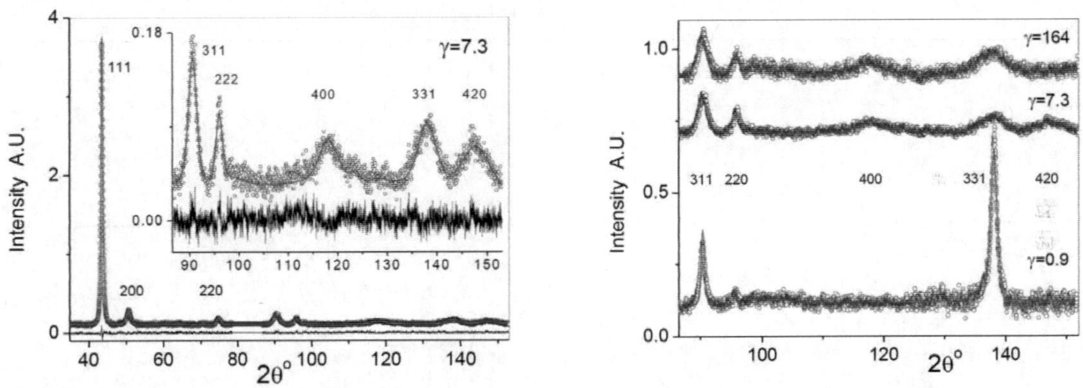

(a) (b)

Figure 1. Typical diffraction patterns of HPT deformed CrMnFeCoNi HEA specimens. (a) Measured (open circles) and CMWP calculated (red lines) patterns for shear strain γ = 7.3 are shown using logarithmic intensity scales. The inset is the enlarged part of the pattern at higher diffraction angles. (b) Comparison of diffraction patterns after different shear strains: γ = 0.9, 7.3 and 164.

The results of the CMWP analysis show that during HPT of CrMnFeCoNi HEA a very fast refinement of the microstructure takes place. The crystallite size ($\langle x\rangle_{area}$) reaches a very low steady state value of 24 nm after a shear strain of about 20 (Fig. 2a). Simultaneously, the dislocation density saturates at 3×10^{16} m^{-2} (Fig. 2b). While the dislocation character q changes from near edge- (q_{edge} = 1.4) to near screw-type (q_{screw} = 2.4) and then slightly decreases again (Fig. 2c), the dipole character M is quite weak (M >1) and saturates at about 6 (Fig. 2d). The twin density has a maximum of 2% at $\gamma \approx$ 20 (Fig. 2e) leading to a mean twin separation distance d_{Tw} smaller than $\langle x\rangle_{area}$ for $10 < \gamma < 60$ (Fig. 2f).

The texture observed after HPT is the typical shear texture observed in FCC metals and alloys. Because of the initial coarse grain structure the texture measured by microdiffraction is statistically

IOP Conf. Series: Materials Science and Engineering **194** (2017) 012028 doi:10.1088/1757-899X/194/1/012028

reliable only after a shear strain of about 5. As shown in Fig. 3a the texture of CrMnFeCoNi HEA is quite weak (below 2.5 mrd). The main texture components developed by dislocation slip and mechanical twinning are $\{111\}\langle 112\rangle$ (A_1^*/A_2^*) and $\{112\}\langle 110\rangle$ ($B/\bar{B}$), with ($B/\bar{B}$) dominating (Fig. 3b). With increasing twinning activity the volume fraction of these components decreases. When twinning ceases texture intensity initially increases but then decreases again at high shear strains.

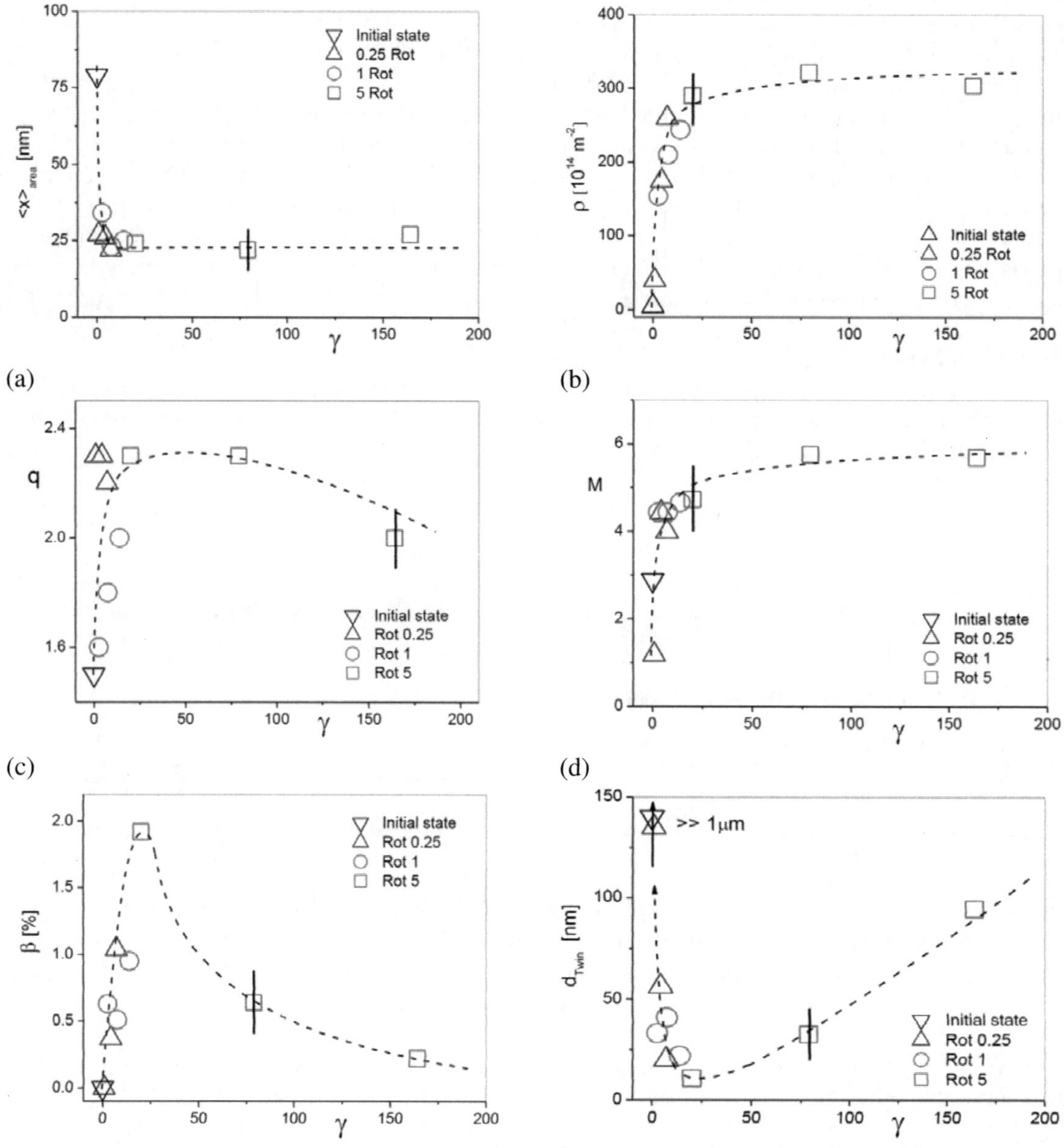

Figure 2. (a) Grain size $\langle x_{area}\rangle$, (b) dislocation density ρ, (c) dislocation character q, (d) dislocation arrangement parameter M, (e) twin density β and (f) average distance between adjacent twin boundaries d_{Tw} versus shear strain γ.

The strong grain refinement is correlated to the high mechanical twinning activity in the medium SFE ($\leq$ 30 mJm^{-2} [21 - 23]) alloy [5, 10]. Based on TEM investigations a mechanism of fast grain refinement via primary and secondary twinning has been proposed by Wang et al. [24] for brass. Primary twins by accumulation of a high density of dislocations evolve into curved high angle grain boundaries from which secondary twins are emitted. The emission of secondary twins further refines the grains and transforms the elongated grains into equiaxed nanograins. At a certain shear strain twinning ceases, may be because grain boundary sliding takes over as the deformation mechanism. A supporting indication of this is the further randomization of texture. The generally low texture strength may be also caused by twinning. Twinning may have led to the almost equal intensities of the twinning related A_1* and A_2^* shear components. Texture simulations to check this are planned.

The medium SFE of this alloy leads to widely dissociated dislocations (screws: ≅ 4 nm, edges: ≅ 6 nm) [21]. After relatively low compressive strains at 77 K, long smoothly curved dislocations having mixed character are seen on the {111} planes suggesting that the mobilities of edge and screw dislocations are not significantly different [21]. Here we find that at RT the dislocation character tends to become more screw-like with increasing shear strain, peaking at a value of $q \approx 2.3$ at a shear strain $\gamma \approx 25$ and then decreasing thereafter to $q \approx 2$ at $\gamma \approx 170$. It is somewhat surprising that the dipole character of dislocations observed is low ($q > 1$) and decreases further with increasing dislocation density. This may be caused by the wide dislocation dissociation suppressing edge dislocation climb and screw dislocation cross slip.

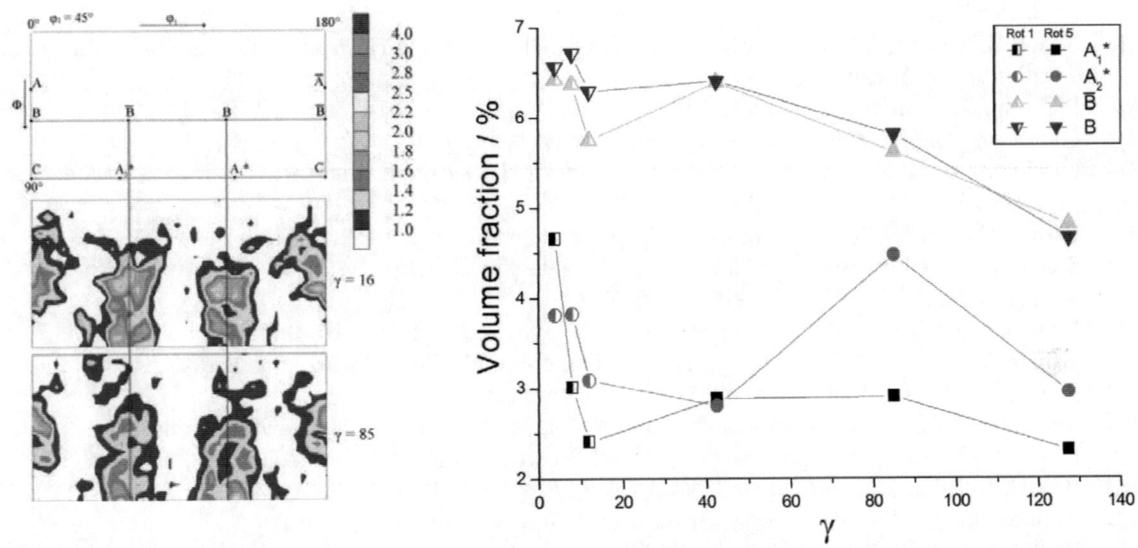

(a) (b)

Figure 3. (a) Texture represented as $\varphi_2 = 45°$ ODF sections at different shear strains γ. Intensities are given in multiples of a random orientation distribution. The key figure shows the main texture components that develop during simple shear of FCC metals. (b) Volume fraction of texture components (15° FWHM) versus shear strain γ.

The brass-type texture in shear with strong B/$\bar{B}$ components agrees with results of texture simulations on HPT samples of nanocrystalline Pd-10at.%Au alloy [25] and rolling of nanocrystalline Ni-18.7 at.%Fe alloy [26]. These simulations strongly suggest that nanoplasticity is determined by slip of partial dislocations emitted from the grain boundaries also leading to twinning. Recent simulations also show that grain boundary sliding leads to randomization of the texture while maintaining the texture signature [27].

4. Conclusions

(1) HPT leads to strong grain refinement of the CrMnFeCoNi HEA. This may be caused by severe mechanical twinning in addition to dislocation slip.

(2) Correlated with the low grain size is a high remnant dislocation density. These dislocations are predominantly screw-type and show a weak dipole character due to a wide dissociation into Shockley partials.
(3) The weak texture is a dominant brass-type shear texture indicating the deformation during SPD occurs mainly by partial dislocation slip accompanied by twinning and grain boundary sliding.

Acknowledgments
Funding of this work (A.H. and R.P.) has been provided by the European Research Council under ERC Grant Agreement No. 340185 USMS and by the Austrian Science Fund (FWF) in the framework of research project P26729-N19. E.P.G is sponsored by the U.S. Department of Energy, Office of Science, Basic Energy Sciences, Materials Sciences and Engineering Division.

References
[1] Yeh J W, Chen S K, Lin S J, Gan J Y, Chin T S, Shun T T, Tsau C H and Chang S Y 2004 *Adv Eng Mater* **6** 299
[2] Cantor B, Chang I T H, Knight P and Vincent A J B 2004 *Mater Sci Eng* A **375-377** 213
[3] Otto F, Yang Y, Bei H and George E P 2013 *Acta Mater* **61** 2628
[4] Otto F, Dlouhy A, Pradeep K G, Kubenova M, Raabe D, Eggeler G and George E P 2016 *Acta Mater* **112** 40
[5] Schuh B, Mendez-Martin F, Völker B, George E P, Clemens H, Pippan, R and Hohenwarter A 2015 *Acta Mater* **96** 258
[6] Pickering E J, Munoz-Moreno R, Stone H J and Jones, N G 2016 *Scripta Mater* **113** 106
[7] Gali A and George E P 2013 *Intermetallics* **39** 74
[8] Gludovatz B, Hohenwarter A, Catoor D, Chang E H, George E P and Ritchie R O 2014 *Science* **345** 1153
[9] Otto F, Dlouhy A, Somsen Ch, Bei H, Eggeler G and George E P 2013 *Acta Mater* **61** 5743
[10] Stepanov N, Tikhonovsky M, Yurchenko N, Zyabkin D, Klimova M, Zherebtsov S, Efimov A and Salishchev G 2015 *Intermetallics* **59** 8
[11] Laplanche G, Kostka A, Horst O M, Eggeler G and George E P 2016 *Acta Mater* **118** 152
[12] Laplanche G, Horst O, Otto F, Eggeler G and George E P 2015 *J Alloys Comp* **647** 548
[13] Sathiaraj G D and Bhattacharjee P P 2015 *J Alloys Comp* **647** 82
[14] Pippan R, Scheriau S, Hohenwarter A and Hafok M 2008 *Mater Sci Forum* **584-586** 16
[15] Ungar T, Ott S, Sanders P G, Borbely A and Weertman J R 1998 *Acta Mater* **10** 3693
[16] Balogh L, Ribarik G and Ungar T 2006 *J Appl Phys* 100 023512
[17] Multex Manual, Multex 3, 2008, Bruker AXS GmbH (Karlsruhe, Germany).
[18] Pawlik K 1986 *Phys Stat Sol* (b) **134** 477
[19] Bunge H-J 1965 *Z Metallkde* **56** 872
[20] Tóth L S and Molinari A 1994 *Acta Metall Mater* **42** 2459
[21] Okamoto N L, Fujimoto S, Kambara Y, Kawamura M, Chen Z M T, Matsunshita H, Tanaka K, Inui H and George E P 2016 *Sci Reports* DOI: 10.1038/srep35836
[22] Yeh J W 2015 *JOM* **67** 2254
[23] Huang S, Li W, Lu S, Tian F, Shen J, Holmström E and Vitos L 2015 *Scripta Mater* **108** 44
[24] Wang Y B, Liao, X Z, Zhao Y H, Lavernia E J, Ringer S P, Horita Z, Langdon T G, Langdon T G and Zhu Y T 2010 *Mater Sci Eng* A **527** 4959
[25] Skrotzki W, Eschke A, Joni B, Ungar T, Toth L S, Ivanisenko Yu and Kurmanaeva L 2013 *Acta Mater* **61** 72
[26] Li L, Ungar T, Toth L S, Skrotzki W, Wang J D, Ren J, Choo H, Fogarassy Z, Zhou X Z and Liaw P K 2016 *Metall Mat Trans* A **47** 6632
[27] Zhao Y, Toth L S, Massion R and Skrotzki W 2017 *Adv Eng Mater*, submitted

Deformation and recrystallization behavior of super high-purity niobium for SRF cavity

Y Yamaguchi[1], H Doryo[1], M Yuasa[2], H Miyamoto[2] and M Yamanaka[3]
[1] Graduate School of Science and Engineering, Doshisha University, Japan.
[2] Department of Mechanical Engineering, Doshisha University, Japan.
[3] Mechanical Engineering Center, High Energy Accelerator Research Organization (KEK), Japan.

E-mail: duq0576@mail4.doshisha.ac.jp

Abstract. Deformation and recyrstallization behavior of pure niobium was investigated in order to clarify the origin of its low hydro-formability despite of its high ductility comparable with pure iron. It was found that pure niobium exhibits lower strain hardening in cold rolling compared with pure iron. Furthermore, in post-deformation annealing, the hardness of niobium decreased monotonously with an increase of temperature, and the typical sharp drop by recrystallization was not evident. This softening behavior was contrasted with the high-purity iron. It is suggested that niobium exhibit the so-called in-situ recrystallization possibly because of low elastic modulus and low accumulative plastic strain energy in spite of high melting temperature. The low hydro-formability of pure niobium sheets or tubes is caused by its low strain hardening and its unique plastic anisotropy which is associated with this recovered residual rolled texture.
Keywords: texture, in-situ recrystallization, elastic modulus

1. Introduction

Fabrication of superconducting radio frequency (SRF) cavities of high-purity niobium have been tried by hydroforming using a seamless tube for the International Linear Collider (ILC) project[1]. While the conventional fabrication methods of accelerator cavity consist of press forming of half cells from sheets and coupling of an electron beam welding, the hydroforming has advantage in high productivity. However the low strain hardening and related low yield ratio with about 30% is the main reason of insufficient formability for bulging formation. In order to commercialize the system by hydroforming, it is necessary to improve the formability of the niobium materials. The insufficient formability of niobium is due not only to low strain hardening but also to low plastic anisotropy related with non-uniform microstructure and texture. The non-uniform texture is also responsible for low *r*-value and ridging[2-5]. Thereofore, it is necessary to improve the non-uniform microstructure and texture for uniform strain distribution for hydroforming to cover the disadvantage of low strain hardening, but there are little studies on this viewpoint. In this study, we therefore aimed to investigate the origin of low strain hardening and non-uniform microstructure and texture formation of high-purity niobium after the cold rolling and annealing focusing on dislocation structure and recrystallization behavior.

IOP Conf. Series: Materials Science and Engineering **194** (2017) 012029 doi:10.1088/1757-899X/194/1/012029

2. Experimental Procedure

A high-purity niobium of RRR 464 was used as a test material. In addition, we used a high-purity iron with 99.99% purity as a comparison material. We performed cold rolling process using a two-stage rolling mill for these materials. Maximum rolling reduction was to 95%. After cold rolling, we annealed these specimens in a vacuum using a image furnace equipment (MILA3000UHV ; ADVANCE RIKO, Inc.). After these processes, we conducted hardness test using a micro Vickers hardness tester (HMV-1 ; SHIMADZU Co.) to examine the relationship between rolling reduction and hardness and the recrystallization temperature. Subsequently, we observed the microstructure using an optical microscople (OPTIPHOT-100 ; NIKON Co.) after rolling and annealing. We measure crystal orientation by electron back scatter diffraction (EBSD) equipped in a scanning electron microscope (SEM, JSM-7001FD ; JEOL Ltd.) and INCA Crystal (Oxford Instuments). Then, we conducted X-ray diffraction (XRD) using a X-ray diffractometer (SMARTLAB ; Rigaku Co.) to evaluate the dislocation density of these specimens.

3. Result and Discussion

3.1 Evaluation of work hardening rate

Figure 1 shows the relationship between the rolling reduction and the Vickers hardness of cold rolled specimens. Both of the cold rolled specimens became harder as the rolling reduction increased, which suggested that work hardening due to cold rolling occureed. In addition, comparing the slope of the hardening curves, the inclination of the high-purity niobium was smaller than that of the high-purity iron. This result suggests that work hardenability of pure niobium is lower than that of pure iron.

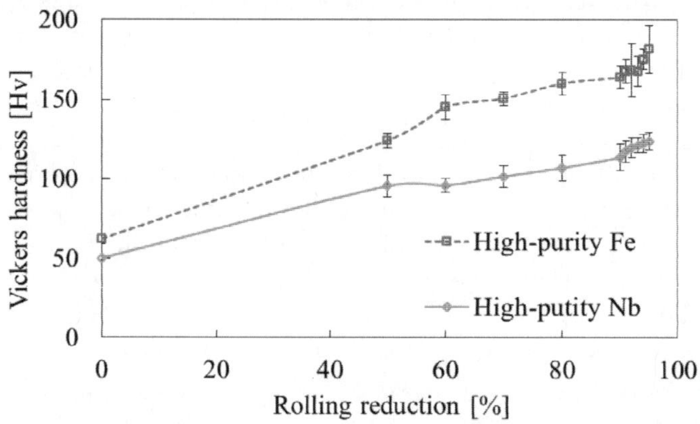

Figure 1. Relationship between rolling reduction and Vickers hardness of the specimens.

3.2 Recrystallization behavior and change of texture

In order to evaluate the influence of the cold rolling process on the recrystallization behavior, we annealed the specimens after cold rolling with reduction of 60% and 95% at various temperatures. Figure 2 shows the hardness of the annealed specimens after heat treatment at each temperature. When the rolling reduction was 95%, the high-purity iron rapidly softened at 773K and softening of the high-purity niobium was completed at 1123K. In other words, the high-purity niobium softened at a higher temperature than the high-purity iron. It is suggested that this difference of softening temperature is due to the melthing point. It is known that a recrystallization temperature of a metal material is approximately $0.4T_m$ to $0.5T_m$, where T_m is a melting point of a material expressed in absolute temperature. The melting points of niobium and iron are 2741K and 1809K, respectively and the recrystallization temperatures are expected to be 1096K to 1371K and 724K to 905K, respectively. Because the recrystallization temperature deduced from the softening curve in figure 2 is generally within this range, the difference of the softening temperature is mainly decided by the melting point. When the rolling reduction was 60%, the high-purity iron and the high-purity niobium softened at 873K and 1173K respectively. In this case also, there was the difference of softening temperature depending on the melting point. However, the hardness of niobium decreased monotonously with an increase of temperature, and the typical sharp drop by recrystallization was not evident. This softening behavior was contrasted with the high-purity iron. It was suggested that in-situ recrystallization of the high-purtiy niobium occurred during the annealing process due to factors other than the melting point. Then we observed the texture of the annealed specimens. Figure 3 shows the texture of the specimens after cold rolling with reduction of 60% in the heat treatment process. It was found that the recrystalized microstructure of high-purity iron was equi-axial crystals over the softening temperature. On the other hand, the texture of the high-purity niobium remained elongated in the annealing process. At the completion of softening grain growth of crystals was observed. This suggests that the occurrence of in-situ recrystallization prevents recrystallization of the high-purity niobium.

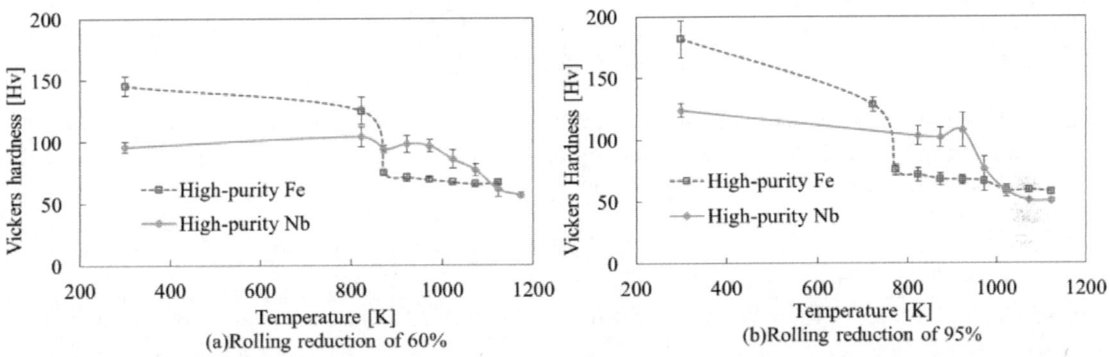

Figure 2. Relationship between the annealing temperature and Vickers hardness of the specimens after cold rolling with reduction of (a) 60% and (b) 95%.

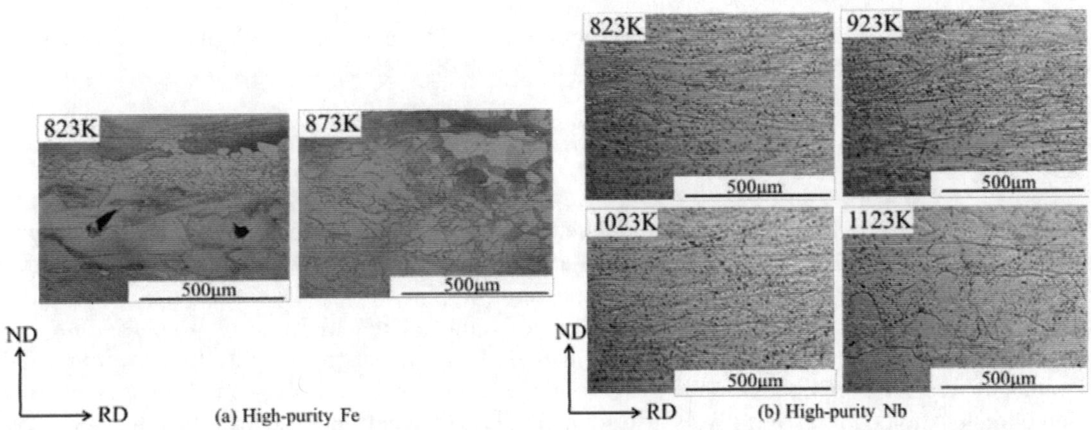

Figure 3. Micrographs of the specimens after cold rolling with reduction of 60% in the heat treatment process with temperature hold time of 120 s. (a) is the high-purity Fe. (b) is the high-purity Nb.

3.3 Texture after cold rolling and annealing

Figure 4 shows orientation image maps and invers pole figures of the high-purity niobium and the high-purity iron in cold rolling and heat treatment process. The dominant orientation of both cold rolled specimens was the {100} <110> texture, the {111} <110> texture and the {111} <112> texture. It was found that the rolling texture of niobium and iron is consistent. Next, we focused on the annealed specimens. From the invese pole figures, it was found that the high-purity niobium has less orientation variation and <100>//ND texture dominated compared to the high-purity iron. Furthermore, in the orientation image map of the high-purity niobium after annealing, elongated <100>//ND texture was observed. It was suggested that <100>//ND texture formed by cold rolling remained after annealing, which caused in-situ recrystallization of the high-purity niobium.

3.4 Evaluation of dislocation density

Dislocation energy, becomes driving force for recovery and recrystallization. Thus, it is presumed that the recrystallization temperature is influenced by the pile-up structures of dislocations. Dislocation densities of the cold rolled specimens were studied by the following two methods.

Dislocation densities of the cold rolled specimens were calculated by the peak value obtained from XRD and the Williamson-Hall method[6]. It is known that the diffraction angle θ_B and the half-value width $\Delta 2\theta$ in XRD have the folling rerationship.

$$\Delta 2\theta \cdot \cos\theta_B/\lambda = ks/D + 2\varepsilon \cdot \sin\theta_B/\lambda \tag{1}$$

λ is the wavelength of the X-ray, *ks* is the Schuler constant, *D* is the crystallite diameter, ε is the lattice strain. *D* and ε can be obtained from the intercept and slopep of the Williamson-Hall plot. Dislocation density is calsulated using the following equation.

$$\rho = \sqrt{43.2}\varepsilon/Db \tag{2}$$

b is the Burgers vector. Figure 5 (a) shows the calculated dislocation densities of the cold rolled specimens.

In addition, the dislocation densities were calculated using the Bailey-Hirsch equation[7] shown below.

$$\Delta\tau = \alpha\mu b\sqrt{\rho} \tag{3}$$

α is the constant (0.5), μ is the rigidity ratio (niobium: μ = 37 GPa, iron: μ = 80 GPa), *b* is the Burgers vector size (niobium: *b* = 0.29 nm, iron: *b* = 0.25 nm), ρ is the dislocation density. $\Delta\tau$ is calculated by the following equation.

$$\Delta\sigma = M\Delta\tau, \quad \Delta\sigma = 3\Delta HV \tag{4}$$

M is the Taylor factor (3.3), and ΔHV is calculated from the difference in Vickers hardness when changing the rolling rate from Figure 1. The calculated dislocation densities of the cold rolled specimens were shown in Figure 5 (b).

Figure 5 showed that in both of the high-purity niobium and the high-purity iron, the pile-up of dislocations increased as the rolling reduction increased. It was suggested that the lowering of the softening temperature accompanying the increase in the rolling reduction shown in FIG. 2 is due to the pile-up of dislocations. Moreover, it was also found that the dislocation density of the cold rolled high-purity niobium is higher than that of the high-purity iron. In other words, dislocations of the high-purity niobium tended to accumulate. This result showed that the difficulty of recrystallization of the high-purity niobium doesn't result from dislocation density.

3.5 Factors impeding recrystallization

It was suggested that dislocation density of the high-purity niobium is higher than that of the high-purity iron. We speculated that this may be due to the low modulus of elasticity of niobium. Iron with high modulus of elasticity has dislocations with strong repulsive force and causes cross slip. Therefore, it is conceivable that dislocation density decreases due to annihilation of dislocations.

Recrystallization depends on the dislocation density as well as the dislocation energy. Since dislocation density is, high in high purity niobium, it is presumed that dislocation energy is affected. The equation of dislocation energy is shown below.

$$E = 1/2\mu b^2 \tag{5}$$

Since the elastic modulus of niobium is lower than that of iron, the dislocation energy of niobium is also lower than that of iron. It is suggested that the residual of rolled texture of niobium after annealing result from the difficulty of recrystallization due to low dislocation energy.

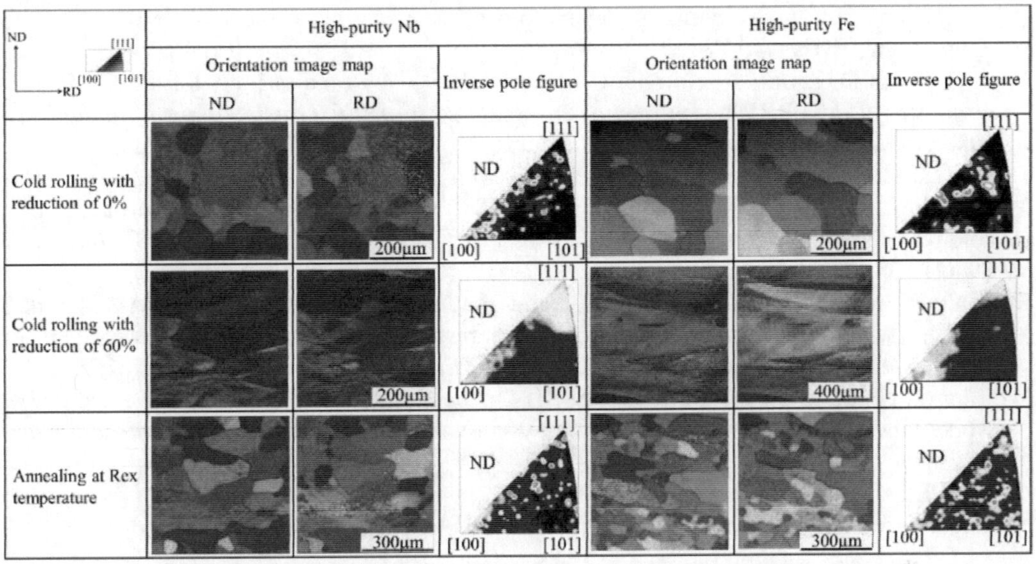

Figure 4. Orientation image maps and invers pole figures of the high-purity niobium and the high-purity iron in cold rolling and heat treatment process.

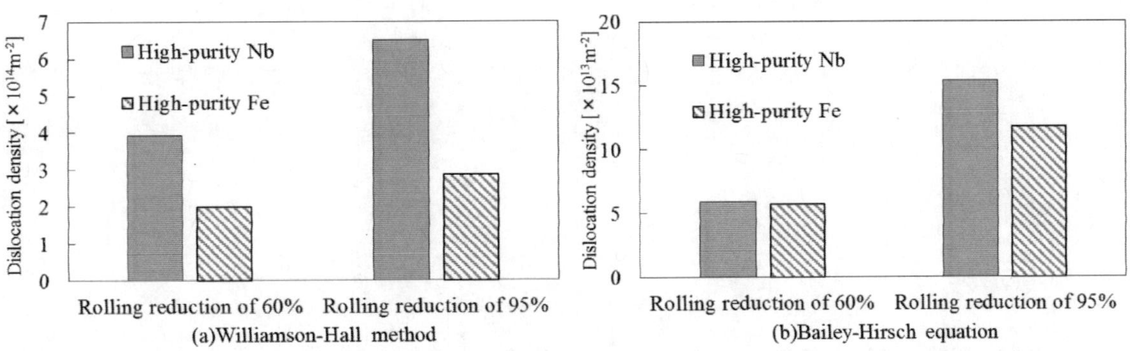

Figure 5. Dislocation densities by (a) Williamson-Hall method and (b) Bailey-Hirsch equation of the specimens after cold rolling with reduction of 60% and 95%.

4. Conclusion

As a result of investigating the texture of the high-purity niobium and the high-purity iron after cold rolling and annealing, the following was clarified. From the rolling hardening curves and softening curves, it was confirmed that high-purity niobium was harder to work hardening than high-purity iron. It was suggested that the cold rolled material of high purity niobium causes in-situ recrystallization due

IOP Conf. Series: Materials Science and Engineering **194** (2017) 012029 doi:10.1088/1757-899X/194/1/012029

to residual rolling texture composed of <100> // ND texture even after annealing. Comparing the dislocation densities of the high-purity niobium and the high-purity iron after cold rolling calculated by the Williamson-Hall method and Bailey-Hirsch equation, it was confirmed that the pure niobium has higher dislocation density than pure iron. It was suggested that the residual of rolled texture of niobium after annealing result from the difficulty of recrystallization due to low dislocation energy.

Reference

[1] Nagata T, Abe N, Masui H, Sshinozawa S, Nagakubo J, Murakami H, Inoue H, Yamanaka M and Kano E 2015 Studies on material properties regarding plastic deformation of niobium product for superconducting accelerator cavity –aiming at production of HOM coupler and other components by press forming technology *Proceedings of the 16th KEK Mechanical Engineering Workshop* 77

[2] Jiang H, Baars D, Zamiri A, Antonie C, Bauer P, Bieler T R and Grimm T L 2007 Mechanical properties of high RRR niobium with different texture *IEEE Transactions on Applied Superconductivity* **17(2)** 1291

[3] Furubayashi E 2000 *Recrystallization and material texture* (Japan: Uchidarokakuho) p16, 99-119

[4] Miyamoto H, Xiao T, Uenoya T and Hatano M 2010 Effect of simple shear deformation prior to cold rolling on texture and ridging of 16% Cr ferritic stainless steel sheets *ISIJ international* **50(11)** 1653-1659

[5] Fuiiwara H, Adachi M, Miyamoto H and Hatano M 2015 Effect of One-Pass ECAP prior to Cold Rolling on the Deformation Structure and Recrystallization in Ferritic Stainless Steel Sheets *Tetsu to Hagane-Journal of the Iron and Steel Institute of Japan* **101(11)** 619-625

[6] Williamson G K and Smallman R E 1956 III Dislocation densities in some annealed and cold-worked metals from measurements on the X-ray debye-scherrer spectrum *Philosophical Magazine* **1(1)** 34-46

[7] Bailey J E and Hirsch P B 1960 The dislocation distribution, flow stress, and stored energy in cold-worked polycrystalline silver *Philosophical Magazine* **5(53)** 485-497

Acknowledgement

This work was financially supported by High Energy Accelerator Research Organization (KEK), ULVAC, Inc. and NEURON JAPAN CO.,LTD. and these supports are gratefully appreciated.

IOP Conf. Series: Materials Science and Engineering **194** (2017) 012030 doi:10.1088/1757-899X/194/1/012030

The effect of severe plastic deformation on the microstructure and mechanical properties of (TiB+TiC) reinforced titanium matrix composites

W Lu, Y Han, J Xiang, J Li, J Mao, D Zhang

State Key Laboratory of Metal Matrix Composites, Shanghai Jiao Tong University, Shanghai, 200240, China

Email: luweijie@sjtu.edu.cn, hyuf1@hotmail.com

ABSTRACT. Equal-channel angular pressing (ECAP) was successfully applied to in-situ synthesized particulate reinforced titanium matrix composites. The results prove that the ultrafine-grained structure can be effectively obtained in the titanium matrix composite by ECAP at different temperatures and passes. Formation of homogenous TiB and TiC particles with ultrafine grains and sub-grains were obtained after 4 ECAP passes, which indicates that the reinforcement-induced ultrafine grains. The lamellar α phases are also twisted obviously due to the shear stress, which can reduce the average thickness of lamellar α phases to less than 1μm. After ECAP at relatively low temperature, dislocation pile-ups, dislocation tangling and plenty of cell structures with size of about 500nm can be observed in the matrix. While many newly formed ultrafine grains with clear grain boundaries and size of 100~500 nanometers have formed. The tensile test results show that the yield strength is improved by ECAP pressing and saturate after 4 ECAP passes to a value of 1150MPa, and the ductility is much higher compared with the first ECAP pass.

Keywords: Titanium Matrix Composite, Microstructure, Severe Plastic Deformation, Mechanical Property

1. Introduction

There is considerable current interest in controlling of materials structures by severe plastic deformation (SPD) [1-4]. An important attractive point of this method is grain refinement to the sub-micro scale, or even the nanometer scale that can be attained in bulk billets, in a cost effective manner and for different metal and alloys [2]. Among a few known methods of SPD, equal channel angular pressing (ECAP) is presently considered as the most popular approach for industrial applications [3, 4]. As high strains are known to introduce larges misorientations within subdivided grains, it has been expected that multi-pass ECAP should lead to ultra-fine grained structures, usually in the range from a few micros to 0.2 micro, provide a reasonable compromise between high strength and satisfactory ductility that is especially attractive for industrial applications. However, realization of ECAP still remains imperfective [5, 6]. Despite of extensive activity in the ECAP, majority of papers published in the journal. This approach was restricted on hard deformed materials. It is difficult to use the method in titanium alloys and still there are few reports on the application of this method to particulate reinforced titanium matrix composites [7, 8].

The incorporation of high strength and high stiffness ceramic reinforcements can improve dramatically mechanical performances of particulate reinforced titanium matrix composites [9], which are currently being widely used in aerospace, commercial automotive engineering, and military applications [10]. Meantime, this material exhibits poor plasticity due to the microstructural heterogeneity. The plasticity is related to the metal matrix behaviour, and the damage is generally due to the particle breaking and interfacial debonding between the metal matrix and particles [11, 12]. Achieving good grain size, particle size, particle distribution and interface bonding is vital if the composites experience severe plastic deformation. It can eliminate the agglomeration phenomenon, and stronger adhesion at the particle interface improves load transfer, further increasing the yield strength and plasticity [13].

Our studies were devoted to developing a process for fabricating titanium matrix composite bars with ultrafine grain structure. The aims focused on the changes in the microstructure and mechanical properties of deformed composites in the process of their production by severe plastic deformation and subsequent heat treatment. Further investigation of the influence of deformation degree applied on the morphology and distribution of TiB short fibers and TiC particles, and to evaluate the deep processing and property improvement of particulate reinforced titanium matrix composite.

2. Experimental material and procedures

(TiB +TiC)/Ti6Al4V titanium matrix composite billet with a diameter of 25mm and length of 1000mm was preliminarily annealed at 600℃ for 30min and air cooled to room temperature. Then the billet was cut into specimens of 10mm×10mm×140mm for equal channel angular pressing (ECAP) on a die-set with the channels' intersection angle of φ=120° and at a temperature of 550℃, while the subsequent extrusion temperature was 800℃. Route B_c is turned by 90° about its longitudinal axis after every pass, and number of passes n=4. After every ECAP pass, the specimen should be removed quickly from the dies and water quenched to keep the high temperature microstructure. Fig.1 shows the schematic illustration of the ECAP process. The appearance of the extruded specimens in intermediate stages is presented in Fig.2.

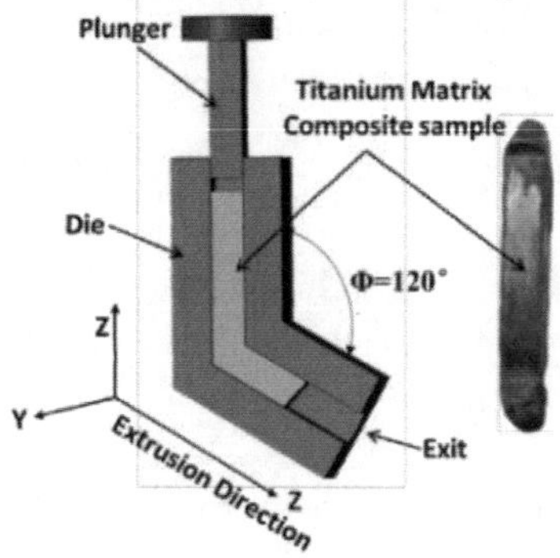

Fig.1 Schematic diagram of ECAP process

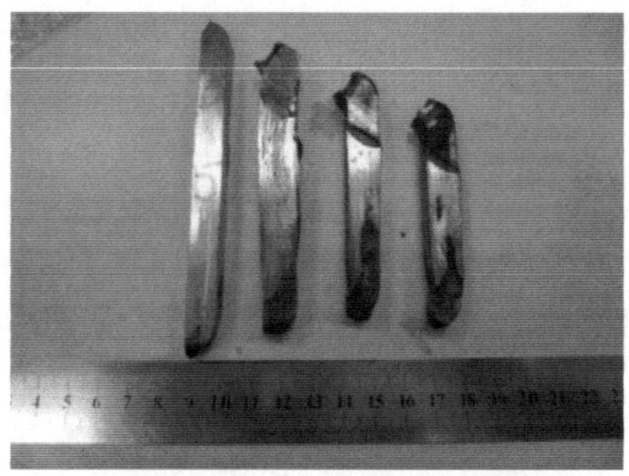

Fig.2 Appearance of extruded specimens

The microstructure was studied by using optical microscopy, scanning electron microscopy and transmission electron microscopes (TEM). All observed surfaces were parallel to the extrusion axis. Tensile tests were performed at room temperature in air using Zwick Z100/SN3A. Flat samples with a gage length of 18 mm and a cross-section of 3 mm×1.6 mm were cut by wire Electrical Discharge Machining (EDM) from the titanium matrix composite billets. A strain rate of 10^{-3} mm/s was used for the all tests.

3. Another section of your paper

3.1. Original microstructure

The microstructure of the as-received material, which had been produced via hot forging followed by annealing, was observed to be equiaxed structure with a volume fraction of approximately 85% equiaxed α phase (Fig.3). On closer examination, it was found that α phases is about ~10μm in diameter. Additionally, SEM analysis of the surface of the material after mechanical polishing revealed that the material had a distribution of reinforcements in the matrix. TiB fibers and TiC particles in these image are platelet and spherical in shape, respectively. All reinforcements are homogenously distributed. In quantitative analysis, there are about 10% long TiB fibers and large TiC particles. TiB fibers are about 20 mm in length and width of ~5 mm, TiC particles are about 15 mm in diameter with an irregular shape, suggesting that the reinforcements are long and large distributed in the metal matrix.

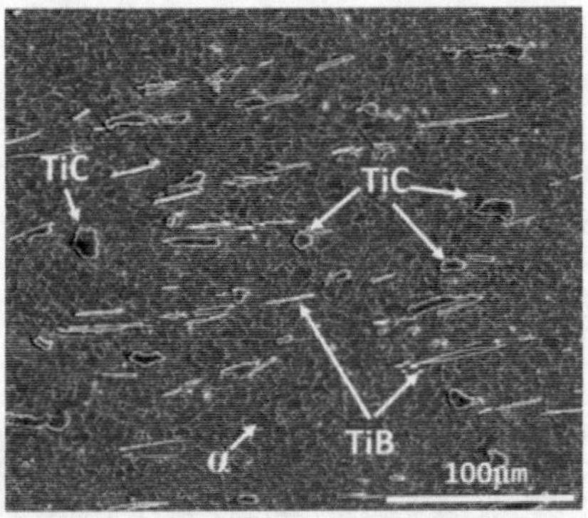

Fig.3 SEM image of original forged samples

3.2. *Microstructure of the composite processed by ECAP*

Fig.4 shows the appearance of the billets after pressing through one and four passes of ECAP at the same pressing speeds, which showed that the surface of the titanium matrix composite samples were smooth and there was some small metal flash after pressing four passes.

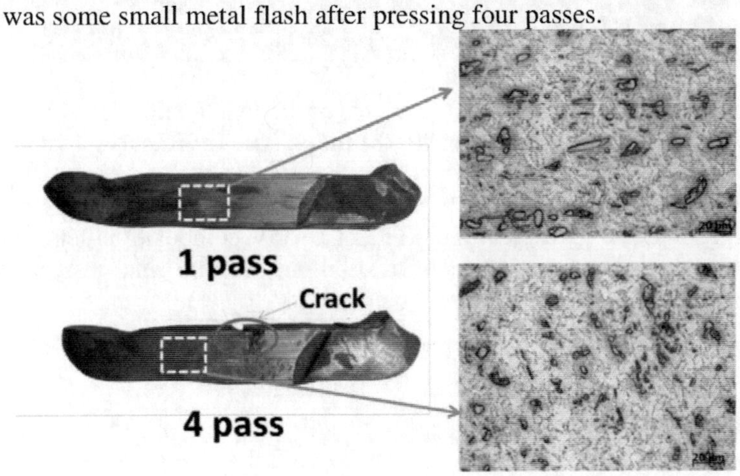

Fig.4 Appearance of the ECAP extruded billet and corresponding microstructure

Nevertheless, close inspection of the sample processed by four passes revealed the presence of cracking in this sample, and the microstructure of the samples processed by four passes showed that the reinforcements dispersed in a homogeneous manner in the composite. After one pass of ECAP, the matrix exhibited a near-fully equiaxed microstructure, and the α phases were elongated in the direction of extruding with average grain size of about 20mm in length, suggesting that the α phases were elongated to be platelet in shape. Meanwhile, TiB fibers in titanium matrix composite showed good alignment along the extrusion direction, and the TiC large particles disperse in the specimens with diameter of about 14 mm. As the ECAP pass number increased, more homogenous structure was formed after four pass of ECAP. Elongated α phases were refined and rearranged significantly in the metal matrix, with the estimated grain size of about ~14mm in diameter, which was a little larger than the original microstructure due to the grain growth at a relative high extruding temperature. The TiB fibers and TiC particle were smaller and distributed more homogenous comparing with the one pass of ECAP.

At higher magnification, SEM micrographs in Fig. 5(a) and Fig.5(b) show the microscopic structure of (TiB+TiC)/Ti6Al4V titanium matrix composite in ECAP billets after one pass and four passes, respectively. After ECAP, the initial TiC large particles and long TiB fibers that are severely fragmented and agglomerated TiC have been observed. Distinct α grains were still visible. However, all the grains were highly elongated, as seen in Fig.4. The evolution of the grain size was determined by quantitative metallographic methods and it varied around 20μm at the beginning ECAP process, and around 14μm at the end of ECAP process, i.e. after four passes of ECAP. The evolution of the reinforcements were investigated by measuring the aspect ratio and particles length by using image analysis software. It was found that the aspect ratio of TiB fibers and TiC particles in all specimens decreased with the number of ECAP passes increasing. Simultaneously, the length of TiB fibers decreased heavily when the billets were deformed by ECAP, which was around 12μm in length at the last of ECAP process. The size of TiC particles were also reduced with the increase of ECAP numbers, refined and near-spherical TiC particles were obtained at the end of the ECAP process.

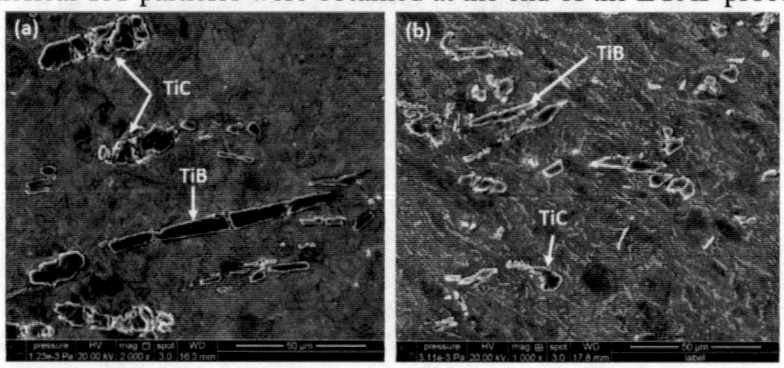

Fig.5 The evolution of the reinforcements of the ECAP extruded composite
(a) one pass of ECAP, (b) four passes of ECAP

3.3. Mechanical properties

Mechanical properties of ECAP extruded (TiB+TiC)/Ti6Al4V composite billets were investigated in the tensile tests at room temperature. The deformation curves after one pass, two passes and four passes of ECAP were plotted as the strain versus the strength (Fig. 6).

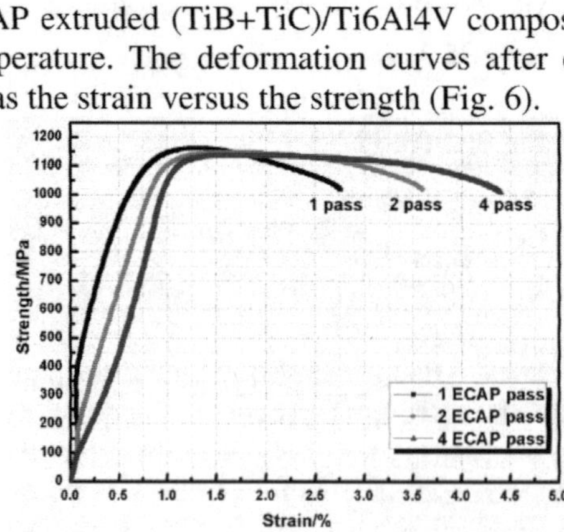

Fig.6 Changes of tensile properties of ECAPed (TiB+TiC)/Ti6Al4V composite billets

IOP Conf. Series: Materials Science and Engineering **194** (2017) 012030 doi:10.1088/1757-899X/194/1/012030

The increase of the ductility was evident, and the ductility increased with the numbers of passes through the 120° ECAP dies. The reason of the improvement was attribute to the refinement of grains after four ECAP passes. In regarded to the tensile strength of the examined samples, it was about 1160 MPa after one pass of ECAP, which decreased slightly to about 1150MPa after two and 4 four passes. The decrease of strength and the increase of ductility could be the interaction of refinement and working hardening. It can be concluded that equal channel angular pressing provides an effective procedure for improving the ductility of particulate titanium matrix composite. The ECAPed (TiB+TiC)/Ti6Al4V composite billets after the second and the fourth pass exhibited higher elongation but similar strength compared with the sample ECAP processed after the first pass.

4. Discussion

The experimental results show that the mechanical properties of TiB+TiC/Ti6Al4V composite depend essentially on its microstructural state, including the small grain size attained after ECAP and the homogeneity of the reinforcements. Since ECAP processing involves the rapid imposition of an exceptionally high strain, the initial microstructural change associated with ECAP is the introduction of very high density of dislocations, and the high pressure involved in the pressing operation may lead to the breaking and fragmentation of internal reinforcements. The observed ductility increase of the four passes of ECAP, in comparison with the one pass of ECAP, is caused by the smaller grain size, and also by the presence of the optimized reinforcements in the matrix. Such changes in microstructure of the composite lead to a vibration of the multiplication, annihilation and reversion of dislocations [14, 15].

Fig. 7 shows the TEM micrographs of (TiB+TiC)/Ti6Al4V composite billets. According to the equivalent strain accumulating formula [16, 17], the values of accumulating strain increases with the increasing of the ECAP pass numbers, which is better to the refinement of grain size and reinforcements, even improving the mechanical properties. However, that does not accord with the strength results. The strength has a slight decrease with the increase ECAP pass numbers. It is observed that the highest strength obtained in the first ECAP pass, and high dislocations are in a high density and tangling together in the matrix (Fig.7a). Some stacking faults are found in the long TiB fibers. As known the low value of dislocation density in the composite before ECAP, it could be known that the dislocation multiplication, which results in dislocation density increasing rapidly in ECAP. As a result, the velocity value of dislocation annihilation is lower than that of dislocation multiplication, and the reversion of dislocations is not evident. Therefore, the strength performs the best in the first ECAP pass number.

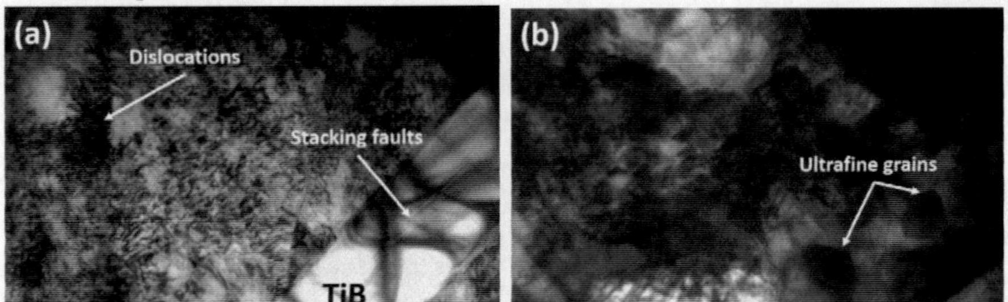

Fig.7 TEM micrographs of TiB+TiC/Ti6Al4V composite billets after (a) one pass of ECAP, (b) four passes of ECAP

Strength characteristics of the titanium matrix composites, measured after two and four passes of ECAP, practically coincide, and their values are slight lower than that for one pass of ECAP. The deformation strain increases to a high value, the dislocation multiplication and annihilation attain a dynamic equilibrium. Therefore, the strength has a slight changes with the increase of ECAP pass numbers. Tendency towards increase of ductility was also observed, which can be associated with accumulation of high dislocation density in an ultrafine grain (~200nm) and presence of a large number of low-angle subgrain boundaries (Fig.7b), which is explained by retention of grain and subgrain sizes. The dislocation in grains migrated to grain boundaries and the interface of TiB and TiC reinforcements. Dislocation density among grain boundaries reached a high value, which formed dislocation cell or wall and then gradually formed subgrain structure. It should be noted that there is a tendency towards enhancement of ductility after four passes. This can be explained with increase of a fraction of high-angle boundaries in the ultrafine grains structure, which are capable of grain boundary sliding during plastic deformation [18].

5. Conclusions

(1)During ECAP processing, bands of elongated α phases are refined and rearranged significantly. The TiB fibers and TiC particles are smaller and distributed more homogenous in the metal matrix.
Ultrafine grains, with relatively homogeneous reinforcement and microstructure, in grain size of about 200nm, can be obtained after four passes of ECAP.
(2)The strength of ECAPed (TiB+TiC)/Ti6Al4V composite billets are closely related to the number of pass. As the following ECAP passes are applied, the value of increase becomes smaller. The enhancement of ductility is dominated by the decreasing of reinforcements' size, and the homogeneity of the reinforcements' distribution through four passes.
(3)With the increasing of the ECAP pass numbers, the small size reinforcements tend to acquire high dislocation density surrounding the TiB fibers and TiC particles, which forms dislocation cell or wall and then gradually forms ultrafine grain structure.

Acknowledgments
The authors are grateful to the National Nature Science Foundation of China (Grant No: 51371114, 51501112, 51504151, U1602274), the supported by the National Engineering and Research Center for Commercial Aircraft Manufacturing (SAMC14-JS-15-047), the fund of the State Key Laboratory of Solidification Processing in NWPU (No. SKLSP201620) and the supported by the 111 Project (Grant No. B16032).

References

[1]. Valiev R 2004 *Nature Materials*. **3** 511-6.
[2]. Shankar M R, Rao B C, Lee S, Chandrasekar S, King A H and Compton W D 2006 *Acta. Mater* **54** 3691-700.
[3]. Azushima A, Kopp R, Korhonen A, Yang D Y, Micari F, Lahoti G D, Groche P, Yanagimoto J, Tsuji N, Rosochowski A and Yanagida A 2008 *Cirp. Ann. Manuf. Techn* **57** 716-35.
[4]. Valiev R Z, Estrin Y, Horita Z and Langdon T G 2006 *JOM* **58** 33-9.
[5]. Zhao X, Yang X, Liu X,Wang X and Langdon T G 2010 *Mater. Sci. Eng.* A **527** 6335-39.
[6]. Shin D H, Kim I, Kim J, Kim Y S and Semiatin S L 2003 *Acta. Mater* **51** 983-96.
[7]. Han Y, Li J, Huang G, Lv Y, Shao X, Lu W and Zhang D 2015 *Mater. Des* **75** 113-9.
[8]. Wang L, Wang X, Zhang L C and Lu W 2015 *Mater. Sci. Eng.* A 645: 99-108.
[9]. Geng K, Lu W, Yang Z and Zhang D 2003 *Mater. Lett* **57** 4054-7.
[10]. Lu W, Zhang D, Zhang X, Wu R and Sakata T 2001 *J. Alloys. Compounds* **327** 240-7.
[11]. Gorsse S, Le Petitcorps Y, Matar S and Matar S 2003 *Mater. Sci. Eng.* A **340** 80-7.
[12]. Ranganath S. J. Mater. Sci 1997, **32** 1-16.
[13]. Tjong S C and Mai Y W. *Compos Sci. Technol*, 2008, **68** 583-601.
[14]. Zhao X, Fu W, Yang X and Langdon T G 2008 *Scripta Mater* **59** 542-5.
[15]. Zhao X, Yang X, Liu X, Wang C T and Huang Y 2014 *Mater. Sci. Eng.* A **607** 482-9.
[16]. Shin D H, Kim I, Kim J, Kin Y S and Semiatin S L 2003 *Acta. Mater* **51** 983-96.
[17]. Sergueeva A V, Stolyarov V V, Valiev R Z and Mukherjee A K 2001 *Scripta Mater* **45** 747-52.
[18]. Arab M S, Mahallawy N E, Shehata F and Agwa M A. *Mater. Des* 2014; **64** 280-6.

IOP Conf. Series: Materials Science and Engineering **194** (2017) 012031 doi:10.1088/1757-899X/194/1/012031

Microstructures and hardness of stir zone for friction stir processed and post-processed heat treatment 7B04-O aluminium alloy

Y Chen[1], H Ding[1] and J Z Li[2]

[1] School of Material Science and Engineering, Northeastern University, Shenyang 110819, PR China

[2] AVIC Beijing Aeronautical Manufacturing Technology Reaserch Institute, Beijing 100024, PR China

E-mail address: hding2013@163.com

Abstract. Friction stir processing (FSP) with different rotation speeds was conducted on 7B04-O aluminium alloy thin-sheet, then the post-processed heat treatment (PPHT) was carried out. The microstructures and hardness of the stir zone (SZ) under as-processed (AP) and PPHT states were investigated. The result showed that the hardness of the SZ was enhanced under the AP state, attributing to the grain refinement. The effect of PPHT on the hardness of the material was different, compared with the AP state, the hardness of the SZ with a low rotation speed under PPHT state was deteriorated while it was enhanced when high rotation speed was adopted. When FSP was conducted on the O-temper base metal (BM), the initial precipitates in the matrix were coarse and stable, reducing the dissolution rate of the precipitates. High heat input was required in order to induce sufficient dissolution of the precipitates, which was beneficial to elevating the impact of PPHT on hardness. The amount and morphology of the precipitates played important roles in the SZ of improvement in hardness.

1. Introduction

Friction stir processing (FSP) was developed as a generic tool for microstructure modification based on the basic principles of friction stir welding. This new solid-state processing technique has been developed by Mishra and coworkers [1]. FSP causes intense plastic deformation and elevates temperatures in stir zone (SZ) [2]. SZ, consisting of recrystallized and fine-grained microstructure, is called dynamic recrystallized zone [3]. And FSP has been proved to be an efficient, cheap and reliable method to produce fine-grained materials [4].

In order to improve the mechanical properties of the SZ produced by FSP, many different methods including optimizing FSP parameters, applying post-processed heat treatment (PPHT), expediting cooling speed, and using submerged FSP, have been explored to minimize the dissolution or coarsening of the strengthening precipitates [5]. Due to the convenience and efficiency, more and more studies have been concentrated on PPHT.

High-strength, precipitation-hardening 7xxx-series aluminum alloy are used extensively in aircraft primary structures. And FSP has been proved to be a successful method in producing fine grains in

7xxx aluminum alloy [6]. Normally FSP and PPHT are often applied in aluminum alloys which are in the hardened condition. However, conducting the FSP on the base metal (BM) in this condition reduces tool life because of the high resistance to deformation, increasing the cost of production. By contrast, there are some advantages in carrying out the FSP with BM in the soft condition (O-temper) as the FSP tool forces are found to be low, which means the FSP system need not to be very expensive [7]. Thus it is necessary to research the effect of FSP and PPHT on the microstructures and mechanical properties in this alloy system in the soft condition. In the present work, FSP was conducted on 7B04-O aluminum alloy with different rotation speeds, and PPHT was applied in order to investigate the effect of PPHT on the microstructure and hardness of 7xxx-series aluminum alloy at O temper.

2. Experimental

Rolled 7B04 aluminum alloy sheets of 2 mm in thickness were used. Annealing heat treatment comprised of initial heating to 400 °C, soaking for 1 hr, followed by furnace cooling to 150 °C and air cooling down to ambient temperature for obtaining the O-temper. The sheets were then processed by FSP with different rotation speeds, which were 1600 rpm and 800 rpm. The traverse speed was 200 mm/min. A tool with a concave shoulder 10 mm in diameter and a M4 cylindrical pin was employed. The tool was made of H13 tool steel with a hardness of HRC 45. After FSP, the artificial aging was carried out at 165 °C for a soaking period of 6 hrs in an electric oven. Microstructural characterization was performed on the cross-section of plates using optical microscopy (OM), electron back-scattered diffraction (EBSD) and transmission electron microscopy (TEM). Differential scanning calorimeter (DSC) test samples were cut from the stir zone under AP state. Samples were heated at a constant heating rate of 10 K/min from room temperature to 550 °C. The Vickers micro-hardness was carried out along the centerlines of the cross-section with a distance between neighboring measured points of 0.5 mm under a load of 50 g for 10 s.

3. Result

3.1 Microstructures in the stir zone

Optical and transmission electron microscopy images of the BM are shown in figure 1. It can be seen that the elongated grains of the BM resulting from the rolling process has an aspect ratio of 3.0-4.0 (figure 1a). The result of TEM (figure 1b) indicates that the particles in the matrix are coarse (larger than 50 nm). In the SZ, the microstructure is characterized by the fine and equiaxed recrystallized grains caused by the severe plastic deformation and thermal exposure during FSP (figure 2). The result of EBSD shows that the average grain size of the SZ in the sample processed at 800 rpm and 200 mm/min (800/200 sample) is 1.15 μm, and 1.07 μm in the sample processed at 1600 rpm and 200 mm/min (1600/200 sample). There is not much difference in grain size when different rotation speeds were adopted in the present case.

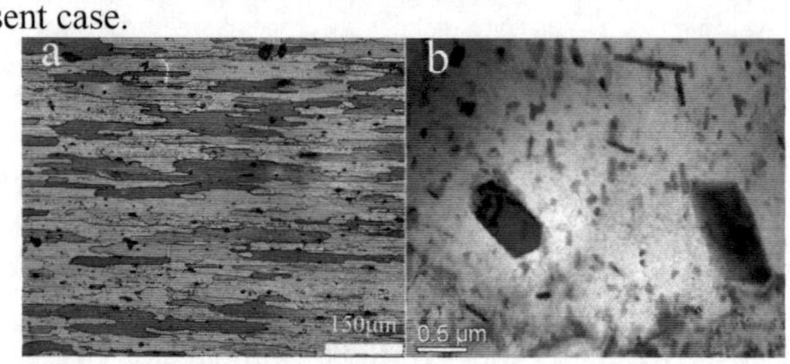

Figure 1. Optical micrograph of BM (a) and TEM micrograph of precipitates in the BM (b).

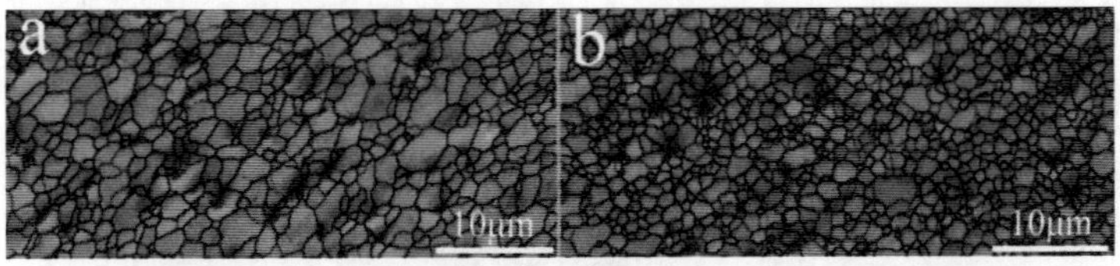

Figure 2. EBSD map of the SZ under AP state (a) 800/200, (b) 1600/200 sample.

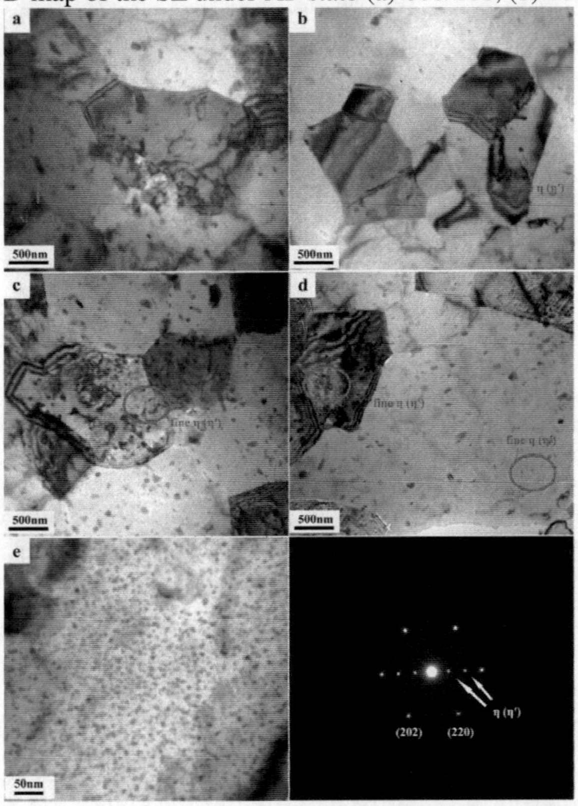

Figure 3. TEM micrograph of the SZ under AP state (a) 800/200, (b) 1600/200 and PPHT state (c) 800/200, (d) 1600/200 and (e) the high-magnification image of the η (η′)-$MgZn_2$.

Figure 3 shows the TEM images of the SZ with different tool rotation speeds under AP and PPHT states. Under AP state (figure 3a and b), two kinds of particles are observed in the matrix. The rod-shaped particles are identified as Al_7Cu_2Fe phase, which remains during FSP because of its high dissolution temperature. And the sphere particles rich in Zn and Mg atoms are identified as η (η′)-$MgZn_2$ phase, the strengthening precipitates in the 7xxx-series aluminum alloy. Under AP state, the amount of the precipitates decreases compared with that of the BM (figure 1b) due to the dissolution of the precipitates during FSP. Apart from the coarsened precipitates, some reformed fine precipitates are also detected. Based on the aging temperature (165 °C in this work), the reformed fine precipitates should be η (η′)-$MgZn_2$. The high-magnification observation is shown in figure 3e, the reformed fine precipitates on the order of 10-20nm distribute homogeneously in the matrix, and the selected area diffraction pattern (SADP) verifies these particles are η (η′)-$MgZn_2$.

3.2 DSC analysis of strengthening precipitates

The results of DSC analysis of the SZ with different rotation speeds under AP state are shown in figure 4. An exothermic peak was found below 300 °C during DSC heating for 1600/200 and 800/200 samples. The appearance of this exothermic peak indicates the formation of η (η′)-$MgZn_2$. It is well known that the area under the peaks is related to the volume fraction of the precipitates. From the calculation about area η (η′) peak, there is less η (η′)-$MgZn_2$ formed in 800/200 sample compared with that in 1600/200 sample.

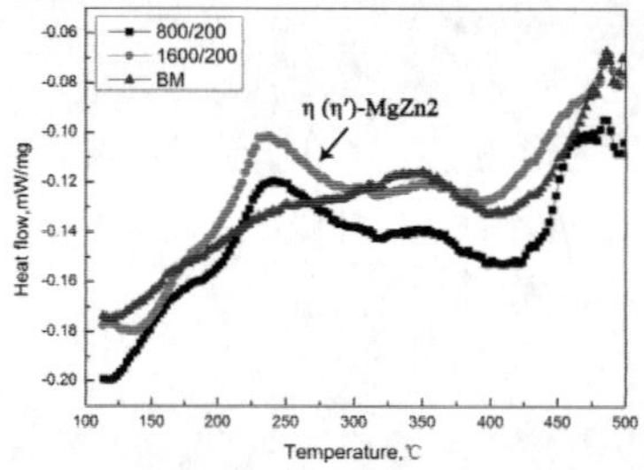

Figure 4. DSC analysis of the precipitates in the SZ and the BM under AP state.

3.3 Hardness

The hardness profiles on the cross-section under AP and PPHT states are shown in figure 5, exhibiting a reverse "U" shape. The hardness in the SZ of 1600/200 and 800/200 samples under AP state are higher than that of the BM. Compared with the un-processed BM, which has an average hardness of 65 HV, the average hardness in the SZ is 115 HV and 124 HV, respectively, corresponding to the 800/200 and 1600/200 samples. The increment in hardness of FSP samples under AP state is mainly caused by the difference in precipitation and grain refinement strengthening effects, which will be quantitatively calculated in this paper. Under PPHT state, the hardness in the SZ of both 1600/200 and 800/200 samples also increase compared with the hardness of the BM. Figure 6 shows the average hardness in SZ with different rotation speeds. It can be seen that the effects of PPHT on hardness are different between 1600/200 sample and 800/200 sample. Compared with AP state, the hardness of 1600/200 sample increases signally, while the hardness of 800/200 decreases under PPHT state.

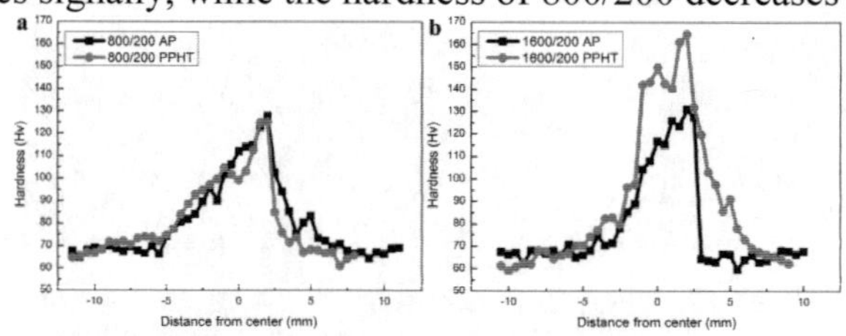

Figure 5. Hardness distribution across the cross-section of plates under different states (a) 800/200 and (b) 1600/200 samples.

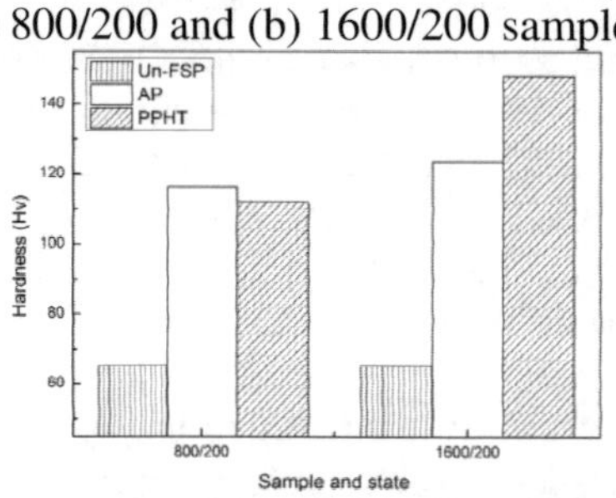

Figure 6. The average hardness in the SZ with different rotation speeds at different heat treatment conditions.

3.4 Effect of the BM condition on the dissolution of the precipitates during FSP

The PPHT did not enhance the hardness of 800/200 sample in the present study. Under PPHT states, the strengthening precipitates reform and its quantity is in proportion to the amount of the dissolved precipitates during FSP. Sufficient reformed precipitates contribute to the elevation in the hardness. Under the soft (O-temper) condition, the initial precipitates in the BM are the coarse stable phases which are difficult for dissolution because of their high dissolution points and larger sizes. And it suggests that the dissolution rate of the precipitates is reduced due to the existence of the coarse stable precipitates. As mentioned above, for 800/200 sample, the low dissolution rate restrains the dissolution of the precipitates, leading to a negative effect of PPHT on the hardness. More heat input has been provided in 1600/200 sample than in 800/200 sample, which is beneficial to induce sufficient dissolution of the precipitates and elevates the hardness of the SZ during PPHT.

3.5 Strengthening effect of grain refinement and precipitation

FSP drastically changed the microstructure in the SZ. And the main differences in microstructure between the BM and the SZ in this study are the grain size, the amount and morphology of the precipitates. Consequently, the hardness is mainly affected by the grain refinement and precipitation strengthening. Based on the report of Feng [8], the hardness of a material is roughly proportional to the yield strength, therefore, the yield strength model is also suitable for predicting the variation in hardness.

Table 1. The volume fraction and the mean radius of the precipitates.

Samples	BM	800/200	1600/200
f	5.86	2.93	2.40
r	75	60	55

Grain refinement strengthening σ_{GB} is described well by the Hall-Petch relationship as

$$\sigma_{GB} = \sigma_0 + kd^{-1/2}$$

where σ_{GB} is the increase in yield strength due to grain boundary strengthening, σ_0 is the yield strength for pure aluminum is about 16 MPa, k is the Hall-Petch constant about 0.065 MPa $m^{-1/2}$, d is the average grain size [9].

The precipitation strengthening σ_p can be given as

$$\sigma_p = \frac{M}{b} \cdot \sqrt{\frac{3f}{2\pi}} \cdot (2\beta G b^2) \cdot \frac{1}{r}$$

where r is the mean precipitates radius, G is the shear modulus (27 GPa for aluminum), b is Burgers vector (2.84 Å for aluminum), and f is the volume fraction of the precipitates, β is a constant (0.36 by Ref [10]). The initial precipitates in the BM are coarse, and in the SZ, there is insufficient time for dissolved precipitates to reform under AP state. The detailed information of the precipitates is listed in table 1. The summary of the grain refinement and the precipitation effects, and the hardness in the SZ and BM is shown in table 2. For the O-temper BM, the precipitation strengthening of the BM is small due to the existence of initial coarse precipitates, and the decrease of the precipitation strengthening caused by the FSP is limited. Since the hardness of the SZ is enhanced by the grain refinement. Besides, more precipitates dissolve in 1600/200 samples during FSP, leading to a higher solid solution strengthening, and increase the hardness further compared with that of 800/200 sample.

Table 2. Summary of the grain refinement and precipitation strengthening effects.

Samples	σ_{GB} (MPa)	σ_p (MPa)	Hardness (HV)
BM	21	37	65
800/200 AP	76	32	115
1600/200 AP	79	33	124

Under PPHT state, the precipitation strengthening rises and the solid solution strengthening drops as the η (η′)-$MgZn_2$ reforms. Sufficient reformed fine η (η′)-$MgZn_2$ are required in order to offset the loss of the solid solution strengthening. As to 1600/200 samples, there are enough fine η (η′)-$MgZn_2$ reforming, and the precipitation strengthening elevates the hardness sharply. On the contrary, the reformed fine precipitates in 800/200 sample are insufficient on account of its small amount of the dissolved precipitates during FSP.

IOP Conf. Series: Materials Science and Engineering **194** (2017) 012031 doi:10.1088/1757-899X/194/1/012031

4. Conclusion

(1) FSP was demonstrated to be very effective in producing fine grains in the SZ, and a same level grain size in the SZ was obtained with rotation speeds of 800 rpm and 1600 rpm. The hardness of the SZ under the AP state was enhanced, which was higher than that of the BM. After PPHT, the hardness of SZ under high rotation speed condition had been enhanced, while the hardness was deteriorated when the low rotation speed was adopted.
(2) For FSPed 7B04-O Al, the strengthening effect of grain refinement overwhelmed the softening effect of precipitate dissolving, and the elevation in the hardness of the SZ under the AP state was mainly caused by the grain refinement. Under the PPHT states, the amount and morphology of the precipitates played important roles in the enhancement in the hardness, sufficient fine reformed precipitates contributed to elevating the hardness.

References

[1] Mishra R S, Mahoney M W, Mvfadden S X, Mara N A, and Mukherjeel A K 2000 *Scripta. Mater* **42** 163
[2] Cui G R, Ma Z Y and Li S X 2008 *Scripta. Mater* **58** 1082
[3] Jata K V and Simiatin S L 2000 *Scripta. Mater* **43** 743
[4] Liu F C and Ma Z Y 2008 *Scripta. Mater* **58** 667
[5] Liu H J, Zhang H J and Yu L 2011 *Mater. Des* **32** 1548
[6] Sharma C, Dwivedi D K and Kumar P 2013 *Mater. Des* **43** 134
[7] Hu Z, Yuan S, Wang X, Liu G and Huang Y 2011 *Mater. Des* **32** 5055
[8] Dixit M, Mishra R S and Sankaran K K 2008 *Mater. Sci. Eng. A* **19** 1256
[9] Zander J and Sandstorm R 2008 *Mater. Des* **65** 1057
[10] Feng X, Liu H and Lippold J C 2013 *Mater. Charact* **29** 97

High Strength and Wear Resistance of Tantalum by Ultrasonic Nanocrystalline Surface Modification Technique at High Temperatures

A Amanov[1]*, Y S Pyun[1] and V K Vasudevan[2]
[1]Mechanical Engineering Department, Sun Moon University, Asan 31460, Korea
[2]Department of Mechanical and Materials Engineering, University of Cincinnati, Cincinnati, OH 45221, USA
E-mail: amanovtribo@gmail.com

Abstract. Enhancing the strength of materials is of great concern in that the failures such as wear, corrosion and fatigue occur on the surface. A tantalum (Ta), which is a refractory metal, was subjected to severe plastic deformation induced by ultrasonic nanocrystalline surface modification (UNSM) technique at room- and high-temperatures of 200^{0}C and 800^{0}C. The mechanical properties and wear resistance of Ta before and after UNSM treatment were characterized by X-ray diffraction (XRD), hardness tester, and ball-on-disk tribometer, respectively. The UNSM treatment at 800^{0}C led to the intensity reduction and broadening, and the diffraction peaks were shifted to lower angles, which means the presence of compressive residual stress and grain size refinement. The hardness of the UNSM-treated sample at room temperature was higher than that of the untreated one, and it was further increased with increasing the UNSM treatment temperature. In addition, the effect of UNSM treatment at high-temperature on wear resistance was significant than that of the UNSM treatment at room-temperature. We demonstrate the possibility of further increasing the mechanical properties and wear resistance of Ta by UNSM treatment at high-temperatures in comparison with the untreated and UNSM-treatment at room temperature, which would be beneficial for applications of Ta.

1. Introduction

In the past few decades, a new era of severe plastic deformation (SPD) technologies have attracted considerable both scientific and technological interests in order to increase the strength of materials by producing a nanocrystalline surface layer with a thickness of several hundreds of microns [1]. A number of newly developed SPD techniques such as surface mechanical attrition treatment (SMAT), equal channel angular pressing (ECAP), etc. have been developed [2, 3]. These technologies are based on imposing high deformation and refining the grain size of coarse-grained (CG) structure of bulk materials into refined nano-grained (NG) structure [4]. In general, an NG structure has a plenty of advantages over the CG one, and besides improving the mechanical properties, while no change in chemical composition after SPD processes, it gives a possibility of applying superplastic deep drawing process to material [5].

Tantalum (Ta) is a refractory metal, which has a wide range of applications in several industries such as electrical devices, aircraft engines, surgical implants, heat exchangers in boilers, nuclear power plants and missiles due to its strength, ductility, toughness, wear resistance, corrosion resistance, thermal conductivity, etc. [6]. In particular, Ta has excellent resistance to corrosive attack by acids and liquid metals, where it displays an oxide film on its surface [7]. Despite the fact that excellent corrosion resistance makes Ta ideal for use in the manufacture and repair of chemical process equipment, heat exchangers in boilers and nuclear power plants, but the wear resistance and fatigue strength are still on demand to be improved. In this regard, there is a need to increase the mechanical properties of Ta by producing a nanocrystalline surface layer with an NG structure. Limited studies can be found in the literature involving the nanostructured Ta by surface modification techniques though it has drawn great attention from industries, which may be due to its high plastic resistance. Zhang et al. have studied the formation of nanocrystalline structure in Ta by sliding friction treatment (SFT) [8]. A nanocrystalline structure was produced successfully with an average longitudinal grain size of about 20 nm. Moreover, Ta sample was subjected to equal channel angular extrusion (ECAE), where an ultrafine grain size was obtained [9].

Ultrasonic nanocrystalline surface modification (UNSM) technology is also an SPD technology, which induces a plastic deformation at the surface and sub-surface regions in order to produce a nanocrystalline surface layer a thickness of several hundreds of microns [10]. The process utilizes an ultrasonic vibration superimposed on a combination impact of static and dynamic loads, which is a core advantage of the UNSM technology over other SPD technologies, to induce high-strain-rate plastic deformation on a material near-surface region. It is also an effective and economical method to produce a nanocrystalline surface layer with NG structure. The UNSM treatment has been used to improve the friction behavior, wear resistance, corrosion resistance, and fatigue strength of various materials through the presence of NG structure having a high hardness and compressive residual stress in the near-surface region [11, 12].

In the present study, we propose a newly developed high-temperature UNSM technique, by means of which a further high strength and smaller grain size were obtained compared to the existing room temperature UNSM technique. It is believed that this high temperature UNSM technique, which is a simple and more efficient holds great potential for various industries, and plays an important role in further improving the overall properties of materials. Hence, so far, the main objective of this present study is to make a comparison between the effects of room- and high-temperature UNSM technique on mechanical properties and wear resistance. Detail microstructural evolution as a function of depth from the top surface by electron backscattered diffraction (EBSD) and transmission electron microscopy (TEM) is still in progress.

2. Experimental procedures

2.1. Sample preparation

Refractory Ta with a body-centred cubic (bcc) crystal structure is a high-density, ductile metal that exhibits exceptional corrosion resistance and good high-temperature strength. Cold worked Ta has a purity of 99.95% (metal base) with a surface of about 194 HV was employed in this study. The round bar Ta was cut into disk samples with a diameter of 20 mm and a thickness of 5 mm. The samples were ground and then were polished with silicon carbide (SiC) abrasive papers down to 2000 grit. Some physical and mechanical properties of Ta are listed in Table 1.

Table 1. Some physical and mechanical properties of Ta employed in this study.

Density (g/cm^3)	UTS (MPa)	Elastic modulus (GPa)	Elongation (%)	Hardness (HV)	Poisson`s ratio
16.6	900	186	5	200	0.35

2.2. Ultrasonic nanocrystal surface modification (UNSM) technique

A UNSM is an ultrasonic-based technology, where a ball (tip) made of tungsten carbide (WC) or silicon nitride (Si_3N_4) attached to a horn strikes the surface of a workpiece at a frequency of 20 and 40 kHz, at an amplitude of up to 100 μm with 1000 to 10000 shots per square millimetre. A static load is exerted on the tip against the surface to control the pressure with which the tip contacts the workpiece surface. In comparison with other SPD technologies, such as shot peening (SP), SMAT, etc. the UNSM technology offers a higher repeatability and reliability because of its capability for precise control of the strike intensity and density. Ta samples were subjected to UNSM technique at room- (25^0C) and high-

(200⁰C and 800⁰C) temperatures under the parameters listed in Table 2. A high temperature of 800⁰C can be reached within 90 s and cooling up to 25⁰C reaches in 15 min. The distance between the sample and the halogen lamp was about 15 cm. Details of the UNSM technique can be found in our previous publication [10, 11].

Table 2. UNSM technique treatment parameters of Ta employed in this study.

Frequency (kHz)	Amplitude (μm)	Static load (N)	Speed (mm/min)	Tip material	Tip diameter (mm)	Interval (μm)
20	40	30	2000	WC	2.38	70

2.3. Friction and wear test performance

The friction coefficient and wear resistance of the untreated and UNSM-treated at both room and high temperatures were investigated using a ball-on-disk tribometer (CSM Instruments, Switzerland) against a bearing steel with a diameter of 6 mm in dry test conditions listed in Table 3. The counterface ball was fixed to the sample holder, while the lower sample was fixed to the main stage, which reciprocated backward and forward.

Table 3. Ball-on-disk wear test conditions.

Normal load (N)	Reciprocating speed (cm/s)	Sliding distance (m)	Temperature (^{0}C)	Relative humidity (%)
10	2.51	30	25	52

2.4. Surface and sub-surface characterizations

The surface hardness was measured with the Vickers hardness tester (Mitutoyo, MVK E3, Japan) at a load of 300 gf and a dwell time of 10 s, while the *P-h* curves were obtained by nano-indentation tester (CSM Instruments, Switzerland) with a Berkovich indenter at a maximum load or 100 mN and a holding time of 10 s. The average six measurements are reported for each data point. The residual stress was measured with Cu-Kα radiation (λ=1.5418Å) using a laboratory based X-ray diffraction (XRD) residual stress analyser (LXRD, Proto, Canada) at an accelerating voltage of 20 kV and a current of 10 mA. The stress was determined using the $\sin^2\psi$ method along two orthogonal directions ($\varphi 0^0$ and $\varphi 90^0$). Full XRD patterns in the range of 30 to 90 degree were obtained using a Rigaku RU-300 with Cu-Kα radiation. The cross-sectional profiles of the wear tracks were measured using a two-dimensional (2D) surface profilometer (Surfcom 1500 SD3, Accretech, Japan). The specific wear rate and the corresponding standard deviation were quantified as the ratio of wear volume loss over the normal load multiplied by total reciprocating sliding distance.

3. Results and discussions

3.1. Surface hardness by the Vickers hardness tester and P-h curve by nanoindentation

The variation in surface hardness of Ta samples before and after UNSM treatment is shown in Figure 1. It can be seen that the surface hardness of Ta was increased by about 19% after UNSM treatment at 25⁰C as shown in Figure 1(a). However, no significant increase in surface hardness was found after heat treatment at 200⁰C, while the surface hardness was increased by about 26% using the combination of heat treatment with UNSM treatment at 200⁰C. Furthermore, the surface hardness was increased significantly up to 489 HV when the heat treatment performed at 800⁰C, which was more increased by about 11% by combination of heat treatment with UNSM treatment at 800⁰C. In addition, the *P-h* curves of the samples was obtained by nanoindentation as shown in Figure 1(b). It was found that the sample subjected to the combination of heat treatment and UNSM treatment at 800⁰C performed the shallowest penetration depth of about 786 nm, which is nearly less than two times than the corresponding penetration depth of the untreated sample. In case of the UNSM-treated samples, the high hardness values may be attributed to a significant grain size refinement and a relatively high dislocation density and the mechanism of increase in hardness can be explained by the Hall-Petch relationship [13].

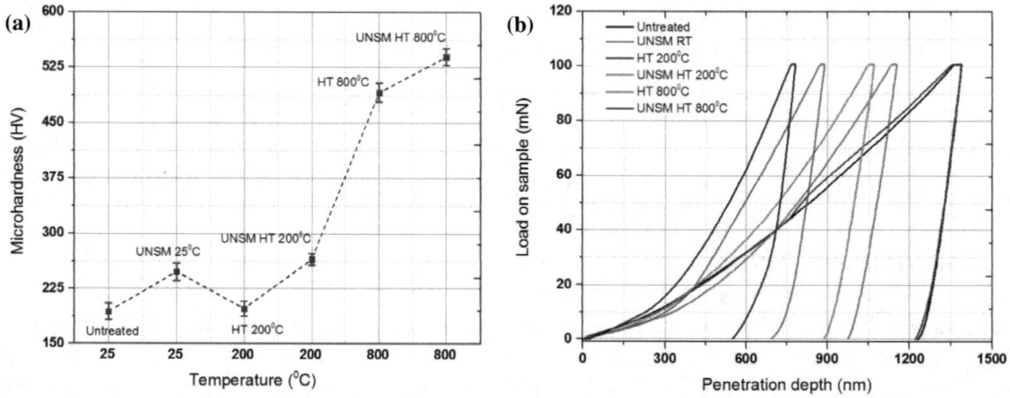

Figure 1. Variation in surface hardness (a) and *P-h* curves (b) of the untreated and UNSM-treated Ta samples, respectively.

3.2. Surface compressive residual stress and XRD patterns

The variation in surface residual stress and XRD patterns of Ta samples before and after UNSM treatment is shown in Figure 2. Residual stresses perpendicular and along to the UNSM treatment direction are represented as $\varphi 0^0$ and $\varphi 90^0$, respectively. The untreated and heat treatment samples at both 200^0C and 800^0C with no UNSM treatment exhibit a tensile residual stress at the surface as shown in Figure 2(a). The UNSM treatment induced a deep compressive residual stress of about -600 MPa, which is quite higher than the corresponding residual stress value of the untreated sample. The residual stress value of the UNSM-treated sample decreased with increasing the temperature to 200^0C perpendicular orthogonal direction of $\varphi 0^0$ and it increased along orthogonal direction of $\varphi 90^0$. It further increased up to -976 MPa and -1057 MPa by the combination of heat treatment and UNSM treatment at 800^0C perpendicular and along orthogonal directions, respectively. Notably, the compressive residual stress of the samples along orthogonal direction of $\varphi 90^0$ was found to be higher than that of the along orthogonal direction of $\varphi 0^0$ UNSM treatment at both high temperatures, which may be attributed to the number of strikes along the two different UNSM treatment directions as determined by the treatment speed and interval.

Summarized XRD patterns obtained from the surface of the samples are plotted in Figure 2(b). Obviously, the intensities of diffraction peaks reduced and broadened after UNSM treatment at 25^0C and 800^0C. However, no significant difference in intensities of diffraction peaks was found for the sample heat treatment and UNSM treatment at 200^0C in comparison with the untreated sample. Increasing the temperature from 200^0C to 800^0C resulted in again reduction and broadening in intensity of diffraction peaks. Moreover, a new diffraction peak was detected at a diffraction angle of 36.7 degree, which is believed due to oxidation during high-temperature treatment. A qualitative estimation of the plastic strain, work hardening and grain size refinement by the UNSM treatment can be made by analysing the intensity and full-width at half maximum (FWHM) of the diffraction peaks [14]. The diffraction peaks of the samples treated at both 200^0C and 800^0C shifted to a slightly lower angle, which indicates the presence of compressive residual stress induced by micro-strain [15], while no shift for the UNSM-treated sample at 25^0C. It is also known that a reduction in intensity and broadening in FWHM after UNSM treatment may also be attributed to the internal stresses, especially stacking faults or twinning [12]. Moreover, when the grains are subjected to a uniform compressive strain, diffraction peaks shift to the lower angles [15]. The grain size quantified according to the Scherrer equation of the UNSM-treated sample at 25^0C was found to be 84 nm, which was further refined by high-temperature UNSM treatment at 800^0C up to 36 nm.

IOP Conf. Series: Materials Science and Engineering **194** (2017) 012032 doi:10.1088/1757-899X/194/1/012032

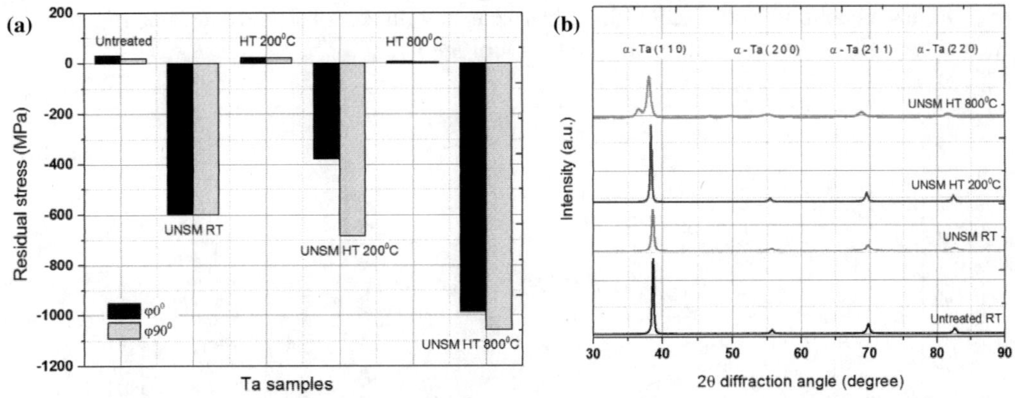

Figure 2. Variation in surface residual stress (a) and XRD patterns (b) of the untreated and UNSM-treated Ta samples.

3.3. Friction coefficient and wear resistance

The variation in friction coefficient of the Ta samples before and after UNSM treatment is shown in Figure 3(a). Clearly, it can be seen that the friction coefficient of the UNSM-treated samples at both room and high temperatures exhibited a lower friction coefficient compared to that of the untreated ones as shown in Figure 3(a). First, in case of the untreated sample, the friction coefficient was found to be about ~0.67 for the first 8 m of sliding, and with continuing the testing time the friction coefficient started reducing to a value of about ~0.58 up to a sliding time of about 17 m, and finally got stabilized to a value of about ~0.47 in the last stage of testing. In case of the UNSM-treated sample at room temperature, the friction coefficient abruptly increased to a value of about ~0.60 at the initial stage of the testing, and then with continuing the testing time the friction coefficient started reducing to a value of about ~0.52 up to a sliding time of about 17 m, which is a similar friction behaviour (trend) occurred for the untreated sample, and finally, it also got stabilized to a value of about ~0.42 in the last stage of testing. Second, in case of the heat treated and UNSM-treated samples at 200^0C, the friction behaviour was found to be similar to the friction behaviour (trend) untreated and UNSM-treated samples at 25^0C, but the UNSM treated sample at 200^0C exhibited a bit lower friction coefficient than that of the UNSM-treated sample treated at 25^0C, while no difference in friction coefficient for the untreated sample at 25^0C and 200^0C. Finally, in case of the heat treated and UNSM-treated samples at 800^0C, the heat treated sample exhibited a similar friction behaviour, which is increased first and then reduced for a while and again reduced and finally got stabilized to a value of about ~0.37 by the end of test. However, the UNSM-treated sample had a different friction behaviour unlike other samples, where the friction coefficient gradually increased up to a sliding time of about 17 m, where friction coefficient increased sharply and got stabilized to a value of about ~0.51 by the end of test. Interestingly, even though the heat treated sample at 800^0C had a higher hardness than other untreated, UNSM-treated at 25^0C, heat treated and UNSM-treated at 200^0C samples could not improve the friction behaviour of Ta at the initial stage of the test, but it exhibited the lowest friction coefficient value of about ~0.37, which is about 19% lower than that of the friction coefficient of the UNSM-treated sample at 800^0C. Ta exhibited a higher friction coefficient with increasing the temperature [16]. The reduced friction coefficient after UNSM treatment is mainly due to the surface morphology modification and reduced surface roughness.

Figure 3(b) shows the comparison in wear rate of the of Ta samples before and after UNSM treatment. It was found that the variation in wear resistance of the UNSM treated at 800 ^{0}C is significant in comparison with the untreated one, but no remarkable change in between the untreated and UNSM treated at 200 ^{0}C. The UNSM treatment at 25^0C increased the wear resistance by about 29%, while the UNSM treatment at 800^0C increased by about 94%. The increase in wear resistance of the UNSM-treated samples is directly related to the increased surface hardness and induced compressive residual stress. Moreover, a modified microstructure having a reduced surface roughness of Ta subjected to UNSM treatment at both room and high temperatures is also responsible for the increase in wear resistance. It has been reported earlier that the wear resistance of NG materials is related to the subsurface deformation behaviour and surface morphology [17].

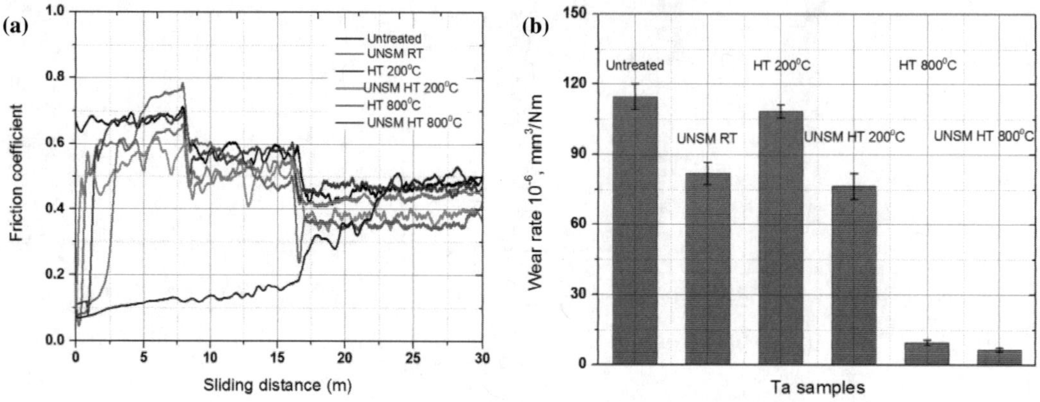

Figure 3. Variation in friction coefficient (a) and comparison in wear rate (b) of the untreated and UNSM-treated Ta samples.

4. Conclusions

In this study, the effect of UNSM treatment at room and high temperatures on the mechanical and tribological properties of Ta was studied. The surface hardness value for the UNSM-treated samples at 25^0C and 800^0C temperatures increased by about 19% and 65% compared to that of the untreated one. The presence of induced compressive residual stress at the surface after UNSM treatment was reached to less and more -1000 MPa for the UNSM-treated at 800^0C sample, which was significantly higher than that of the untreated sample. The grain size refinement after UNSM treatment at both 25^0C and 800^0C was confirmed by diffraction patterns, where the intensity reduced and broadening occurred. The friction and wear performance of the UNSM-treated sample at 800^0C improved considerably in comparison with those of the untreated and UNSM-treated samples at 25^0C. These findings would be beneficial for high-temperature applications of Ta in various industries.

5. References

[1] Estrin Y and Vinogradov A 2013 *Acta Mater.* **61(3)** 782.
[2] Basu S, Wang Z and Saldana C 2016 *Proc. Math. Phys. Eng. Sci.* **472(2187)** 20150486.
[3] Raab G 2014 *IOP Conf. Series: Mater. Sci. Eng.* **63** 012009.
[4] Valiev R, 2004, *Nature Mater.* **3** 511.
[5] Valiev R, Estrin Y, Horita Z, Langdon T G, Zechetbauer M J and Zhu Y T 2006 *JOM* **58(4)** 33.
[6] Shankar M R, Rao B C, Lee S, Chandrasekar S, King A H and Compton W D 2006 *Acta Mater.* **54(14)** 3691.
[7] Huang Y, Maury N, Zhang N X and Langdon T G 2014 *IOP Conf. Series: Mater. Sci. Eng.* **63** 012100.
[8] Zhang Y S, Wei Q M, Niu H Z, Li Y S, Chen C, Yu Z T, Bai X F and Zhang P X 2014 *Int. J. Refract. Metals Hard Mater.* **45** 71.
[9] Mathaudhu S N and Hartwig K T 2007 *Mater. Sci. Eng. A* **463** 94.
[10] Amanov A, Cho I S, Pyoun Y S, Lee C S and Park I G 2012 *Wear* **286-287** 136.
[11] Amanov A, Cho I S, Pyun Y S 2016 *Appl. Surf. Sci.* **388** 185.
[12] Amanov A, Cho I S, Park I G and Pyun Y S 2016 *Mater. Lett.* **174** 142.
[13] Ghamarian I, Samimi P, Telang A, Vasudevan V K, Collins P C 2017 *Mater. Sci. Eng. A* **688** 524.
[14] Amanov A, Pyun Y S and Sasaki S 2014 *Appl. Surf. Sci.* **72** 187.
[15] Amanov A, Urmanov B, Amanov T and Pyun Y S 2017 *Mater. Lett. (In Press).*
[16] Scharf T W, Prasad S V, Kotula P G, Michael J R and Robino C V 2015 *Wear* **330-331** 199.
[17] Chen X, Han Z, Li X and Lu K 2016 *Sci. Adv.* **2(12)** e1601942.

IOP Conf. Series: Materials Science and Engineering **194** (2017) 012033 doi:10.1088/1757-899X/194/1/012033

Optimization of RT superplasticity of UFG Zn-22Al alloy by applying ECAP at different temperatures and phase regions

M Demirtas[1,2], H Yanar[2], G Purcek[2]

[1]Bayburt University, Faculty of Engineering, Department of Mechanical Engineering, 69000, Bayburt, Turkey
[2]Karadeniz Technical University, Faculty of Engineering, Department of Mechanical Engineering, 61080, Trabzon, Turkey

E-mail : purcek@ktu.edu.tr

Abstract. Zn–22Al alloy was subjected to either one-step or two-step equal channel pressing (ECAP) to investigate the effect of processing temperature on its microstructure and room temperature (RT) superplasticity. In one-step ECAP processes, 4 passes ECAP were applied to the alloy at different temperatures: RT, 100°C and 250°C in two-phase region below eutectoid temperature and 350°C in single-phase region above eutectoid temperature. In two-step ECAP processes, one-step ECAP-processed samples were subjected to four more passes ECAP at RT. Considering the one-step ECAP processing, RT superplasticity increased with decreasing ECAP temperature as expected, and the highest RT superplasticity was achieved as 350% after 4 passes ECAP at RT. On the other hand, application of 4 more passes ECAP at RT to the sample showing the lowest superplastic elongation after one-step ECAP (the sample processed at 350°C) resulted in the maximum RT elongation of 400% at a high strain rate of $5x10^{-2}$ s^{-1}. These results suggest that first step temperature of two-step ECAP process is needed to increase above the eutectoid point of Zn-22Al alloy to achieve high RT superplasticity. These results were attributed to the changes in microstructure inside the single-phase and two-phase regions during the processes.

1. Introduction

Superplastic behavior can be explained as the high neck-free tensile elongation to failure which occurs at some polycrystalline materials. High tensile test temperature (about 0.5Tm where Tm is the absolute melting point of the material), low strain rate (between $1x10^{-5}$ s^{-1} and $1x10^{-3}$ s^{-1}) and small grain size (typically below than 10 μm) are the main requirements that should be fulfilled in order to achieve superplasticity [1, 2]. Among these requirements, grain size can be considered as the most important one since any change in grain size also affects two other requirements. Decreasing grain size decreases the temperature and increases strain rate at which superplasticity are achieved [3, 4]. Therefore, grain refinement techniques have been among the main interests of scientist and engineers who deal with the superplastic behavior of metals.

In order to achieve fine grained (FG) and/or ultrafine grained (UFG) microstructures, some thermal and thermomechanical processes have been applied to superplastic materials [5-18]. Considering the thermomechanical processes, severe plastic deformation (SPD) techniques have been developed and performed successfully [5, 9-18] besides the conventional plastic deformation processes like rolling, extrusion and forging. Furthermore, SPD techniques have been found to be more effective to achieve smaller grain size and thus higher superplastic elongation [14] comparing to the conventional grain refinement techniques. Equal channel angular pressing (ECAP) [5, 9-14], friction stir processing (FSP) [15], torsional straining (TS) [16,17] and cross-channel angular extrusion (CCAE) [18] are the main SPD techniques which have been used for grain refinement in superplastic materials. Among them, ECAP is the most commonly used one and it has been applied to many different superplastic materials in order to achieve FG and UFG microstructures. For this purpose Zn-22Al alloy as a model superplastic material is one of the most commonly studied superplastic alloy and ECAP has been applied to this alloy at different temperatures as a grain refinement tool [5, 9-14].

Effects of some ECAP parameters on the final microstructure and mechanical properties of the materials have been studied [19] including processing route, numbers of passes, pressing speed, back pressure and pressing temperature. Pressing temperature is considered as a key factor and it has been shown that lower pressing temperature leads to lower final grain size [19]. On the other hand, two-step ECAP in which 4 passes were applied to Zn-22Al alloy at 350 °C followed by 4 more passes at RT (8 total passes) resulted in higher superplastic elongation [14] compared to 8 passes ECAP performed at room temperature (RT) [5]. Thus, two step ECAP seems to be more suitable to achieve higher superplastic elongations than ECAP in which all passes are performed at low temperature. Regarding the dependency of final grain size of superplastic materials to the ECAP temperature, it will be beneficial to determine whether it is possible to achieve higher superplastic elongations by decreasing the first step ECAP temperature (like performing first step ECAP at 100 °C or 250 °C) of two-step ECAP processes or not. Therefore, the main purpose of this study is to analyze the effect of first step temperature of two-step ECAP on the RT superplasticity of Zn-22Al alloy.

2. Experimental procedure

As-cast Zn-22Al ingot was homogenized at 375 °C for 24 h and billets with dimensions of 13 x 13 x 130 mm^3 were machined from the ingot for the subsequent ECAP. After then, the billets were homogenized at 375 °C for 48 h for the second time and quenched into ice-water. ECAP was performed as either one-step or a two-step processes. In the one-step ECAP, each billet was subjected to 4 passes at RT, 100 °C and 250 °C in the two phase region below the eutectoid temperature and at 350 °C in the single phase region above the eutectoid transformation temperature of the alloy. In the two-step ECAP, 4 more passes were applied to the billets at RT followed by the 4 passes performed in the first step. ECAP processes were conducted at 1mm.s^{-1} pressing speed using route Bc where the billets were rotated along their longitudinal axis among each passes.

Scanning electron microscopy (SEM) was utilized for the microstructural examination after the applied processes. For this purpose, SEM examination samples were cut from the ECAP-processed billets as so examination plane was perpendicular to the extrusion direction (ED) of the billets using wire electro-discharge machining (wire-EDM). The samples were ground, polished and etched in a solution containing 5 g CrO_3, 0.5 g Na_2SO_4 and 100 ml H_2O. SEM was conducted using a JEOL-6400 microscope.

In order to evaluate the RT superplasticity after the applied processes, tensile tests were performed at strain rates ranging between $1x10^{-3}$ and $1x10^{0}$ s^{-1}. Dog bone shaped test samples having 2 mm x 3 mm x 5 mm gauge section dimensions were extracted from the billets with their tensile axis aligned with the ED. All tests were repeated at least three times in order to confirm the validity of the tensile test results and mean of these tests results are given in the manuscript.

3. Results and discussion

SEM micrographs showing the microstructures of the alloy after the applied processes are given in figure 1. In these micrographs, the bright and dark contrast correspond to Zn-rich η- and Al-rich α-phases, respectively. 4 passes ECAP at RT resulted in a UFG microstructure with 350 nm grain sizes (Figure 1(a)-(b)). In one-step ECAP processes, increasing the ECAP temperature up to 250 °C also increased the final grain size, and grain sizes after 4 passes ECAP performed at 100 ºC and 250 ºC were measured to be 500 nm ((Figure 1(c)-(d))) and 1 µm (Figure 1(e)-(f)), respectively. On the other hand, application of ECAP above the eutectoid temperature brought about more refined microstructure with 250 nm grain size (Figure 1(g)-(h)). However, the microstructure consists of some regions with lamellar structure (LS) after that process (Figure 1(h)).

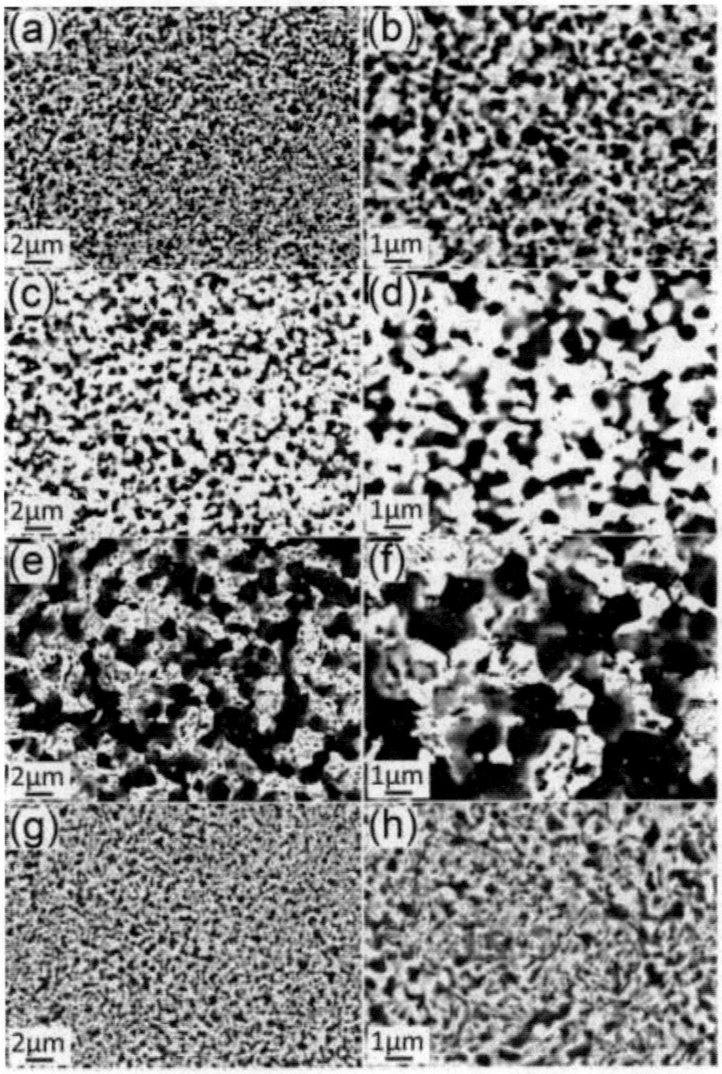

Figure 1. SEM micrographs showing the microstructures after one-step ECAP performed at: (a)-(b) RT, (c)-(d) 100 ºC (e)-(f) 250 ºC and (g)-(h) 350 ºC.

Application of four more passes ECAP at RT to the one-step ECAP-processed samples resulted in more refined microstructures compared to one-step ECAP processes. 8 total passes at RT brought about 250 nm grain sized microstructure (Figure 2(a)-(b)). Grain sizes after 4 passes ECAP at RT following one-step ECAP at 100 ºC and 250 ºC were measured as 400 nm (Figure 2(c)-(d)) and 700 nm (Figure 2(e)-(f)), respectively. Lamellar structure formed after 4 passes ECAP at 350 ºC was completely eliminated by 4 more passes ECAP at RT (Figure 2(g)-(h)). Lamellar-free UFG microstructure with 200 nm grain size was achieved instead [14].

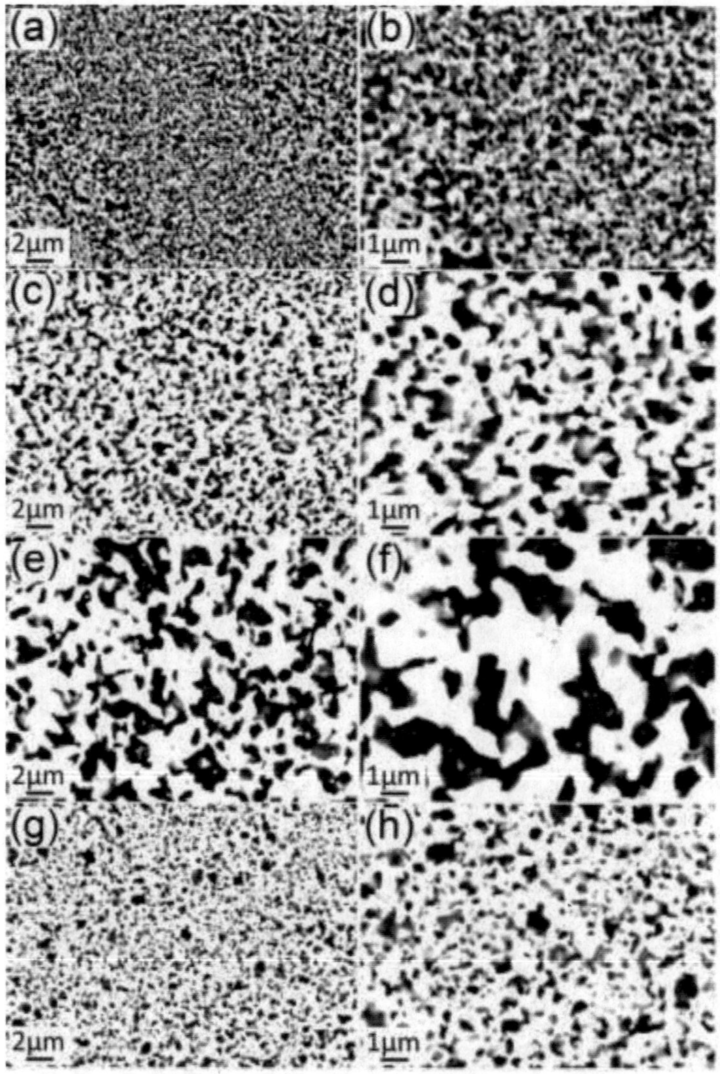

Figure 2. SEM micrographs showing the microstructures after two-step ECAP processes in which first step was performed at: (a)-(b) RT, (c)-(d) 100 ºC (e)-(f) 250 ºC and (g)-(h) 350 ºC.

It was stated previously that decreasing pressing temperature of ECAP resulted in finer grain sizes [19]. The final grain sizes obtained after one-step ECAP processes are in a good agreement with that conclusion except for the 4 passes ECAP applied at 350 ºC; i.e. below the eutectoid temperature the smallest grain size was achieved after 4 passes ECAP at RT. Regarding all one-step ECAP processes, on the other hand, although 350 ºC is the highest process temperature, the smallest grain size was achieved after 4 passes ECAP at 350 ºC. Considering ECAP temperatures of RT, 100 ºC and 250 ºC, the alloy was processed at two phase region, and grain refinement occurred due to the only ECAP. However, eutectoid transformation occurred at 350 ºC before the ECAP process, and ECAP performed in one phase region in that process. In this process, the ECAP billets were quenched between each passes. It is known that, quenching above the eutectoid temperature results in refined microstructure in Zn-22Al alloy [20]. Thus, grain refinement effects of both ECAP and quenching are combined, and the smallest grain size was achieved after 4 passes ECAP at 350 ºC.

Considering the grain sizes achieved after the applied processes, it can be concluded that decreasing ECAP temperature below the eutectoid temperature also decreases final grain size. However, processing the alloy above the eutectoid temperature results in more refined microstructure with some lamellar structure which is not suitable for achieving high superplastic elongation [20]. Thus, in one-step ECAP processes it is beneficial to keep ECAP temperature as low as possible for desired microstructure for superplasticity. On the other hand, in two-step ECAP, the smallest grain size was achieved after 4 passes ECAP at 350 ºC followed by 4 more passes at RT. Therefore the alloy should be processed above the eutectoid temperature in the first step of two-step ECAP processes in order to achieve more refined microstructure.

The RT superplastic elongations after the applied processes with respect to the initial strain rates are shown in figure 3. Considering the results of one-step ECAP processes, the highest RT elongation was achieved after 4 passes ECAP at RT as 350% at a strain rate of $1x10^{-1}$ s^{-1} (Figure 3(a)). 4 passes ECAP at 350 ºC, 250 ºC and 100 ºC resulted in maximum elongations of 110%, 195% and 315%, respectively (Figure 3(a)). Further grain refinement in two-step ECAP processes by means of 4 more passes ECAP at RT increased elongation to failure comparing to all one-step ECAP processes. In general, the sample subjected to 4 passes ECAP at 350 ºC + 4 passes ECAP at RT reflected to the highest RT elongation to failure as 400% at a strain rate of $5x10^{-2}$ s^{-1} (Figure 3(b)) [14]. 8 total ECAP passes at RT also resulted in high superplastic elongation of 375% [1]at $1x10^{-1}$ s^{-1}(Figure 3(b)

Decreasing grain size increases the maximum superplastic elongations in superplastic materials due to the effective grain boundary sliding (GBS) which occurs as the main deformation mechanism at more grain boundaries [3, 4]. Considering the maximum elongations after one-step ECAP processes, the results are consistent with this conclusion except for the maximum elongation obtained after 4 passes ECAP at 350 ºC. Although the alloy has the smallest grain size after 4 passes ECAP at 350 ºC among all one-step ECAP-processed samples, the lowest superplastic elongation was achieved after this process. Similar observation was also achieved in Zn-22Al alloy in [20]. It was shown that, lamellar structure inside the microstructure makes difficult the accommodation process of GBS by intragranular dislocation slip process and causes stress concentration especially near the lamellar structure. Thus, in the early stage of superplastic deformation (about 50% superplastic elongation) some micro-cracks nucleate near the lamellar structure and premature failure occurs [20]. Similarly, the unexpected low superplastic elongation after 4 passes ECAP at 350 ºC can be attributed to the partially lamellar structure which causes stress concentration and thus premature failure.

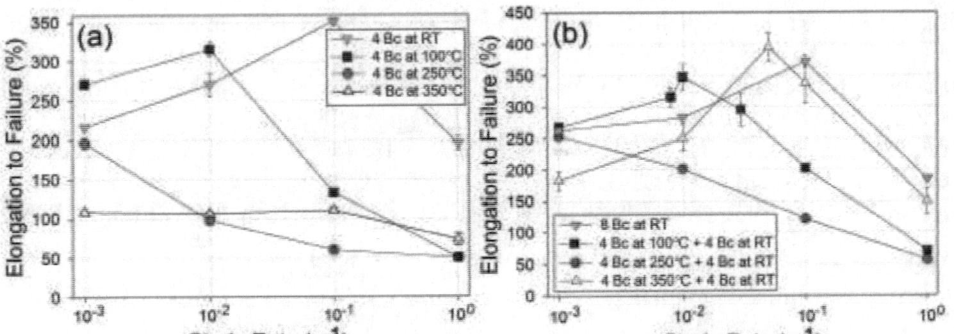

Figure 3. Variations of elongation to failure with the initial strain rates after: (a) one-step ECAP, and (b) two-step ECAP processes.

4. Conclusions

In this study, the effects of one-step and two-step ECAP temperatures on the final grain size and RT superplasticity of Zn-22Al alloy were investigated. The main findings and conclusions can be listed as below.

1. Decreasing ECAP temperature below the eutectoid temperature also decreases final grain size. However, processing the alloy above the eutectoid temperature results in more refined microstructure with some lamellar structure.

2. In one-step ECAP processes, it is beneficial to keep ECAP temperature as low as possible. On the other hand, the alloy should be processed above the eutectoid temperature in the first step of two-step ECAP processes in order to achieve more refined microstructure.
3. In one step ECAP, 4 passes at 350 °C resulted in the lowest superplastic elongation although it brought about the lowest grain size. This unexpected low superplastic elongation was attributed to the partially lamellar structure which causes stress concentration and thus premature failure. The maximum RT elongation was achieved to be 400% at a strain rate of $5x10^{-2}$ s^{-1} after two-step ECAP process which resulted the lowest grain size among all applied processes.

Acknowledgments
This research was supported by Scientific Research Projects of Karadeniz Technical University, Turkey, under Grant no10501.

References
[1] Kawasaki M and Langdon T.G 2014 *J. Mater. Sci.* **49** 6487.
[2] Kaibyshev O.A 1992 *Superplasticity of Alloys Intermetallides and Ceramics*, (Berlin: Springer-Verlag).
[3] Langdon T.G 1994 *Mater. Sci. Eng. A* **174** 225.
[4] Mayo M.J 1997 *Nanost. Mater.* **9** 717.
[5] Xia S.H, Wang J, Wang J.T and Liu J.Q 2008 *Mater. Sci. Eng. A* **493** 111-115.
[6] Tanaka T, Makii K, Ueda H, Kushibe A, Kohzu M and Higashi K 2003 *Inter. J. Mech. Sci.* **45** 1599.
[7] Tanaka T, Makii K, Kushibe A and Higashi K 2002 *Mater. Trans.* **43** 2449.
[8] Tanaka T, Makii K, Kushibe A, Kohzu M and Higashi K 2003 *Scr. Mater.* **49** 361.
[9] Tanaka T, Watanabe H and Higashi K 2003 *Mater. Trans.* **44** 1891.
[10] Tanaka T and Higashi K 2004 *Mater. Trans.* **45** 1261
[11] Kumar P, Xu C and Langdon T.G 2006 *Mater. Sci. Eng. A* **429** 324.
[12] Huang Y and Langdon T.G 2002 *J. Mater. Sci.* **37** 4993.
[13] Yang C.F, Pan J.H and Chuang M.C 2008 *J. Mater. Sci.* **43** 6260.
[14] Demirtas M, Purcek G, Yanar H, Zhang Z.J and Zhang Z.F 2014 *Mater. Sci. Eng. A* **620** 233.
[15] Hirata T, Tanaka T, Chung S.W, Takigawa Y and Higashi K 2007 *Scr. Mater.* **56** 477.
[16] Furukawa M, Horita Z, Nemoto M, Valiev R.Z and Langdon T.G 1996 *J. Mater. Res.* **11** 2128.
[17] Kawasaki M and Langdon T.G 2011 *Mater. Sci. Eng. A* **528** 6140.
[18] Chou C.Y, Lee S.L, Lin J.C and Hsu C.M 2007 *Scr. Mater.* **57** 972.
[19] Valiev R.Z and Langdon T.G 2006 *Prog. Mater. Sci.* **51** 881.
[20] Demirtas M, Purcek G, Yanar H, Zhang Z.J and Zhang Z.F 2016 *J. Alloy. Compd.* **663** 775.

IOP Conf. Series: Materials Science and Engineering **194** (2017) 012034 doi:10.1088/1757-899X/194/1/012034

Precipitates studies in ultrafine-grained Al alloys with enhanced strength and conductivity

V.D. Sitdikov[a,b], **M.Yu. Murashkin**[a,b], **R.Z. Valiev**[a,b*]

[a]Institute of Physics of Advanced Materials, Ufa State Aviation Technical University, Ufa, 450000, Russia

[b]Laboratory for Mechanics of Bulk Nanomaterials, Saint Petersburg State University, Saint Petersburg, 198504, Russia

*ruslan.valiev@ugatu.su, svil@ugatu.su

Abstract. The paper analyzes precipitation of nanosized particles in dynamically aged Al-Mg-Si alloy subjected to high pressure torsion in the temperature range from 25°C to 230°C. Precipitate types are identified and quantified within the frames of the modified X-ray phase analysis (XPA) technique. The generality and difference in the mechanisms and kinetics of precipitation during artificial and dynamic strain aging are analyzed, which opens new ways to further increase properties in promising aluminum conductors.

1. Introduction

Today aluminum alloys of the Al-Mg-Si system are widely applied as material for electrical conductors. The set balance of strength and electrical conductivity in them is achieved through thermal and thermomechanical treatment [1, 2]. In particular, in order to improve the balance of mechanical and performance properties, conducting alloys 6101 and 6201 are subjected to artificial aging (AA) after quenching. As a result of AA secondary phase particles precipitate in the Al matrix [1, 2]. Decomposition of supersaturated solid solution (SSS) controlled by temperature and ageing time occurs in the following order:

$$SSS \rightarrow GP\ zones \rightarrow \beta'' \rightarrow \beta' \rightarrow \beta.$$

Here β'' – needle-like precipitates of Mg_5Si_6 with a diameter of 3…6 nm and length of 20…45 nm, β' - rod-like precipitates of Mg_9Si_5 with a diameter of 8…15 nm and length of 3…100 nm, β – cube-like Mg_2Si precipitates [3, 4].

Our recent studies demonstrated that a significant increase of the properties of the Al-Mg-Si alloys was possible through refinement of grain structure via severe plastic deformation (SPD) techniques [5-7]. It was discovered that formation of an ultrafine-grained (UFG) structure in the alloys during SPD was accompanied by dynamic strain aging (DSA). Transmission electron microscopy (TEM) and atom probe tomography (APT) demonstrate that sphere-shaped precipitates with a diameter of 5-50 nm form in 6201 alloy after high pressure torsion (HPT) performed in the range of temperatures 27…230°C [6, 7].

Recent progress in the XPA is promising for such precipitates studies. The analysis consists in identification and quantification of precipitates with an improved accuracy on the basis of the analysis of the transmission diffraction patterns obtained using a parallel X-ray beam [8]. In this work this approach is used for phase analysis of precipitates and statement of microstructure parameters of the

Al-Mg-Si alloy after DSA fulfilled at different HPT temperatures. As a result of application of the new technique for 6201 alloy all the precipitate types were identified. The kinetics and mechanisms of precipitation during AA and DSA are discussed on the basis of the obtained data.

2. Materials and methods

Aluminum alloy 6201 (Mg 0.81, Si 0.79, Fe ≤ 0.1, base Al (wt. %)) is chosen to study DSA. Prior to AA and DSA, an initial rod is subjected to solution treatment at 550°C for 2 h and water quenching. AA is performed at 180°C for 24 hours (T6 regime). In order to implement DSA, billets are subjected to HPT [6,9] at room temperature (RT) and at 130°C, 180°C and 230°C.

The microstructure of the AA and DSA alloys is analyzed with the help of X-ray scattering on a diffractometer Rigaku Ultima IV using Cu$K\alpha$ radiation. Unlike the conventional techniques of XPA, in this work the diffraction patterns are obtained on foils with a thickness of 80 μm using a parallel beam in the transmission regime [8]. To identify the diffraction lines from precipitates, such exposure regimes were optimized as the distance from the detector to the studied sample, size and intensity of a beam, sample surface. As a result, unlike the conventional exposure in the reflection regime, the maximums of all the three types of precipitates were identified during AA, they are shown in the right corner in Fig. 1. Precipitates are identified in the PDXL software [10], and the quantitative estimation of their volume fraction f is performed by the Rietveld method [11]. As demonstrated below, this enabled studying precipitates with a volume fraction of about 0.1%. The microstructure of samples is also characterized by TEM to provide information about a grain size and shape of precipitates.

3. Results and discussion

Figure 1 displays the diffraction patterns for the samples after AA via regime T6 and DSA induced by HPT at different temperatures. The comparison of angles positions and peaks intensity with those stored in the base of diffractometric data shows that after AA precipitates of metastable β'', β' and stable β modifications form in the Al solid solution (Fig. 1a). β''-type precipitates are identified as a Mg_5Si_6 phase [4], β'-type precipitates are identified as a Mg_9Si_5 [12]. Besides, β-type precipitates are identified as a Mg_2Si phase [12]. The quantitative estimations show that after AA the volume fraction of the β'' phase in the alloy is 0.41 %, that of the β' phase is 0.28%, and that of the β phase is not over 0.11 % (Table 1).

Table 1. Microstructure parameters obtained by X-ray analysis and TEM studies.

Treatment	f, %			a, nm	ρ, 10^{14} m^{-2}	m_{ixp}, %	d_{gr}, nm	r_p, nm
	β''	β'	β					
T6	0.41(6)	0.28(4)	0.11(5)	0.405108(7)	0.08(3)	43(4)	6*10^4	
DA RT	0.18(5)	0.11(4)	0.05(3)	0.405234(5)	4.5(4)	95(2)	140	-
DA 130 °C	0.29(8)	0.26(6)	0.16(4)	0.405125(6)	2.1(3)	86(3)	275	5
DA 180 °C	0.23(4)	0.37(7)	0.53(7)	0.405052(7)	1.1(5)	79(4)	430	15
DA 230 °C	0.10(4)	0.24(4)	0.82(6)	0.405017(5)	0.3(4)	71(4)	980	25

The obtained results correlate well with the data obtained earlier as a result of the study of aged alloys of the Al-Mg-Si system [13-16]. In particular, TEM studies revealed β'' precipitates with a volume fraction of 0.35% in Al-Mg-Si after AA at 180°C for 17 hours [13], in 6011 alloy after T6 the volume fraction of β'' precipitates was 0.48% [14]. On the other hand, the APT technique [16] enabled revealing β'' phase of 0.25%, β' phase of 0.15% and β phase of 0.07% in Al-Mg-Si after AA at 180°C for 30 minutes. At the same time, small-angle neutron scattering [16] enabled stating that in Al-Mg-Si after AA (180°C for 18 hours) the volume fraction of β'' phase is ~0.25%. Thus, the XPA in the transmission regime unlike the TEM techniques and APT, on the example of AA of Al alloy with enhanced statistics and improved accuracy makes possible to identify and estimate all the three types of precipitates simultaneously, as a larger volume is analyzed. Besides, in this technique the estimates and sample preparation are performed faster. Thus, this technique is applied to evaluate precipitates after DSA.

In case of DSA at RT, judging by the XPA data, the microstructure contains only β" and β' precipitates with fractions of ~0.18% and ~0.11% (Fig. 1b, Table 1). However, DSA occurred during HPT at a temperature of 130°C (Fig. 1c) results in intensification of diffraction peaks corresponding to β", β' and β precipitates, which is connected with increase of the volume fraction of β" phase precipitates to 0.29%, the fraction of β' increases to 0.26%, and the fraction of the β-phase increases to 0.16% (Table 1). DSA at 180°C leads to reduction of the fraction of β" phase precipitates to 0.23% and increase of β' and β type precipitates to 0.37 and 0.53%, respectively. Further growth of the DSA temperature to 230°C leads to change of the maximums on the diffraction patterns (Fig. 1e), which is connected with growth of fractions of β precipitates to 0.82% and decrease of β' and β" precipitates (Table 1).

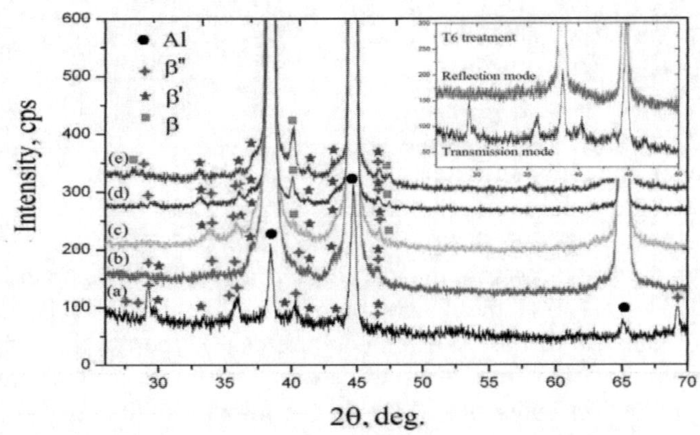

Figure 1. X-ray transmission diffraction patterns after AA (a) and after DSA at RT (b), at 130°C (c), at 180°C (d) and at 230°C (e).

Unlike AA, during HPT at RT the growth of dislocation density and lattice parameter is observed (Table 1). The same tendency is observed in the Al alloy of the Al-Mg-Sc-Zr system subjected to HPT [17]. The DSA temperature increase results in decrease of the ρ and a (Table 1). It is also established that the fraction of edge dislocations m_{ixp} after AA is 43%, and after DSA at RT it has the maximum value, and with the temperature increase it reduces monotonously (Table 1).

TEM analyses demonstrate also the formation of UFG structure and nano-sized sphere-shaped precipitates during HPT (Fig. 2). The HPT temperature increase leads to the grain size d_{gr} growth (Fig. 2) and increase of precipitates. In particular, after DSA at 130°C the average radius of precipitates r_p is ~5 nm, and after 180°C and 230°C it is about ~15 nm and ~25 nm, respectively.

The performed comparative analysis shows that the states of the alloy after DSA and AA differ in the quantity as well as in the morphology of precipitates. In particular, after AA mainly β" precipitates are observed in the alloy microstructure, which have a needle-like shape [4]. The main mechanism of AA is activation of mobile vacancies transmitting Mg and Si atoms across the Al matrix [18]. The GP zones capture migrating impurity atoms and grow along the directions <100> of the Al matrix acquiring the shape of needles. Further holding at 180°C results in growth of needles, which leads to formation of β'-type rod-shaped precipitates [4]. As it is seen from Table 1, AA ensures only partial SSS decomposition, as mainly β" and β' precipitates are found in the microstructure. After DSA, unlike AA, occurred during HPT at RT and during a short time (20 minutes), β" and β' precipitates form. With the HPT temperature increase above 130°C, the fractions of β" and β' precipitates decrease, and that of the β type grows. This fact testifies to the completeness and acceleration of decomposition of SSS during DSA. All the precipitate types, which are observed in this work and in recent publications [6, 7, 19], have a sphere shape (Fig. 2a,b).

The observed increase of the fraction of precipitates during DSA at 180°C as compared to AA testifies to accelerated depletion of the Al matrix from alloying elements. Such process should lead to a significant change of the Al lattice parameter (Table 1). When the DSA temperature increases to 230°C, the lattice parameter value tends to but does not achieve 0.40494 nm (Table 1), the typical value of pure Al. This indicates that some excess Mg or Si atoms partially remain in the Al matrix after precipitation. Increase of the DSA temperature results in growth of precipitates. As is seen from Table 1, the precipitates with an average radius of 5 nm relate to β'' type, those with a radius of ~15 nm and ~25 nm to β' and β types respectively.

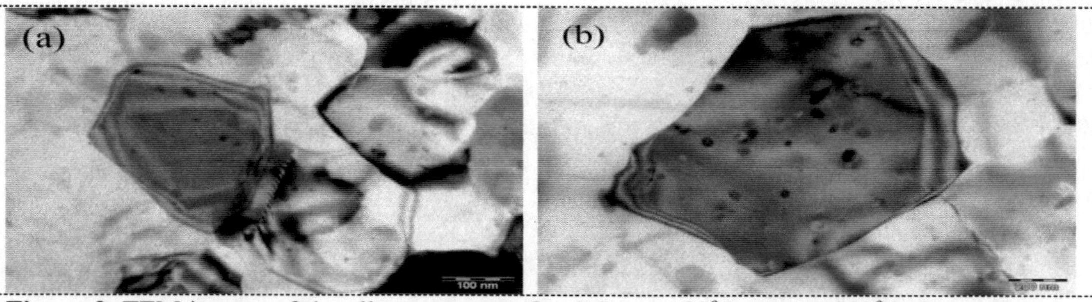

Figure 2. TEM images of the alloy structure after DSA at 130°C (a) and 230°C (b) testifying to only sphere-shaped precipitates in the microstructure.

Thus, after DSA at RT mainly β''-type precipitates located along grain boundaries are found in the microstructure. During DSA at 130°C they form both in grain interiors (GI) and (GB), and their fraction increases. Further increase of the DSA temperature causes the fraction of β''-type precipitates to reduce due to their transformation into β' modification. The fraction of β' precipitates grows till 180°C, and further their reduction takes place due to the transformation β'→β. The transformation β'→β starts at 130°C, and at a subsequent DSA temperature increase the activity of such transformations grows, which results in a monotonous increase of their fraction. Thus, the scheme of SSS decomposition during DSA remains though occurs much faster. At the same time the temperature increase also accelerates the decomposition process. The completeness of decomposition is determined by the chemical composition of Mg and Si in the alloy.

As it is mentioned above, the migration of mobile vacancies lies in the basis of the AA mechanism. In case of SPD the concentration of vacancies in the deformed material grows considerably [20]. At the same time, the applied pressure during HPT can reduce the activity of vacancy migration. In this connection, other decomposition mechanisms such as diffusion of impurity atoms over dislocation nuclei, their segregation and migration of GBs are more probable to take place. During HPT at RT dislocations relate to edge ones (Table 1), which according to [21] are located in GB. Dislocations in GBs contribute to segregation of impurity atoms in GBs and formation of β'' precipitates in them. When the HPT temperature increases, the fraction of screw dislocations grows (Table 1), which are mainly located in GIs [21]. This leads to Mg and Si atoms segregating not only in GBs but also GIs gradually transforming the GP zones into β'' precipitates. At the same time dislocations during HPT go around precipitates, and the atmosphere formed around precipitates is isotropic for capturing Mg and Si atoms by precipitates, which leads to formation of sphere-shaped precipitates. Due to transmission of impurity atoms across dislocation nuclei, precipitates increase in size and change in stoichiometry first transforming in β' modification and then in β one.

4. Conclusions

In summary, use of XPA in the transmission regime enabled performing precision studies of precipitates in the aged Al alloy, where the volume fraction of these particles does not exceed 1%. The precipitation rate during DSA is several orders higher than that during AA at the same temperature conditions. The mechanism of nucleation and growth of precipitates during DSA is conditioned by the diffusion along dislocations and GBs. During DSA as well as in case of AA transformation of precipitates $\beta''\rightarrow\beta'\rightarrow\beta$ is observed, but this transformation occurs only at achievement of some critical sizes by particles, which are 10 nm and 20 nm, respectively. The control over nucleation and growth of precipitates during DSA opens new possibilities for achieving high strength and electrical conductivity in advanced Al conductors.

Acknowledgments
This work was partially supported by the Russian Federal Ministry for Education and Science (Grant No. 14.B25.31.0017) and in part through the President of the Russian Federation grant № NSh-7996.2016.8 providing the support for leading scientific schools.

References

1. Lin G, Zhang Z, Wang H, Zhou K, Wei Y 2016 *Mater. Sci. Eng.* **A650** 210.
2. Karabay S 2006 *Mater. Des.* **27** 821.
3. Yang W, Wang M, Zhang R, Zhang Q, Sheng X 2010 *Scr. Mater.* **62** 705.
4. Huis MA, Chen JH, Sluiter MHF, Zandbergen HW 2007 *Acta Mater.* **55** 2183.
5. Sabirov I, Murashkin MYu, Valiev RZ 2013 *Mater. Sci. Eng.* **A560** 1.
6. Valiev RZ, Murashkin MYu, Sabirov I 2014 *Scr. Mater.* **76** 13.
7. Sauvage X, Bobruk EV, Murashkin MYu, Nasedkina Y, Enikeev NA, Valiev RZ 2015 *Acta Mater.* **98** 355.
8. León-Reina L, García-Maté M, Álvarez-Pinazo G, Santacruz I, Vallcorba O, De la Torre AG, Aranda MAG 2016 *J. Appl. Cryst.* **49** 722.
9. Valiev RZ, Islamgaliev RK, Alexandrov IV 2000 *Prog. Mat. Sci.* **45** 103.
10. *The Rigaku Journal* 2010 **26(1)**.
11. Bish DL, Howard SA 1988 *J. Appl. Cryst.* **21** 86.
12. van Huis MA, Chen JH, Zandbergen HW, Sluiter MHF 2006 *Acta Mater.* **54** 2945.
13. Marioara CD, Andersen SJ, Zandbergen HW, Holmestad R 2005 *Metall. Mater. Trans.* **A36a** 691.
14. Esmaeili S, Wang X, Lloyd DJ, Poole WJ 2003 *Metall. Mater. Trans.* **A34** 751.
15. Zandbergen MW, Xu Q, Cerezo A, Smith GDW 2015 *Data in Brief* **5** 626.
16. Chang CST, De Geuser F, Banhart J 2015 *J. Appl. Cryst.* **48** 455.
17. Fatay D, Bastarash E, Nyilas K, Dobatkin S, Gubicza J, Ungar T 2003 *Z. Metallkd.* **94(7)** 1.
18. Pogatscher S, Antrekowitsch H, Leitner H, Ebner T, Uggowitzer PJ 2011 *Acta Mater.* **59** 3352.
19. Sha G, Tugcu K, Liao XZ, Trimby PW, Murashkin MYu, Valiev RZ, Ringer SP 2014 *Acta Mater.* **63** 169.
20. Setman D, Schafler E, Korznikova E, Zehetbauer MJ 2008 *Mater. Sci. Eng.* **A493** 116.
21. Zehetbauer M, Seumer V 1993 *Acta Metall. Mater.* **41** 577.

IOP Conf. Series: Materials Science and Engineering **194** (2017) 012035 doi:10.1088/1757-899X/194/1/012035

Enhanced fatigue properties of Ti-6Al-4V alloy turbine blades via formation of ultra-fine grained structure and ion implantation of surface

***I P Semenova, M K Smyslova, K S Selivanov, R R Valiev, Yu M Modina**
Ufa State Aviation Technical University, Ufa, Russia

*Corresponding author, E-mail: semenova-ip@mail.ru

Abstract. An overall approach to enhance service properties of Ti alloys is based on material nanostructuring in its volume and surface layer. In this work ultrafine-grained (UFG) structure is formed in Ti-6Al-4V alloy through equal channel angular pressing via the Conform scheme with subsequent drawing. Samples modelling the shape of blades are prepared. Their surface is subjected to ion implantation with N^+. Fatigue tests are performed in the conditions imitating the stress-strain state of blades. It is shown the increasing the resistance of a high-cycle fatigue of blades due to formation of the UFG structure and subsequent surface modification in the material. The fatigue behaviour of the UFG alloy subjected to ion implantation is discussed.

1 Introduction

Development of modern branches of mechanical engineering imposes high requirements to service properties of structural materials and items produced out of them. In many cases it is important to ensure a set of properties, for example a combination of high fatigue strength, corrosion and erosion resistance along with a low specific weight of such items as high-load parts of power units and aircrafts. Ti alloys are widely used in practice for this purpose [1]. Ti alloys have a much higher specific strength as compared to refractory steels and Ni alloys [1, 2]. Today Ti-6Al-4V alloy is the most widely used one due to favourable alloying [1, 2].

As is known, there is generally a correlation between the fatigue endurance limit and the ultimate tensile strength for different materials [3] Recent studies testify to an increase in the fatigue properties of metals and alloys through formation of ultrafine-grained (UFG) structures using severe plastic deformation (SPD) processing [3, 4]. In most cases, this improvement is connected with enhancement of the monotonic strength as a result of microstructural refinement. The formation of a bulk UFG structure in Ti alloys by SPD techniques allows increasing their specific strength, fatigue resistance, fatigue life, which creates possibilities to enhance service properties of items made out of these materials [5-7]. Alongside with that, the surface condition plays an important role in resistance of parts operating in extreme conditions (high cycle loads, aggressive environment, elevated temperature, erosion impact, etc.) to external loads. Among successful and promising techniques of surface treatment in structural materials are the techniques connected with impact of concentrated energy flows, to which ion implantation belongs [8-13].

The physical entity of ion implantation is introduction (implantation) of an alloying element in the surface layer of a part as a result of bombardment by ions with high kinetic energy [9, 10]. In particular, the surface modification of conventional Ti-6Al-4V alloy by ion implantation with N^+ enables improving its tribological [11], corrosion properties [12], creep rupture strength at 600°C [13].

As is known that ionized atoms or molecules of an alloying substance in the surface layer form a complex dislocation substructure, and ions introduced in the crystalline lattice of a metal base form solid solutions or new chemical compounds [9, 10]. Unlike coarse-grained (CG) materials, nucleation and accumulation of new dislocations are constrained in ultrafine grains with a high dislocation density [14]. Probably, this can affect the formation mechanisms and the modified layer parameters in the UFG Ti alloy surface and, therefore, the level of its service properties.

Thus, the aim of this work is to study the influence of consecutive impact of SPD-processing and ion modification of the surface on the fatigue properties of Ti-6Al-4V alloy. To form a bulk UFG structure in the alloy, equal channel angular pressing via the Conform scheme (ECAP-C) is used. To study the fatigue resistance, samples modelling a gas turbine engine (GTE) compressor blade are produced. The surface of the samples is subjected to low-energy implantation with N^+ ions.

2 Material and experimental procedure

The investigations are carried out on Ti-6Al-4V alloy (Ti-basis, Al – 6.6 %; V – 4.9 %; Zr – 0.02 %; Si – 0.033 %; Fe – 0.18 %; C – 0.007 %; O_2– 0.17 %). The temperature of polymorphic transformation (TPT) is 975±5 °C. The initial billets with a diameter of 15 mm and 1000 mm long are subjected to heat treatment (HT) by quenching from T=950°C (heating during 20 min) and consequent 4-hour annealing at 675°C [15]. Further the billets are subjected to 6 passes of ECAP-C on a die-set with the channels intersection angle ψ=120° (e = 4.5) at 550°C (the die-set heating is 350°C). The subsequent drawing at 450°C is performed till a rod diameter of 12 mm (ε~35%), from which samples are produced for fatigue tests (Fig. 1). After the treatment, the samples are subjected to annealing at 500°C C for 1 hour to reduce the total density of dislocations, internal elastic stresses [7].

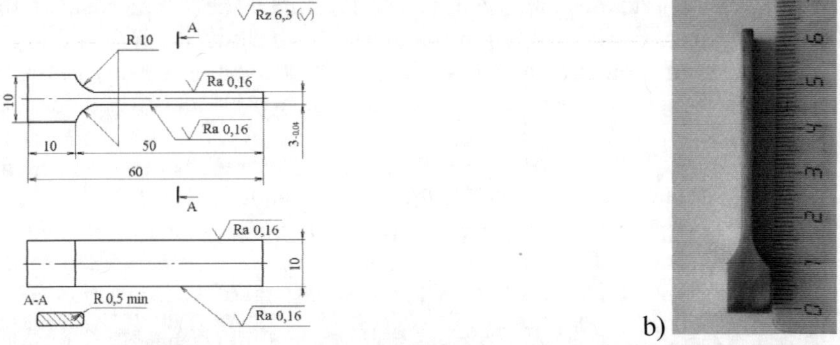

a) b)

Figure 1 The scheme (a) and the image (b) of the fatigue sample.

The microstructure is studied by optical and transmission electron microscopy (TEM) on a JEM-200B microscope with an accelerating voltage of 200 kV. The experimental X-ray data are produced using a diffractometer DRON-4-07, copper radiation was employed. The distribution of elements in the surface layer is studied with the help of Auger electron spectroscopy [16]. Auger profiling of samples in depth is performed via surface sputtering with ions of Ar^+ with energy of ~ 3 keV. The quantitative analysis of the surface composition is performed by comparing the intensities of sorted peaks of elements with account of the data on their atomic sensitivity factors [16].

Cylindrical samples with a gauge length of 15 mm and diameter of 3 mm are subjected to tensile testing on an Instron tensile machine at room temperature and the strain rate $\dot{\varepsilon} = 10^{-3}$ s^{-1}. The fatigue sample surface is subjected to electrolyte-plasma treatment, which is performed at a voltage of 250-270 V and T=95°C in the electrolyte on the basis of fluoride salts [17] to ensure the required surface roughness. The subsequent ion implantation (ii) with nitrogen N^+ is carried out at an accelerated voltage of E= 300 eV and a dose of $2x10^{19}$ ions/cm^2 [18]. The fatigue tests are carried out on the electrodynamic stand VEDS-400A on the basis of $2x10^7$ cycles under the symmetric (R=-1) bending in the first vibration mode. The vibration frequency varies from 930 to 1020 Hz, the maximum stress of the cycle varies from 500 to 640 MPa.

3. Results

3.1 Microstructure and mechanical properties of Ti-6Al-4V after SPD processing

Figure 2 displays the microstructure images of the initial (CG) Ti-6Al-4V alloy and that after the combined SPD - processing (HT+ECAP-C+drawing+annealing). The microstructure of the initial rod is a typical bimodal one with a size of primary α-phase grains of (10±5) μm and areas with a plate (α+β) structure. The primary α-phase fraction is about 65% (Fig.2a). The β-phase fraction does not exceed 12 % according to the X-ray results. After the combined SPD-processing the fraction of primary α-phase grains decreases to 25%, and its grain size reduces down to 5 μm (Fig. 2b).

a) 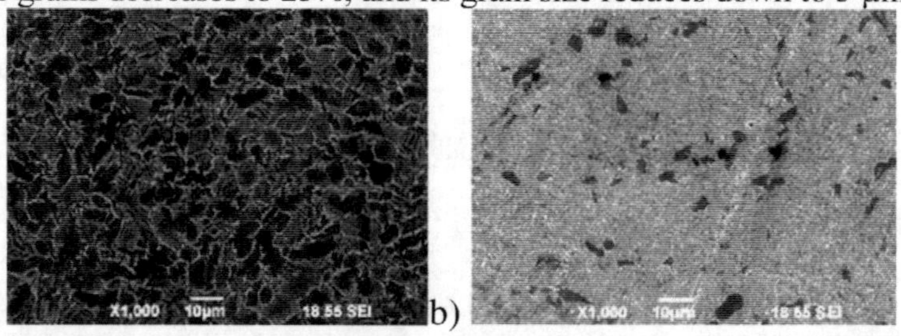 b)

Figure 2 Microstructure of the Ti-6Al-4V alloy: (a) in as-received state and (b) after SPD-processing: transversal section.

Figure 3 displays TEM images of the billet microstructure in the transversal and longitudinal sections after SPD processing. The microstructure is mainly of grain character with clearly observed boundaries, which is connected with redistribution of the dislocation structure and formation of more defect-free equilibrium grain boundaries after the final annealing. The average size of structural elements is 200±20 nm (Fig. 3a). Grain elongation is observed along the strain direction in the ratio 1:2 in the longitudinal section of a billet (Fig. 3 b, c). The total dislocation density in the billet after SPD is about $4.5x10^{15}$ m^{-2}, which is much higher than the dislocation density in the initial alloy ($0.19x10^{15}$ m^{-2}).

It is seen from Table 1 that the UFG structure formation in the alloy results in enhancement of the ultimate tensile strength (UTS) from 1030 to 1460 MPa due to considerable reduction of the α-phase grain/subgrain size from 10 to 0.2 μm, which correlates well with the previously obtained results for this alloy subjected to conventional ECAP with subsequent thermo-mechanical treatment [5]. Some reduction of the relative elongation (from 15 to 8%) is also typical of Ti alloys after SPD [5].

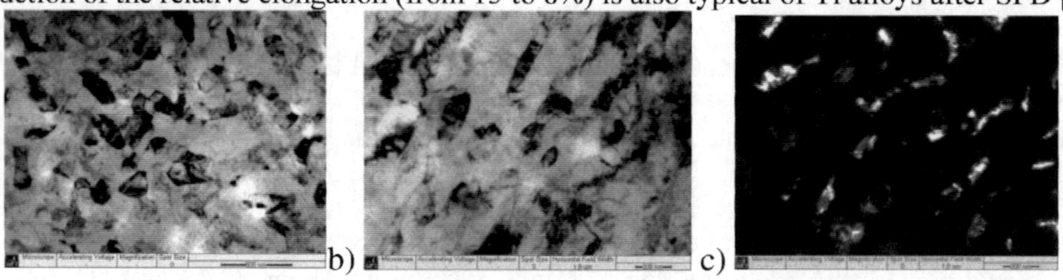

Figure 3 TEM images of the Ti-6Al-4V alloy microstructure after SPD-processing in transversal (a) and longitudinal (b, c) sections.

Table 1 Mechanical properties of the Ti-6Al-4V alloy before and after SPD-processing

State	Average grain size, D, μm	UTS, MPa	YS 0.2, MPa	Elong.,%
As-received	10±5	1030	850	15,0
SPD-processing	0.20±0.02	1460	1380	8,5

3.2 Ion implantation of the surface of UFG Ti-6Al-4V alloy blade samples.

Atomic N^{+}, introduced in the surface of UFG samples, is confirmed by an *X*-ray diffraction pattern (Fig. 4). The value of the modified layer is estimated using the Auger spectroscopy [16]. Fig. 5

displays the concentration profiles of alloying elements and *N* in at. % in depth of the sample surface. On the average, the depth of N^+ penetration in the surface does not exceed 6 μm. As is known that the value of the layer modified with N^+ in Ti alloy at low-energy ion implantation can vary from 4 to 10 μm depending on the structure state of the surface and radiation dose [8-10].

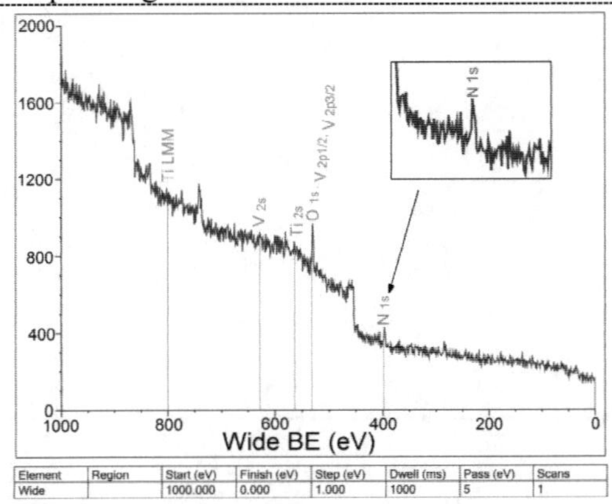

Figure 4 *X*-ray diffraction pattern of the UFG samples surface after N^+ ion implantation.

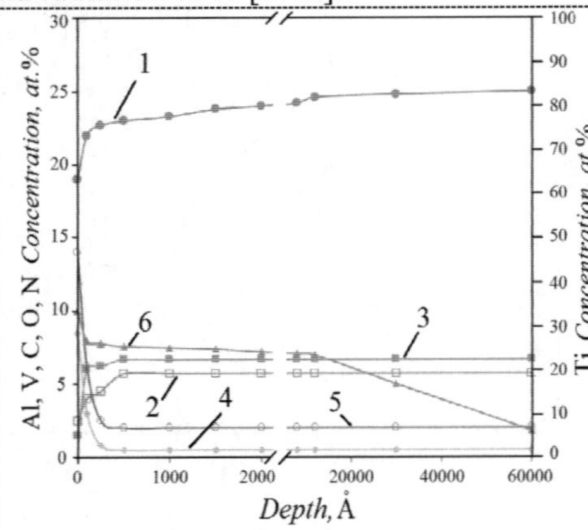

Figure 5 Concentration profiles of alloying elements in depth of the sample surface: **1-** Ti; **2–**Al; **3–**V; **4–**C; **5–**O; **6-**N

3.3 Fatigue tests of UFG Ti-6Al-4V alloy samples with ion implantation of the surface.

Fig. 6 displays the fatigue test results of sample-items from CG and UFG Ti-6Al-4V alloys on the basis of $2x10^7$ cycles. The main increment in the fatigue strength is observed as a result of formation of the UFG structure in it (from 500 to 570 MPa), i.e. approximately by 15%. The similar result was achieved in [15], where the fatigue endurance limit of CG Grade 5 with a bimodal structure and UFG states were 600 and 690 MPa respectively on the basis of 10^7 cycles of standard samples with a round section. Preliminary tensile tests of UFG samples with N^+ ion implantation show that the mechanical properties at room temperature practically do not change and are within the statistical spread. The surface modification of sample by ion N^+ gave an additional increment in the fatigue strength, which was 600 MPa.

4 Discussions

In this work the ECAP-Conform technique applied in combination with drawing and annealing allows achieving effective refinement of α-grains in the Ti-6Al-4V alloy and forming a UFG structure with an average size of 200 nm and, thus, enhancing the UTS from 1030 to 1460 MPa. As it is mentioned above, the strength enhancement in Ti-6Al-4V allows increasing the fatigue endurance limit at room temperature, which is testified in previous works [5, 15]. The main difference of the current work from the previous ones is evaluation of the fatigue resistance of samples imitating the shape of a GTE blade (Fig. 1). The results of bench tests of such UFG alloy samples on the enhanced basis up to $2x10^7$ cycles in the conditions close to operating ones regarding frequency and stresses, also showed improvement of fatigue resistance (from 500 to 570 MPa) (Fig. 6).

The influence of UFG Ti alloy surface modification with N^+ ions on the fatigue strength should be noted. On the one hand, the mechanical properties at room temperature practically did not change. However, after ion implantation the fatigue strength tends to improvement and it is 600 MPa (Fig. 6). This can be explained by the fact that the positive effect of surface modification is conditioned by alteration of the crack initiation mechanism. In particular it was shown in [10] that fatigue crack initiation in Ti-6Al-4V alloy samples occurred on the surface. In case of radiated samples fatigue cracks initiated at a distance of 25-150 μm from the surface. In this work the main contribution in the fatigue strength of UFG alloy is made by grain refinement, i.e. grain boundary strengthening mechanism.

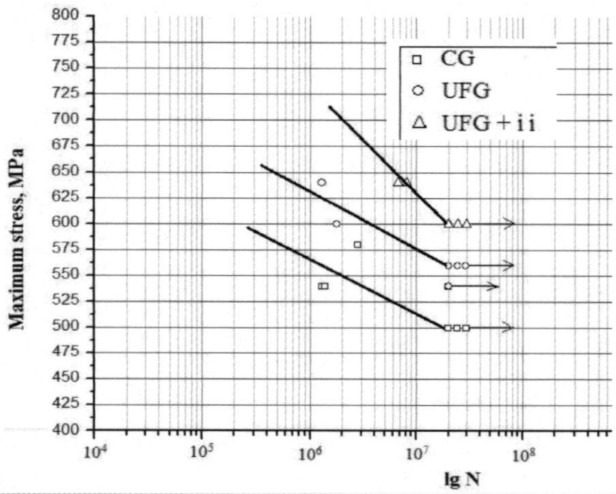

Figure 6. Fatigue curves of CG, UFG and UFG+ii N^+ samples.

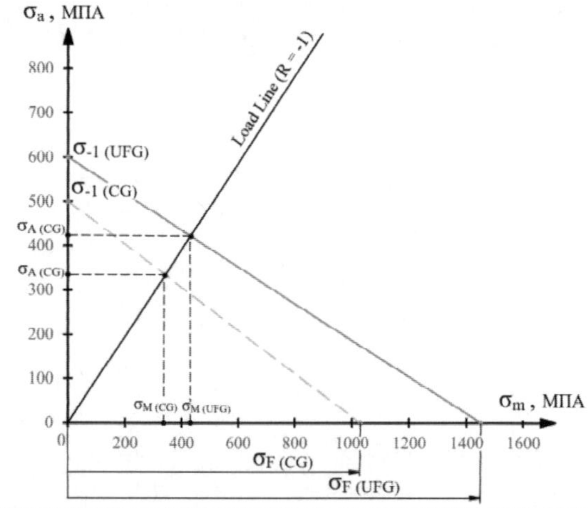

Figure 7 Haigh diagrams: $\sigma_{A(CG)}$, $\sigma_{A(UFG)}$ and $\sigma_{M(CG)}$ $\sigma_{M(UFG)}$ - ultimate values of the amplitude and mean stresses, corresponding for CG and UFG states.

In engineering calculations of the design integrity and evaluation of the fatigue behavior of an item, probabilistic methods based on the Goodman line [19] in Haigh diagrams and corresponding fatigue safety factors are used. These diagrams allow seeing what stress amplitude material and/or an item can suffer without failing at the given mean stress. With use of the experimental data obtained in this work, the diagrams are plotted to determine the ultimate values of the amplitude σ_A and mean stress σ_M, shown in Fig. 7. Table 2 lists the values of σ_F (UTS) and fatigue endurance limit σ_{-1} for the CG and UFG alloys studied in this work. Fig. 7 shows that the Goodman lines for the UFG sample and item are higher than those for the CG state. This testifies to the fact that the fatigue safety factor at the given mean stress σ_m will be higher in the item with the UFG structure. For example, for the σ_m =300 MPa, the amplitude σ_a is close to the ultimate value $\sigma_{A\ (CG)}$ for the CG alloy as against the UFG alloy, for which the value σ_a is much lower than the ultimate one $\sigma_{A(UFG)}$ (500 MPa) (Fig. 7). The fatigue safety factor S_f is determined via the formula [19]: $S_f = \frac{1}{\frac{\sigma_a}{\sigma_{-1}} + \frac{\sigma_m}{\sigma_f}}$, where σ_a is the amplitude stress, σ_m – the mean stress, σ_f is the material strength (in this case, it is UTS); σ_{-1} - the fatigue endurance limit at the stress ratio R=-1. It is seen from Table 2 that at the mean stress σ_m=300 MPa, S_f is 0.64 for the CG alloy and 1.41 i.e. higher by 30% for the UFG one.

Table 2 Evaluation fatigue safety factor S_f at $\sigma_a = \sigma_m$=300 MPa for CG and UFG Ti-6Al-4V alloy

	Endurance limit σ_{-1}, MPa	σ_f (UTS), MPa	Mean stress σ_m, MPa	Stress amplitude σ_a, MPa	fatigue safety factor, S_f
CG	500	1030	300	300	1.12
UFG	600	1460	300	300	1.43

Thus, the above considered experimental data testify to a principal possibility for enhancement of service properties of items manufactured from Ti-6Al-4V with an UFG structure with subsequent ion modification of the surface with N^+. It is known that parts of GTE from Ti alloys are exploited at temperatures up to 450°C. One can assume that in these exploitation conditions, creation of a layer modified with N^+ in the UFG alloy surface at elevated temperatures can cause improvement of the creep strength. A similar effect was observed in [13], where ion implantation N^+ of Ti-6Al-4V alloy with resulted in improvement of the creep strength at 600°C. However, it requires more careful studies, which are planned to be performed by the authors in the future. Such a comprehensive approach to Ti-6Al-4V alloy hardening is promising for manufacturing of GTE items, for example, compressor blades, to which high requirements on operating properties at elevated temperatures are imposed.

5 Conclusions

1. It is established that formation of the UFG structure in Ti-6Al-4V with the average size of grains/subgrains of 200 nm by ECAP-Conform technique with subsequent drawing results in enhancement of the UTS at room temperature to 1460 MPa with a relative elongation of about 8 %.
2. It is shown that consecutive impact of SPD and N^+ ion implantation of Ti-6Al-4V sample surface ensures increase of the fatigue resistance in the conditions imitating the stress-strain state of GTE blades by no less than 20%. The fatigue safety factor S_f for the given acting stress increases by 30%.
3. The produced results demonstrate the perspectives for overall hardening of Ti-6Al-4V, which includes formation of a UFG structure in the volume by SPD techniques and subsequent modification of the surface with N^+ ions to increase the service properties of GTE items.

Acknowledgments
This work was funded by the Russian Science Foundation of the Russian Federal Ministry of Education and Science (grant №16-19-10356).

References

[1] Peters M, Kumpfert J, Ward C H, Leyens C 2003 *Adv. Eng. Mater.* **5** (6) 419
[2] *Materials Properties Handbook: Titanium Alloys* 1998**,** ed Boyer R, Welsch G, Collings E W, (ASM International: USA), p 1048
[3] Höppel H W, Kautz M, Xu C, Murashkin M, Langdon T G, Valiev R Z, Mughrabi H, 2005 *Int. J. Fatigue* **28**, 1001
[4] An X H, Wu S D, Wang Z G, Zhang Z F 2014 *Acta Materialia* **74** 200
[5] Semenova I P, Raab G I, Valiev R Z 2014 *Nanotechnologies in Russia* **9** (5-6) 311
[6] Zherebtsov S, Salishchev G, Galeyev R, Maekawa K 2005 *Mater.Trans.* **46** (9) 2020
[7] Polyakov A, Semenova I, Huang Yi, Valiev R, Langdon T 2014 *Adv.Eng.Mater.***16** (8) 1038
[8] Carter G, Katardjiev I V, Nobes M I 1989 In: *Materials modification by high-fluence ion beams*. ed Kelly R, Fernanda da Silva M (Dordrecht: Kluwer) pp 3-27
[9] Guseva M I, Martynenko Yu V, Smyslov A M 1996 *In Proc. 3-rd Inter Workshop on Plasma-Based Ion Implantation (*Rossendorf, Germany) pp 101-106
[10] Guseva M I, Smyslov A M, 2000 Surface (X-ray, synchrotron and neutron researches) **6** 68 (in Russian)
[11] Budzynski P, Youssef A A, Sielanko J 2006 *Wear* **261** (11-12) 1271
[12] Zhecheva A, Sha W, Malinov S, Long A 2005 *Surface & Coatings Technology* **200** (7) 2192
[13] Castagnet M, Yogi L M, Silva M M, Ueda M, Couto A A, Reis D A P, M Neto C 2012 *Mater. Sci. Forum* **727 - 728** 50
[14] Valiev R, Estrin Y, Horita Z, Langdon T, Zehetbauer M, Zhu Y 2015 *Mater Res Lett.* **4** 1
[15] Semenova I, Polyakov A, Polyakova V, Huang Yi, Valiev R and Langdon T 2016 *Adv. Eng. Mater.* **18** (12) 2057
[16] Briggs D and Sikh M P (Eds.) 1987 *Surface Analysis by Auger and X-Ray Photoelectron Spectroscopy* (Russian translation: Mir, Moscow) p 600
[17] Patent RF 2006 *No* 2552203
[18] Patent RF 1998 *No* 2117073
[19] Jelaska D 2011 *Adv. Eng.* **6**, ISSN1846

Optimized layer architecture for an extended fatigue life of ultrafine-grained AA1050/AA5005 laminated metal composites

Frank Kümmel, Thea-Simone Tegtmeier, Heinz Werner Höppel and Mathias Göken

Institute I: General Materials Properties, Department of Materials Science and Engineering, Friedrich-Alexander-University Erlangen-Nürnberg , Martensstr. 5, Erlangen 91058 , Germany

E-mail: Frank.Kuemmel@fau.de

Abstract. The influence of interfaces on the fatigue life in AA1050/AA5005 ultrafine-grained laminated metal composites was investigated. At constant sheet thickness, the layer thickness and the number of material interfaces was varied by performing different cycles of accumulative roll bonding. It is found that crack deviation occurs if the crack approaches the soft to hard (AA1050/AA5005) material interface resulting in strongly enhanced fatigue lives compared to the mono-material. The most favourable layer architecture for enhanced fatigue life strongly depends on the applied stress amplitudes. At intermediate stress amplitudes a large layer thickness of 500 µm (N4) and 8 material interfaces gives the longest fatigue life. In contrast, at low stress amplitudes the sheets with 32 material interfaces and a layer thickness of 125 µm (N6) shows the longest fatigue life.

1. Introduction

Ultrafine-grained (UFG) materials are widely known for their high specific monotonic strength compared to their coarse grained (CG) starting material, see i.e. refs. [1–3]. But also the cyclic mechanical properties can be strongly enhanced [4–6]. The accumulative roll bonding (ARB) process enables the production of UFG reinforced [7] and/or layered [8–10] metallic structures with tailored or graded properties. As the cyclic properties are regarded only limited data is available in the literature for such UFG laminated metal composites up to now. Previous publications by the authors [11,12] revealed that the fatigue life of AA1050/AA5005 UFG laminated metal composites can be significantly enhanced compared to their CG and their UFG mono-material counterparts. This is caused by an arresting of the plastic zone in front of the crack tip at the material interface and thus a crack branching if the crack approaches the soft to hard material interface. Consequently, the composites show significantly longer fatigue lives compared to their mono-material sheets [11]. Fatigue tests on AA1050/AA5005 and AA1050A/AA2024, in which the interlayer material was varied, revealed that an increasing hardness gradient at the material interface enhances this crack deviation mechanism if the crack approaches the interface [12]. This is associated with a strongly enhanced fatigue life at medium stress amplitudes. This cracking behavior is quite similar to the results reported by Sugimura et al. [13] and Pippan et al. [14], who investigated the cracking behavior at coarse grained bimetallic interfaces. Both authors found that, if the fatigue crack approaches the layer interface from a plastically softer to a stronger material, the crack propagation rate diminishes in the vicinity of the interface [13,14]. In the current work the influence of the number of material interfaces and layer thickness on the fatigue properties of UFG laminated metal composites is investigated by producing AA1050/AA5005 composites with different numbers of ARB cycles.

2. Experimental Procedure

2.1. Processing of the UFG laminated metal composites

The laminated metal composites in this work, made of the technical pure aluminum AA1050 (Al99.5) and the solution hardening alloy AA5005 (AlMg1), were produced by the accumulative roll bonding process using a four high rolling mill (BW 300, Carl Wezel, Germany) at a nominal thickness reduction of 50 % per cycle. The sheets had an initial geometry of length × width × height of 300 × 200 × 4 mm³. Prior to roll bonding the surfaces were cleaned by acetone and then wire brushed in order to remove the oxide layer and to ensure a sufficient bonding. Subsequently, the treated surfaces were stacked on each other and roll bonded at room temperature. The bonded aluminum sheets were air cooled and halved before performing the next cycle. In the first ARB cycle (N1), an AA1050A sheet was roll-bonded with an AA5005 sheet. In the second cycle (N2), the two AA5005 surfaces were stacked together and roll-bonded to receive a sandwich like structure with AA1050A outer layers. Consequently, in the following ARB cycles (N > 3), always two AA1050A surfaces were roll-bonded. In order to prevent necking and rupture of the harder AA5005 layer recrystallization by annealing for 10 min at 500 °C was used in addition after different ARB cycle numbers. Details on the processing can be found in previous papers published by the authors [11,15]. Hence, all final composites exhibit an UFG condition, which is very similar to an UFG sheet processed by only 4 ARB cycles. The number of layer interfaces with dissimilar materials and the theoretical layer thickness of the processed composites are given in Table 1 a). For reference, AA1050A with 4 ARB cycles and AA5005 mono-material sheets with 0 and 4 ARB cycles were also investigated.

2.2. Microstructural and mechanical characterization

The layer architecture of the composites were characterized by scanning electron microscopy (Crossbeam 1540 EsB, Zeiss, Germany) using secondary electron imaging. All sheets were analyzed in side view (rolling direction × normal direction). Therefore, the specimens were mechanically ground down to a grit size of 6 µm and then mechanically polished up to 1 µm. In the final step, the specimens were chemo-mechanically polished using an OPS suspension.

The local mechanical properties of the layers were measured by nanoindentation experiments (Nanoindenter XP, MTS Nano Instruments, USA) using a three-sided Berkovich pyramid with the continuous stiffness method. In each sample the hardness profile across the material interface was measured in an angle of 30° to the rolling direction. The indentation depth was 500 nm and the distance between the indents was always the twentyfold of the indentation depth.

The fatigue tests were carried out on a vibrophore testing machine (HFP 5100, Roell Amsler, Germany) in 3-point bending mode. The specimen had a geometry of 20 × 9.5 × 4 mm³ in length, width and height and were ground mechanically down to a grid size of 6 µm before testing. All tests were conducted under force control with an R-value of 0.1 leading to a permanent tension stress at the lower part of the sample. The maximum bending stress amplitudes, which are in the middle at the bottom of the sample, ranged in between 100 MPa and 200 MPa. Furthermore, a frequency observation enabled the detection of macro-crack nucleation and propagation. After fatigue testing the crack path was analyzed by scanning electron microscopy under secondary electron contrast conditions.

3. Results and Discussion

3.1. Characterization of the initial state after ARB processing

All composites consist of continuous layer architecture up to 12 ARB cycles. Due to the intermediate recrystallization steps at higher numbers of ARB cycles pronounced necking of the harder AA5005 layer could be suppressed effectively. The layer thicknesses of the two different materials are very similar, see Table 1 b). The AA5005 layer thickness is only slightly higher than that of the AA1050 layer. This is due to the higher hardness of these layers leading to a slightly preferred deformation of the softer AA1050 layers, especially during the first ARB cycles. Furthermore, the experimental determined thickness is in good accordance to the theoretical values, c.f. Table 1 a) and b).

In Table 1 c) the hardness of the materials in the AA1050/AA5005 composites after different numbers of ARB cycles are listed. It is clearly visible that the hardness of the AA5005 layer is higher compared to the AA1050 layer. In the AA5005 layers, the hardness is quite constant at 1.2 GPa after all ARB cycles. In the AA1050 layers, the hardness stays almost constant in between ARB cycle 4 and 8 at about 0.9 GPa. After 12 ARB cycles the hardness of the AA1050 layer slightly rises to about 1.0 GPa. This is probably a result of the strong increase in the number of material interfaces. Previous investigations on ARB processed laminated metal composites revealed that the grain refinement near the material interface is strongly enhanced compared to the evolution in the middle of the layers leading to a higher hardness in these areas [16]. This leads to an improvement of the macroscopic mechanical properties at higher ARB cycles, where the fraction of the material that is affected by the interfaces is significantly larger, see refs. [15,17]. The hardness of the AA1050 and AA5005 ARB processed N4 mono-material sheets are very similar to the one in the composites, see Table 1 c). This was also observed in many other investigations, especially at lower numbers of ARB cycles, see for example refs. [15,18].

Table 1: Calculated and experimentally determined parameters of the AA1050/AA5005 LMCs and the AA1050 and AA5005 mono-material sheets after different numbers of ARB cycles.

	Composite				Mono-material
	N4	N6	N8	N12	N4
a) Calculated number of material interfaces and material layer thickness					
Material interfaces	8	32	128	2048	0
Layer thickness / μm	500	125	32	2	4000
b) Mean layer thickness ± standard deviation of the materials / μm					
AA1050	510 ± 18.5	127 ± 17.6	33.6 ± 1.6	2.0 ± 0.7	-
AA5005	539 ± 34.9	137 ± 10.9	34.8 ± 4.7	2.2 ± 0.7	-
c) Mean hardness ± standard deviation / GPa					
AA5005	1.19 ± 0.09	1.22 ± 0.05	1.21 ± 0.07	1.24 ± 0.06	1.14 ± 0.07
AA1050	0.91 ± 0.02	0.91 ± 0.03	0.93 ± 0.04	1.04 ± 0.05	0.86 ± 0.05

3.2. Fatigue life and damage mechanism of the ARB processed samples

The fatigue life curves of the AA1050/AA5005 composites after different numbers of ARB cycles and, for reference, of the AA1050 after 4 ARB cycles and AA5005 mono-material sheets after 0 (as received) and 4 ARB cycles are shown in Figure 1 a). In this Wöhler S-N curve the maximum stress amplitude $\Delta\sigma/2$, which is calculated by the linear-elastic bending theory, is plotted versus the number of fatigue cycles to failure N_f. It should be noted that the UFG microstructure and the hardness of all composites are very similar among each other and also comparable to the N4 mono-material reference sheet, as all samples were subjected to 4 ARB cycles after a final recrystallization step. Therefore, differences in the fatigue lives have to be related to the different number of material interfaces and/or layer thicknesses. The fatigue life curves of all samples can be divided by characteristic threshold stresses $\sigma_{threshold}$ into two regimes with clearly different slopes. Furthermore, the fraction of macro-crack initiation on the total fatigue life, which was determined by a drop in the testing frequency, strongly differentiates in between these areas. At higher stresses, macro-crack initiation spans about 90 % of N_f in all composites and 95 % of N_f in the mono-material sheets. At stresses lower than the threshold stresses macro-crack initiation

IOP Conf. Series: Materials Science and Engineering **194** (2017) 012036 doi:10.1088/1757-899X/194/1/012036

periods above 98 % of N_f are found for all samples. The fatigue life of all UFG samples is clearly enhanced compared to the CG AA5005 N0 mono-material sheet. The threshold stress of the UFG AA5005 N4 mono-material sheet lies about 50 MPa above its CG counterpart. This is related to the much higher strength of the UFG samples and was already observed by many other authors on uniaxial fatigue tests, see i.e. refs. [4–6]. A comparison of the fatigue lives of the composites and the AA1050 N4 mono-material sheets is only possible at lower stresses ($\Delta\sigma/2 < 160$ MPa). As all tests were performed at a R-value of 0.1 a mean stress of about 200 MPa is achieved, which exceeds the ultimate tensile strength of UFG AA1050, see i.e. refs. [15,19,20], leading to early crack initiation and failure of the AA1050 N4 mono-material sheets. The fatigue lives are increased by a factor of 5 for the N4 composite and of 4-18 for the N6 composite in comparison to that of the UFG AA1050 N4 mono-material sheet. In contrast, the fatigue lives of the N8 and N12 composites are very similar to the AA1050 N4 mono-material sheets for the whole testing regime. At higher stress amplitudes the N4 composite shows a fatigue life even in the range of the AA5005 N4 mono-material sheet. The most favorably LMC architecture strongly depends on the applied stress amplitudes. At higher maximum stress amplitudes ($\Delta\sigma/2 > 125$ MPa), where macro-crack propagation extends a larger period of the fatigue life, an increasing number of ARB passes leads to a decreasing fatigue life. Concurrently, at lower maximum stress amplitudes, where the fatigue life is almost entirely determined by the macro-crack initiation period, the longest fatigue life is achieved for the N6 composite followed by the N4 composite and, at last, the N8 and N12 composites. This is related to the highest threshold stress in the N8 composite, which means that the onset of the formation of macro-crack is shifted to higher stress amplitudes. The same trend is also found for the endurable fatigue strength for $5\cdot10^6$ cycles of the different samples, see Figure 1 b). Again, it is shown in this diagram that the fatigue properties at lower stress amplitudes are clearly enhanced in the N6 composite compared to the other composites and the AA1050 mono-material sheet.

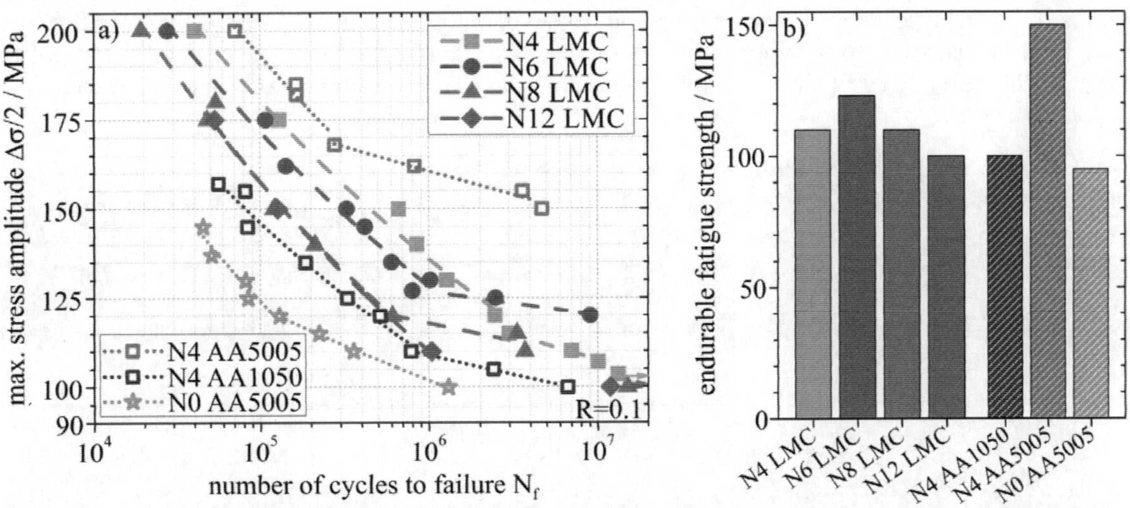

Figure 1: a) Fatigue life curves of the AA1050/AA5005 composites after 4, 6, 8 and 12 ARB cycles. For comparison, the fatigue life for AA1050 and AA5005 mono-material sheets after 0 and 4 ARB cycles are also plotted. b) Endurable fatigue strength for $5\cdot10^6$ cycles of the different samples.

The difference in fatigue life and in the threshold stresses of the composites at higher and lower stress amplitudes can be directly correlated to the different cracking behavior of the samples. In Figure 2 the fatigue crack paths of the N4, N6 and N8 composites are depicted that are typical for samples fatigued at an intermediate stress amplitude ($\Delta\sigma/2 = 200$ MPa) or at a low stress amplitude ($\Delta\sigma/2 = 120$ MPa). For N8 composites, where the layer thickness is rather small, a straight crack path forms for the whole testing regime. Therefore, the reinforcement by the AA5005 layers has no influence on the cracking behavior in this composite. This applies also for the N12 composite, where the layer thickness is even smaller than in the N8 composites. The fatigue lives of these composites are very similar to the AA1050 mono-material sheets. Thus, it becomes obvious, that a rather thin layer thickness is detrimental for the fatigue life of the composites at intermediate stress amplitudes. Contrary to that, the N4 composite shows crack branching when the crack approaches the soft to hard material interface for all stress amplitudes. This leads to a drastically increased fatigue life compared to the N8 and N12 composites. The cracking behavior of the N6 sample shows a strong dependency on the applied stress in the testing range. At higher stress amplitudes the fatigue crack growths very straight across the layer thickness, whereas at lower stress amplitudes a pronounced crack branching can be observed. This is associated with a lower fatigue life at intermediate stress amplitudes and an increased fatigue life at lower stress amplitudes compared to the N4 composite.

This different behavior can be explained by a shielding of the damage zone in front of the crack tip when the crack approaches the soft to hard layer interface. At high stress amplitudes the size of the damage zone in front of the crack tip is relatively large. Consequently, a high AA5005 layer thickness is necessary to shield the plastic zone in front of the material interface and to hinder crack bridging in between the softer AA1050 layers. Thus, crack propagation is effectively delayed. For this reason, only the N4 composite with a layer thickness of about 500 µm shows clear crack deviation in front of the material interfaces leading to the highest fatigue life at these amplitudes. Concurrently, the crack propagates very straight across the material interfaces in the N6, N8 and N12 composites, as the layer thickness is too thin. Furthermore, crack initiation and propagation at discontinuities at the bonding interfaces is observed at higher stress amplitudes. Both factors lead to a decrease in fatigue life with an increasing number of ARB cycles, see Figure 1. At lower stress amplitudes the plastic zone in front of the crack tip is significantly smaller. At these amplitudes, the fatigue crack is deviating from the straight crack path in front of the material interface in both the N4 and N6 composites. However, due to the much higher number of material interfaces in the N6 composite crack deviation is more pronounced leading to the delayed macro-crack propagation in this composite. In the N8 and N12 composites the crack again spreads very straight across the whole sheet thickness. As a consequence, the N6 composite shows the highest fatigue life at lower stress amplitudes, as crack deviation at the material interfaces is most pronounced.

Figure 2: Crack paths of the N4, N6 and N8 AA1050/AA5005 composites fatigued at a stress amplitude of 200 MPa (top row) and 120 MPa (bottom row). Please note that all pictures were taken at the side view of the sample. At the left margin, the different layers are indicated by red (AA5005) and blue (AA1050) bars. The maximum tensile bending stresses are located at the top in the middle of each micrograph.

4. Conclusion

In this study the influence of the layer architecture (number of material interfaces and layer thickness) on crack propagation and the fatigue life of ultrafine-grained AA1050/AA5005 laminated metal composites (LMCs) was investigated. A clear correlation between the crack propagation mechanisms and the fatigue life was observed: strongly enhanced fatigue lives are achieved in the LMCs that showed pronounced crack deviation at the material interface. The most favorably LMC architecture strongly depends on the applied stress amplitudes. At higher stress amplitudes, a high layer thickness is necessary to effectively block the plastic zone at the AA1050-AA5005 material interface leading to a deviation of the crack path. Therefore, the LMC after 4 ARB cycles shows the longest fatigue life. Concurrently to that, at lower stress amplitudes the layer thickness of the N6 composite is already sufficient to effectively hinder straight crack propagation. However, due to the much higher number of material interfaces in the N6 composite, crack path deviation is much more pronounced leading to a strongly enhanced fatigue life. By optimizing, the number of material interfaces and layer thickness in the composite the fatigue life is significantly enhanced in comparison to the UFG AA1050 mono-material sheets. At higher stress amplitudes, the fatigue lives of the composites almost reach the ones of the UFG AA5005 sheets although the volume fraction of this harder material is only 50 % in the composites. Compared to the GC AA5005 mono-material sheets, strongly enhanced fatigue lives were achieved in all UFG samples.

References

[1] Saito Y, Tsuji N, Utsunomiya H, Sakai T and Hong R G 1998 *Scr. Mater.* **39** 1221–7
[2] Tsuji N, Ito Y, Saito Y and Minamino Y 2002 *Scr. Mater.* **47** 893–9
[3] Höppel H W, May J and Göken M 2004 *Adv. Eng. Mater.* **6** 781–4
[4] Patlan V, Vinogradov A, Higashi K and Kitagawa K 2001 *Mater. Sci. Eng. A* **300** 171–82
[5] Mughrabi H, Höppel H W and Kautz M 2004 *Scr. Mater.* **51** 807–12
[6] Kunz L, Lukáš P and Svoboda M 2006 *Mater. Sci. Eng. A* **424** 97–104
[7] Schmidt C W, Knieke C, Maier V, Höppel H W, Peukert W and Göken M 2011 *Scr. Mater.* **64** 245–8
[8] Quadir M Z, Al-Buhamad O, Bassman L and Ferry M 2007 *Acta Mater.* **55** 5438–48
[9] Hausöl T, Höppel H W and Göken M 2011 *Mater. Sci. Forum* **667–669** 217–22
[10] Roy S, Nataraj B R, Suwas S, Kumar S and Chattopadhyay K 2012 *Mater. Des.* **36** 529–39
[11] Kümmel F, Hausöl T, Höppel H W and Göken M 2016 *Acta Mater.* **120** 150–8
[12] Kümmel F, Höppel H W and Göken M 2017 *Submitt. Mater. Sci. Eng. A*
[13] Sugimura Y, Lim P G, Shih C F and Suresh S 1995 *Acta Metall. Mater.* **43** 1157–69
[14] Pippan R, Flechsig K and Riemelmoser F O 2000 *Mater. Sci. Eng. A* **283** 225–33
[15] Kümmel F, Kreuz M, Hausöl T, Höppel H W and Göken M 2016 *Metals* **6** 56
[16] Hausöl T, Höppel H W and Göken M 2010 *J. Mater. Sci.* **45** 4733–8
[17] Ma X, Huang C, Moering J, Ruppert M, Höppel H W, Göken M, Narayan J and Zhu Y 2016 *Acta Mater.* **116** 43–52
[18] Su L, Lu C, Tieu A K, Deng G and Sun X 2013 *Mater. Sci. Eng. A* **559** 345–51
[19] Topic I, Höppel H W and Göken M 2008 *J. Mater. Sci.* **43** 7320–5
[20] Ruppert M, Böhm W, Nguyen H, Höppel H W, Merklein M and Göken M 2013 *J. Mater. Sci.* **48** 8377–85

IOP Conf. Series: Materials Science and Engineering **194** (2017) 012037 doi:10.1088/1757-899X/194/1/012037

Effect of stress relieving treatment on low cycle fatigue behavior of USSP treated 7075 aluminium alloy

Vaibhav Pandey*, K. Chattopadhyay, N. C. Santhi Srinivas, Vakil Singh

Department of Metallurgical Engineering, Indian Institute of Technology (Banaras Hindu University), Varanasi – 221005, India

email: vaibhavmech12@gmail.com

Abstract. The effect of ultrasonic shot peening (USSP) on low cycle fatigue (LCF) behavior of the 7075 aluminium alloy was studied at room temperature. There was grain refinement approximately to 20 nm size, appreciable increase in micro hardness, and inducement of residual compressive stress in the surface region, due to the USSP treatment. The modified microstructure in surface region of the specimens subjected to USSP was characterized by X-ray diffraction and transmission electron microscopy. There was marked increase in LCF life of the specimens due to USSP, however, LCF life of the USSP specimens was reduced due to subsequent treatment of stress relieving. The results are discussed in terms of the process of crack initiation and propagation in the different conditions.

1. Introduction

Precipitation-hardening 7075 aluminum alloys are used in aerospace applications due to good combinations of their specific strength, corrosion resistance and formability [1]. In these structural applications, the components undergo cyclic loading. Since nearly 90 % failures of such components occur due to fatigue it is important to improve their fatigue resistance.

Fatigue cracks are known to initiate mostly from surface of the components, therefore there is much scope of surface modification in improving fatigue resistance of such components. Introduction of compressive residual stress in surface region of components through shot peening is a well-established process of surface treatment for improving fatigue life [2-4]. Ultrasonic shot peening (USSP) is a novel process of surface modification. It introduces compressive residual stress along with grain refinement in the surface region to nano level. As compared to conventional shot peening, USSP leads to development of higher compressive residual stress to a larger depth and better surface finish due to usage of steel balls of smoother surface. There have been several reports on the effect of USSP on microstructural modifications of various metallic materials like Al [5], Cu [6], stainless steels [7-8] and Ti [9], however, there are very limited data related to effect of USSP treatment on their low cycle fatigue (LCF) behavior.

In USSP, the specimen is bombarded with hard steel balls, causing plastic deformation in the surface region. The work hardening in the surface region leads to increase in micro hardness and development of high compressive residual stress. It is now well established that materials with fine grains have greater resistance against fatigue-crack initiation whereas coarse-grained materials offer high resistance to fatigue crack propagation [10-12]. The advantage of gradient structures lies in maximizing physical and mechanical properties and minimizing the material cost. The materials with gradient structure are fundamentally different from their conventional coarse-grained counterparts, possessing high strength and hardness and enhanced physical properties. Therefore, it is expected that a structure with fine grain surface region and coarse grained interior should possess superior fatigue properties.

In the present investigation, we studied he effect of grain refinement in surface region combined with induced compressive residual stress, on LCF properties of the 7075 aluminium alloy and also the effect of relieving the associated residual compressive stress of the USSP specimens, without altering the nanostructure in the surface region, on the LCF behavior.

2. Experimental

The 7075 aluminium alloy used in the present investigation was procured from the Hindalco Industries, Renukot, India, in form of a cylindrical bar. The chemical composition of the alloy is presented in Table 1. The material was used in retrogressed and re-aged condition (RRA). It was solution treated at 470°C for 0.5 h, pre-aged at 120°C for 24 h and subjected to retrogression at 200°C for 10 mins followed by secondary ageing at 120°C for 24 hrs.

Table 1 Chemical Composition of the 7075 aluminium alloy.

Element	Zn	Mg	Cu	Si	Fe	Mn	Al
Wt %	4.89	2.12	1.52	0.33	0.007	0.09	Bal.

Ultrasonic shot peening treatment of the alloy was performed with steel balls of 3mm diameter at an amplitude of 80μm for 180 seconds. Stress relieving treatment of the USSPed samples was done at 90°C for 4h, followed by furnace cooling. The specimen without USSP is designated as un-USSP, treated by USSP as USSP, and stress relieved after USSP as USSP+SR. A Rigaku X-ray diffractometer with Cu Kα radiation was used to determine phase constitution, average crystallite size and mean micro strain of the USSP specimen. Crystallographic structure of the USSP samples was characterized by XRD in 2Θ range of 30° to 90°. Residual stress analysis was also carried out using PANanalytical X-ray diffractometer. Microhardness was measured using Shimadzu micro-hardness tester at an applied load of 100 g with dwell time of 5s Electrochemical polishing of TEM foil was done in an electrolyte containing 20% nitric acid in methanol at -28°C at applied voltage of 20V. Transmission electron microscopy was done using TECNAI 20 G2 operating at 200 kV. LCF tests were conducted on a servo hydraulic MTS™ of 50kN (model 810) under total strain control mode.

3. Results and Discussion

3.1. *X-ray Diffraction Analysis*

X-ray diffraction profiles of the un-USSP, USSP and USSP+SR samples are shown in Fig. 1 (a). Peaks of only α-Al were observed for all the conditions. Thus, it was obvious that there was no phase transformation, neither due to USSP nor from the stress relieving treatment. Fig.1(b) shows diffraction peak corresponding to (220) for the USSP condition, broadening and shifting of peak to the side of lower angle is quite evident. Broadening was due to increase in lattice strain and refinement of surface grains.

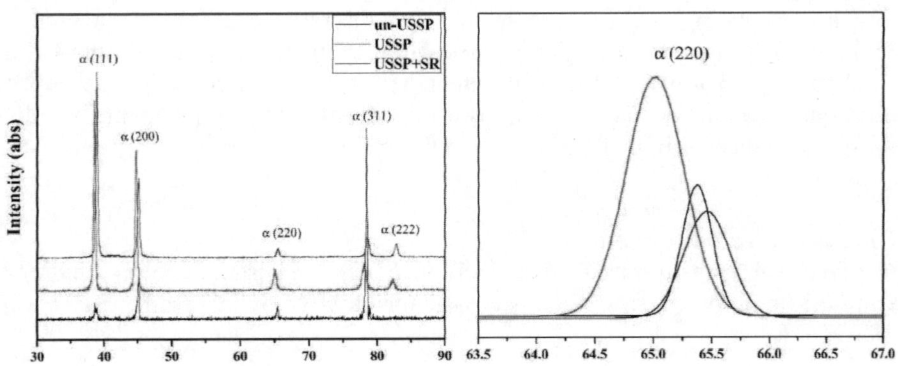

Figure 1. (a) XRD profiles of un-USSP, USSP and USSP+SR samples and (b) (220) peaks of the base α-matrix.

The average grain size was calculated using Scherrer and Wilson equation [13] from the five diffraction peaks and was found to be 18nm and 20nm for the USSP and USSP+SR conditions respectively. The micro strain resulting from the USSP treatment was calculated using the Williamson Hull equation [14] and was found to be 0.320% and 0.285% for the USSP and USSP+SR conditions respectively. Substantial decrease in micro strain was due to stress relieving treatment and it is obvious from Fig. 1 (b) that after stress relieving the peak again shifts towards the higher angle side which may be attributed to decrease in micro strain and the peak is still broad due to presence of nano grains in the surface region.

3.2. Microstructural Characterization

Fig 2 (a) shows bright field TEM micrographs of 7075 aluminium alloy in the un-USSP condition. Fine precipitates with some coarse intermetallic particles are distributed homogeneously throughout the grains with some η ($MgZn_2$) precipitates segregated at the grain boundary. The coarse intermetallic precipitates have been identified as Al_3Fe and Al_7Cu_2Fe [15-16].

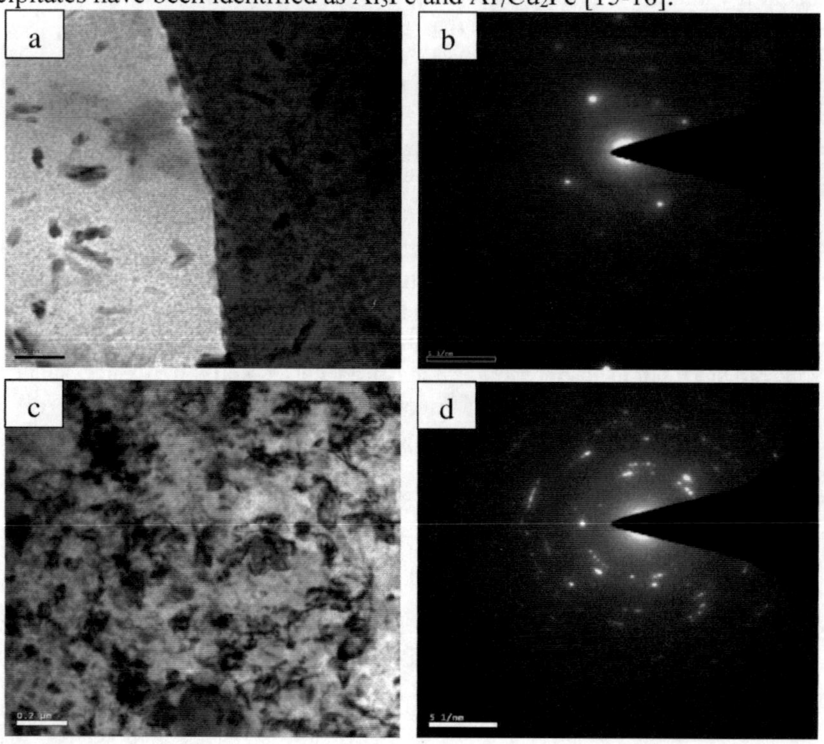

Figure 2 TEM micrographs of the 7075 aluminium alloy: (a) un-USSPed, (b) corresponding SAD pattern, (c) USSP, (d) corresponding SAD pattern.

Fig. 2 (c) shows bright field TEM micrograph of the USSP treated sample and its corresponding SAD pattern. Nano grains of approximately 20-30nm size are formed in surface region of the USSP specimen. Some of the grain boundaries are visible but many of them are not seen clearly. The formation of ring in the corresponding SAD pattern with some discretion indicates formation of nanograins in the USSP region with random orientations. The high strain and strain rates from USSP in the surface region led to subdivision of original coarse grains to nanograins. High density of dislocation network and dislocation tangles may be seen inside the grains, likewise the contrast within the grains is not uniform and indicates high level of internal stresses and elastic distortion in the crystal lattice.

3.3. Microhardness & Residual Stress variation

The microhardness in surface regions of the un-USSP, USSP and USSP+SR samples was measured. The hardness of the un-USSP sample was 154H_V and was increased to 172H_V after the USSP

IOP Conf. Series: Materials Science and Engineering **194** (2017) 012037 doi:10.1088/1757-899X/194/1/012037

treatment due to severe plastic deformation caused by repeated multiple impacts of hard steel balls and the consequent work hardening in the surface region. The increase in dislocation density and formation of nanograins increased the hardness in surface region. After the stress relieving treatment, the hardness was reduced to 157H_V and was comparable to that of the un USSP specimen.

The residual stresses resulting from USSP treatment were found compressive in nature in surface regions of the USSP and USSP+SR samples. For the USSP treated specimen the residual stress was -261 MPa and was reduced to -111 MPa following the stress relieving treatment. It may be attributed to decrease in dislocation density for the USSP+SR sample that caused decrement in the compressive residual stress in the surface region. The main aim was to relieve the associated residual stress to bring out the effect of only the nanostructured surface layer on low cycle fatigue behavior of the 7075 aluminium alloy.

3.4. Low Cycle Fatigue Behavior

The variation of cyclic stress with number of cycles for the un-USSP, USSP and USSP+SR conditions at different strain amplitudes is shown in Fig. 3. LCF tests were conducted at three different total strain amplitudes of ±0.6%, ±0.45% and ±0.38% at a strain rate of 5 X 10^{-3} s^{-1}. It may be seen that there was continuous cyclic softening in the un-USSP specimen tested at the highest strain amplitude of ±0.60% till failure, whereas cyclic hardening was observed for the two lower strain amplitudes. In the case of USSP and USSP+SR conditions, cyclic hardening was observed at all the three strain amplitudes. Irrespective of strain amplitude, there was rapid cyclic hardening in the initial 10 cycles. The degree of cyclic hardening increased with increase in strain amplitude for the USSP samples. The initial rapid hardening was mainly from the work hardened surface layer and interaction of dislocations generated during cyclic loading, with the initial dislocations resulting from USSP treatment. The degree of cyclic hardening was more in the USSP+SR condition as compared to that in the USSP.

The dependence of fatigue life as reversals to failure (2Nf), on plastic strain amplitude ($\Delta\varepsilon_p/2$), was analyzed using the Coffin–Manson relationship $\Delta\varepsilon_p/2 = \varepsilon'f(2N_f)^c$, where ε'f and c are fatigue ductility coefficient and fatigue ductility exponent respectively (Fig. 4). A linear trend in the variation may be observed for the un-USSP, USSP and USSP+SR conditions.

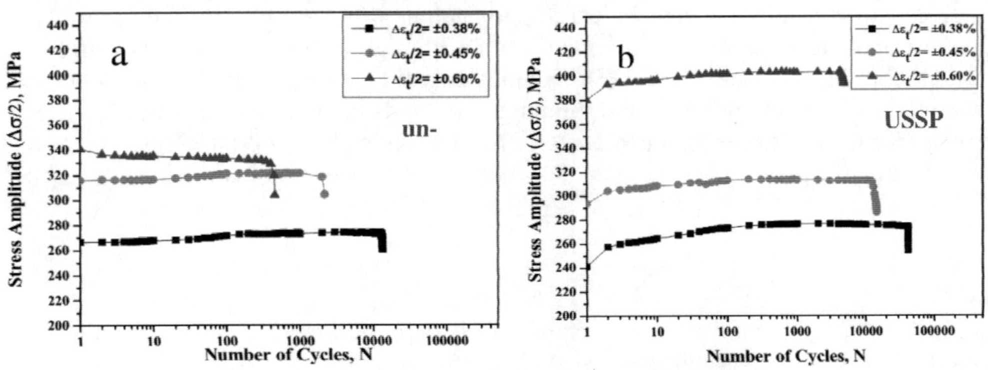

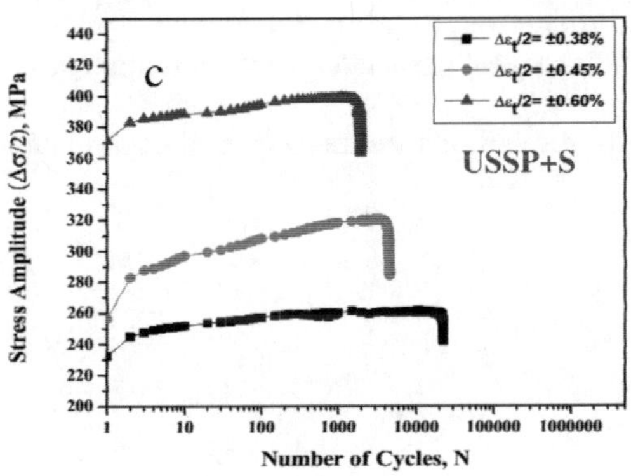

Figure 3 Variation of cyclic stress in (a) un-USSP, (b) USSP and (c) USSP+SR conditions.

IOP Conf. Series: Materials Science and Engineering **194** (2017) 012037 doi:10.1088/1757-899X/194/1/012037

LCF life was found to increase for the USSP treatment and the increment in life increased with decrease in strain amplitude. The increment in fatigue life was due to combined effect of nanostructured surface layer and the induced compressive residual stresses from the USSP treatment [17]. The nanostructured surface layer resisted the initiation of fatigue crack whereas the compressive residual stress reduced the rate of fatigue crack propagation and led to increase in LCF life. After the stress relieving treatment LCF life was lower as compared to that of the USSP condition but was still higher than that of un-USSP.

the

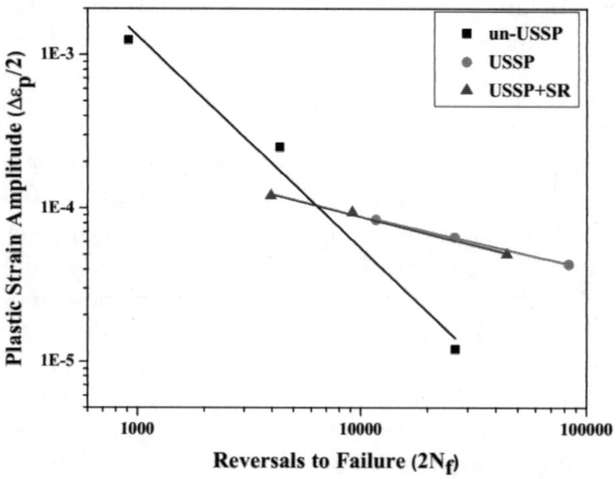

Figure 4 Coffin-Manson plots of 7075 aluminium alloy in the un-USSP, USSP and USSP+SR conditions.

Due to the severity of deformation resulting from USSP treatment there was increase in surface hardness and formation of nanograins in the surface region. The deformation was relatively more homogeneous in the USSP treated surface layer whereas it was heterogeneous in the un-USSP condition and concentrated along slip bands in coarse grains. The homogeneity of deformation decreased the degree of stress concentration hence the process of crack initiation was delayed. The surface and sub-surface compressive residual stresses reduced the tensile component of stress responsible for crack growth, thus the rate of crack propagation was reduced and fatigue life was increased. Thus, the nanostructured surface in combination with high compressive residual stress led to increase in fatigue life. In the case of USSP+SR samples only the nanostructured surface layer was there to delay the process of crack initiation and the fatigue life was higher than that of the un-USSP condition. However, due to absence of high compressive residual stress in the surface region there was no effect on the process of fatigue crack propagation and LCF life was less than that of the USSP condition.

4. Conclusions

i. A nanostructured layer with grain size of ~20nm was formed close to surface of the 7075 aluminium alloy subjected to ultrasonic shot peening.
ii. There was no phase transformation in the alloy due to the USSP treatment.
iii. The nanostructured surface layer associated with compressive residual stress led to enhancement of LCF life.
iv. Stress relieving treatment of the USSP specimens caused decrement in LCF life.

References

[1] Payne J. MIAC Newsl 1995; December:4-6.

[2] De los Rios, E. R., Walley, A., Milan, M. T., & Hammersley, G. Fatigue crack initiation and propagation on shot-peened surfaces in A316 stainless steel. International Journal of Fatigue, (1995); 17(7), 493-499.

[3] Almer, J. D., Cohen, J. B., & Moran, B. The effects of residual macrostresses and microstresses on fatigue crack initiation. Materials Science and Engineering: A, (2000); 284(1), 268-279.

[4] Altenberger, I. Mikrostrukturelle Untersuchungen mechanisch randschichtverfestigter Bereiche schwingend beanspruchter metallischer Werkstoffe. Univ.-Bibliothek, (1999).

[5] Wu, X., Tao, N., Hong, Y., Xu, B., Lu, J., & Lu, K. Microstructure and evolution of mechanically-induced ultrafine grain in surface layer of AL-alloy subjected to USSP. Acta materialia, (2002); 50(8), 2075-2084.

[6] Wang, K., Tao, N. R., Liu, G., Lu, J., & Lu, K. Plastic strain-induced grain refinement at the nanometer scale in copper. Acta Materialia, (2006); 54(19), 5281-5291.

[7] Zhang, H. W., Hei, Z. K., Liu, G., Lu, J., & Lu, K. Formation of nanostructured surface layer on AISI 304 stainless steel by means of surface mechanical attrition treatment. Acta materialia, (2003); 51(7), 1871-1881.

[8] Lin, Y., Lu, J., Wang, L., Xu, T., & Xue, Q. Surface nanocrystallization by surface mechanical attrition treatment and its effect on structure and properties of plasma nitrided AISI 321 stainless steel. Acta Materialia, (2006); 54(20), 5599-5605.

[9] Zhu, K. Y., Vassel, A., Brisset, F., Lu, K., & Lu, J. Nanostructure formation mechanism of α-titanium using SMAT. Acta Materialia, (2004); 52(14), 4101-4110.

[10] Suresh S., Fatigue of Materials, Cambridge University Press, New York. (1991)

[11] Hanlon, T., Kwon Y.N., Suresh S., Grain size effects on the fatigue response of nanocrystalline metals, Scripta Mater., (2003); 49, 675-680.

[12] Mughrabi, H., Hoppel, H.W., Kautz, M., Fatigue and microstructure of ultrafine-grained metals produced by severe plastic deformation, Scripta Mater., (2004); 51, 807-812.

[13] YE, H. Q., & FAN, X. M. Surface nanocrystallization of 7A04 aluminium alloy induced by circulation rolling plastic deformation. Transactions of Nonferrous Metals Society of China, (2006); 16, 656-660.

[14] Williamson, G. K., & Hall, W. H. X-ray line broadening from filed aluminium and wolfram. Acta metallurgica, (1953); 1(1), 22-31.

[15] Li, J.F., Peng, Z.W., Li, C.X., Jia, Z.Q., Chen, W.J., Zheng, Z.Q., Mechanical properties, corrosion behaviors and microstructures of 7075 aluminium alloy with various aging treatments, Trans. Nonferrous Metals Soc. of China, (2008); 18(4), 755-762.

[16] Birbilis N., Cavanaugh M.K., Buchheit R.G. et al., Proceedings Symposium Applications of Materials Science to Military System, Materials Science and Technology, Pisttsburgh, PA; (2005).

[17] Pandey, V., Rao, G.S., Chattopadhyay, K., Srinivas, N.C.S., Singh, V., Effect of surface Nanostructuring on LCF behavior of aluminum alloy 2014, Mater. Sci. Eng. A, (2015); 647, 201-211.

IOP Conf. Series: Materials Science and Engineering **194** (2017) 012038 doi:10.1088/1757-899X/194/1/012038

Warm deformation behaviour of UFG CP-Titanium produced by I-ECAP

Muhammad Jawad Qarni[1,*], Andrzej Rosochowski[2] and Sonia Boczkal[3]

[1]Advanced Forming Research Centre (AFRC), University of Strathclyde, Inchinnan, Renfrewshire, PA4 9LJ, UK

[2]Design Manufacture and Engineering Management (DMEM), University of Strathclyde, Glasgow, G1 1XJ, UK

[3]Institute of Non-Ferrous Metals in Gliwice, Light Metals, Division, ul. Pilsudskiego 19, 32-050 Skawina, Poland

*Corresponding autor: jawad.qarni@strath.ac.uk

Abstract. The objective of the present study is to investigate the deformation behaviour of Ultrafine-grain (UFG) commercial purity Titanium (CP-Ti) at warm temperatures. Firstly, CP-Ti billets were processed through six passes of incremental equal channel angular pressing (I-ECAP) at 300° C using a die channel angle (Φ) of 120°. Uniaxial compression tests were then performed under isothermal conditions on cylindrical samples obtained from the UFG CP-Ti billets. A series of these tests were conducted at different temperatures of 400, 500 and 600 °C and at varying strain rates of 0.01, 0.1 and 1.0 s^{-1}. In each test, the original height of the sample was deformed by ~50% of its original value. The true stress-strain curves obtained, revealed that the flow stress was sensitive to both temperature and strain rate. In general, the flow stress was higher for lower temperatures and higher strain rates. For tests conducted at 400 and 500 °C, the flow stress quickly reaches a peak value, beyond which it exhibits a steady state response where there is no appreciable change in flow stress with increasing strain. The 600 °C tests however shows a strain hardening behaviour. Microstructure of the sample deformed at 600 °C and 0.01 s^{-1}, exhibited significant grain growth.

1. Introduction

Owing to its high specific strength, low density, outstanding corrosion resistance and excellent bio-compatibility; Titanium and it's alloys are a material of choice in many aerospace, military, chemical and biomedical applications [1]. Ti-6Al-4V is most widely used for medical device applications such as in total replacement implants, where higher strength characteristics is generally a requirement [2]. The addition of alloying elements such as Al and V in Titanium, significantly enhance the mechanical performance, but these are considered to be toxic and therefore undesirable for full bio-integration [3]. Commercial purity Titanium (CP-Ti) is biologically more compatible but cannot satisfy the higher strength level requirements. An effective solution for improving the mechanical strength and performance of CP-Ti is by grain refinement through severe plastic deformation (SPD) processes. To date, strengthening of various metals including CP-Ti via grain refinement has been demonstrated by numerous studies [4, 5].

Equal channel angular pressing (ECAP) is by far the most promising SPD technique available. It is capable of producing bulk UFG materials, large enough for practical applications [6]. Developed by Segal et al. [7], the technique involves pressing a billet through a die consisting of two channels with equal cross-sections, intersecting at an angle (Φ). As the billet passes through the intersection, it is subjected to simple shear, while retaining the original cross-sectional area. The same billet can therefore be passed through the die multiple times, in order to impart desired level of plastic strain. The billet is usually rotated along its longitudinal axis between the passes, creating different ECAP routes [8]. ECAP has been widely applied on CP-Ti for the purposes of refining the grain structure and subsequently improving its strength. Research has shown that grain refinement in CP-Ti, not only improves its yield and tensile strength [9-12] but also improves fatigue behaviour and corrosion resistance considerably [13, 14].

CP-Ti has a hexagonal closed packed (HCP) structure at room temperature. Compared to face-centred cube (FCC) and body-centred cube (BCC) metals, it is regarded as 'hard-to-deform' because of its poor formability. The inferior deformability, is mainly due to the limited slip systems available in HCP structure. In order to improve the formability in CP-Ti, it is necessary to carry deformation at elevated temperatures [15]. However, use of excessively high temperatures can cause increased recovery, leading to significant grain growth. This will result in considerable loss of hardness and strength, especially if the deformation process is carried at very slow strain rates. Deforming CP-Ti in the warm temperature range is therefore desirable, not only in terms of retaining the strength characteristics of UFG but also in terms of reducing the forming process energy requirements. Finding a good balance between formability and loss in strength is essential by developing a deeper understanding into the warm temperature range deformation behaviour of UFG CP-Ti. It is generally accepted that compression flow stress data can be used to simulate the deformation behaviour during metal forming processes. The effective use of accurate flow stress data in simulations combined with the knowledge of grain growth at different temperatures and strain rates, will greatly aid in establishing the optimal processing regimes of UFG CP-Ti.

The present study is therefore concerned with studying the warm temperature deformation behaviour and subsequent grain growth in UFG CP-Ti during compression. Firstly, the incremental equal channel angular pressing (I-ECAP) process, which is a recognized SPD processing technique [16] was employed to obtained UFG structure in CP-Ti. Uniaxial compression tests were then performed on cylindrical samples obtained from the processed billets. To obtain a comprehensive flow stress dataset, a series of these tests were conducted for a temperature range of 400 to 600 °C and for varying strain rates. Moreover, by using the electron backscatter diffraction (EBSD) technique, the microstructure was analysed to observe grain growth following compression tests.

2. Material and method

2.1. Material

The material used in the present study was commercial purity titanium, grade 2 (here after referred to as CP-Ti). The material was received in the form of 12.5 mm thick plate from Dynamic Metals Ltd (UK), having an average grain size of ~20 μm. The reported chemical composition (max wt. %) was 0.08% C, 0.03% N, 0.18% O, 0.015% H, 0.20% Fe and balance Ti. Square cross-section billets measuring 10 mm^2 and 120 mm in length were EDM wire-cut from the plate for I-ECAP processing. Billets were cut such that the longitudinal axis of the billet was parallel to the rolling direction of the plate.

2.2. I-ECAP experimental procedure

Despite the success of ECAP process, it is not an ideal option from the viewpoint of commercialization. Its inability to process very long or continuous billets, due to very high contact pressure, is the main limitation. I-ECAP process developed by Rosochowski and Olejnik [17], is a possible solution. The schematic illustration of the process is shown in Figure 1(a). Here, material pressing and deformation, takes place in two separate stages as opposed to ECAP [18]. By separating the pressing and deformation stages, reduces friction substantially. This is turn lowers the force required to press the material; thereby enabling processing of very long or continuous billets.

There are three main tools in I-ECAP: (1) die, (2) plunger and (3) punch; denoted by A, B and C respectively. During processing, punch tool (a working die) is oscillating with a certain frequency and therefore comes cyclically in contact with the billet top surface. In material pressing stage, while the punch tool is retracting, the billet material is pressed into the deformation zone in increments of 'a' by the plunger tool known as pressing stroke (see Figure 1(a)). In the deformation stage, the punch comes down and deforms the billets. The blue colour outline of the billets in Figure 1(a) represents the pressing stage whereas the red colour outline represents the deformation stage with the dashed outline representing the plastically deformed zone. The mode of deformation is similar to that in ECAP i.e. simple shear, provided the pressing stroke is not too large.

In the present study, the double-billet variant of the I-ECAP process, with a channel intersection angle (Φ) of 120° was used. This technique enables processing two square cross-section billets simultaneously, the two billets are marked 1 and 2 in Figure 1(a). A detailed description of the I-ECAP apparatus used in the study can be found elsewhere [19].Experiments were performed at 300 °C, on a 1000 KN servo-hydraulic press shown in Figure 1(b), at a pressing rate of 0.2 mm/cycle (0.1 mm/s), punch oscillatory movement of 0.5 Hz and with a peak to peak amplitude of 1.6 mm. The die configuration lead to an imposed strain of ~0.67 per pass through the die. The billets were processed repeatedly and were subjected to six passes, leading to an induced total strain of ~4.02. Processing route B_C was followed, as it is considered to be the most effective in achieving homogenous microstructure of grains separated by high angle boundaries [20, 21].

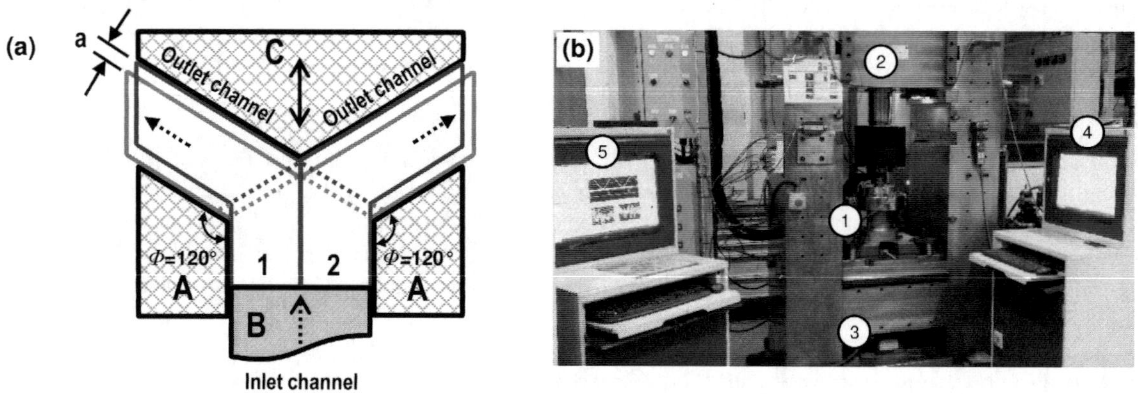

Figure 1. (a) Schematic illustration of double-billet variant of I-ECAP process (A=Die, B=Plunger and C=Punch) and (b) the I-ECAP experimental rig (1=I-ECAP die, 2=1000 KN servo-hydraulic press, 3=servo motor coupled to screw jack for material pressing, 4=Zwick Controller for controlling the press and 5=LabVIEW application for data acquisition)

2.3. Compression testing

For investigating the warm deformation behaviour of CP-Ti, uniaxial compression tests were performed following the ASTM E209-00 on the Zwick/Roell HA250 servo hydraulic testing machine with a 250 kN load cell. Compression samples, measuring ∅8 mm in diameter and 8 mm in height, were precision turned on a lathe machine, from the sixth pass processed billets such that the compression axis was parallel to longitudinal axis of the billet. To avoid end effects, samples were taken from the central region only, after cutting away the head and tail region of the billets.

Compressions tests were conducted at constant strain rate and under isothermal conditions. A series of these tests were conducted, at three deformation temperatures of 400, 500 and 600 °C and at three different strain rates of 0.01, 0.1 and 1.0 s^{-1}. A minimum of two tests were performed at each test conditions to ensure repeatability of results. Temperature was monitored using two calibrated N-Type thermocouples, which was located near the deformation zone. When the temperature value was stable by ±2 °C, compression test was initiated and the height of cylindrical sample was deformed by ~50% in a single loading step, achieving a true strain of ~0.80. After the test, each sample was immediately quenched in cold water to arrest the microstructure and to stop any grain growth.

2.4. *Microstructure*
The deformed compression samples were cut along the compression axis from the centre. The surface of each sample was polished using standard mechanical polishing techniques and was then ion milled on Leica RES101 as a final step. The microstructure was characterized by scanning electron microscopy (SEM) using electron backscatter diffraction (EBSD) technique. The SEM used was a FEI Inspect F50 with an EDAX TSL EBSD detector. The sample was tilted 70° from the horizontal for EBSD data collection, at a 20 kV accelerating voltage, 200 mA beam current and with a 80 nm step size.

3. Results and discussion

3.1. *Flow stress behaviour*
Figure 2, shows the true stress and true strain curves obtained from the compression testing, representing the flow stress and deformation behaviour of CP-Ti studied in the present work at various temperature and strain rate conditions. Room temperature compression tests at 0.01s^{-1} were also performed on unprocessed (0P) and for severely deformed (6P) material, to observe the increase in strength and strain hardening behaviour following I-ECAP processing. Figure 2(a) shows this comparison, here the solid line represents the processed material displays a much higher yield strength compared to the unprocessed material. The results for the unprocessed material is consistent with the findings by Gray [22]. He concluded that the rate of strain hardening increases with increase in grain size due to pronounced increase in deformation twinning.

Figure 2(b-d), shows the flow stress behaviour of UFG CP-Ti at warm deformations temperatures of (b) 400, (c) 500 and (d) 600 °C. As expected, the flow stress increases with decreasing temperature and increasing strain rate. At same temperature, the flow stress decreases with decreasing strain rate. It is evident from these graphs that compared to the room temperature result, the flow stress at warm temperatures for all strain rates is considerably lower. These figures also displays low levels of strain hardening behaviour of UFG CP-Ti at warm deformation temperatures. For test temperatures of 400 and 500 °C, the flow stress quickly reaches a peak value near ~0.02 true strain, beyond which it exhibits a steady state response and where there is no appreciable strain hardening. Except for the test conducted at 500 °C and 0.10 s^{-1}, which actually displays a flow softening behaviour. Flow stress obtained at 600 °C for all strain rates, shows slightly increasing levels of strain hardening, compared to at 400 and 500 °C. However, here the flow stress has been dropped by half compared to 500 °C.

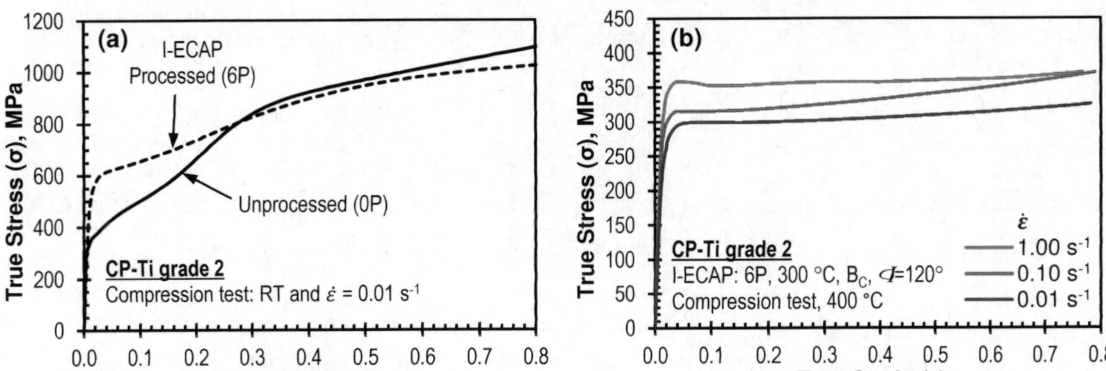

IOP Conf. Series: Materials Science and Engineering **194** (2017) 012038 doi:10.1088/1757-899X/194/1/012038

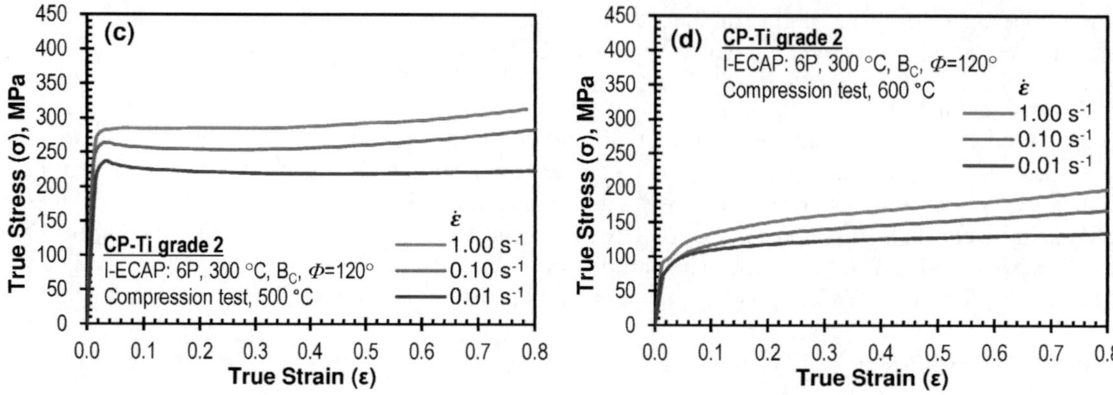

Figure 2. Representative true stress-true strain curves obtained from uniaxial compression testing of CP-Ti at (a) room temperature testing of both unprocessed (0P) and processed material (6P) and in the warm temperature range testing of processed material (6P) at (b) 400 °C (c) 500 °C and (d) 600 °C

3.2. Deformed microstructure

Figure 3(a and b) shows the colour inverse pole figure (IPF) maps obtained from the SEM based EBSD analysis, shows the microstructure after 50% compression at (a) 400 °C and 1.0 s^{-1} and (b) 600 °C and 0.01 s^{-1}. These two extremes conditions have been selected to shows the extent of grain growth. The average grain size shown in Figure 3(a) is 0.85 μm, and it dramatically increases to ~4.0 μm as shown in Figure 3(b). The IPF map in Figure 3(b), also shows the presence of some fraction of twins within the matrix of large grains. Analysis of the grain misorientation histogram (not shown here), shows a noticeable peak around 65° and a significant peak at 85°, this strongly suggests existence of $\{11\bar{2}2\}$ and $\{10\bar{1}2\}$ twins.

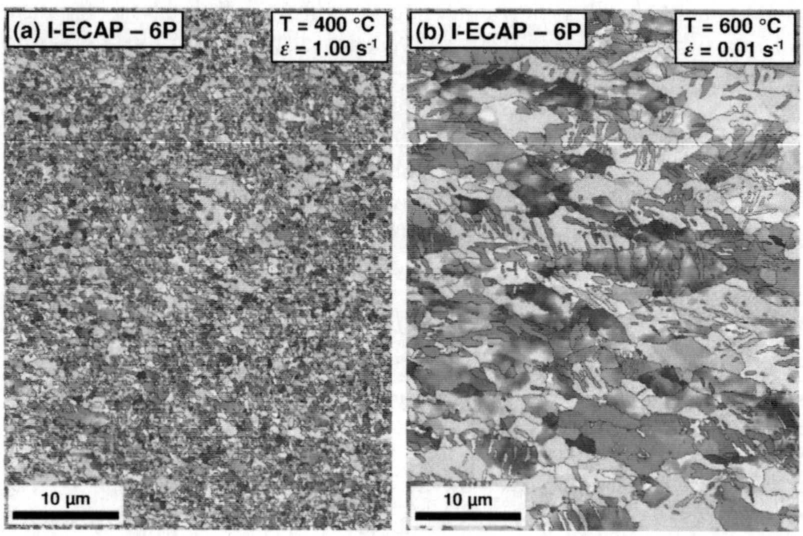

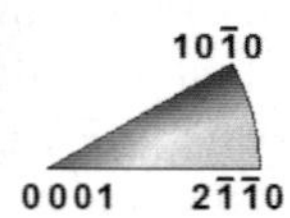

Figure 3. Coloured inverse pole figure (IPF) maps obtained from EBSD analysis showing the grain growth and deformation characteristics during compression testing at (a) 400 °C, $\dot{\varepsilon}$ = 1.00 s^{-1} and (b) 600 °C, $\dot{\varepsilon}$ = 0.01 s^{-1}. The colour corresponds to the crystallographic orientation shown in the standard stereographic triangle. Colour variations within the grains qualitatively represent differences in internal misorientations.

4. Conclusion

The following conclusions are drawn from the study:

1. CP-Ti grade 2 billets were severely deformed to achieve ultrafine grain (UFG) structure, after six passes of I-ECAP processing at 300 °C using a die angle of 120° following route B_C.
2. Room temperature compression tests of unprocessed and I-ECAP processed material, revealed significant increase in strength due to processing. However, the unprocessed material exhibited higher levels of strain hardening compared to processed material.

3. Compression testing of UFG CP-Ti at warm temperatures of 400, 500 and 600 °C and at varying strain rates of 0.01, 0.1 and 1.0 s^{-1}, revealed that the flow stress was sensitive to temperature and strain rate. The flow stress increases with decreasing temperatures and increasing strain rates.
4. In compressions tests conducted at 400 and 500 °C, the flow stress quickly reaches a peak value, beyond which it exhibits a steady state response where there is no appreciable change in flow stress with increasing strain. The 600 °C tests however shows a strain hardening behaviour.
5. Microstructural observations using EBSD, exposed significant grain growth in the sample deformed at 600 °C and 0.01 s^{-1}. This also led to a considerable drop in flow stress during compression.

References

[1] Donachie MJ 2000 Titanium - A technical guide Second edition ASM International ISBN: 0-87170-686-5
[2] Geetha M, Singh AK, Asokamani R and Gogia AK 2009 *Progress in Materials Science* **54** 397-425
[3] Lugowski SJ, Smith DC, McHugh AD and Van Loon JC 1991 *Journal of Biomedical Materials Research* **25** 1443-1458
[4] Meyers MA, Mishra A and Benson DJ 2006 *Progress in Materials Science* **51** 427-556
[5] Langdon TG 2013 *Acta Materialia* **61** 7035-7059
[6] Segal VM, Dobatkin SV and Valiev R. Z. 2004 *Russian Metall* **1** 1-102
[7] Segal VM, Reznikov VI, Drobyshevskiy AE and Kopylov VI 1981 *Russ Metally* **1** 99-104
[8] Iwahashi Y, Wang J, Horita Z, Nemoto M and Langdon TG 1996 *Scripta Materialia* **35** 143-146
[9] Purcek G, Yapici GG, Karaman I and Maier HJ 2011 *Materials Science and Engineering: A* **528** 2303-2308
[10] Sergueeva AV, Stolyarov VV, Valiev RZ and Mukherjee AK 2001 *Scripta Materialia* **45** 747-752
[11] Zhao X, Yang X, Liu X, Wang CT, Huang Y and Langdon TG 2014 *Materials Science and Engineering: A* **607** 482-489
[12] Stolyarov VV, Zhu YT, Alexandrov IV, Lowe TC and Valiev RZ 2003 *Materials Science and Engineering: A* **343** 43-50
[13] Balyanov A, Kutnyakova J, Amirkhanova NA, Stolyarov VV, Valiev RZ, Liao XZ, Zhao YH, Jiang YB, Xu HF, Lowe TC and Zhu YT 2004 *Scripta Materialia* **51** 225-229
[14] Semenova IP, Salimgareeva GK, Latysh VV, Lowe T and Valiev RZ 2009 *Materials Science and Engineering: A* **503** 92-95
[15] Semiatin SL, DeLo DP, Segal VM, Goforth RE and Frey ND 1999 *Metallurgical and Materials Transactions A* **30** 1425-1435
[16] Azushima A, Kopp R, Korhonen A, Yang DY, Micari F, Lahoti GD, Groche P, Yanagimoto J, Tsuji N, Rosochowski A and Yanagida A 2008 *CIRP Annals - Manufacturing Technology* **57** 716-735
[17] Rosochowski A and Olejnik L 2007 *AIP Conference Proceedings* **907** 653-658
[18] Rosochowski A and Olejnik L 2011 *Materials Science Forum* **674** 19-28
[19] Qarni MJ, Sivaswamy G, Rosochowski A and Boczkal S 2017 *Materials & Design* **122** 385-402
[20] Oh-Ishi K, Horita Z, Nemoto M, Furukawa M and Langdon TG 1998 *Metallurgical and Materials Transactions A* **29** 2011-2013
[21] Stolyarov VV, Zhu YT, Alexandrov IV, Lowe TC and Valiev RZ 2001 *Materials Science and Engineering: A* **299** 59-67
[22] Gray G, T. 1997 *Le Journal de Physique IV* **07** 423-428

IOP Conf. Series: Materials Science and Engineering **194** (2017) 012039 doi:10.1088/1757-899X/194/1/012039

The effect of high-pressure torsion on the microstructure and properties of magnesium

Roberto B Figueiredo[1,2], Shima Sabbaghianrad[2], Terence G Langdon[2,3]

[1] Department of Materials Engineering and Civil Construction, Universidade Federal de Minas Gerais, Belo Horizonte, MG 31270-901, Brazil
[2] Departments of Aerospace & Mechanical Engineering and Materials Science. University of Southern California, Los Angeles, CA 90089-1453, USA
[3] Materials Research Group, Faculty of Engineering and the Environment, University of Southampton, Southampton SO17 1BJ, UK

E-mail:figueiredo@demc.ufmg.br, ssabbagh@usc.edu, langdon@soton.ac.uk

Abstract. High-pressure torsion provides the opportunity to introduce significant plastic strain at room temperature in magnesium and its alloys. It is now established that this processing operation produces ultrafine-grained structures and changes the properties of these materials. The present paper shows that the mechanism of grain refinement differs from f.c.c. and b.c.c. materials. It is shown that fine grains are formed at the grain boundaries of coarse grains and gradually consume the whole structure. Also, the processed material exhibits unusual mechanical properties due to the activation of grain boundary sliding at room temperature.

1. Introduction

Magnesium exhibits poor formability at room temperature due to the limited number of slip systems readily available. Basal slip and twinning play a major role in accommodating plastic deformation but these mechanisms can only withstand a few percent strain. It has been shown that the maximum strain during compression of pure magnesium with grain size of ~24 μm is in the range of 15%~20% [1]. The maximum elongation of as-cast magnesium in tension is ~5% [2]. EBSD *in situ* tensile [3] and compression [4] testing have shown that slip is accommodated by basal slip and twinning and very limited non-basal slip takes place at low temperatures.

However, recent papers have shown a change in deformation behavior of pure magnesium when the grain size is reduced below a certain level. Hall-Petch breakdown was reported in pure magnesium with grain sizes of several microns [5]. An increase in strain accommodation during low strain-rate compression of magnesium takes place in samples with ~2.7 μm [1]. Finally, a recent paper has shown that grain refinement of magnesium processed by high-pressure torsion leads to a remarkable increase in elongation during tension [2]. This change in deformation behavior is attributed to the onset of grain boundary sliding at room temperature in fine-grained pure magnesium [1, 2, 5].

Thus, high-pressure torsion provides the opportunity to produce magnesium samples with extreme ductility at room temperature. For this, a homogeneous distribution of ultrafine grains must be achieved throughout the sample. The mechanism of grain refinement in pure magnesium processed by severe plastic deformation at high temperatures differs from other metallic materials with b.c.c. and f.c.c. structures [6-8]. However, very limited information is available on microstructure evolution of magnesium processed by severe plastic deformation at room temperature. A recent paper reported the grain structure of pure magnesium [2] at different stages of processing but the area analysed was limited. The present paper aims to evaluate the grain structure at different stages of HPT processing considering larger areas and evaluate the texture evolution.

2. Experimental material and procedures

The material used in this research was commercial purity magnesium. Cylindrical billets with 10 mm diameter were machined from a cast slab. Discs with ~1.0 mm thickness were cut from billets and ground to ~0.85 mm. High-pressure torsion was carried out using quasi-constrained anvils. The nominal pressure was 6.0 GPa and the samples were processed at room temperature at 1 rpm. Samples were processed to 1/8, 1/2 and 10 turns.

The processed discs were ground and polished to a mirror-like finish using diamond suspensions. A final polishing step with colloidal silica polishing was required to obtain clear EBSD images. EBSD characterization was carried out using a JEOL JSM-7001F scanning electron microscope operating at 7 kV. The microstructure was characterized at mid-radius position of the discs.

Miniature tensile specimens with ~1.0 mm gauge length were machined by spark-erosion from discs processed by 10 turns of HPT. The gauge areas of the specimens were located near the mid-radius positions of the discs. Tensile tests were carried out at constant cross-head displacement rate with initial strain rates in the range $10^{-5} - 10^{-1}$ s^{-1}. Engineering stress vs. strain curves were determined from the load vs. displacement data.

3. Results and discussion

Figure 1 shows the inverse pole figures of the grain structure at the mid-radius positions of discs processed to different numbers of turns of HPT. Coarse grains of ~10 μm and ultrafine grains are observed in the material processed to 1/8 turn. This multimodal grain size distribution is typical of magnesium and magnesium alloys processed by equal-channel angular pressing at high temperatures and is attributed to a different mechanism of grain refinement in these materials [6-8]. Inhomogeneous grain size distribution has also been reported in Zr processed by HPT [9] which suggests this might be a common feature in h.c.p. metals. The microstructure of the sample processed to 1/2 turn is largely composed of ultrafine grains although a small number of grains with a few microns are still observed. Processing by 10 turns of HPT does not change significantly the structure but decreases the number of grains larger than 1 μm.

Figure 2 shows the frequency distribution of the grain boundary misorientations at different stages of HPT processing. A high frequency of low-angle boundaries is observed at 1/8 turn but the overall fraction of these boundaries is significantly smaller than observed in b.c.c. or f.c.c. metals processed to low numbers of turns in HPT [10-12]. Further processing leads to an increase in high-angle boundaries and a pronounced peak in distribution is observed around 30°. A high frequency of grain boundaries with this misorientation range is predicted in magnesium with a basal fiber texture [13]. It is worth noting a high frequency of grain boundaries with misorientation in the range 85°-90° which is associated with twinning [14].

IOP Conf. Series: Materials Science and Engineering **194** (2017) 012039 doi:10.1088/1757-899X/194/1/012039

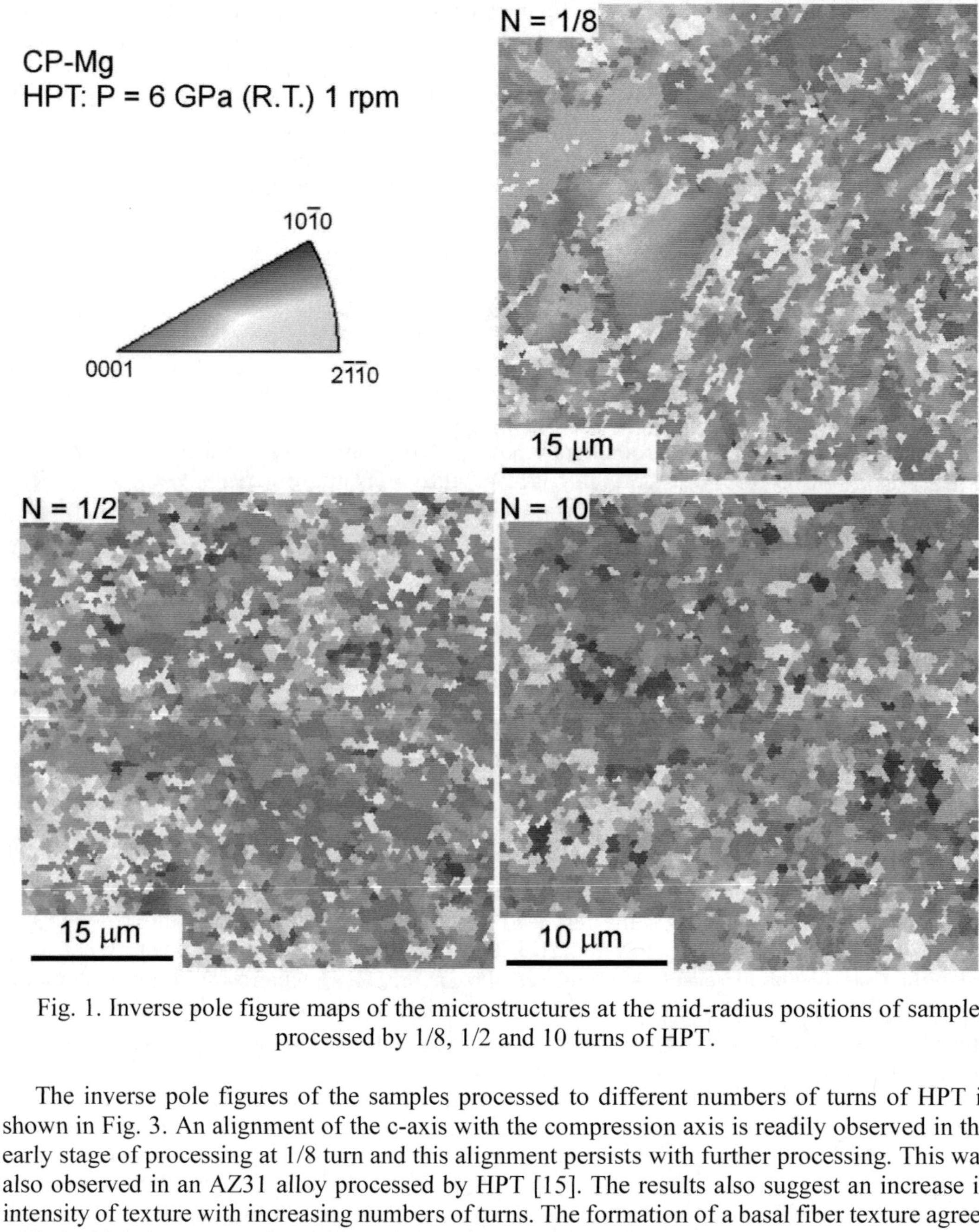

Fig. 1. Inverse pole figure maps of the microstructures at the mid-radius positions of samples processed by 1/8, 1/2 and 10 turns of HPT.

The inverse pole figures of the samples processed to different numbers of turns of HPT is shown in Fig. 3. An alignment of the c-axis with the compression axis is readily observed in the early stage of processing at 1/8 turn and this alignment persists with further processing. This was also observed in an AZ31 alloy processed by HPT [15]. The results also suggest an increase in intensity of texture with increasing numbers of turns. The formation of a basal fiber texture agrees with the high frequency of grain boundaries with misorientations of 30°.

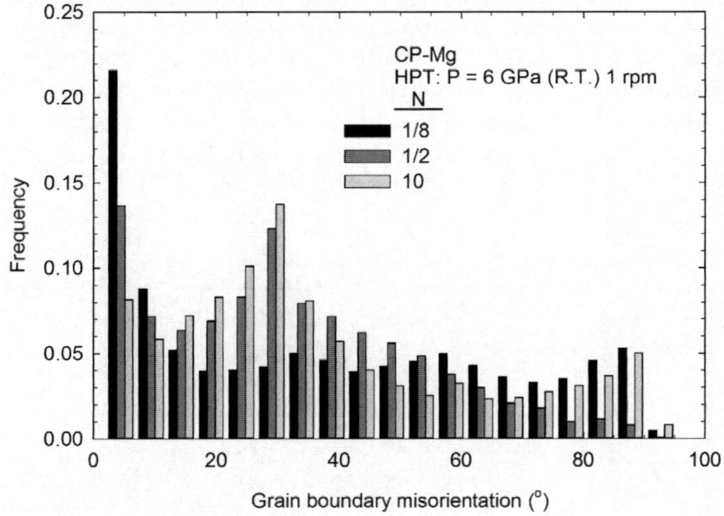

Fig.2. Frequency of grain boundary misorientation distribution at different stages of HPT.

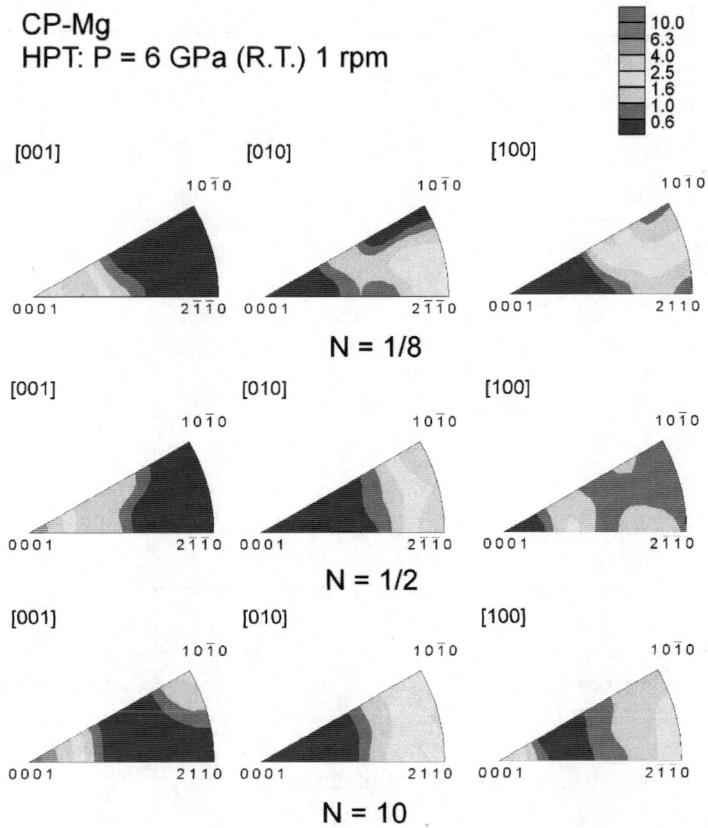

Fig. 3. Inverse pole figures at different stages of HPT processing.

The engineering stress vs. engineering strain curves of the material processed by 10 turns of HPT are shown in Fig. 4. The curves obtained during testing at different strain rates show that this parameter plays a major role in the deformation response of this material. The maximum flow stress observed during testing at 10^{-5} s^{-1} is ~50 MPa and increasing the strain rate to 10^{-1} s^{-1} leads to a 4-fold increase in flow stress. This agrees with the high strain rate sensitivity reported for fine-grained magnesium [1, 2, 5]. Also, the curves show a pronounced increase in elongation with decreasing strain rate which confirms that HPT processing enhances the ductility of magnesium.

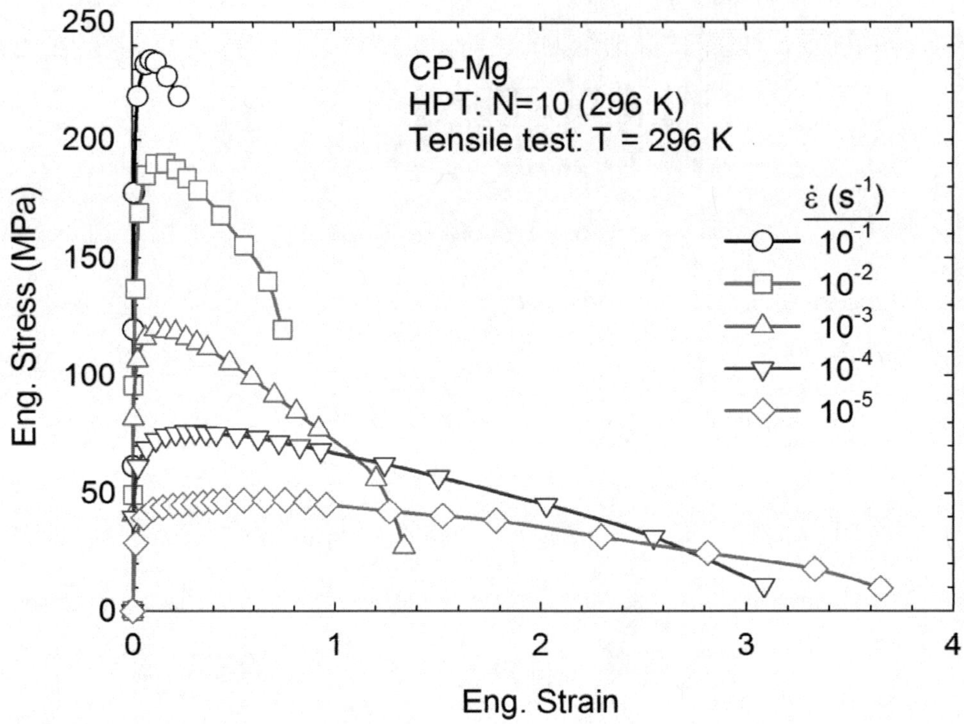

Fig. 4. Engineering stress versus engineering strain for the CP-Mg alloy processed by HPT.

4. Conclusions

The mechanism of grain refinement of magnesium differs from that observed in f.c.c. and b.c.c. metals. A multi-modal grain size distribution is observed in the early stage of deformation and gradually evolves to a homogeneous distribution of ultrafine grains.

The distribution of grain boundary misorientations shows significantly lower frequencies of low-angle boundaries in the early stage of processing compared to aluminium and iron. A high frequency of misorientations around 30° is observed at 1/2 and 10 turns of HPT.

A basal fiber texture with the c-axis aligned to the torsion axis develops in the early stage of HPT processing and persists with increasing numbers of turns.

HPT processing improves the ductility of magnesium.

Acknowledgements

The authors acknowledge the support of CNPq, CAPES, FAPEMIG and PPGEM. This work was supported by the European Research Council under Grant Agreement No. 267464-SPDMETALS.

References

[1] Figueiredo RB, Poggiali FSJ, Silva CLP, Cetlin PR, Langdon TG. 2016 *J. Mater. Sci.* **51** 3013-24.
[2] Figueiredo RB, Sabbaghianrad S, Giwa A, Greer JR, Langdon TG. 2017 *Acta Mater.* **122** 322-31.
[3] Cepeda-Jiménez CM, Molina-Aldareguia JM, Pérez-Prado MT. 2015 *Acta Mater.* **84** 443-56.
[4] Cepeda-Jiménez CM, Molina-Aldareguia JM, Pérez-Prado MT. 2015 *Acta Mater.* **88** 232-44.
[5] Somekawa H, Mukai T. 2015 *Metall. Mater. Trans. A* **46A** 894-902.
[6] Figueiredo RB, Langdon TG. 2010 *J. Mater. Sci.* **45** 4827-36.
[7] Figueiredo RB, Langdon TG. 2009 *Int J Mater Res* **100** 1638-46.
[8] Figueiredo RB, Langdon TG. 2009 *J. Mater. Sci.* **44** 4758-62.
[9] Wang YB, Louie M, Cao Y, Liao XZ, Li HJ, Ringer SP, Zhu YT. 2010 *Scripta Mater.* **62** 214-217.
[10] Ivanisenko Y, Valiev RZ, Fecht HJ. 2005 *Mater. Sci. Eng. A* **390** 159-65.
[11] Ito Y, Horita Z. 2009 *Mater. Sci. Eng. A* **503** 32-6.
[12] Xu C, Horita Z, Langdon TG. 2010 *Mater. Trans.* **51** 2-7.
[13] del Valle JA, Pérez-Prado MT, Ruano OA. 2002 *Rev. Metall. Madrid* **38** 353-357.
[14] Partridge PG. 1967 *Metall. Rev.* **12** 169-94.
[15] Huang Y, Figueiredo RB, Baudin T, Helbert AL, Brisset F, Langdon TG. 2013 *Mater. Res.* **16** 577-85.

IOP Conf. Series: Materials Science and Engineering **194** (2017) 012040 doi:10.1088/1757-899X/194/1/012040

The potential for achieving superplasticity in high-entropy alloys processed by severe plastic deformation

Hamed Shahmir[1,2], Megumi Kawasaki[3,4], Terence G Langdon[1,4,*]

[1] Materials Research Group, Faculty of Engineering and the Environment, University of Southampton, Southampton SO17 1BJ, UK
[2] School of Metallurgy and Materials, College of Engineering, University of Tehran, Tehran, Iran
[3] Division of Materials Science and Engineering, Hanyang University, Seoul 04763, South Korea
[4] Departments of Aerospace & Mechanical Engineering and Materials Science. University of Southern California, Los Angeles, CA 90089-1453, USA

* langdon@soton.ac.uk

Abstract. High-entropy alloys (HEAs) are now becoming important because they offer unique combinations of solid solution strengthening and good ductility at low temperatures. Only limited information is at present available on the high temperature mechanical properties of these materials. Nevertheless, it is evident that, as in conventional metallic alloys, processing through the application of severe plastic deformation can reduce the grain size to the nanometer range and this provides a potential for achieving good superplastic elongations. The superplastic data available to date are examined in this review and a comparison is made between the behaviour of HEAs and conventional superplastic alloys.

1. Introduction

High-entropy alloys (HEAs) are a new class of material containing five or more principal elements with each elemental concentration between 5 at.% and 35 at.% but producing a relatively simple structure based on solid solution phases [1-3]. Besides the principal elements, HEAs can also contain minor elements with each below 5 at.%. These alloys are designated HEAs because their liquid or random solid solution states have significantly higher mixing entropies than those in conventional alloys [4]. Generally, when the number of alloying elements increases beyond five, the contribution of configurational entropy to the total free energy becomes sufficiently significant that it can overcome the enthalpies of compound formation and phase separation, thereby stabilizing the solid solution state relative to the multi-phase microstructure [1-3]. In practice, there is an opportunity for achieving a combination of high solid solution strengthening and good ductility if the solid solution phase possesses a simple crystal structure such as a face-centred cubic (fcc) lattice where there are a large number of slip systems [1-4]. In practice, HEAs are very attractive materials due to their potential beneficial mechanical, magnetic, and electrochemical characteristics, such as high strength, high thermal stability and oxidation resistance. Thus, these promising properties offer many potential applications in various fields, such as tools, molds and in magnetic films [5–7].

2. The significance of superplasticity in tensile testing

When metals are pulled in tension under a constant rate of straining, they generally break after only a limited amount of strain. But in some circumstances the material may pull out essentially uniformly and exhibit exceptionally high strains prior to failure. This process is known as superplasticity and it is the major feature of materials that are used for industrial superplastic forming where complex curved shapes are fabricated for use in a range of applications in the aerospace, automotive and commercial product sectors [8]. It is now well established that there are two essential requirements for attaining superplasticity in bulk metals [9]. First, the grain size of the material must be very small and typically <10 μm. Second, since superplastic flow is a diffusion-controlled process, it requires a temperature of the order of at least ~$0.5T_m$ where T_m is the absolute melting temperature. In the superplastic forming industry these very small grains are achieved through thermo-mechanical treatments which are generally capable of reducing the grain sizes to ~2-5 μm.

Over the last 25 years it has become clear that even smaller grains, typically within the submicrometer or even the nanometer range, may be attained by processing metals through the application of severe plastic deformation (SPD). The first demonstration of the potential for achieving superplasticity in these ultrafine-grained (UFG) materials was in 1988 [10] but subsequently the approach of processing through SPD has been adopted in many research institutes around the world and it is now recognized as a viable and useful procedure for achieving exceptional grain refinement [11-15]. Although there are several different SPD processing techniques, most attention to date has focussed on the two different procedures of equal-channel angular pressing (ECAP) [16] where a rod or bar is pressed through a special die containing a channel bent through a sharp angle and high-pressure torsion (HPT) [17] where a disk is subjected to a high applied pressure and concurrent torsional straining.

In order to provide a definitive criterion for the occurrence of superplastic flow, the advent of superplasticity is now defined as a measured tensile elongation of at least 400% [18]. In superplastic flow it has been shown that the strain arises from the occurrence of grain boundary sliding [19] and this sliding must be accommodated by the glide of intragranular dislocations that cross the grains, pile up at the opposite grain boundaries and then climb into the boundaries [20,21]. A theoretical model based on grain boundary sliding accommodated by intragranular glide and climb leads to a superplastic strain rate, $\dot{\varepsilon}_{sp}$, which is given by a relationship of the form [22]

$$\dot{\varepsilon}_{sp} = \frac{AD_{gb}G\mathbf{b}}{kT}\left(\frac{\mathbf{b}}{d}\right)^2\left(\frac{\sigma}{G}\right)^2 \quad (1)$$

where A is a dimensionless constant having a value of ~10, D_{gb} is the coefficient for grain boundary diffusion, G is the shear modulus, **b** is the Burgers vector, k is Boltzmann's constant, T is the absolute temperature, d is the grain size and σ is the applied stress. Equation (1) provides an excellent description of the superplastic flow of conventional metals with grain sizes of a few micrometers but recently it was shown by analyses that the equation applies equally well to bulk ultrafine-grained materials with submicrometer grain sizes produced by either ECAP or HPT [23-25]. Specifically, it was shown that there is very good agreement between published data and the predictions of equation (1) for both aluminum-based and magnesium-based alloys. Thus, it is interesting to determine whether the same agreement between theory and experiment applies also to superplastic flow in HEAs.

3. An examination of superplasticity in HEAs

Some limited results are now available documenting the occurrence of superplasticity in a number of HEAs. For convenience, these results are summarized in Table 1 where information is provided on the composition of the HEA, the processing procedure used to attain a superplastic grain size, the gauge dimensions of the tensile specimens, the temperature and strain rate of the tensile testing and the measured elongations to failure. Data are presented for AlCoCrCuFeNi processed by multiaxial foging [26-28], CoCrFeNiMn [29] and $CoCrFeNiMnTi_{0.1}$ [30] processed by HPT and CoCrFeNiMn processed by rolling [31].

Table 1. Experimental results for superplasticity in high-entropy alloys [26-31].

Composition	Processing	Gauge dimensions (mm^3)	Temperature (K)	Strain rate (s^{-1})	Elongation (%)	Reference
AlCoCrCuFeNi	Multiaxial forging	16 × 3 × 1.5	1073	1.0×10^{-3}	160	Shaysultanov *et al.* [26] and Stepanov *et al.* [27]
			1073	1.0×10^{-4}	325	
			1173	1.0×10^{-2}	350	
			1173	1.0×10^{-3}	585	
			1173	1.0×10^{-4}	490	
			1273	1.0×10^{-1}	600	
			1273	1.0×10^{-2}	1240	
			1273	1.0×10^{-3}	850	
AlCoCrCuFeNi	Multiaxial forging	16 × 3 × 1.5	1073	1.0×10^{-3}	604	Kuznetsov *et al.* [28]
			1173	1.0×10^{-3}	405	
			1273	1.0×10^{-1}	442	
			1273	1.0×10^{-2}	858	
			1273	1.0×10^{-3}	864	
			1273	1.0×10^{-4}	753	
CoCrFeNiMn	HPT	1.1 × 1.0 × 0.6	773	1.0×10^{-3}	160	Shahmir *et al.* [29]
			873	1.0×10^{-1}	330	
			873	1.0×10^{-2}	400	
			873	1.0×10^{-3}	520	
			973	1.0×10^{-1}	410	
			973	1.0×10^{-2}	500	
			973	1.0×10^{-3}	570	
			1073	1.0×10^{-1}	310	
			1073	1.0×10^{-2}	360	
			1073	1.0×10^{-3}	390	
$CoCrFeNiMnTi_{0.1}$	HPT	1.1 × 1.0 × 0.6	873	1.0×10^{-2}	460	Shahmir *et al.* [30]
			973	1.0×10^{-1}	630	
			973	1.0×10^{-2}	830	
			973	1.0×10^{-3}	650	
			1073	1.0×10^{-2}	570	
CoCrFeNiMn	Rolling	3 × 1 × 0.3	1023	1.0×10^{-1}	160	Reddy *et al.* [31]¶
			1023	1.0×10^{-3}	290	
			1023	1.0×10^{-4}	320	

¶Superplasticity is defined formally as a tensile elongation of at least 400% and therefore the results for this CoCrFeNiMn HEA are strictly outside of the range required for true superplastic flow. Nevertheless, the experimental results were interpreted by Reddy *et al.* [31] as evidence for "superplastic-like flow" with additional evidence that the ductility may be limited due to the occurrence of cavitation. Accordingly, based on this interpretation and in view of the rather limited results available to date for superplasticity in HEAs, these results are included in this tabulation and in the subsequent analysis.

IOP Conf. Series: Materials Science and Engineering **194** (2017) 012040 doi:10.1088/1757-899X/194/1/012040

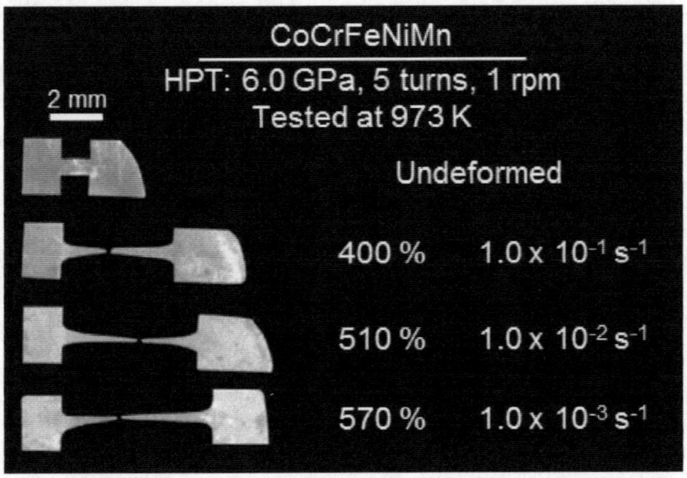

Figure 1. Examples of superplasticity in a CoCrFeNiMn HEA after processing by HPT and then testing in tension at a temperature of 973 K.

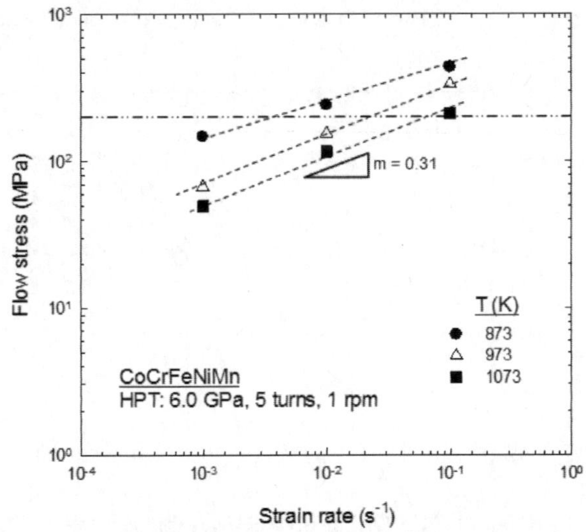

Figure 2. Flow stress versus strain rate for the superplastic CoCrFeNiMn HEA showing a strain rate sensitivity of $m \approx 0.31$ [29].

Inspection of Table 1 shows several excellent results for superplasticity in HEAs. It is necessary to also mention two important features of the data in Table 1. First, the results were obtained using different processing conditions including multiaxial forging [26-28] and HPT [29,30] which are SPD processing methods and rolling [31] which is not an SPD method. Second, noting that superplasticity requires an elongation of at least 400% [18], the results from rolling give elongations of up to only 320% which are not within the true superplastic range. Nevertheless, these results are included in Table 1 because the data were interpreted as indicative of superplastic-like flow but with the overall elongations limited by the development of cavitation.

Examples of true superplasticity are shown in Figure 1 where results are recorded for a CoCrFeNiMn HEA tested in tension at 973 K and exhibiting elongations up to 570% in samples where, as required for true superplasticity, there is no evidence for necking within the gauge length [32]. Figure 2 shows a plot of flow stress against initial strain rate for these samples where the results fall along reasonably straight lines for three different temperatures giving an average strain rate sensitivity of $m \approx 0.31$. This value of m is not consistent with equation (1) where m corresponds to the inverse of the stress exponent so that the anticipated strain rate sensitivity is $m = 0.5$. A value of $m \approx 0.3$ suggests control by an intragranular dislocation glide process [33] but samples deforming by dislocation glide are not capable of achieving elongations up to 500%. Accordingly, the data were interpreted in a different way by noting that an anomalously low strain rate sensitivity may be attained due to the occurrence of massive grain growth during tensile testing where the grains in the CoCrFeNiMn alloy grew from ~10 nm after HPT processing to ~1.0 μm after tensile testing at 973 K [29].

4. An analysis of the flow mechanism in superplastic HEAs

To evaluate the flow mechanism in these superplastic HEAs, equation (1) was re-arranged and experimental data from each set of results were plotted in the form of the temperature and grain size compensated strain rate against the normalized stress. The results are shown in Figure 3 with $D_{gb} = D_o \exp(-Q_{gb}/RT)$ where D_o is a frequency factor (19.4×10^{-4} m^2 s^{-1} for pure Ni [34]), Q_{gb} is the activation energy for grain boundary diffusion (113 kJ mol^{-1} for superplastic deformation of UFG HEA [29]), R is the gas constant, $\mathbf{b} = 2.55 \times 10^{-10}$ m [34] and $G = 85-\{16/[\exp(448/T)-1]\}$ GPa [35]. The plot was constructed taking grain sizes of $d = 2.1$ μm [26,27], 1.5 μm [28], 1.0 μm [29], 0.5-3.0 μm [30] and 1.4 μm [31]. The solid line labelled $\dot{\varepsilon}_{sp}$ shows the theoretically predicted rate for superplasticity occurring by grain boundary sliding. Thus, the experimental data for HEAs are in excellent agreement with the theoretical prediction and this is consistent with conventional alloys processed by SPD.

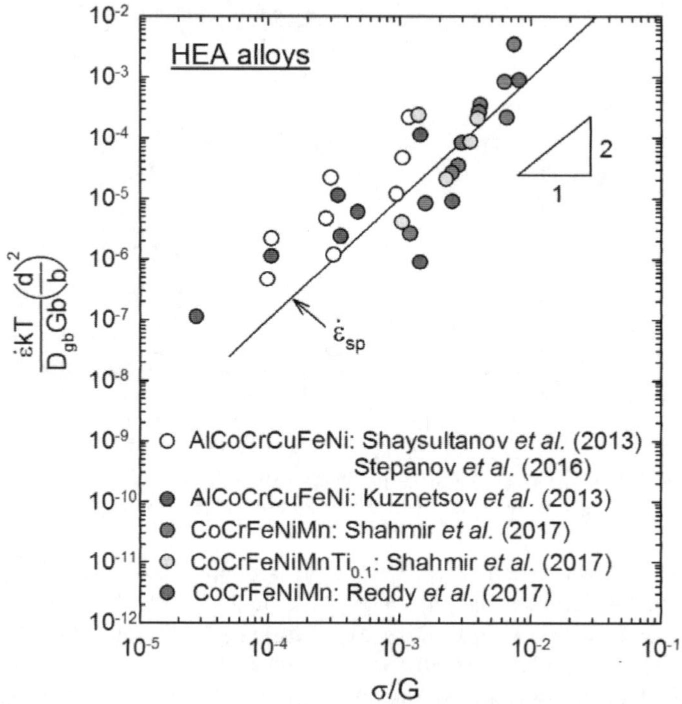

Figure 3. Temperature and grain size compensated strain rate versus normalized stress showing excellent agreement with the theoretical prediction for conventional superplasticity.

5. Summary and conclusions

1. Superplastic data are now available for several HEAs with tensile elongations up to >1000%.
2. An analysis of the experimental data shows good agreement between the measured strain rates and the predictions for conventional superplastic alloys and superplastic alloys processed by SPD.

Acknowledgements

This work was supported by the European Research Council under ERC Grant Agreement No. 267464-SPDMETALS. One of the authors acknowledges support from the NRF Korea funded by MoE under Grant No. NRF-2016R1A6A1A03013422 and by MSIP under Grant No. NRF-2016K1A4A3914691 (MK).

References

1. Yeh JW, Chen SK, Lin SJ, Gan JY, Chin TS, Shun TT, Tsau CH and Chang SY 2004 *Adv. Eng. Mater.* **6** 299.
2. Cantor B, Chang ITH, Knight P and Vincent AJB 2004 *Mater. Sci. Eng. A* **375-377** 213.
3. Yeh JW, Chen YL, Lin SJ and Chen SK 2007 *Mater. Sci. Forum* **560** 1.
4. Tsai MH and Yeh JW 2014 *Mater. Res. Lett.* **2** 107.
5. Huang PK, Yeh JW, Shun TT and Chen SK 2014 *Adv. Eng. Mater.* **6** 74.
6. Tong CJ, Chen YL, Chen SK, Yeh JW, Shun TT, Tsau CH, Lin SJ and Chang SY 2005 *Metall. Mater. Trans. A* **36A** 881.
7. Hsu US, Hung UD, Yeh JW, Chen SK, Huang YS and Yang CC 2007 *Mater. Sci. Eng. A* **460-461** 403.
8. Barnes AJ 2007 *J. Mater. Eng. Perform.* **16** 440.
9. Langdon TG 1982 *Metall. Trans. A* **13A** 689.
10. Valiev RZ, Kaibyshev OA, Kuznetsov RI, Musalimov RSh and Tsenev NK 1988 *Dokl. Akad. Nauk SSSR* **301** 864.
11. Valiev RZ, Islamgaliev RK and Alexandrov IV 2000 *Prog. Mater. Sci.* **45** 103.
12. Valiev RZ, Estrin Y, Horita Z, Langdon TG, Zehetbauer MJ and Zhu YT 2006 *JOM* **58(4)** 33.
13. Estrin Y and Vinogradov A 2013 *Acta Mater.* **61** 782.
14. Langdon TG 2013 *Acta Mater.* **61** 7035.
15. Valiev RZ, Estrin Y, Horita Z, Langdon TG, Zehetbauer MJ and Zhu YT 2016 *JOM* **68** 1216.
16. Valiev RZ and Langdon TG 2006 *Prog. Mater. Sci.* **51** 881.
17. Zhilyaev AP and Langdon TG 2008 *Prog. Mater. Sci.* **53** 893.
18. Langdon TG 2009 *J. Mater. Sci.* **44** 5998.
19. Langdon TG 1994 *Mater. Sci. Eng. A* **174** 225.
20. Falk LKL, Howell PR, Dunlop GL and Langdon TG 1986 *Acta Metall.* **34** 1203.
21. Valiev RZ and Langdon TG 1993 *Acta Metall. Mater.* **41** 949.
22. Langdon TG 1994 *Acta Metall. Mater* **42** 2437.
23. Kawasaki M and Langdon TG 2007 *J. Mater. Sci.* **42** 1782.
24. Kawasaki M and Langdon TG 2014 *J. Mater. Sci.* **49** 6487.
25. Kawasaki M and Langdon TG 2016 *J. Mater. Sci.* **51** 19.
26. Shaysultanov DG, Stepanov ND, Kuznetsov AV, Salishchev GA and Senkov ON 2013 *JOM* **65** 1815.
27. Stepanov ND, Shaysultanov DG, Salishchev GA and Senkov ON 2016 *Mater. Sci. Forum* **838-839** 302.
28. Kuznetsov AV, Shaysultanov DG, Stepanov ND, Salishchev GA and Senkov ON 2013 *Mater. Sci. Forum* **735** 146.
29. Shahmir H, He J, Lu Z, Kawasaki M and Langdon TG 2017 *Mater. Sci. Eng. A* **685** 342.
30. Shahmir H, Nili-Ahmadabadi M, Shafie A and Langdon TG 2017 To be submitted.
31. Reddy SR, Bapari S, Bhattacharjee PP and Chokshi AH 2017 *Mater. Res. Lett.*.Submitted.
32. Langdon TG 1982 *Metal Sci.* **16** 175.
33. Mohamed FA and Langdon TG 1974 *Acta Metall.* **22** 779.
34. He JY, Zhu C, Zhou DQ, Liu WH, Nieh TG and Lu ZP 2014 *Intermetallics* **55** 9.
35. Laplanche G, Gadaud P, Horst O, Otto F, Eggeler G and George EP 2015 *J. Alloys. Compds* **623** 348.

IOP Conf. Series: Materials Science and Engineering **194** (2017) 012041 doi:10.1088/1757-899X/194/1/012041

Superplasticity of nanostructured Ti-6Al-7Nb alloy with equiaxed and lamellar initial microstructures processed by High-Pressure Torsion

Jorge M. Cubero-Sesin [1,2,*], Joaquin E. Gonzalez-Hernandez[1], Elena Ulate-Kolitsky[1], Kaveh Edalati[3,4] and Zenji Horita[3,4]

[1]Escuela de Ciencia e Ingeniería de los Materiales, Instituto Tecnológico de Costa Rica, Cartago 159-7050, Costa Rica.
[2]Centro de Investigación en Ciencia e Ingeniería de los Materiales, Instituto Tecnológico de Costa Rica, Cartago 159-7050, Costa Rica.
[3]Department of Materials Science and Engineering, Kyushu University, Fukuoka 819-0395, Japan.
[4]WPI, International Institute of Carbon-Neutral Energy Research (WPI-I2CNER), Kyushu University, Fukuoka 819-0395, Japan.

E-mail: jcubero@itcr.ac.cr

Abstract. Microstructural modifications of a biomedical Ti-6Al-7Nb alloy were accomplished via heat treatment in 3 different quenching mediums and then processed by High-Pressure Torsion (HPT) at room temperature. The microstructure of the as-received condition is composed of an equiaxed duplex (α+β) structure. After the heat treatment, a combination of primary α phase and lamellar structures was obtained with an increasing fraction of the martensitic lamellar with increasing cooling rate. After HPT processing, refinement of the microstructures was observed for N=5 revolutions. Transmission electron microscopy (TEM) of the sample quenched in liquid nitrogen confirmed the nanostructure with grain sizes below 100 nm and high density of lattice defects after HPT processing for N=5 revolutions. High-temperature tensile tests were carried out at 800 °C with an initial strain rate of $2x10^{-3}$ s^{-1} on specimens with different combinations of heat treatment and HPT straining. The test in the as-received condition presented a maximum elongation to failure of ~400% after HPT processing for N=5 revolutions. The highest elongation to failure in the heat-treated samples was ~580% in the sample quenched in liquid nitrogen and processed for N=5 revolutions.

Keywords: *Titanium-aluminum-niobium, quenching, microstructure evolution, high-pressure torsion, superplasticity.*

1. Introduction

Titanium-based alloys have been used as a biomedical material for decades due to their excellent mechanical properties, corrosion resistance and high biocompatibility [1-3]. Ti-6Al-7Nb is a practical option for medical devices and implants because of its excellent biological compatibility and good biomechanical properties. Several research works have demonstrated further potential of this alloy in the field of bioscience, by experimenting with surface modifications in the nanoscale to improve the corrosion resistance and biocompatibility [4-8]. This suggests that the combination of grain refinement in the bulk with surface modification could improve in higher proportion both the mechanical properties and biocompatibility of this material for medical applications [9].

The microstructural evolution of duplex α+β type Ti-6Al-4V alloy has been deeply studied with different initial microstructures and methods for grain refinement [11-13]. Also, Ti-6Al-7Nb exhibited superplastic properties after HPT due to grain refinement [14,15]. It is well known that the regime for superplastic behaviour requires grain sizes < 10 μm and that this structure is stable at elevated temperatures. Also, it is necessary work with a high testing temperature, usually above ~0.5 T_m, where T_m is the absolute melting temperature of the material [14-16]. There are several conditions and other important variables during the test, such as, tensile specimen geometry, strain rate and initial microstructure of the alloy.

The objective of the current study is focused on the effects of the initial microstructure of a Ti-6Al-7Nb alloy processed by HPT over its superplasticity. The microstructure evolution due to the heat treatments and the grain refinement have been described. Finally, the deformation mechanism of the lamellae and the recrystallization of the structure has been proposed.

2. Experimental procedures

Discs of 10 mm diameter and ~1 mm thickness were cut from a commercial rod of Ti-6Al-7Nb using a wire-cutting electric discharge machine (EDM). Then, the discs were sealed in quartz tubes in an Ar atmosphere and treated at 960 °C for 2 h, followed by quenching in different mediums: (1) air at ~28°C, (2) ice water at ~5°C and (3) liquid nitrogen $N_{2(l)}$ at ~ -196°C. The specimens were further ground to a final thickness of ~0.8 mm, and then processed by HPT under a pressure of 6 GPa at room temperature, with a rotational speed of 1 rpm for N=1 and N=5 revolutions. After HPT processing, the discs were polished to a mirror-like surface and etched using Kroll's reagent for metallographic observations at several distances from the center of the discs. High-temperature tensile tests were performed after HPT processing from tensile specimens were extracted using EDM. The length and width of the gage portion of the specimens were 1.85 mm and 0.6 mm respectively, and the thickness was kept within 0.65±0.05 mm. The tensile testing was conducted at 800 ± 1 °C with an initial strain rate of $2.0 \times 10^{-3}\ s^{-1}$.

Microstructural studies were carried out firstly by X-ray diffraction (XRD) with the Cu-Kα (λ = 1.54016 Å) radiation in the range 2θ = 30°-80°, with a scan step of 0.01° and a scan speed of 0.06 s/step. The microstructure after HPT processing of the sample with highest elongation to failure was studied by transmission electron microscopy (TEM) with an acceleration voltage of 200 kV. For this purpose, a 3mm disc was extracted from the HPT-processed discs, and thinned to electron transparency by twin-jet electropolishing at 18°C, using a solution of 90% acetic acid and 10% perchloric acid under an applied voltage of 20 V.

3. Results and discussion

The microstructural observations before and after HPT processing are shown first. Similar metallographic observations and microhardness measurements were carried out for the as-received and the heat-treated conditions to understand the initial microhardness levels and the microstructure before HPT processing. Figure 1 shows optical micrographs of the initial microstructures, prior to HPT. Figure 1(a) shows the as-received condition, while Figure 1(b), 1(c) and 1(d) show the resulting structures after quenching from 960°C in air, ice water and liquid nitrogen N_2, respectively. The as-received microstructure was composed of an equiaxed α+β duplex microstructure, the areas in bright contrast correspond to α-phase grains, with β-phase particles along the grain boundaries shown in dark contrast. After heat-treating the alloy at 960°C with different cooling rates, a bimodal microstructure composed of an equiaxed α grain matrix surrounded by very fine lamellae was produced. At low cooling rates, these lamellae should be alternating α and β lamellae, and with increasing cooling rate the lamellae are replaced by martensitic α′ and metastable β structures, because the quenching is to the temperature above the martensitic transformation (Ms) temperature of ~910°C [2]. It can be seen from the comparison of Figure 1(b) and 1(c) that the volume fraction of lamellar phase increases with the increase of the cooling rate by quenching in ice water with respect to air cooling, but not much change occurs in the volume fraction by quenching in liquid nitrogen, as in Figure 1(d). However, it is expected that the phase fraction of α in the lamellae is replaced by α′ in a higher fraction by quenching in liquid nitrogen in comparison with cold water.

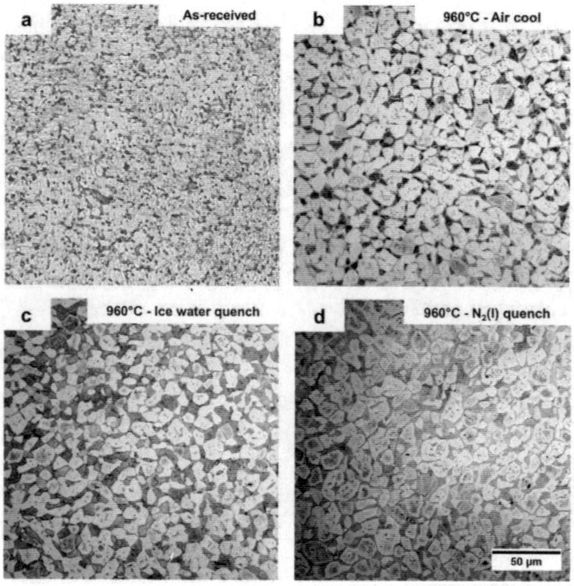

Figure 1. Optical micrographs of Ti-6Al-7Nb prior to HPT processing: (a) as-received condition. (b), (c) and (d) heat-treated samples quenched from 960°C in air, ice water and N_2, respectively.

Figure 2 shows the XRD profiles of the as-received and the heat-treated samples with their corresponding profiles before and after being processed by HPT for N=5 revolutions. The α/α′ and β phase reference diffraction positions are shown as dotted lines. The α′ phase has the same space group as the α phase, and shares the fundamental crystallographic diffraction peaks, so it is not possible to differentiate them by XRD. The α/α′ phases have an HCP crystal structure and their most prominent crystallographic planes are (100) at 2θ = 35.09°, (101) at 40.17° and (110) at 62.94°. The β phase has a BCC crystal structure with fundamental peaks: (110) at 2θ = 38.84° and (200) at 56.11°. These peaks are consistent with the result for the as-received condition. The XRD profiles of the samples quenched from 960 °C show a significant increase in the intensity of the (110) peak (2θ ~63°) which suggests that it corresponds to a preferred orientation in the lamellae, associated with the increase in the fraction of the (α+β) and (α′+β) at the expense of the α phase in the microstructure of these samples. The sample cooled in air has the lowest fraction of lamellar phase because the cooling rate was not enough to transform the whole β phase, and thus the (110) peak corresponding to the β phase is still visible, which is clearly appreciable in the profile of the as-received condition at 2θ~38.9°. The profiles from the samples quenched in ice water and liquid nitrogen, respectively, do not exhibit this peak, probably because the β phase in the lamellae is very fine in structure. On the other hand, the α phase peak at 2θ~38° is more prominent in these two profiles, corresponding to the faster cooling, than in the air-cooled sample, where it is hardly appreciable. This suggests α/α′ is dominant in the lamellar structure. In general, the XRD profiles from HPT processed specimens for N=5 revolutions present significant peak broadening, associated with the grain refinement in the material and the formation of large number of crystallographic defects such as vacancies and dislocations [17]. Additionally, due to the peak broadening there is a peak overlap between the α and β phases around 2θ~38-39° as a result of the high strain imposed by the HPT process. In addition, the preferred orientations defined by the initial heat treatments seem to have evolved to more random orientations. However, the significant peak broadening suggests that refinement of the microstructure is large.

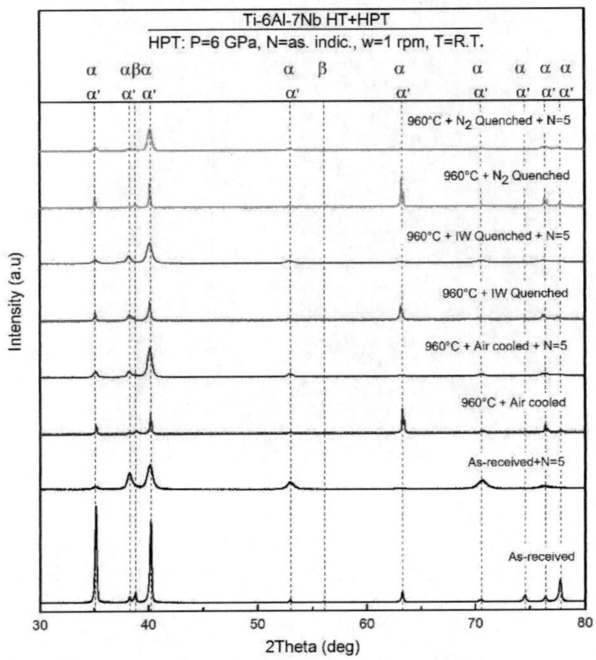

Figure 2. XRD profiles of Ti-6Al-7Nb for heat-treatments before and after HPT processing for N=5 revolutions.

Therefore, TEM observations were carried out to understand the extent of grain refinement in the HPT-processed samples. Figure 3 shows TEM images from the sample quenched in liquid nitrogen from 960 °C and processed by HPT for N= 5 revolutions. Figure 3(a) shows a bright-field image, where the presence of a nanostructured lamellae with thicknesses below 10 nm can be observed, along with fine crystallites with large amount of lattice distortion and dislocations at grain boundaries. Thus, in Figure 3(b) a dark-field image was taken from the diffracted beam pointed with an arrow in the inset SAED pattern, which corresponds to the (101) crystallographic plane, consistent with the highest intensity peaks at 2θ~40.2° in the XRD profiles after HPT processing, shown in Figure 2. From the dark-field image, the size of crystallites of α phase is below 100 nm. The SAED pattern in the inset corresponds to a ring pattern characteristic of a nanostructured material with a high fraction of high angle grain boundaries and lattice defects, commonly associated to metallic materials subjected to the HPT process [18].

Figure 4 show the stress-strain curves from the high temperature tensile tests. Figure 4(a) corresponds to the samples processed by HPT for N=1 and Figure 4(b) for N=5 revolutions. The comparison shows that HPT straining results in a higher elongation to failure of the samples. The total elongation to failure measured from the samples shown in Figure 4(a) was of ~220%, ~380%, ~345% and ~325% for the as-received, air, ice water and liquid N_2 quenched samples respectively. These results are all under the threshold value of 400% which is considered as superplastic behavior [16]. Also, there is an improvement in the elongation in terms of the bimodal microstructure achieved by quenching with respect to the as-received condition, but a slight decrease in the total elongation with respect to the increase in the cooling rate. On the other hand, the samples processed for N=5 revolutions exhibited higher elongations to failure as ~400%, ~455%, ~480% and ~580% for the as-received, air, ice water and liquid N_2 quenched samples, respectively. In this case, the trend is clear so that the total elongation increases with the increase in the cooling rate. These elongations are all considered in the superplastic regime, achieving the best performance in the sample quenched in liquid N2.

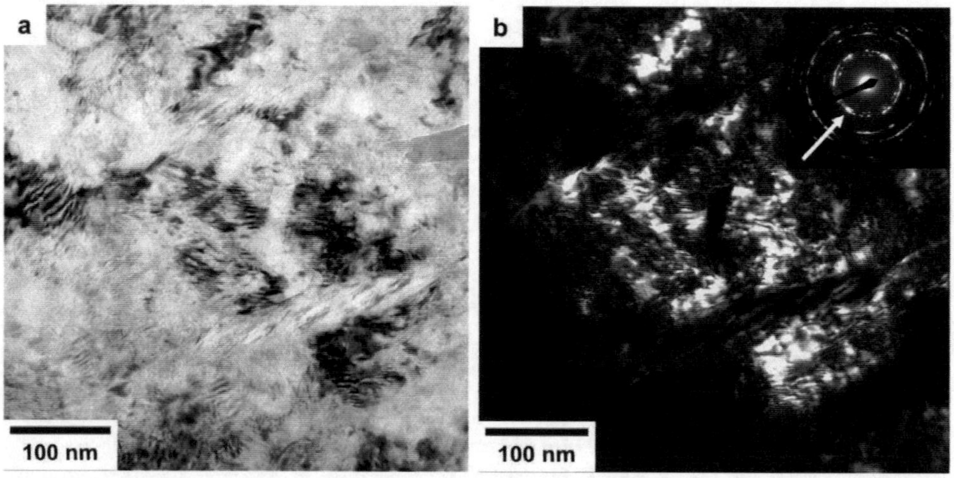

Figure 3. TEM (a) bight-field and (b) dark-field images of sample quenched from 960°C in liquid nitrogen N_2 and processed by HPT for N= 5 revolutions. Selected-area electron diffraction (SAED) pattern shown as inset in (b).

The results in Figure 4(b) show that quenching from 960 °C in liquid N_2 in combination with HPT processing for N=5 revolutions is the best condition for superplasticity. The reason for the high elongation to failure with respect to the as-received condition can be explained from the microstructural point of view in terms of the level of microstructure refinement of an already fine microstructure that forms as a result of quenching, which favors key mechanisms for superplasticity, such as grain boundary sliding [19]. The lamellar structures achieved by quenching are already in the ultrafine-grained level, as opposed to the equiaxed structures in the as-received condition, which after HPT are further refined to form a truly ultrafine-grained structure consisting of α phase grains below 100 nm and deformed lamella of (α+β) / (α′+ β) with spacings below 10 nm, as shown in Figure 3.

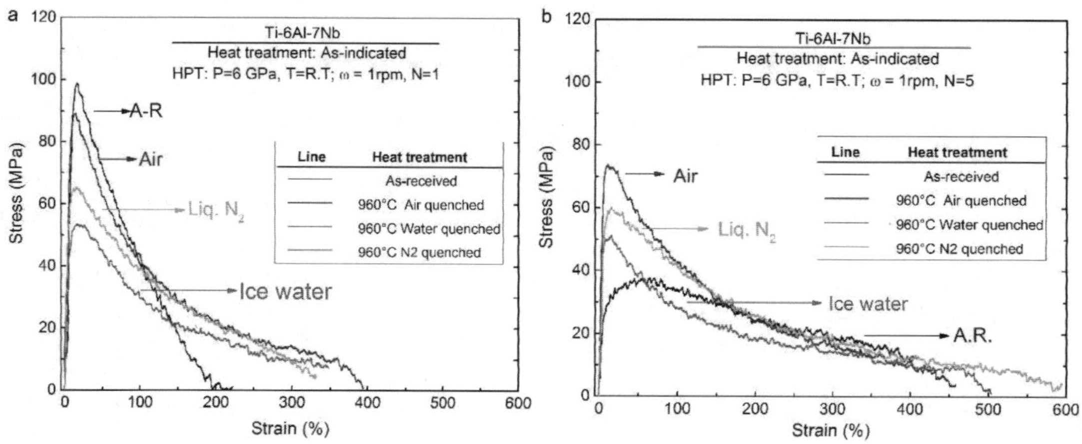

Figure 4. *Stress-strain curves from samples tested at 800°C processed by HPT in as-received condition and quenched from 960°C in different mediums for (a) N=1 and (b) N=5 revolutions*

Other factors that play a role in superplastic behaviour should be considered, which are not present in the as-received condition, such as the bimodal structure and the reverse phase transformation to the equilibrium structure, as effective mechanisms to accommodate the deformation during the tensile test. Further microstructural analyses of the failed tensile specimens will allow a better understanding of the microstructural behaviour during the high temperature tensile test, so it will be the focus of the future work.

4. Conclusions

Microstructural modifications were achieved on the Ti-6Al-7Nb alloy having bimodal structures by heat treatment and HPT processing to examine the effect of the initial microstructure on mechanical properties of interest. The following conclusions were obtained:

(1) Quenching from 960°C provides a bimodal structure: α'+β lamellae surrounding the α phase grains.

(2) Grain refinement occurred by application of HPT processing. The prior heat treatment giving a higher fraction of (α'+β) lamellae led to a smaller grain size after processing by HPT.

(3) TEM observations and SAED analyses revealed that the microstructure consists of nanograins with α grains and (α'+β) lamellae after quenching to liquid N_2 and HPT processing. Such an ultrafine-grained structure results in higher superplastic elongation than the as-received state consisting of an equiaxed duplex (α+β) structure.

References

[1] Leyens C and Peters M 2003 *Titanium and Titanium Alloys: Fundamentals and Applications,* Wiley–VCH Verlag, USA.
[2] Donachie M 2000 *Titanium: A Technical Guide.* ASM International, Ohio, 2nd ed.
[3] Lutjering G and Williams J 2007 *Titanium: Engineering materials and processes*. Springer Berlin Heidelberg, New York, 2nd ed.
[4] Akahori T, Mitsuo K, Niimomi M, and Suzuki A 2002 *Metal. Mater. Trans. A*. **33A** 503
[5] Huang H, Wu C, Sun Y and Lee T 2013 *Thin Solid Films* **528** 157
[6] Mohan L, Anandan C and Rajendran N 2015 *Mater. Sci. Eng. C.* **50** 394
[7] Huang H, Wu C, Sun Y, Yang W and Lee T 2014 *J. Alloys Comp.* **615** S648
[8] Rafieerad A, Bushroa A, Nasiri-Tabrizi B, Kabol S, Khanahmadi S, Amiri A, Vadivelu J, Yusof F, Basiruni W and Wasa K 2017 *J. Mech. Behav. Biomed. Mater.* **69** 1
[9] Oliveira D, Prokofiev E, Sanches L, Polyakova V, Valiev R, Botta W, Junior A and Bolfarini C 2015 *J. Alloys Comp.* **643** S241.
[10] Murray J 1987 *ASM Handbook: Nb (Niobium) Binary Alloy Phase Diagrams.* **3** 2304
[11] Matsumoto H, Nishihara T, Yohei Iwagaki Y, Tohru Shiraishi T, Ono Y and Chiba A 2016 *Mater. Sci. Eng. A*. **661** 68
[12] Nakahigashi J and Hirofumi Y 2002 *Materials Trans*. **43** No.11 2768
[13] Kral P, Dvorak J, Blum W, Kudryavtsev E, Zherebtsov S, Salishchev G, Kvapilova M and Sklenicka V 2016 *Mater. Characterization.* **116** 84
[14] Cui W, Ji Z, Guo A and Zhou L 2009 *Mater. Sci. Eng.* A. **499** (1-2) 252.
[15] Ashida M, Chen P, Doi H, Tsutsumi Y, Hanawa T and Horita Z 2015 *Mater. Sci. Eng. A.* **640** 449
[16] Kawasaki M, Figueiredo R, Xu C and Langdon T 2007 JOM *Metal. Mater. Trans. A* **38A** 1891
[17] Estrin Y and Vinogradov A 2013 *Acta Mater.* **61** 782
[18] Valiev R, Islamgaliev R and Alexandrov I 2000 *Prog. Mater Sci.* **45** 103
[19] Zehetbauer M and Zhu Y 2009. *Bulk Nanostructured Materials*. WILEY-VCH Verlag GmbH & Co. KGaA, Weinheim

IOP Conf. Series: Materials Science and Engineering **194** (2017) 012042 doi:10.1088/1757-899X/194/1/012042

Effect of equal-channel angular pressing on the mechanical behavior of a Bi-Sn eutectic alloy

Fariba Naghdi[1], Roberto B Figueiredo[2,3], Terence G Langdon[1,3]

[1] Materials Research Group, Faculty of Engineering and the Environment, University of Southampton, Southampton SO17 1BJ, UK

[2] Department of Materials Engineering and Civil Construction, Universidade Federal de Minas Gerais, Belo Horizonte, MG 31270-901, Brazil

[3] Departments of Aerospace & Mechanical Engineering and Materials Science. University of Southern California, Los Angeles, CA 90089-1453, USA

E-mail: fnaghdi@ut.ac.ir, figueiredo@demc.ufmg.br, langdon@soton.ac.uk

Abstract. Equal-channel angular pressing (ECAP) was utilized to process a Bi–42Sn eutectic alloy at room temperature using processing route A for 1, 2, 4, and 8 passes. Tensile testing was performed at room temperature under initial strain rates in the range of 1.0×10^{-5} to 1.0×10^{-2} s^{-1}. The results demonstrate that processing by ECAP improves the ductility in this material and the elongations to failure increased with decreasing strain rate in all samples. The largest elongation to failure, ~1300%, was recorded in the sample processed by ECAP for 8 passes at an initial strain rate of 1.0×10^{-5} s^{-1}. This elongation is much improved by comparison with the as-cast Bi–42Sn alloy.

1. Introduction

The ability of a polycrystalline material to exhibit, in a generally isotropic manner, very high elongations prior to failure is known as superplasticity. The measured elongations in superplasticity are generally at least 400% and the measured strain rate sensitivities are close to ~0.5. Superplasticity is a diffusion-controlled process and therefore it requires a relatively high testing temperature, typically at or above $0.5T_m$ where T_m is the absolute melting temperature of the material [1]. In addition to high testing temperatures, the materials should have a small and stable grain size, typically smaller than ~10 μm, to exhibit superplastic flow [2]. Superplastic behaviour is usually observed either in two-phase eutectic/eutectoid materials or materials containing finely dispersed second phase particles to inhibit grain growth at high temperatures [1]. The reason is that it is hard for both of the two necessary conditions of superplasticity to be satisfied simultaneously in pure metals and solid solution alloys since grain growth occurs at high temperatures.

Materials processed by severe plastic deformation (SPD) methods are also capable of exhibiting high elongations to failure, and in some cases superplastic behaviour, when tested at high temperatures, typically more than $0.5T_m$. Large amounts of strain may be imposed to the working material using SPD techniques since the imposed strain makes no significant change in the overall dimensions of the material. Therefore, exceptional grain refinements may be achievable in metals through SPD methods. Equal-channel angular pressing (ECAP) is one of the well-known SPD methods which uses relatively large bulk samples and has other advantages such as simplicity in operation. For this reason it is generally considered superior to most other SPD methods [3,4]. Studying the superplastic behaviour in materials processed by SPD methods, including ECAP, has been the subject of much research and it has been established that ECAP-processed materials can exhibit excellent superplasticity. For example, an elongation of 3050% was achieved in a magnesium ZK60 alloy after processing for 2 passes of ECAP and then testing in tension at 473 K using a strain rate of 1.0×10^{-4} s^{-1} [5]. Similarly, a superplastic elongation of 3060% was achieved in a Pb-62% Sn eutectic alloy processed by ECAP for 16 passes under a strain rate of 1.0×10^{-3} s^{-1} at 413 K [6].

True superplastic behaviour, with an elongation to failure of more than 400%, was reported for the first time in an extruded Bi-Sn eutectic alloy after ageing for 7 days [7]. This material showed an elongation to failure of ~1950% when tested under a constant stress of ~1.7 MPa. However, there are very few investigations of the effect of SPD processing on the superplastic behaviour of this material. In the only study conducted to date, the Bi-Sn eutectic alloy was processed by high-pressure torsion (HPT) and the ductility of the material was significantly improved in tensile testing at room temperature. A superplastic elongation of ~1200% was achieved after processing by HPT through 10 turns and pulling to failure at an initial strain rate of 1.0×10^{-4} s^{-1} [8]. Nevertheless, the effect of processing by ECAP on the eutectic Bi-Sn alloy has not been studied. Therefore, the aim of this research was to investigate the superplastic behaviour of this material after ECAP processing.

2. Experimental material and procedures

The alloy was cast Bi-42% Sn from which rods with 60 mm lengths and cross-sectional dimensions of 10×10 mm^2 were machined for ECAP processing. This alloy has a low melting temperature of 411 K and experiences self-annealing when held at room temperature of 298 K. Therefore, no heat treatment was performed on the alloy before ECAP processing. The ECAP was conducted at room temperature using a die with an internal channel angle, φ, of 135°. The pressing was conducted using route A in which the sample is not rotated between consecutive passes [9]. The pressing rate was 1 $mm.s^{-1}$ and the ECAP was conducted for 1, 2, 4 and 8 passes. The processed samples were aged at room temperature for 40 days.

Two tensile specimens were cut from the longitudinal sections of the billets parallel to the pressing direction using electro-discharge machining with gauge dimensions of $4.0 \times 3.0 \times 2.0$ mm^3. Tensile tests were performed under constant cross-head speeds with initial strain rates from 1.0×10^{-5} to 1.0×10^{-2} s^{-1} at room temperature. The stress-strain curve was plotted for each specimen and the ultimate tensile strength was measured from the curve. The elongations to failure were estimated carefully by measuring the gauge lengths before and after tensile testing.

3. Results and discussion

Representative plots of engineering stress against elongation to failure are demonstrated in Fig. 1 for the samples processed by ECAP for 8 passes, aged at room temperature for 40 days and pulled to failure at room temperature under different initial strain rates. The values of flow stress and elongation to failure for the samples processed by ECAP through 1, 2, 4 and 8 passes followed by aging at room temperature for 40 days were derived from such plots and the results are presented in Table 1. The corresponding data for the as-cast condition is also included. It is seen in Table 1 that processing by ECAP leads to a decrease in the flow stress of the as-cast material and by decreasing the strain rate the flow stress decreases implying that this material has positive strain rate sensitivity, *m*. Furthermore, processing by ECAP leads to an increase in the elongation to failure and samples processed to higher numbers of ECAP passes show higher elongations to failure. The total elongation to failure increases with decreasing imposed stain rate.

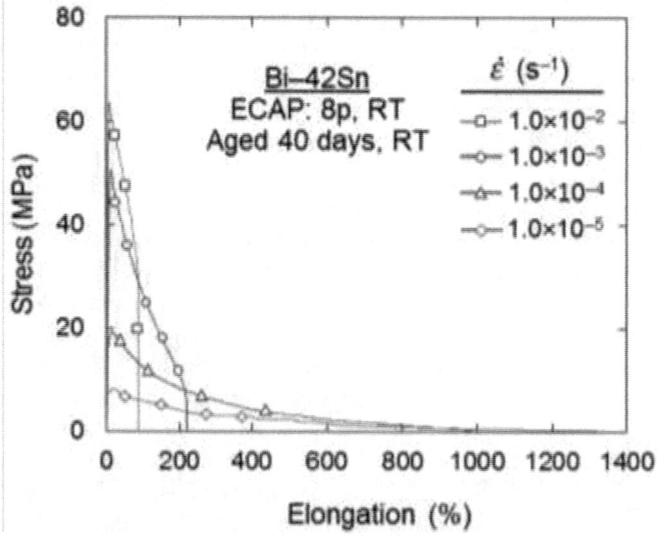

Fig. 1. Engineering stress-elongation curves of the Bi-42Sn alloy processed by ECAP for 8 passes at different initial strain rates in the range of 1.0×10^{-5} to 1.0×10^{-2} s^{-1}.

Table 1. Mechanical properties of the Bi-42Sn alloy in the as-cast condition and after ECAP processing for different numbers of passes

Condition	Strain rate (s^{-1})	Flow stress (MPa)	Elongation to failure (%)
As-cast	1.0×10^{-2}	81	32
	1.0×10^{-3}	71	53
	1.0×10^{-4}	54	92
	1.0×10^{-5}	38	112
1p	1.0×10^{-2}	83	45
	1.0×10^{-3}	65	112
	1.0×10^{-4}	40	150
	1.0×10^{-5}	24	175
2p	1.0×10^{-2}	79	51
	1.0×10^{-3}	61	143
	1.0×10^{-4}	36	172
	1.0×10^{-5}	16	245
4p	1.0×10^{-2}	78	80
	1.0×10^{-3}	55	172
	1.0×10^{-4}	24	389
	1.0×10^{-5}	15	596
8p	1.0×10^{-2}	72	88
	1.0×10^{-3}	57	225
	1.0×10^{-4}	25	988
	1.0×10^{-5}	14	1325

IOP Conf. Series: Materials Science and Engineering **194** (2017) 012042 doi:10.1088/1757-899X/194/1/012042

Figure 2 shows the variations of the elongation to failure versus the imposed initial strain rate for samples processed for different ECAP pass numbers. This plot more effectively demonstrates the effects of number of ECAP passes and initial strain rate on the elongation to failure of different samples. As mentioned earlier, the Bi-42Sn eutectic was the first alloy in which real superplasticity with an exceptional elongation to failure of ~1950% was recorded after extrusion and ageing at room temperature for 7 days [7]. With such a background, it is expected that by grain refinement through SPD techniques higher superplastic elongations may be achievable in this alloy. It is apparent from Fig. 2 that Bi-42Sn has a very low elongation to failure in the as-cast condition which is not strongly dependent on the initial strain rate. Under a strain rate of 10^{-2} s^{-1}, the as-cast Bi-42Sn is very brittle and has an elongation to failure of 32%. Under the same strain rate, the samples processed for 1, 2, 4 and 8 passes of ECAP possess elongations of 45, 51, 80, and 88%, respectively. A significant improvement in the elongation to failure is observed after ECAP for 4 passes, especially at lower initial strain rates. An elongation to failure of ~600% was recorded at the lowest initial strain rate of 1.0×10^{-5} s^{-1}. The elongation to failure in the sample processed for 8 passes is more strain rate sensitive and under the lowest strain rate of 1.0×10^{-5} s^{-1} a total elongation to failure of ~1300% was achieved, where this is a considerable improvement compared to the as-cast material. The as-cast Bi-42Sn eutectic alloy has a typical lamellar structure containing coarse lamellae of Bi and Sn which are broken down during ECAP processing leading to an improved elongation to failure in the processed samples especially after 4 passes of ECAP.

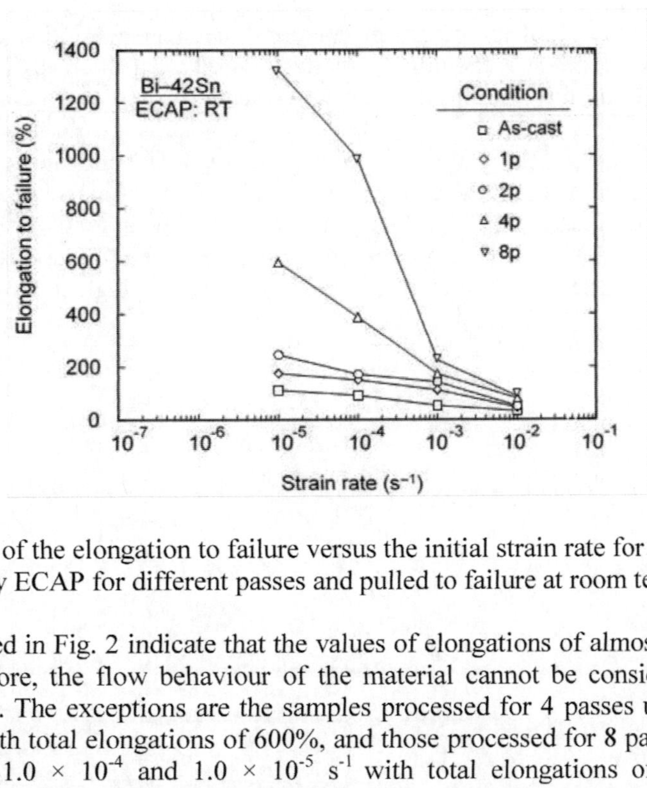

Fig.2. Variation of the elongation to failure versus the initial strain rate for the Bi-42Sn alloy processed by ECAP for different passes and pulled to failure at room temperature

The results presented in Fig. 2 indicate that the values of elongations of almost all samples are less than 400% and therefore, the flow behaviour of the material cannot be considered as superplastic under these conditions. The exceptions are the samples processed for 4 passes under an initial strain rate of 1.0×10^{-5} s^{-1} with total elongations of 600%, and those processed for 8 passes and tested under initial strain rates of 1.0×10^{-4} and 1.0×10^{-5} s^{-1} with total elongations of 990% and 1325%, respectively. These values confirm the occurrence of superplasticity in these conditions. The appearance of the specimens of the Bi-42Sn alloy processed for 8 passes after testing to failure at room temperature with different initial strain rates are demonstrated in Fig. 3. Close inspection shows that the superplastic samples pull out in quite a uniform manner with no evidence for the formation of any necking within the gauge length which is an important additional requirement of true superplastic flow [10].

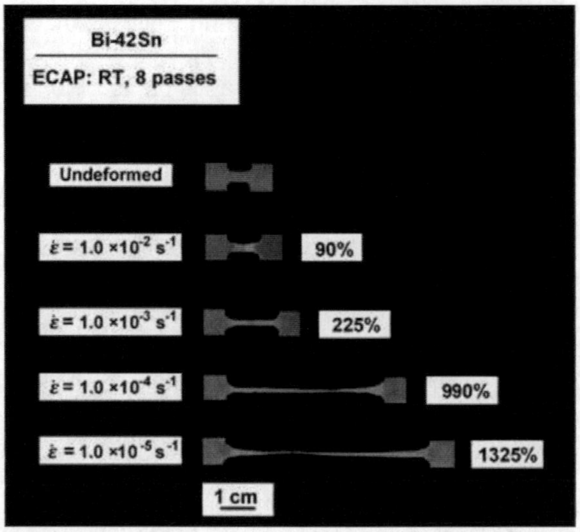

Fig. 3. Appearance of specimens of Bi-42Sn alloy, processed by ECAP for 8 passes, after testing to failure at room temperature with different initial strain rates.

For a better understanding of the dominant operating mechanisms during deformation of the Bi-42Sn alloy after the ECAP process, the flow stress was plotted logarithmically against strain rate using the data from Table 1, and the resulting diagram is presented in Fig. 4. In conventional superplastic alloys, three distinct regions of flow are observed in the flow stress-strain rate curve; over a range of intermediate strain rates high elongations to failure are recorded and the strain rate sensitivity is typically ~0.5. This is region II. At low strain rates in region I and at high strain rates in region III the strain rate sensitivity decreases [11]. However, the experimental range of strain rates in the present study was not sufficient to reveal the three regions of flow generally associated with conventional superplastic alloys where the measured elongations decrease at both high and low strain rates.

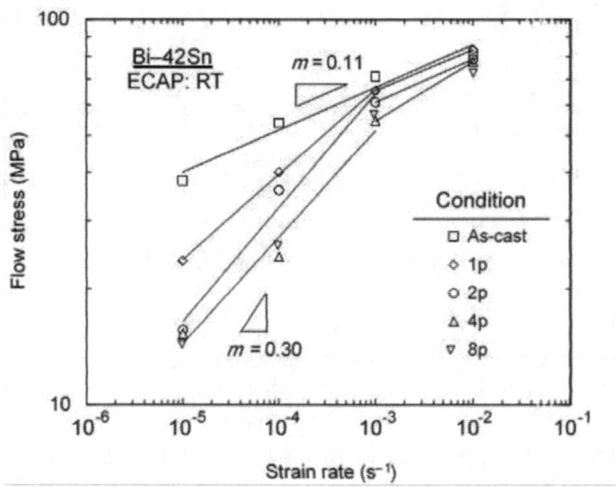

Fig. 4. Flow stress versus strain rate for the Bi-42Sn alloy processed by ECAP. Values of the strain rate sensitivity, *m*, are shown for the as-cast material and the alloy processed for 8 passes.

Values of the strain rate sensitivity, *m*, for the tested samples were derived from the data reported in Fig. 4. The measured *m*-value for the as-cast material was ~0.11 over three orders of magnitude of strain rate. As expected, no superplastic behaviour was observed in this condition which is in agreement with results of tensile tests. The same is true for the samples processed through 1 and 2 passes of ECAP. The variations of flow stress against strain rate in these samples tend to show regions II and III of a typical curve of a conventional superplastic alloy. However, the *m*-value in region II was smaller with values of ~0.22 and ~0.29 for the samples processed by ECAP for 1 and 2 passes, respectively. It is apparent that the values of strain rate sensitivity do not tend to increase significantly by increasing the numbers of passes and the samples processed for 4 and 8 passes have *m*-values of ~0.28 and ~0.3, respectively. These *m*-values are lower than the usual value of $m = 0.5$ associated with superplastic flow [12]. A value of $m = 0.3$ is observed where dislocation glide is the rate-controlling process [13]. However, dislocation glide does not give total elongations of more than 400%. Therefore, the maximum total elongation of ~1300% cannot be associated with dislocation glide. Microstructural observations in an HPT-processed Bi-42Sn alloy showing an elongation to failure of 1220% at an initial strain rate of 1.0×10^{-4} s^{-1} revealed that Bi-Sn and Sn-Sn boundaries have excellent grain boundary sliding behaviour whereas cavities are formed on Bi-Bi boundaries. Therefore, the enhanced superplastic flow behaviour of the fine-grained Bi-Sn alloy was attributed to the homogeneously distributed Bi-Sn and Sn-Sn boundaries after HPT processing [14]. In the present study, it is likely that grain boundary sliding is the dominant operating mechanism and the observed *m*-value of ~0.3 is probably due to the occurrence of grain growth during tensile testing which may lead to an apparent strain rate sensitivity of $m \approx 0.3$ in superplastic alloys [15].

4. Conclusions

ECAP was found to be a very effective way to achieve high elongations in the Bi-42Sn eutectic alloy. A sample processed for 8 passes showed an elongation to failure exceeding 1300% when testing at an initial strain rate of 1.0×10^{-5} s^{-1} and this is a considerable enhancement compared to the coarse-grained as-cast sample that only exhibited an elongation to failure of 112% under the same conditions. By producing more refined microstructure, for example by processing through larger numbers of passes, higher elongations to failure are expected in the Bi-42Sn alloy. The calculated value for strain rate sensitivity for the samples processed for 8 passes was ~0.3 over two orders of magnitude of strain rate. This implies that grain growth may operate during the deformation together with grain boundary sliding.

Acknowledgement

This work was supported by the European Research Council under Grant Agreement No. 267464-SPDMETALS.

References

[1] Langdon TG 2009 *J. Mater. Sci.* **44** 5998.
[2] Langdon TG 1982 *Metall. Trans. A* **13** 689.
[3] Langdon TG 2013 *Acta Mater.* **61** 7035.
[4] Valiev RZ and Langdon TG 2006 *Prog. Mater. Sci.* **51** 881.
[5] Figueiredo RB and Langdon TG 2008 *Adv. Eng. Mater.* **10** 37.
[6] Kawasaki M, Lee S and Langdon TG 2009 *Scripta. Mater.* **61** 963.
[7] Pearson CE 1934 *J. Inst. Metals* **54** 111.
[8] Wang CT and Langdon TG 2014 *IOP. Conf. Series: Mater. Sci. Eng.* **63** 012107.
[9] Furukawa M, Iwahashi Y, Horita Z, Nemoto M and Langdon TG 1998 *Mater. Sci. Eng. A* **257** 328.
[10] Langdon TG 1982 *Metal Sci.* **16** 175.
[11] Ishikawa H, Mohamed FA and Langdon TG 1975 *Phil. Mag.* **32** 1269.
[12] Langdon TG 1994 *Acta Metall. Mater.* **42** 2437.
[13] Mohamed FA and Langdon TG 1974 *Acta Metall.* **22** 779.
[14] Wang CT and Langdon TG 2015 *Lett. Mater.* **5** 301.
[15] Shahmir H, He J, Lu Z, Kawasaki M and Langdon TG 2017 *Mater. Sci. Eng. A* **685** 342.

IOP Conf. Series: Materials Science and Engineering **194** (2017) 012043 doi:10.1088/1757-899X/194/1/012043

The effect of Cu-based core-sheath configurations on the processing of Nd-Fe-B-based permanent magnets via equal-channel angular pressing

L Besley[1], H Zhang[1], A Molotnikov[1], H Kishimoto[2], A Kato[2], C Davies[1], K Suzuki[1]

[1] Department of Materials Science and Engineering, Monash University, Melbourne, Australia
[2] Toyota Motor Corporation, Mishuku, Susono, Shizuoka 410-1193, Japan

E-mail: luke.besley@monash.edu

Abstract. Equal channel angular pressing (ECAP) has been used as an alternative manufacturing route for preparation of $Nd_2Fe_{14}B$-based anisotropic magnets, facilitating processing temperatures much lower than conventional die upsetting. While this method can produce a suitable texture and microstructure in permanent magnetic materials, it still remains novel; involving extremely high pressures which present a high risk of both process failure and die and tooling damage. Powder metallurgical processes frequently incorporate an external layer of secondary material (commonly an outer foil layer or can) for separation between the primary material and die as well as the control of surface effects such as friction through appropriate choice of secondary material. This work implements such modifications to this manufacturing route by incorporation of an outer layer of Cu foil, the addition of which negatively affected both the powder compaction and strength of texture produced via ECAP. Also investigated was the incorporation of solid Cu bar as part of the sample cross section. This modification facilitated processing without any compromise on observed magnetic properties, whilst also reducing damage to both the die and tooling. This type of methodology may aid in improving the reliability of producing bulk anisotropic permanent magnets via ECAP.

1. Introduction

Equal channel angular pressing (ECAP) is a thermo-mechanical processing method capable of developing suitable texture in $Nd_2Fe_{14}B$-based permanent magnets at temperatures low enough to avoid grain coarsening [1]. However, extremely high pressures during pressing [2] of permanent magnet materials, which are highly brittle [3][4], frequently lead to failure of the process and tooling damage. Such a risk of failure presents an obstacle to the upscaling and reliability of this manufacturing process.

It is common practice for powder metallurgical processes to incorporate a secondary material to isolate the primary material from the die and control properties arising from material-die contact such as friction [5]. With a view towards reducing pressure during ECAP and increasing the chance of a successful procedure, a suitably low friction material, Cu, was chosen for incorporation into the ECAP process. Two arrangements of Cu incorporation were investigated: a thin outer layer of foil acting as a sheath insulating the powder from the die wall, and a solid Cu bar at the leading end of the sample.

IOP Conf. Series: Materials Science and Engineering **194** (2017) 012043 doi:10.1088/1757-899X/194/1/012043

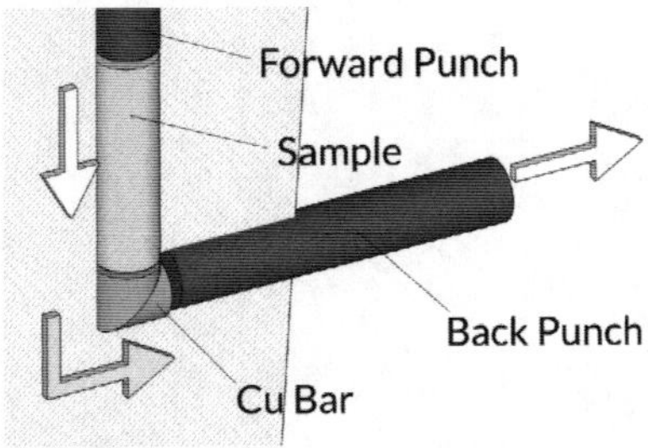

Figure 1. Schematic of ECAP with an incorporated Cu Bar.

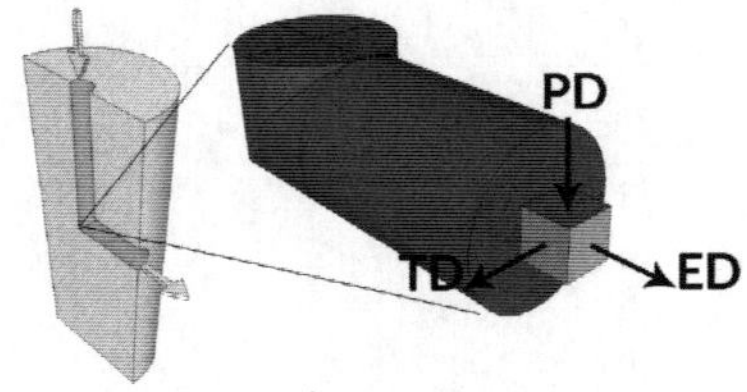

Figure 2. The sample after ECAP, showing three principle directions (PD,TD, and ED).

2. Methods

Commercial melt-spun $Nd_{13.5}Fe_{73.8}Co_{6.7}B_{5.6}Ga_{0.4}$ powders (Magnequench MQU-F1, referred to herein as 'F1') were placed into an ECAP die with channel angle of 90° and channel diameter of 10mm. The sample was then processed via ECAP at a temperature (T) equal to $500°C$ with a backing pressure (σ) approximately equal to 0.5 GPa and a pressing speed of approximately 3 mm/s. This process was then modified by first placing a sheath of Cu foil with a thickness of approximately 76 μm into the die channel and then placing F1 powders into this foil. The foil was folded at each end to prevent powder escaping, and processed via ECAP under the same conditions.

Figure 1 illustrates the experimental setup for the ECAP process modified by incorporating a Cu bar. A solid cylindrical Cu bar with a diameter slightly under 10 mm was first placed into an ECAP die and deformed at room temperature to fill the channel. F1 powder was then inserted above this bar. The top of the bar remains above the channel corner, meaning all powder is confined entirely in the pressing channel. The sample was then processed via ECAP under the same conditions.

Figure 2 shows schematically the sample after ECAP and three principle manufacturing directions, the pressing direction (PD), exit direction (ED) and transverse direction (TD). Following ECAP, cubes with a length of 3 mm were cut from the center of the sample for magnetic measurements via vibrating sample magnetometer along the PD, TD and ED directions. Prior to each directional measurement, the sample was magnetised in a pulsed field of $\mu_0 H = 7T$ using a pulse magnetiser. Disc-shaped samples were cut in the plane normal to ED for texture analysis on a GBC X-ray diffractometer. Finally, samples were cut along the plane normal to TD for optical microscopy.

In the addition to the experimental work, modelling of ECAP process utilising a finite element package QForm VX [6] was carried out. This package is designed to simulate metal forming processes and uses an automated re-meshing procedure which is advantageous in the case of severe plastic deformation processes. The set-up of the simulation included the drawing of the geometry of the ECAP die, forward and backward punches and the workpieces. The values for the pressing speed, temperature, back-punch pressure were adopted from the experiments. In order to simulate the material behaviour of the F1 alloy and Cu, experimental stress-strain curves of both materials were imported into the database of QForm VX as flow stress curves, where the stress is represented as a function of strain, strain rate and temperature. This information was compiled from other published data describing the mechanical behaviour of these materials [7]. The interaction between two workpieces was modelled using high value of friction factor $\mu = 1$.

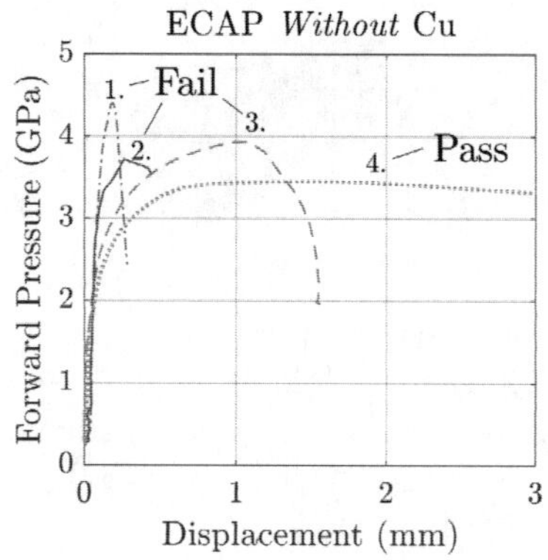

Figure 3. Forward pressure vs. displacement without Cu additions. 1, 2, 3=Failure. 4=Pass.

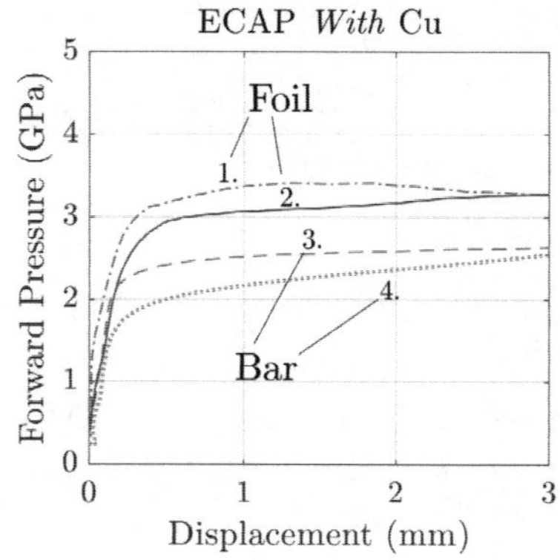

Figure 4. Forward pressure vs. displacement with Cu additions. 1, 2=Cu foil. 3, 4=Cu bar.

3. Results

3.1. Pressure output during ECAP

Figure 3 shows the relationship between the pressure applied to the forward punch (forward pressure) and sample displacement during ECAP for trials without Cu. Trials marked 1, 2, and 3 are trials where the ECAP process failed. In these trials, the forward pressure increase rapidly with little subsequent change in sample displacement. The initial static friction is not overcome, and the sample does not begin to flow. At a critical forward pressure of approximately 4 GPa, punch failure occurs. Trial 4 shows the forward pressure behaviour of a successful passing without any Cu. This successful Cu-free trial was selected for further testing, and forms the basis for measurements in Figs. 6, and 7.

Figure 4 shows forward pressure vs. displacement after addition involving Cu. Trials 1 and 2 included Cu foil, while trials 3 and 4 involved Cu bar. With the addition of Cu foil, the pressure vs. displacement curves behave similarly to trial 4 in Fig. 3 (a successful pass without Cu) where the build up of forward pressure reaches approximately 3 GPa before flow can begin. With the addition of a Cu bar, the build up of forward pressure before the sample overcomes the initial static friction and begins to pass is significantly reduced, allowing deformation to proceed at a forward pressure of around 2 to 2.5 GPa. Trials 1 and 4 from Fig. 4 were chosen for further testing.

3.2. Simulation of deformation behaviour

Figure 5 displays a comparison of the shape of the extruded sample after one passing of ECAP with the predicted simulation. It can be seen that a good agreement between the simulation and the experiment is obtained. The result obtained by finite element simulation indicates that the equivalent strain in the central region of the sample is close to the analytically predicted value of 1.15 [8].

3.3. Magnetic measurements

Figure 6 shows demagnetisation curves taken in manufacturing directions PD, TD and ED (which are defined in Fig. 2). The highest remanent magnetisation value (M_r, i.e. $M(H = 0)$) in the pressing direction suggests a high degree of alignment of the magnetically easy (001) axes parallel to PD; a sign of developed texture [9]. This is the case for F1 processed both without

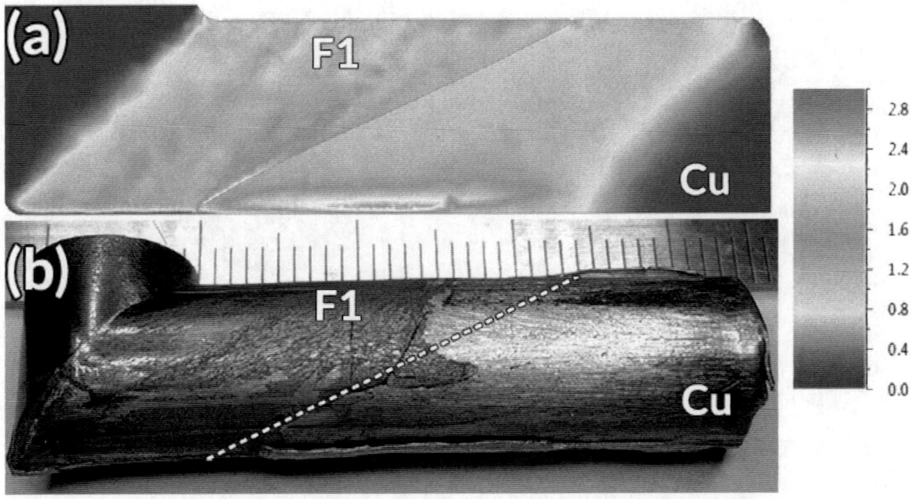

Figure 5. Comparison of the shape of the extruded sample after one pass of ECAP: (a) the distribution of equivalent strain as predicted by simulation and (b) photo of the sample.

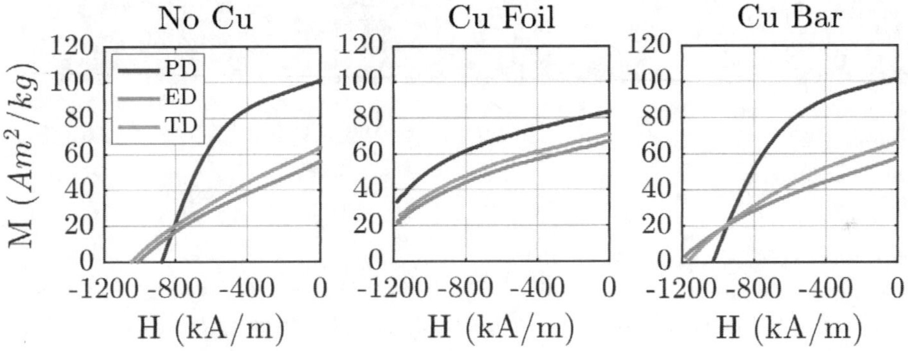

Figure 6. Demagnetisation curves taken parallel to PD, TD and ED (illustrated in Fig. 2) for ECAP with no Cu, Cu foil, and Cu bar respectively.

Cu and with a Cu bar, which show similar M_r values of 100.9 and 101.4 emu/g respectively along PD. F1 processed with Cu foil however shows similar demagnetisation behaviour along all axes, with a significantly decreased M_r along PD. This suggests that the incorporation of the outer Cu foil layer diminishes the degree of texture produced via ECAP.

3.4. Texture analysis

Figures 7 and 8 are (001) pole figures taken from the plane normal to the exit direction (ED) after ECAP for F1 processed without Cu and with Cu bar respectively. The center of each pole figure is parallel to ED, with each ring marking a 5° increase in altitude away from ED. The peak m.r.d (multiple of random distribution) intensity is observed at an altitude of between 50-70° away from ED for both trials. The azimuthal range and maximum m.r.d value of approximately 4.5 remain consistent between F1 without Cu and F1 with Cu bar. This similarity is also consistent with magnetic measurements in Fig. 6, further supporting the notion that an equivalent texture has been produced in these two trials.

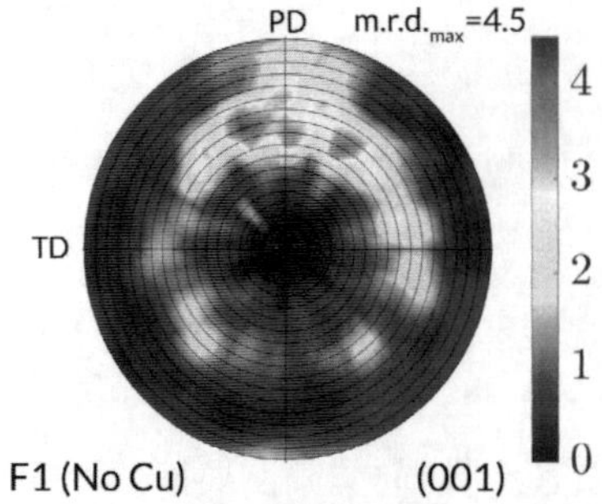

Figure 7. (001) pole figure for F1 without Cu.

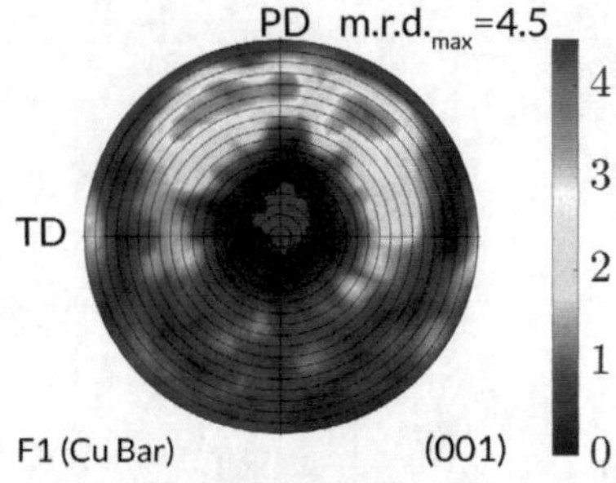

Figure 8. (001) pole figure for F1 with Cu bar.

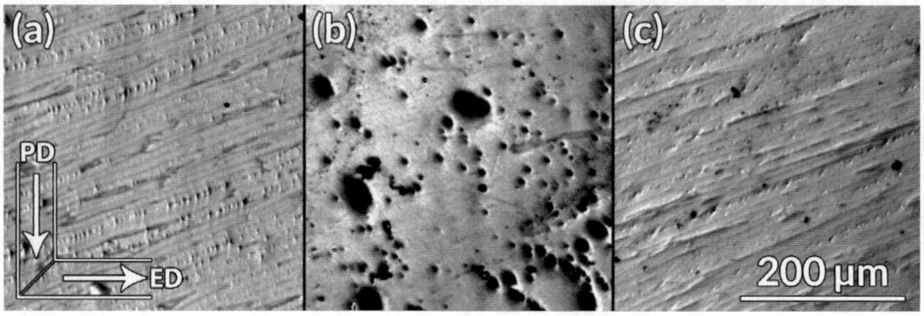

Figure 9. Optical micrographs of transverse faces after ECAP for (a) No Cu (b) Cu foil, and (c) Cu bar.

3.5. Optical microscopy
Figure 9 shows the transverse face of samples after ECAP. Without Cu (Fig. 9(a)), the sample becomes quite dense and particles tend to both elongate and realign at an inclination angle of $\theta \approx 26°$ (consistent with the angle produced via simple shear in ECAP with a die angle of 90° [10]) away from the exit direction, ED. When processed with Cu foil, the microstructure retains a significant amount of porosity (as visible in Fig. 9(b)) and particle alignment and elongation is less prominent. In 9(c), when a Cu bar is incorporated, a dense microstructure with aligned and elongated particles similar to Fig. 9(a) is observed.

4. Discussion and Conclusions
The incorporation of Cu foil successfully reduced the risk of process failure during ECAP and the initial build up of forward pressure before significant deformation occurs is reduced. However, the degree of texture produced via ECAP is significantly reduced. The microstructure retains an appreciable degree of residual porosity. Effects stemming from the interaction between the die wall and primary material, such as friction, may play a key role in successful development of a well-textured, dense microstructure for $Nd_2Fe_{14}B$-based materials via ECAP.

The incorporation of a Cu bar also successfully reduced the risk of process failure during ECAP. The build up of forward pressure before deformation occurs is significantly reduced. Unlike modifications based on Cu foil, the interaction between the sample and die wall is not affected by the Cu bar, and an equivalent degree of texturing and densification as in the case without Cu is achievable, which is supported by results showing an equivalent texture (Figs. 7 and 8), magnetic performance (Fig. 6) and microstructure (Fig. 9). This modification presents a promising technique for improving the reliability of the ECAP process for brittle materials requiring high forces without compromise on material and functional properties.

References

[1] Onal E, Lapovok R, Kishimoto H, Kato A, Davies C H J and Suzuki K 2015 *J. Appl. Phys.* **117**

[2] Segal V M 2004 *Mat. Sci. Eng. A* **386** 269

[3] Horton A and Wright J L 1996 *IEEE Trans. Magn.* **32** 4314–4316

[4] Grünberger W, Hinz D, Kirchner A, Müller K H and Schultz L 1997 *IEEE Trans. Magn.* **33** 3889–3891 ISSN 00189464

[5] Roberts P R and Ferguson B L 1991 *International Materials Reviews* **36** 62–79 ISSN 09506608

[6] Qform simulation software URL `http://www.qform3d.com/`

[7] Yasuda H Y, Kumano M, Nagase T, Kato R and Shimizu H 2011 *Scr. Mater.* **65** 743–746 ISSN 13596462

[8] Segal V 1995 *Mat. Sci. Eng. A* **197** 157–164 ISSN 09215093

[9] Mishra R K 1987 *J. Appl. Phys.* **62** 967–971 ISSN 00218979

[10] Beyerlein I J and Tóth L S 2009 *Progress in Materials Science* **54** 427–510 ISSN 00796425

IOP Conf. Series: Materials Science and Engineering **194** (2017) 012044 doi:10.1088/1757-899X/194/1/012044

3D modeling of unconstrained HPT process: role of strain gradient on high deformed microstructure formation

A Ben Kaabar[1], A Aoufi[2], S Descartes[1], C Desrayaud[2]

[1] Univ Lyon, INSA-Lyon, CNRS, UMR5259 LaMCoS, F-69621 Villeurbanne, France

[2] Univ Lyon, Ecole des Mines de Saint-Etienne, CNRS, UMR 5307 Laboratoire Georges Friedel, F-42023 Saint-Etienne, France

Corresponding author: sylvie.descartes@insa-lyon.fr

Abstract. During tribological contact's life, different deformation paths lead to the formation of high deformed microstructure, in the near-surface layers of the bodies. The mechanical conditions (high pressure, shear) occurring under contact, are reproduced through unconstrained High Pressure Torsion configuration. A 3D finite element model of this HPT test is developed to study the local deformation history leading to high deformed microstructure with nominal pressure and friction coefficient. For the present numerical study the friction coefficient at the interface sample/anvils is kept constant at 0.3; the material used is high purity iron. The strain distribution in the sample bulk, as well as the main components of the strain gradients according to the spatial coordinates are investigated, with rotation angle of the anvil.

1. Introduction

During tribological contact's life, different deformation paths lead to the formation of high deformed microstructure in the near-surface layers of the bodies. Those structural changes occurring at the near surface are one of the natural responses of materials to tribological stresses, i.e. temperature, hydrostatic pressure and shearing gradients. This response is named Tribologically Transformed Structures (TTS) [1–3]. In most cases, the TTS are often described as submicrometric structured near-surface layers produced at contacts [1-5]. Initially attributed to flash temperatures generated by frictional heating at the contact zone, followed by subsequent quenching [5, 6], several studies have shown that the temperature increase was often too low to explain these phase transformations [7-10]. Thus these transformations could result from high strain under high hydrostatic pressure [9-12] and even at low temperatures. In parallel, among the Severe Plastic Deformation (SPD) processes [13], High-Pressure Torsion (HPT) experiments are efficient to obtain materials with ultra-fine structures [14-17] or phase transformations in materials. Thus, carrying out HPT tests seems to be an efficient means of investigating TTS initiation mechanisms under severe shearing conditions and simultaneously analyzing and controlling their formation under conditions that can be controlled with greater precision than in a contact.

Furthermore, in the case of the unconstrained version of HPT, previous experimental work [18, 19] has shown the development of a refined structure after only one revolution, for two carbon steels materials and high purity iron (Fe 99.9999%) tested on such apparatus, at room temperature. The evolution of the microstructure has been correlated with the contact conditions (localization of the sliding or adhesion zones between sample and anvil surfaces) [20, 21]. In the present work, the unconstrained HPT process simulates numerically the HPT tests. The material is a high purity iron (Fe 99.9999%). The 3D finite element model developed allows to study locally the specificity of the stress and strain loading cycles within the areas where TTS are formed during these tests. To calculate significant stress and strain cycles in the right range of tribological conditions, this modeling takes into account both the contact conditions (friction at the sample / anvil interfaces) and the material behavior law in an adequate range of pressure and stresses.

2. 3D finite element model

2.1. Model geometry

In order to investigate the stress and strain results but also the strain gradients in the sample volume, a 3D model (figure 1) was developed with commercial software (Abaqus 6.9) to simulate the cylindrical sample compressed and sheared between two anvils. The two cylindrical tungsten carbide (WC) anvils were considered to be rigid, with mass and inertial moment. The anvil geometry used in these simulations was based on the unconstrained HPT set-up geometry. The anvils were smooth with a diameter of 3.2mm. The sample was 3.0 mm in diameter and 0.5 mm thick. The model parts were meshed with 8 noded-50 μm elements (C3D8R elements) for the samples and 100 μm elements for the anvils. The simulations, through a quasi-static analysis, were used to examine the processing operation up to a quarter revolution of the bottom anvil. The model was assuming an ALE (Arbitrary Lagrangien Eulerien) numerical scheme providing the displacement variable as the sum of both the material and the grid motion in order to:

- satisfy the mechanical balance equilibrium,
- minimize the mesh distortion according to criteria defined on aspect ratios and angles of the elements of the mesh.

Practically in the explicit version of the Abaqus software, a smooth procedure of the mesh is processed at a frequency of 10 increments (default value) to respect the limit defined in terms of distortion of the elements which the topology was constant during whole the calculation. The explicit formulation of the problem implied an averaging procedure of the output variables to avoid numerical noise in the solution. Finally the mass scaling procedure applied in the prescribed range enabled to decrease significantly the CPU time at completion.

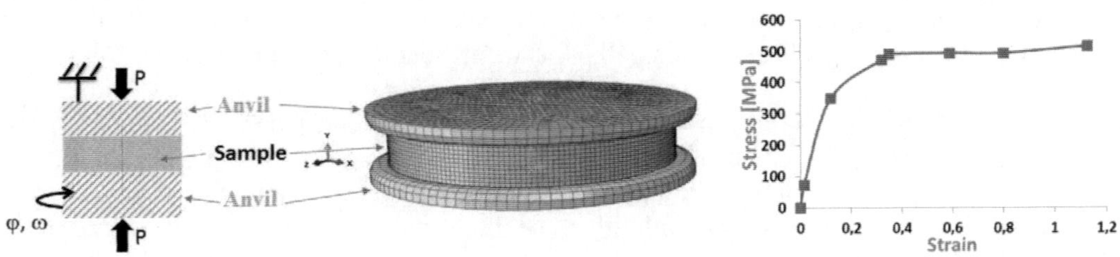

Figure 1: (a) Schematics of HPT model, (b) 3D view, (c) Behaviour law for pure iron

2.2. Material behaviour and boundary conditions

An ultra-high purity iron Fe was studied: the Young's modulus E was 210 GPa, the Poisson ratio υ was 0.3, the density was 8.7 10^{-9} t/mm^3 and the yield stress was 120 MPa. The material plastic behaviour law was not known under the test conditions, i.e. coupling between high pressures and strong shear deformation. Thus, in this work, the material behaviour was elasto-plastic and based on flow curve determined experimentally through compression tests under hydrostatic pressure up to 0.8 GPa. The stress–strain curves used are plotted in figure 2. In real HPT process, slippage occurs between the anvils and the sample. One Coulomb friction coefficient value was chosen and equal to 0.3, based on previous results [20-21].

2.3. Description of a numerical test

In the first load step, the target vertical load (3.5kN) was applied via the bottom anvil through a vertical displacement along Y (in exactly the same way as in the experiment). This load corresponds to a nominal pressure P of 1 GPa, on a circular 3 mm diameter disk. Throughout the simulation, the top anvil was prevented from moving or rotating. In the second step, the vertical displacement was stopped and the normal load was maintained constant while small incremental rotation around the axis (Y) was imposed on the bottom anvil. In the last step the rotation was applied up to the 3 target rotation angles: 1°, 30° and 90° of anvil rotation.

3. Results and discussion

Previous experimental results [18, 20] have shown that the transformed layers exhibiting the finest microstructures were produced in a specific sample volume linked with sliding/adhesion contact conditions (Figures 2a, 2b). The adhesion zone was characterized by a radius rA. The surface conditions thus govern the location of material transformation in the bulk. Different paths (a path is composed of consecutive nodes) were thus defined, in relation to previous experimental work highlighting the TTS location within the sample [18, 20] (Figure 2c). The stresses and strains could thus be calculated during a simulation at different depths and along radius within the sample following these paths, at a given rotation angle.

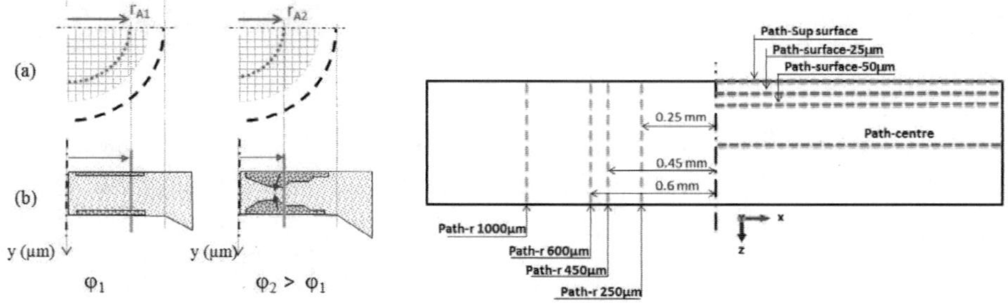

Figure 2: Sliding/adhesion contact. (a) The adhesion radius rA decreases with increased φ (b) Schematics evolution of the microstructure refinement (pink) based on previous experimental results, (c) Location of the chosen paths for the present numerical model

The evolution of contact conditions was calculated at the interface. The contact pressure distribution (CPress in abaqus software) was presented for 90° of anvil rotation (Figure 3a). For 3 different angles of rotation, the relative sliding between the anvil and the sample was quantified along "path-sup surface". An adhesion zone could thus be highlighted (Figure 3b); it decreased with increasing of φ(°).

The history of the Von Mises equivalent plastic strains ε (variable PEEQ used in Abaqus software) was also plotted along radius r and through thickness of the sample, for the different paths and for different anvil rotation angle (Figure 4). For a given depth, these strain were low at the centre of the sample and at the periphery, but reached a maximum value around a radial position close to the transition area between the adhesion and the sliding zones. The radial location of the maximum approached the centre of the sample when the angle increases. For a given radius, the strain values are maximal in the skin.

During the unconstrained HPT process pure shear strain ($\varepsilon_{z\theta}$) occurred but also a shearing strain ($\varepsilon_{r\theta}$) at the frontier between the sliding and the adhesion zones, considering that elongation was possible in the radial direction. Thus those strain components were plotted along the different paths (Figures 5, 6, 7). The strain components reached a maximum value and their radial positions draw closer to the transition area between the adhesion and the sliding zones. While $\varepsilon_{\theta z}$ evolution stabilized between 30° and 90° of anvil rotation (Figure 5), $\varepsilon_{r\theta}$ values increased with increase of φ(°), especially in the near-surface layer (Figure 6). The increase was closely linked to rA evolution. ε_{rr} increased continuously along the different paths r- with increase of φ (Figure 7) in the whole sample.

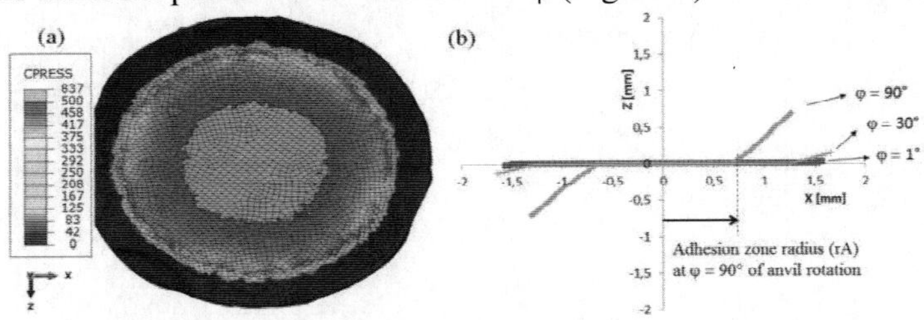

Figure 3: (a) Contact pressure distribution for φ =90°, (b) Evolution of the adhesion zone

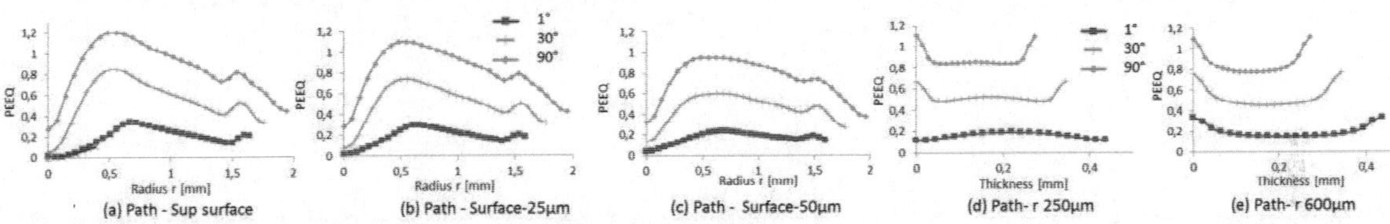

Figure 4: Von Mises equivalent plastic strain for 3 angles of rotation 1°, 30° and 90°, along different paths (see figure 2c)

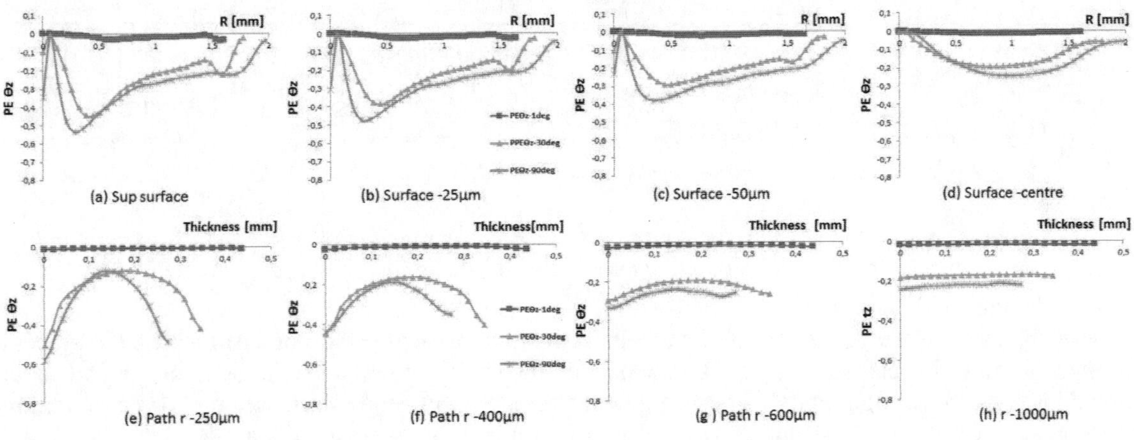

Figure 5: Plastic strain $\varepsilon_{\theta z}$ for 3 angles of rotation 1°, 30° and 90°, along different paths (see figure 2c)

Previous work [20] highlighted strong equivalent plastic strain values coupled with large equivalent strain gradients in both radial and axial directions, where TTS were located. To go further in the mechanical analysis, the main components of the strain gradients according to the spatial coordinates were calculated in the present work (tables in the figure 8). The gradients were obtained as the average values of the strain variation ($\Delta PE_{r\theta}$, $\Delta PE_{z\theta}$, ΔPE_{rr}) along the considered coordinates (r or h). These table provided an estimation of the curvature. The strain gradients (i. e. components of the curvatures tensor) were maximal in the areas in which "TTS" were observed (Figure 8). The gradients of $\varepsilon_{r\theta}$ were the highest ones within whole the sample, especially in the axial direction, and below the adhesion zone. Maximal gradients of ε_{rr} values were linked with the edge sliding/adhesion zones. Nevertheless the accurate values can be obtained by the calculation of second gradients of the displacements (i.e. curvature) [22], $\alpha_{ij} = \varepsilon_{jkl} \dfrac{\partial \beta_{li}^{p}}{\partial x_k}$ (m/m²); this work is in progress.

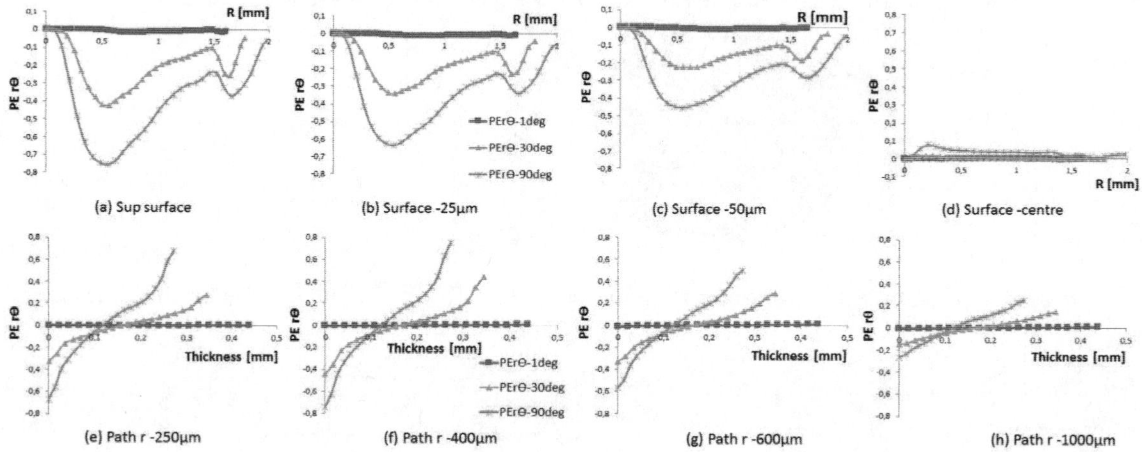

Figure 6: Plastic strain $\varepsilon_{r\theta}$ for 3 angles of rotation 1°, 30° and 90°, along different paths (figure 2c)

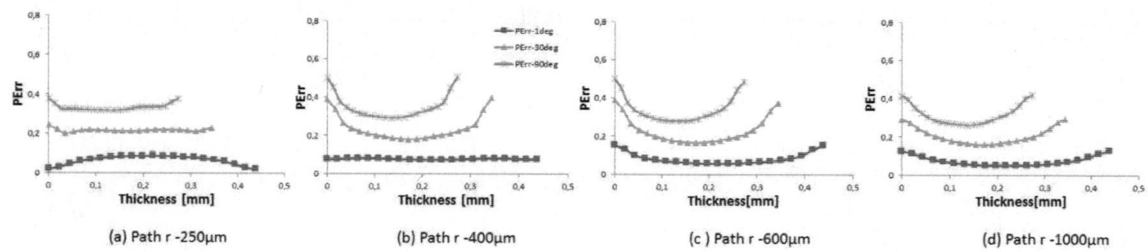

Figure 7: Plastic strain ε_{rr} for 3 angles of rotation 1°, 30° and 90°, along different paths (figure 2c)

4. Conclusion

The objective of the present work was to investigate numerically the unconstrained HPT process through a 3D finite element model. Gradients of the strain components in both radial and axial directions have been highlighted. Those gradients coupled with high strain seem to be a condition necessary to form TTS. This model will allow to investigate in an accurate way the strain paths and curvatures, in relation with the developed microstructures during a test and their localization in the sample.

Acknowledgments
The authors would like to thank A. Eleod and his team from Université Technique de Budapest UTB (Hungary) to give them the stress-strain curves under hydrostatic pressure for pure iron. This work was supported by the LABEX MANUTECH-SISE (ANR-10-LABX-0075) of Université de Lyon, within the program "Investissements d'Avenir" (ANR-11-IDEX-0007) operated by the French National Research Agency (ANR).

IOP Conf. Series: Materials Science and Engineering **194** (2017) 012044 doi:10.1088/1757-899X/194/1/012044

		1°	90°
$\Delta PE_{r\theta}$	/r	-0	-1.7
	/h	-0.01	6.1
$\Delta PE_{z\theta}$	/r	-0.03	-2.9
	/h	0.003	-4
ΔPE_{rr}	/r		
	/h	-0.5	0.01

		1°	90°
$\Delta PE_{r\theta}$	/r	-0.02	-0.9
	/h	0.04	10.2
$\Delta PE_{z\theta}$	/r	-0.07	0.35
	/h	-0.03	-2
ΔPE_{rr}	/r		
	/h	-0.06	1.43

		1°	90°
$\Delta PE_{r\theta}$	/r	0.01	0.5
	/h	0.06	1.7
$\Delta PE_{z\theta}$	/r	0.015	0.2
	/h	-0.02	0.3
ΔPE_{rr}	/r		
	/h	0.4	1.1

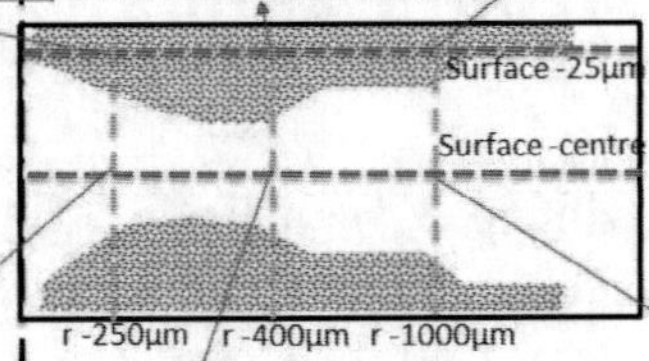

		1°	90°
$\Delta PE_{r\theta}$	/r	0	0.4
	/h	0.02	2.4
$\Delta PE_{z\theta}$	/r	0.001	-0.3
	/h	-0,01	1.85
ΔPE_{rr}	/r		
	/h	-0.1	0.3

		1°	90°
$\Delta PE_{r\theta}$	/r	0	-0.1
	/h	0.02	2.8
$\Delta PE_{z\theta}$	/r	0.003	-0.4
	/h	-0.008	0.92
ΔPE_{rr}	/r		
	/h	0.06	0.5

		1°	90°
$\Delta PE_{r\theta}$	/r	0	-0.007
	/h	0.04	1.3
$\Delta PE_{z\theta}$	/r	0.0007	-0.08
	/h	-0.006	-0.4
ΔPE_{rr}	/r		
	/h	0.09	0.5

Figure 8: Strain gradients components

References

[1] Berthier Y, Vincent L and Godet M 1989 *Tribol. Int.* **22** 235
[2] Sauger E, Fouvry S, Ponsonnet L, Kapsa Ph, Martin J-M and Vincent L 2000 *Wear* **245** 39
[3] Descartes S, Renouf M, Fillot N, Gautier B, Descamps A, Berthier Y, Demanche Ph 2008 *Wear* **265** 1408
[4] Zhou Z R, Sauger E, Liu J J and Vincent L 1997 *Wear* **212** 50
[5] Chou Y. K and Song H 2005 *Int. Journal of Machine Tools & Manufacture* **45** 481
[6] Kalin M and Vizintin J 2001 *Wear* **249** 172
[7] Hurricks P L 1970 *Wear* **15** 389
[8] Sproles E S and Duquette D J 1978 *Wear* **47** 387
[9] Colombié Ch, Berthier Y, Floquet A, Vincent L and Godet M 1984 *Transactions of the ASME* **106** 106
[10] Berthier Y, Colombié Ch, Vincent L and Godet M 1988 *Journal of Tribol.* **110** 517
[11] Rigney A 1997 *Tribol. Int.* **30** 361
[12] Kuhlmann-Wilsdorf D 2001 in: *Proc. of the 27th Leeds-Lyon Symp. on tribology* **39** Elsevier, Lyon, 417
[13] Valiev RZ, Islamgaliev R K and Alexandrov IV 2000 *Progress in Mater. Sci.* **45** 103
[14] Xu C, Horita Z and Langdon T G 2007 *Acta Mater.* **55** 203
[15] Zhilyaev AP, Nurislamova GV, Kim BK, Baro MD, Szpunar JA, Langdon TG 2003 *Acta Mater.* **51** 753
[16] Todaka Y, Umemoto M, Yin J, Liu Z and Tsuchiya K, 2007 *Mater. Sci. and Eng. A*, **462** 264
[17] Rentenberger C, Waitz T and Karnthaler H P, 2004 *Scripta Mater.* **51** 789
[18] Descartes S, Desrayaud C and Rauch E F 2011 *Mater. Sci. Eng. A* **528** 3666
[19] Busquet M, Descartes S and Berthier Y 2009 *Tribology International* **42** 1730
[20] Descartes S, Busquet M and Berthier Y 2011 *Wear* **271** 1833
[21] Halloumi A, Busquet M, Descartes S 2014 IOP Conf. Series: Mater. Sci. and Eng. **63** 012036
[22] Nye J F 1953 *Acta Metall.* **1** 153

IOP Conf. Series: Materials Science and Engineering **194** (2017) 012045 doi:10.1088/1757-899X/194/1/012045

High Pressure Torsion: from Laminar Flow to Turbulence

R Kulagin[1], Y Beygelzimer[2], Y Ivanisenko[1], A Mazilkin[1,3] and H Hahn[1]

[1]Institute of Nanotechnology (INT), Karlsruhe Institute of Technology (KIT), Hermann-von-Helmholtz-Platz 1, 640, 76344 Eggenstein-Leopoldshafen, Germany
[2]Donetsk Institute for Physics and Engineering named after O.O. Galkin, NASU, Vernadsky Str. 46, 03142 Kyiv, Ukraine
[3]Institute of Solid State Physics, RAS, Academician Ossipyan Str.2, 142432 Chernogolovka, Russia

roman.kulagin@kit.edu

Abstract. We analyzed experimentally and theoretically deviations from the simple High Pressure Torsion (HPT) scheme. We showed that the nature of the plastic flow and the deformed state of the sample under HPT depends on the structure and mechanical properties of the material being processed. Both laminar and turbulent flows under HPT were analysed. It was demonstrated that non-laminar flow causes intensive mass transfer and mixing of the deformed material. As an example, mechanical synthesis of the high entropy alloys from elemental powders by HPT is presented.

1. Introduction

High Pressure Torsion (HPT) is a widely used process for modifying the materials structure and properties by means of severe plastic deformation [1]. It is a generally accepted assumption that deformation during HPT occurs by simple shear. The shear is uniform over the sample's thickness and shear strain γ at each point of the sample is proportional to the distance r from the anvil rotation axis and to the rotation angle β:

$$\gamma = \frac{r\beta}{t} \tag{1}$$

where t is the sample thickness.

Experiments and numerical simulations in recent years show that the actual plastic flow during HPT may differ significantly from this simple scheme. The influence of the geometric parameters and conditions of contact friction during the HPT deformation on the spatial strain distribution in the sample's bulk was studied in [2–5]. In [6] the dependence of the strain rate field on the stress-strain curve and rotation angle was investigated.

It was shown that if the strain hardening obeys the power-law and the strain-rate hardening is absent, the strain rate field in the sample volume is independent on the anvil rotation angle. In this case, the value of shear strain at any point of the sample is proportional to the rotation angle. Any deviation from the aforementioned conditions results in the unsteadiness of the strain rate field. The latter means that shear strain ceases to depend linearly on the rotation angle. It was also shown in [6] that when strain hardening is saturated and the strain-rate hardening is absent, then the strain is concentrated in a thin layer of the sample. This effect is not related to the geometry of the anvils or friction, but is rather determined by instability of the material during the plastic shear deformation [7, 8].

We show it in Section 2 of this work that certain adjustments to the conclusions of [6] have to be made if strain-rate hardening is to be considered. On increasing the magnitude of the strain-rate hardening, the thickness of the layer where the deformation is concentrated becomes progressively larger until no localization is observed. When the strain-rate hardening stops to increase, the strain-rate field approaches a steady state and shear deformation depends linearly on the rotation angle β.

We carried out HPT deformation experiments on copper samples with aluminum markers (see Section 3). The experiments showed that at sufficiently large strains the two materials (Cu and Al) were intermixed. We regard this result as an evidence of transition from laminar to turbulent flow in the material under HPT deformation. This transition corresponds to the two-stage simple shear hypothesis [9].

2. Dependence of the strain-rate field on the anvil rotation angle

It was shown in [10] that the saturation of strain hardening during HPT should lead to strain localization in a thin layer of the sample if the strain-rate hardening is absent [6]. We will show by means of FEM computer simulations that in this case strain-rate hardening may stabilize the plastic flow at HPT and prevent strain localization.

All calculations were done with the DEFORM – 2D/3D software using a 2D axisymmetric torsion model [11]. The set-up for the HPT experiment is shown in figure 1a

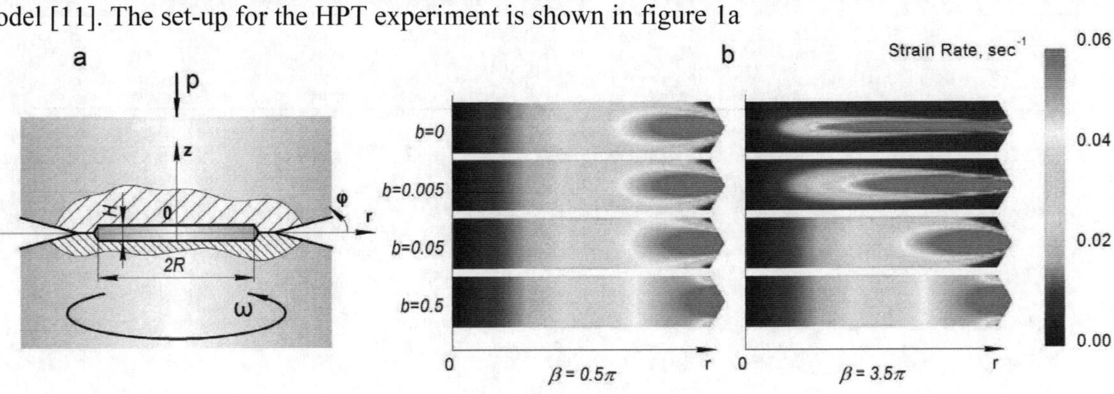

Figure 1. To the calculation of the strain-rate field during HPT: (a) schematic geometry of HPT, (b) von Mises strain rate maps for a horizontal plane at two different anvil rotation angles (FEM simulation); values of the strain-rate are indicated in the figure. The parameter b is defined in the text below.

The stress-strain curve was specified by the following power-law relation:

$$\sigma_s[MPa] = \sigma_s^*(e_M) \cdot \left(\frac{\dot{e}_M}{a}\right)^b \qquad (2)$$

where $\sigma_s^*(e_M)$ is the experimental dependence for powder iron [6] and a and b are strain-rate hardening parameters.

For the calculations the value of a was assumed to be equal to $0.05 s^{-1}$. We studied the evolution of the strain rate field depending on the b value for b equal to 0, 0.005, 0.05 and 0.5. In all experiments, the sample diameter was 10 mm, and its thickness was 1 mm.

The results of the calculations are shown in figure 1b. It is seen that, as the b parameter increases, the area of large shear deformation becomes broader. Ultrafine-grained materials processed by HPT usually have strain rate sensitivity coefficient in the range of 0.01 - 0.1 [13, 14]. This means that equation (1) can still be applied as a rough estimate for the shear strain during HPT. However, it should be noted that the area of large deformation may be reduced. The reason is that the saturation of strain hardening when the value of strain-rate hardening may be incomplete.

3. Turbulent flow of the material during HPT

This section describes the results of HPT experiments on Cu samples containing inserts of Al and brass markers. Samples were prepared as follows. Eight 2 mm channels were drilled in Cu cylinder having the length of 35 mm and the diameter of 20 mm. The channels were drilled at a distance of 4.0 and 7.0 mm from the axis of the workpiece. The Al wire was inserted into four channels which were closer to the axis of the cylinder. Brass wire was put into the peripheral channels. The workpiece was subjected to direct extrusion and its diameter shrank to 12 mm to seal the wires in copper matrix. Disc-shaped specimens 1 mm in thickness were cut from the workpiece. We used these disks for HPT experiments. Deformation was performed at a pressure of 4.5 GPa and an angular velocity of the anvils of 1 revolution per minute. Figure 2 shows sections of the samples normal to the rotation axis after the rotation to angles $\pi/2$ (figure 2a) and π (figure 2b).

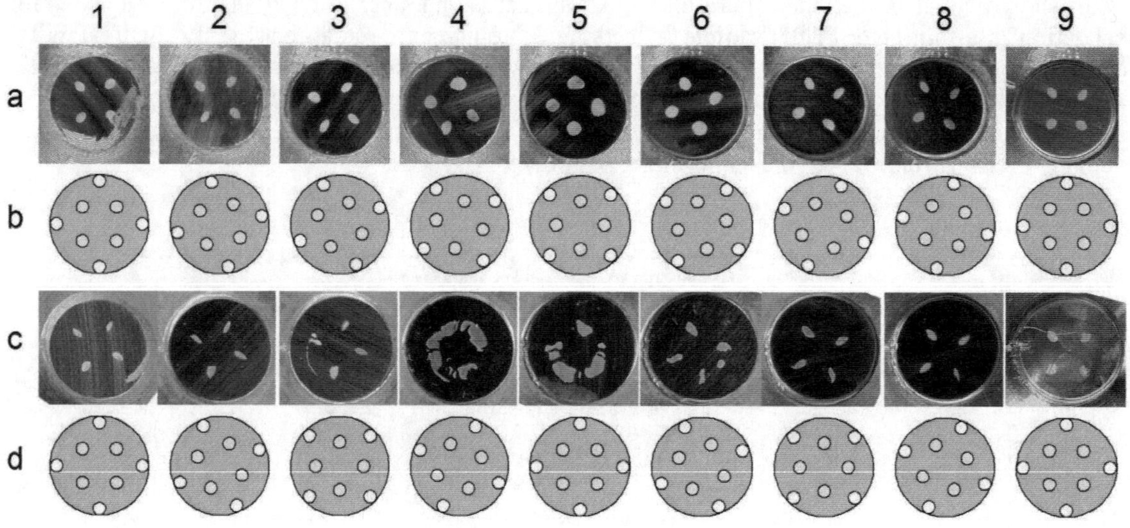

Figure 2. Rotation to 90° (a and b); rotation to 180° (c and d). Numbers in figure correspond to the depth step (each step is 100 μm).

Figure 2a, b shows that when $\beta = \pi/2$ the positions of the markers coincide well with those calculated from the model of the simple shear deformation. This indicates the validity of this model, at least for this value of the rotation angle. The picture changes completely for $\beta = \pi$. The markers are blurred for the depths from 300 to 500 μm (figure 2c, d). This suggests that the major deformation is localized in this 300 μm thick layer. The cross-section of an HPT deformed disk after three revolutions of the anvils is shown in figure 3. It can be seen from the figure that HPT leads to the mixing of the material of the matrix with that of the markers.

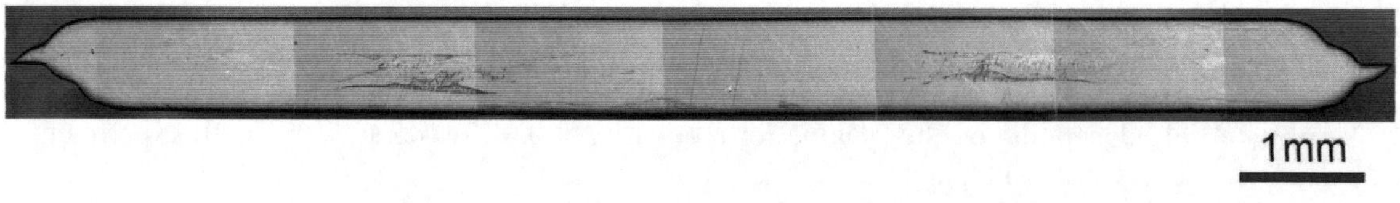

Figure 3. Cross-section of a Cu disk with markers after HPT deformation; three revolutions of the anvils were performed.

This mixing is possible only in the presence of turbulent eddy currents in the sample during HPT. This conclusion is consistent with the suggestions of the authors [9] that turbulent flow takes place in the material at large simple shear strains. We suppose that it is the turbulent plastic flow which explains the efficiency of HPT for the cold fusion of various materials in the first works of Bridgman [15]. Formation of the alloys is associated with the mass transfer by the diffusion mechanism. The diffusion can be rather fast at elevated temperature, but the rate of the diffusion mass transfer at low temperature is not high enough. The rate of the mass transfer necessary for cold fusion can be provided only by a stochastic vortex motion, which is similar to the turbulent flow in liquids or gases. The next section of this paper will show that HPT processing of a powder mixture of a certain composition results in the formation of high entropy alloys at ambient temperatures. We believe that a turbulent flow of the material during the HPT deformation plays the most important role in this process.

In [20] the formation of vortices in the plastically deformed material under simple shear is associated with a local blocking of the shear deformation. Let us show by numerical modeling that the local blocking of the shear deformation in the sample under HPT leads to the twists and turns of the obstacles. We regard this as the physical cause for the turbulent flow during the HPT.

Figure 4 shows the scheme and the results of calculations. Layer of plastic material is located between the two plates, moving in opposite directions. There are obstacles in the layer, locally locking the shear. The obstacle is of the material with the yield stress ten times higher than that in the layers and it has the shape of a cylinder (perpendicular to the plane of the figure).

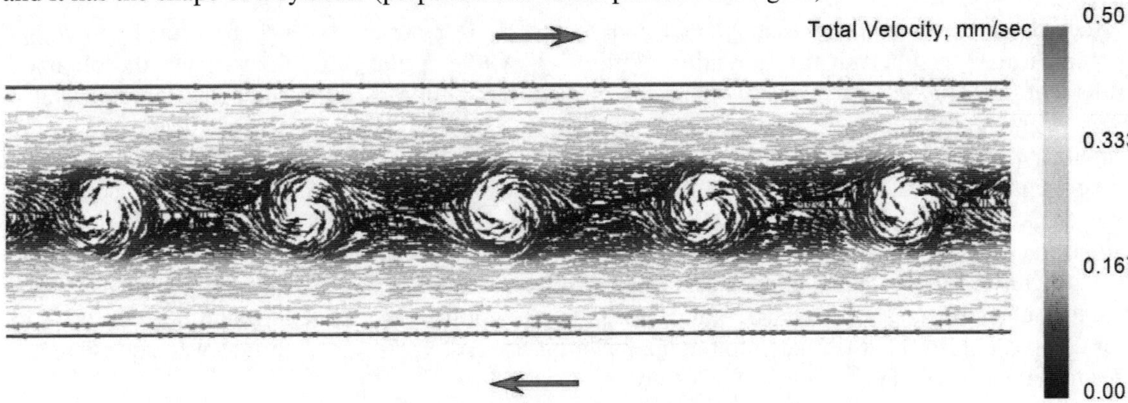

Figure 4. Simple shear blocked locally by the obstacles (FEM simulation).

Figure 4 shows that the local blocking of the shear really causes rotation of obstacles, which leads to the formation of vortices at sufficiently large shear strain. As obstacles during the HPT deformation, non-uniformities in the material of different kinds may be considered.

4. Obtaining high entropy alloy by turbulent flow during HPT

A high entropy CoCrMnFeNi alloy was produced from a mixture of powders by means of high pressure torsion. To prepare this mixture, we used powders of Co(>99.8 wt%), Cr (>99.2 wt%), Mn (>99.6 wt%), Fe (>99.5 wt%) and Ni (>99.9 wt%) with the size of about 1-10 μm. The mixture prepared for the HPT treatment contained equal atomic fractions of all components. The treatment was carried out in a custom built computer-controlled HPT device using flat anvils under the following conditions: the applied pressure P=5 GPa, anvils rotation speed ω=5 rpm and number of rotations N=100. The temperature in the deformation zone was monitored by the laser pyrometer and it was ~200°C.

The obtained sample was investigated in a TITAN 80–300 transmission electron microscope (TEM). The sample for the investigation was prepared using a FEI Strata 400S dual beam facility. The results of the investigations are presented in figure 5.

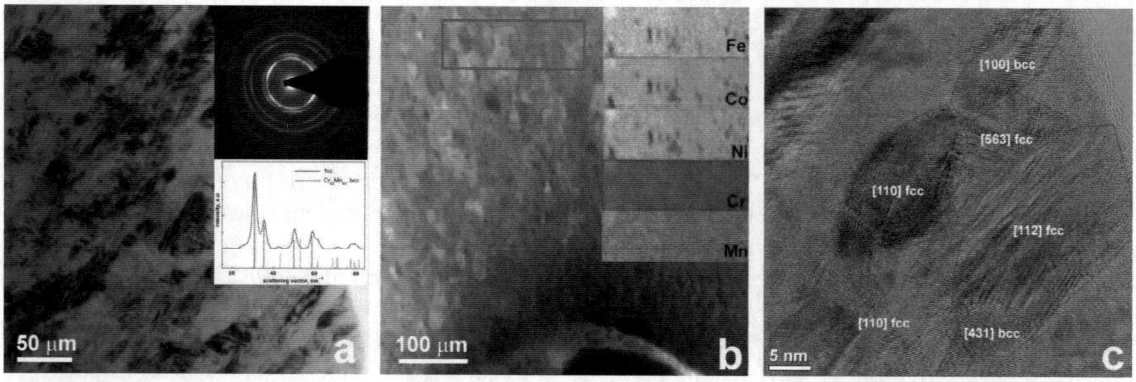

Figure 5. Microstructure analysis of the CoCrMnFeNi alloy after HPT: (a) bright field image (in the inserts are a selected area electron diffraction (SAED) pattern and a corresponding intensity radial distribution plot). Vertical lines denote the positions of the reflections from CoCrMnFeNi fcc phase (red) and $Cr_{60}Mn_{40}$ bcc phase (green) correspondingly); (b) HAADF STEM image with the elemental maps for Fe, Co, Ni, Cr and Mn as inserts; (c) high resolution TEM image of the grain structure (red lines denote the boundaries between the grains of different phases with indexed normal orientations).

According to the results presented in figure 5, the HPT treatment of the mixture of individual powder particles resulted in the formation of the CoCrMnFeNi alloy. It follows from the obtained results that the alloy consists mainly of the fcc phase although the particles of second phase are also observed in the structure. According to the SAED, the second phase is $Cr_{60}Mn_{40}$ bcc phase (W structure type). This conclusion is also supported by the results of elemental mapping (figure 5b). It can be seen that all the elements are distributed homogeneously over the sample, with the exception of small areas about 10 nm in size, which are mostly composed of Cr and Mn. We can estimate the volume fraction of bcc phase to be as high as 7 %. High resolution TEM image (figure 5c) shows the grain structure of the obtained alloy. From this image the grain size can be estimated at about 20 nm. The mechanism of the solid-state HEA alloy formation is not the purpose of this study. Irrespective of what this mechanism is, it is related to the intensive mass transfer. As stated earlier, we suppose that it is the turbulent flow that ensures the mass transfer.

Conclusions

The results obtained indicate that the mechanical properties and the structure of the processed materials can have a significant effect on the velocity field and strain state in a sample process by HPT. At stress saturation in a material with low or zero strain-rate hardening, the deformation is localized in a thin layer of the sample. The thickness of this layer increases with increasing exponent b of strain-rate hardening. For $b > 0.1$ no strain localization occurs. The occurrence of strain non-uniformities inhibiting simple shear leads to a turbulent flow pattern under HPT. This accelerates the mass transfer required for the formation of alloys. We show in our study that the HEA alloys can be obtained by means of this approach.

Acknowledgment

YB would like to express his gratitude to the State Fund for Fundamental Research of the Ukraine for financial support of research on vortex flow under twist extrusion through the competitive grant F71/56-2016.

References

[1] Zhilyaev A P and Langdon T G 2008 *Prog. Mater Sci.* **53** 893
[2] Figueiredo R B, Faria G C V, Cetlin P R and Langdon T G 2013 *J. Mater. Sci.* **48** 4524

[3] Yoon S C, Horita Z and Kim H S 2008 *J. Mater. Process. Technol.* **201** 32
[4] Pippan R, Scheriau S, Hohenwarter A and Hafok M 2008 *Mater. Sci. Forum* **584-586** 16
[5] Lee D J and Kim H S 2014 *J. Mater. Sci.* **49** 6620
[6] Beygelzimer Y, Kulagin R, Toth L S and Ivanisenko Y 2016 *Beilstein J. Nanotechnol.* **7** 1267
[7] Hill R 1958 *J. Mech. Phys. Solids* **6** 236
[8] Beigelzimer Y, Efros B, Varyukhin V and Khokhlov A 1994 *Eng. Fract. Mech.* **48** 629
[9] Beygelzimer Y, Valiev R Z and Varyukhin V 2011 *Mater. Sci. Forum* **667-669** 97
[10] Beygelzimer Y, Toth L S and Jonas J J 2015 *Adv. Eng. Mater.* **17** 1783
[11] DEFORM *Scientific Forming Technologies Corporation* http://www.deform.com/
[12] Zhao Y J, Massion R, Grosdidier T and Toth L S 2015 *Adv. Eng. Mater.* **17** 1748
[13] Chinh N, Szommer P, Horita Z and Langdon T 2006 *Mater. Sci. Forum* **503–504** 1001
[14] Valiev R Z, Murashkin M, Kilmametov A, Straumal B, Chinh N and Langdon T 2010 *J. Mater. Sci.* **45** 4718
[15] Bridgman P W 1935 *Phys. Rev.* **48** 15.
[16] Panah M P 2015 *Shear-Induced Mixing in Metallic Systems, PhD Thesis* (Darmstadt, Germany)
[17] Kim H J, Karthikeyan S and Rigney D 2009 *Wear* **267** 1130
[18] Sundaram N K , Guo Y and Chandrasekar S 2012 *Phys. Rev. Lett.* **109** 106001
[19] Prandtl L and Tietjens OG 1957 *Fundamentals of Hydro - and Aeromechanics*, (New York: Dover Pub. Inc.)
[20] Beygelzimer Y 2011 *Mater. Sci. Forum* **683** 213

IOP Conf. Series: Materials Science and Engineering **194** (2017) 012046 doi:10.1088/1757-899X/194/1/012046

Analytical and Experimental Investigation of Process Loads on Incremental Severe Plastic Deformation

Mehmet Okan Görtan

Mechanical Engineering Department, Hacettepe University, Ankara/Turkey

okangortan@hacettepe.edu.tr

Abstract. From the processing point of view, friction is a major problem in the severe plastic deformation (SPD) using equal channel angular pressing (ECAP) process. Incremental ECAP can be used in order to optimize frictional effects during SPD. A new incremental ECAP has been proposed recently. This new process called as equal channel angular swaging (ECAS) combines the conventional ECAP and the incremental bulk metal forming method rotary swaging. ECAS tool system consists of two dies with an angled channel that contains two shear zones. During ECAS process, two forming tool halves, which are concentrically arranged around the workpiece, perform high frequency radial movements with short strokes, while samples are pushed through these. The oscillation direction nearly coincides with the shearing direction in the workpiece. The most important advantages in comparison to conventional ECAP are a significant reduction in the forces in material feeding direction plus the potential to be extended to continuous processing. In the current study, the mechanics of the ECAS process is investigated using slip line field approach. An analytical model is developed to predict process loads. The proposed model is validated using experiments and FE simulations.

1. Introduction

Ultrafine grained (UFG) materials produced by severe plastic deformation processes (SPD) are known for their unusual properties such as a combination of high strength and ductility or improved fatigue behavior [1, 2]. Among the SPD processes, Equal Channel Angular Pressing (ECAP) is the most investigated and developed one due to its ease of use and scalability [3]. However, ECAP involves several drawbacks in its basis form which are mostly related to the high frictional losses in the production system. Firstly, friction generates an additional resistance against the punch movement in ECAP tool systems and hence increases the axial loads. Increased loads enlarge the buckling risk of punch and limit the ECAP workpiece lengths. Therefore, conventional ECAP is seen as a discontinuous process. Moreover, high friction forces results in a poor surface appearance.

In order to overcome these disadvantages, an incremental ECAP (I-ECAP) process has been proposed [4-6]. I-ECAP tool system is split into three major parts; a stationary body and a moveable wall as well as a feeding unit. During I-ECAP, he moveable die wall is opened before forming so the material can be fed freely. After clamping the sample in the forming position, the moveable wall pushes the material into the groove of the stationary die body. Hence, the workpiece is formed gradually. Since the relative movement between I-ECAP tool walls and the specimen is relatively small for every stroke, the friction is minimized by this process.

A similar incremental SPD process, called as Equal Channel Angular Swaging (ECAS) has been proposed by the author [7-9]. ECAS combines the conventional ECAP and the incremental bulk metal forming process of rotary swaging. Since this process optimizes friction forces during forming, it has a great potential for a continuous production. However, in its first variant, ECAS tool system had channels with a high angle. Therefore, plastic deformation was limited. In order to improve the efficiency of ECAS, channel angle should be reduced. In the current study, the mechanics of ECAS process has been investigated using slip line field theory. An analytical model has been developed to predict process forces during SPD with ECAS. Afterwards, a tool system has been designed with the aid of the proposed analytical model and it is validated with experiments and FE simulations.

2. Principles of ECAS

The ECAS process bases upon the combination of conventional ECAP and the incremental bulk metal forming method of rotary swaging. During the proposed ECAS process, two forming tool halves, which are concentrically arranged around the workpiece, perform high frequency radial movements with short strokes, while samples are pushed through these. Due to this oscillation motion, the contact time and the relative movement between the tool and the workpiece are very small within a stroke cycle. This feature enables the feeding of the material while the tools are open. Moreover, the discontinuous contact enables the permanent lubrication of the tools and the workpiece. Furthermore, ECAS tool systems contain multiple forming sections to ensure that the material exits the process in the same direction as the entry. In the current study, a tool with two forming sections is investigated. Another important aspect of ECAS is that the samples are not pushed against stationary tool walls. Since the oscillation direction nearly coincides with the pushing direction, the workpiece is pulled by the tools during forming. ECAS tool system is shown in 'figure 1' schematically.

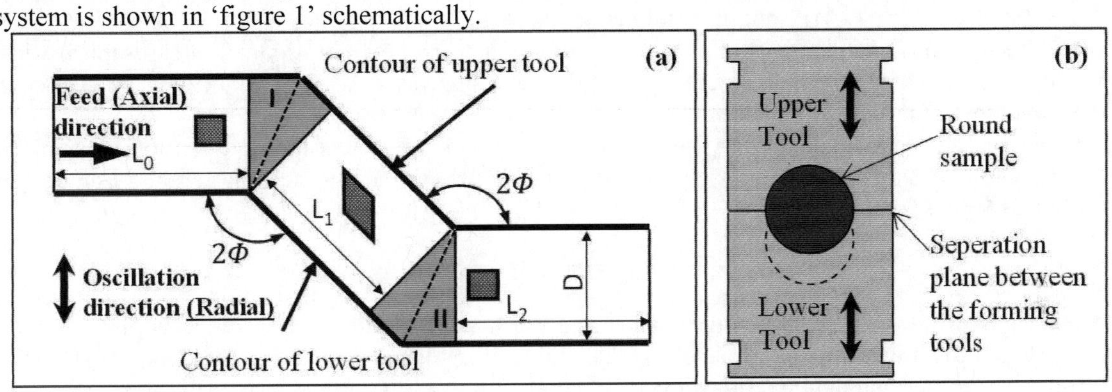

Figure 1. Schematic illustration of ECAS tool (a) cross-section (b) front-view.

In the first studies about ECAS process, a tool system with high channel angles (150°) was used. Although it was possible to refine grain structure of metals with that tool configuration, it required multiple passes [7, 8]. The efficiency of ECAS can be increased by decreasing the channel angle and thus increasing the strain induced in every pass. However, a decrease in the channel angle will cause a significant increase in the forming forces. In such a case, it should be assured that the required force in feeding direction doesn't cause a yielding in the material outside the tool in axial direction. In order to prevent such a failure, first of all, process loads are investigated analytically in the current study.

3. Mechanics of ECAS

Process loads of conventional ECAP has been investigated analytically in many studies [10, 11]. The only force source in conventional ECAP is the punch, which loads workpieces in axial direction. Analytical formulas developed for conventional ECAP process predict the punch force. On the other hand, workpieces are loaded in oscillation direction of the tools during ECAS process. These radial forces in oscillation direction cause the deformation in the material. Forces in feeding direction (axial forces) only compensate the redundant force components in the feeding direction. Therefore, analytical formulas developed for conventional ECAP can't be used for ECAS process.

The forces acting during ECAS process are shown in 'figure 2' schematically. ECAS tool system involves two forming sections. Angles in every channel intersection is the same. Therefore, the workpiece leaves the process in the same direction as it enters. Hence, forming follows the Route A. Radial forces are applied to the workpiece by the tool in oscillation direction. Axial forces are applied in feed direction.

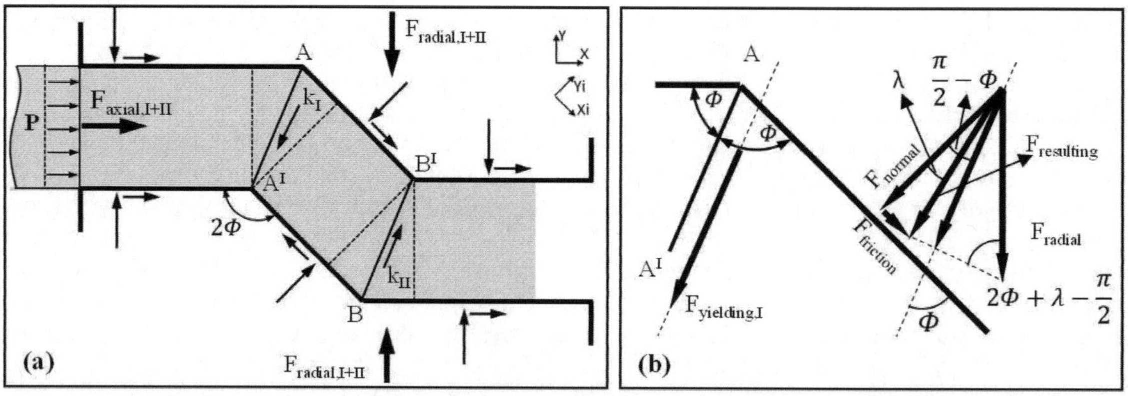

Figure 2. (a) Forces during ECAS process and (b) detail of force distribution.

Unlike conventional ECAP, due to the kinematics of the tools, friction is acting on the sample's feeding direction in ECAS, not on the opposite direction. Thus, as shown in 'figure 2' it eases the feeding and reduces the axial forces. This aspect is seen as the most significant advantage of ECAS process over ECAP concerning the force requirement.

For the analytical calculation of the process forces, it is assumed that the deformation takes place along single lines. As a results, force required to form the material along the first yielding line can be expressed as follows:

$$F_{yielding,I} = \frac{k_I}{\sin\phi} \cdot A \tag{1}$$

where, k_I is the shear strength of the material in the first deformation zone, Φ is the half channel angle and A is the area of the workpiece. In order to be able form the material, the component of the resulting force in the yielding direction should be equal to the yielding force. The resulting force can be represented as follows:

$$F_{resulting,I} = \frac{k_I}{\sin\phi \cdot \sin(\phi+\lambda)} \cdot A \tag{2}$$

where λ is the friction angle. Moreover, this resulting force is a consequence of the radial force. After making the trigonometrical calculations, the radial force required to form the material in the first and second deformation zone can be calculated using the following equation:

$$F_{radial,I+II} = \frac{(k_I+k_{II})}{\sin\phi \cdot \sin(\phi+\lambda) \cdot (-\cos(2\phi+\lambda))} \cdot A \tag{3}$$

where, k_{II} is the shear strength of the material in the second deformation zone. Additionally, the component of the resulting force in the feeding direction should be compensated by the axial force. However, part of that force is compensated by the friction acting on the tool walls. By doing the trigonometrical calculations, the following formula can be derived to predict axial force in the investigated ECAS process, which contains two forming sections:

$$F_{axial,I+II} = \left[\frac{(k_I+k_{II}) \cdot \sin(\phi+\lambda)}{\sin\phi \cdot \sin(\phi+\lambda)} - \frac{(k_I+k_{II}) \cdot (-\cos(2\phi+\lambda))}{\sin\phi \cdot \sin(\phi+\lambda) \cdot} \cdot \mu\right] \cdot A \tag{4}$$

where, μ is the friction coefficient. A common method to demonstrate forces on such a system is to use the dimensionless P/2k value. Assuming that the shear strength of the material is the same in both deformation zones, the change in P/2k value in axial and radial directions dependent on friction coefficient and half channel angle is shown in 'figure 3'.

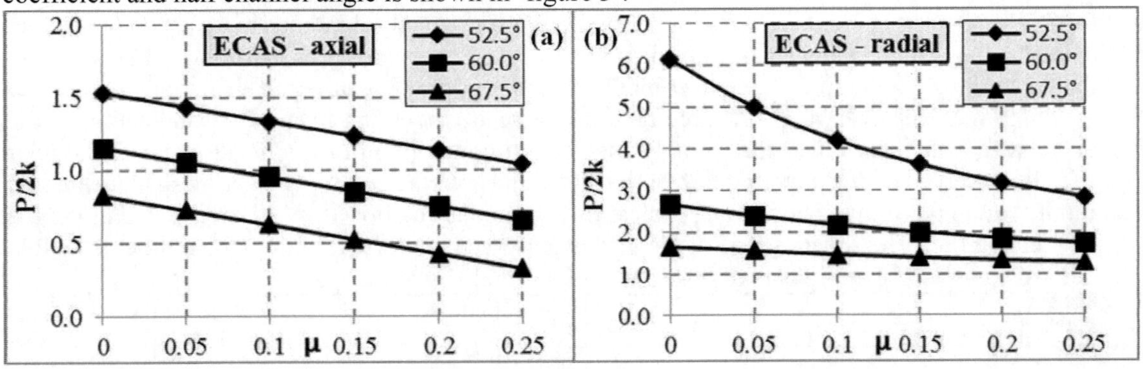

Figure 3. Unitless pressure values of ECAS in (a) axial and (b) radial direction.

The pressures in both axial and radial directions decrease while the friction coefficient increases in ECAS process. Besides the fact that the radial pressures are extremely higher than the axial ones and the ECAS process is inefficient in terms of overall pressure requirement compared to ECAP, the axial pressure in the feeding direction is considerably lower. Since a major prerequisite for a continuous forming operation is a low force requirement in the feeding direction, ECAS process has a high potential for the continuous and thus cost efficient production of UFG materials. An important aspect is that the pressures in the axial direction for a channel angle (2Φ) of 120° is lower than the yield strength of the material by friction coefficients higher than 0.07. Thus, in order to prove the validity of the derived formulas, an ECAS tool system with a channel angle of 120° will be used in this study.

4. Experimental Results

4.1. Process parameters

For the experimental investigations of ECAS process and validation of the derived analytical formulas, a tool system as shown in 'figure 4' with two forming sections at an channel angle 2Φ=120° and a channel diameter of 20 mm was used. Other geometrical parameters of the tools are demonstrated in 'figure 4'. The oscillation frequency and amplitude of the tools was 30 Hz and 0.7 mm, respectively. The feed speed was 1 mm/s. Commercially pure copper (Cu 99.9%) round bar samples with a diameter of 20 mm were used in the experiments. Commercial FE code of MSC.Marc has been used for the numerical investigation of the ECAS process.

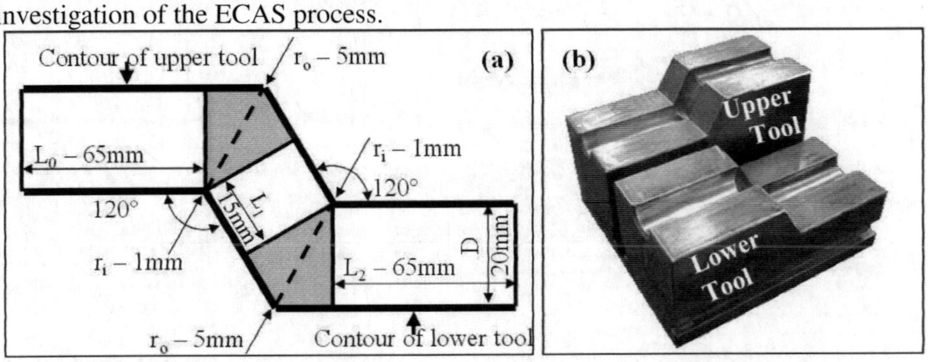

Figure 4. (a) Geometrical parameters and (b) picture of ECAS tool system.

4.2. Strain distribution

In order to verify the feasibility of the developed ECAS tool system, model experiments are conducted with marked copper samples. Therefore, 320 mm long round copper bars are split in the middle surface and then marked with a square grid of 2 mm. Afterwards, these two parts are put together and processed by ECAS. The results of the experiments together with the numerically simulated mesh are demonstrated in 'figure 5'. The entry channel is not fully shown in this figure for a better visualization. The regularly distributed markings in the middle channel indicate a homogeneous shearing of the material along the first yielding line. However, as the sample reaches the second yielding line, the lower section of the workpiece tend to be bent rather than simply sheared. Although in an insignificant manner, a similar effect is also observed on the upper section of the material. Nevertheless, the regular mesh in the middle section of the sample in the exit channel indicates that the deformation is mostly recovered along the second yielding line. The square shape of the grid is preserved after the forming with a little deviation. The deformation in the experiments resembles the one from FE simulations

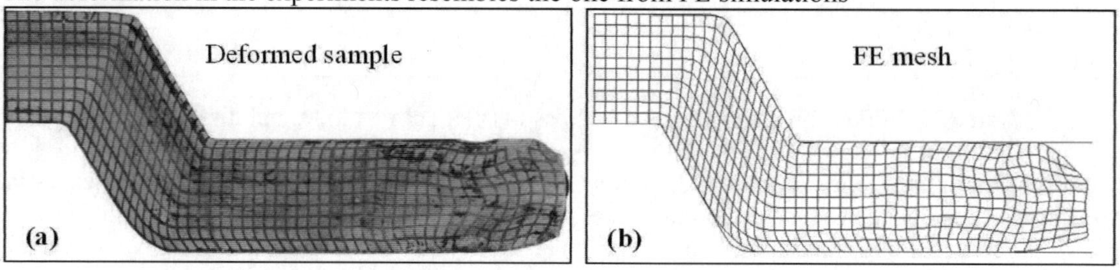

Figure 5. Deformation patterns obtained from (a) experiment and (b) FE simulation.

4.3. Force requirement

In 'figure 6', measured forces are shown with light gray lines, FE results are demonstrated with black lines, analytically predicted ones are displayed with broken lines and their values are written on these. A strain gage based force measurement system was used to measure axial and radial loads during ECAS process. The forming starts with a sudden increase in axial forces and stabilizes shortly after. A similar distribution is observed also when the sample reaches the second forming section. Stabilized forces after the first and second forming sections are 45 kN and 77 kN, respectively. Forces in radial direction after first and second forming section are 74 kN and 143 kN, respectively. It is clear that FE simulations can predict process loads more accurately than the developed analytical formula. It can be explained by investigating the material flow demonstrated in 'figure 5'. By the derivation of the analytical model, it was assumed that the tool geometry is completely filled and the deformation is simple shear along a single line. However, especially along the second deformation section, the samples are rather bent than sheared. Accordingly, an unfilled corner section is present at the second deformation zone.

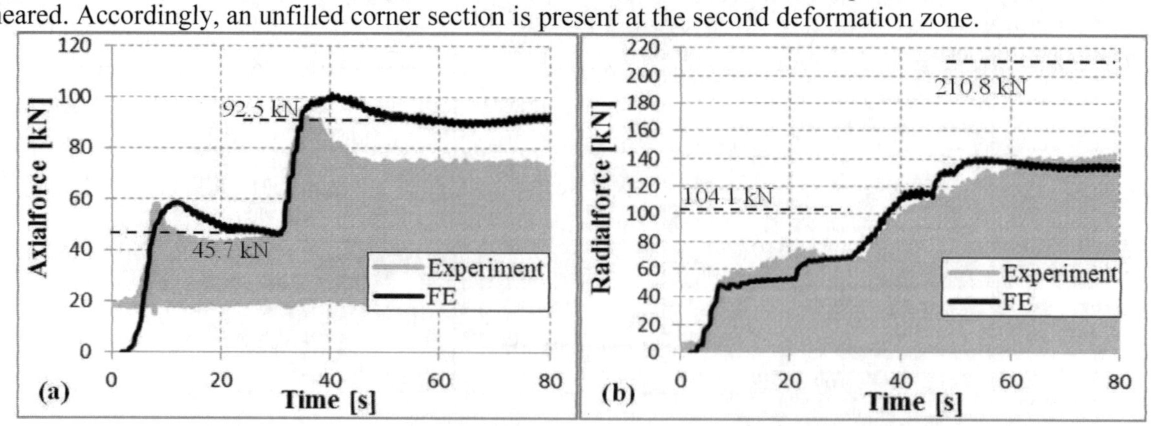

Figure 6. Measured, FE simulated and analytically calculated (a) axial and (b) radial forces.

4.4. Microstructural evolution

In order to investigate the ability of ECAS process to refine grain size, the microstructure of the processed copper samples was analyzed with EBSD in a high resolution SEM after four times ECAS processing using route E. The grain distribution is shown in 'figure 7' as an Image Quality Picture. The dark lines between the grains represent the grain boundaries in this figure. Here, the fine grains dominate the microstructure. The average grain size is slightly less than 0.7 μm. If all grain boundaries are taken into account including the ones in the grains, the proportion of HAGBs is approximately 20 %. On the other hand the proportion of the HAGBs, which separates the grains from each other, is over 80 %.

Figure 7. Microstructure of the copper sample after four ECAS steps.

5. Conclusion and Outlook

A new SPD process called as equal channel angular swaging (ECAS) has been proposed recently which combines the conventional ECAP and the incremental bulk metal forming method of rotary swaging. The most important advantages of ECAS process in comparison to conventional ECAP are a significant reduction in the forces in material feeding direction plus the potential to be extended to continuous processing. In the current study, the mechanics of the ECAS process is investigated using slip line field approach. An analytical model is developed to predict process loads. The proposed model is used to design a new set of ECAS die with low channel angles. Model experiments and FE simulations validated that ECAS process with a low channel angle is feasible. In future, other materials should be tested.

Acknowledgments
This work was supported by Research Fund of the Hacettepe University. Project Number: 13661. The author wishes to thank Dr. Enrico Bruder for his support in the microstructure analysis.

References

[1] Valiev R Z, Islamgaliev R K and Alexandrov I V 2000 *Prog. Mater. Sci.* **45** 103
[2] Valiev R Z, Estrin Y, Horita Z, Langdon T G, Zehetbauer M J and Zhu Y T 2006 *JOM* **58/4** 33
[3] Azushima A, Kopp R, Korhonen A, Yang D Y, Micari F, Lahoti G D, Groche P, Yanagimoto J, Tsuji N, Rosochowski A and Yanagida A 2008 *CIRP Annals* **57/2** 716-735
[4] Rosochowski A and Olejnik L 2007 *Proc. 10th ESAFORM Conf. on Material Forming (Zaragoza, Spain)* 653-658
[5] Rosochowski A and Olejnik L 2008 *Int. J. Mater. Form.* **1** 483-486.
[6] Rosochowski A, Rosochowska M, Olejnik L and Verlinden B 2010 *steel res. int.* **81** 470-473.
[7] Görtan, M O, Bruder E, Grcohe, P and Müller C 2010 *steel res. int.* **81** 478-481
[8] Bruder E, Görtan M O, Groche P and Müller C 2011 *Mater. Sci. For.* **667-669** 103-107
[9] Görtan M O, Bruder E, Groche P and Müller C 2012 *Proc. 18th PLASTICITY'12 (Puerto Rico)* 226-228
[10] Segal V M 2003 *Mater. Sci. Eng. A* **345** 36-46
[11] Segal V M 2004 *Mater. Sci. Eng. A* **386** 269-276

Experimental and numerical analysis of HPTE on mechanical properties of materials and strain distribution

B Omranpour[1,2,3*], **R Kulagin**[1], **Y Ivanisenko**[1],**E Garcia Sanchez**[2]

1- Institute of Nanotechnology (INT), Karlsruhe Institute of Technology (KIT), Germany
2- Facultad de Ingeniería Mecánica y Eléctrica (FIME), Universidad Autónoma de Nuevo León (UANL), Mexico
3- Department of Materials Engineering, Tallinn University of Technology (TTU), Estonia

* Email: babak.o@daad-alumni.de

Abstract. High Pressure Torsion Extrusion (HPTE) is a novel technique which has been recently introduced to the society of Nano-SPD researchers. HPTE exploits the deformation mechanics of HPT but in a larger scale using rod-shape samples and is capable of applying high values of strain to materials in one pass. This research aims to evaluate the effect of HPTE on mechanical properties of materials and also to study the effect of geometry of HPTE die on strain distribution in deformed samples by using Finite Element Method (FEM). Commercial pure Aluminium AA1050 was used for experimental work; and eccentric dies with parallel-misaligned channels were developed for evaluation by numerical modelling. Results of this research will help us better understand the effect of process parameters and also geometry of the die on materials.

Keywords: High Pressure Torsion Extrusion (HPTE); Die design; FEM; Eccentric dies

1. Introduction

Severe Plastic Deformations are very famous techniques in producing Ultra Fine-Grained materials (UFG) by imposing high strains on materials and subdividing original coarse grains into smaller subgrains. This reduction of grain size from micro-meters towards nanometers affects the mechanical properties of materials such as strength, hardness and ductility [1]. Basically in SPD processes, yield strength and hardness are improved, and ductility is reduced; although there are some reports about retaining ductility along with improving other mechanical properties [2].

High Pressure Torsion Extrusion, hereinafter called as HPTE, is a novel technique of Severe Plastic Deformation, introduced in 2015 [3] and has distinctive characteristics in deformation mechanism and properties. HPTE is composed of two separate upper and lower dies rotating relative to each other and a punch will extrude the sample through the die. HPTE approaches a combination of mechanisms of deformation including conventional High Pressure Torsion (HPT) [4] and Cyclic Expansion and Extrusion (CEE) [5] [6]. The major advantage of HPTE over HPT is the possibility of producing rod-shape samples, similar to ECAP; whereas in HPT, using a small disk with an average thickness of less than 1 mm restricts the applications to laboratory research [4] [7]. Plastic strain in this process is composed of torsion, expansion and extrusion (Figure1). Accumulated strain after one pass of HPTE can be calculated by the following equation:

IOP Conf. Series: Materials Science and Engineering **194** (2017) 012047 doi:10.1088/1757-899X/194/1/012047

$$E = 2\ln\frac{D1}{D0} + 2\ln\frac{D1}{D2} + \frac{\omega \cdot R \cdot D1}{\sqrt{3} \cdot v \cdot D2} \tag{1}$$

Where D0, D1, and D2 are diameters of inlet channel, deformation chamber and the outlet channel, respectively; ω is the rotational speed of the lower die, v is extruding speed and R is the desired distance from the center at the section of the sample. The first two terms of this equation correspond to expansion and extrusion, and the third term is referred to the plastic strain resulted by torsion [3].

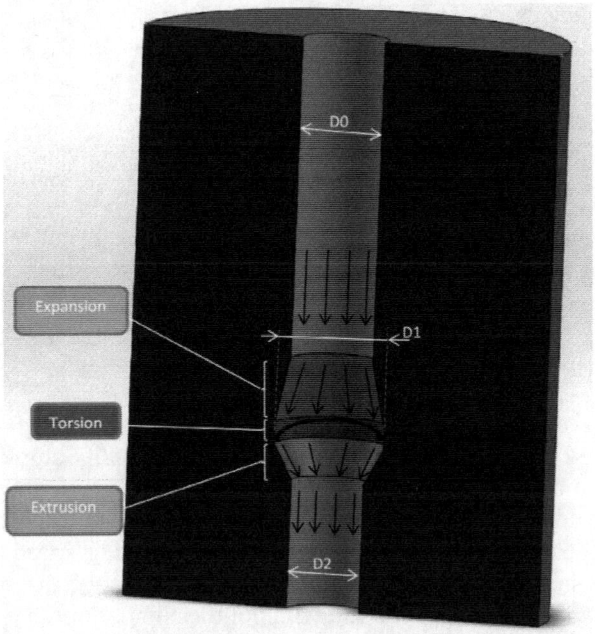

Figure 1. Schematic illustration of HPTE die

A key factor in HPTE die is concerned with the design of inner side of the die at the shear zone. Torsion at this zone is provided by rotating the lower die in order for twisting the workpiece and applying high amounts of shear strain to the specimen. Since the initial state of the material has a cylindrical shape, it is necessary to provide enough friction torque to prevent the materials from slippage. This issue is solved by using jutting surfaces inside the die. These jutting surfaces act as holding elements (resembling hexagon socket head cap screws) to constrain the material and apply torsion when the die is rotating (Figure 2.C).

This research in the first step tries to evaluate the effect of HPTE on mechanical properties of materials and in the second step, intends to propose a new design with eccentric channels using Finite Element Methods (FEM). Numerical modeling methods such as FEM are nowadays very common in metal forming processes either to study different aspects of a new technique or to optimize a process [8][9]. FEM as the most popular technique in numerical modeling is capable of assessing different parameters of a process including tool geometry, process conditions and materials properties [10]. The aim of simulation in this work is to devise a new design of HPTE die with eccentric channels towards manipulating strain distribution and adapting new properties in the materials. A theoretical concept of this idea firstly was introduced in an earlier work [3].

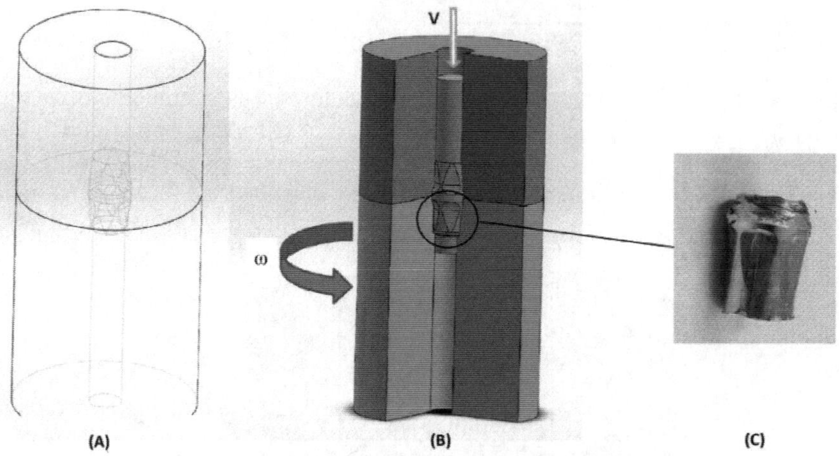

Figure 2. (A) Wireframe demonstration of HPTE die; (B) Process parameters of HPTE: Extruding speed, V and rotational speed, ω; (C) Small section of a workpiece extracted from the deformation zone of the die

2. Methodology

2.1. Experimental testing and evaluation

Processing of HPTE was performed on commercial pure Aluminium AA1050. Samples were prepared in the dimensions of 11.6 mm in diameter and 13.5mm in length. One specimen was considered as a reference and annealed at 350 °c for 30 minutes. Computer-controlled HPTE machine executed the process at room temperature with rotational speed of ω=1 rpm and extruding speed of V=7mm/min. Instant pressing force values together with extruding speed and rotational speed were monitored and recorded in real-time operating system to be used for validation of the simulations.

In order for mechanical testing of materials two testing methods of compression and hardness tests were conducted on the materials. A "SHIMADZU" universal testing machine performed the compression test on the samples with a compressive speed of 2 mm/min; Teflon sheet was used as a solid lubricant. Criterion of maximum length reduction of 66.67% (e=2/3) was established in the test. "Buehler Micromet-5104" tester applied micro-indentation at the cross section of the specimens on 11 points at regular intervals to measure Vickers hardness. Pressing load of 0.2 Kg.F with a dwell time of 15 seconds was conducted into all samples.

2.2. FEM simulations:

In this study, DEFORM-Ver.11 was implemented for simulation of the process. Assumptions of simulation rest on using the same extruding and rotational speed as in the experiments (ω=1 rpm, V=7mm/min), friction coefficient of 0.4, neglecting the deformation heating and considering the die as rigid body.

2.3. Scheme of the new design

First and foremost, an FEM simulation of the base die with the exact dimensions was carried out to verify the simulations; thereafter, a new design with eccentric channels was modeled. This new die includes two channels with the same diameters as the base die, but they are misaligned in parallel lines. Therefore there is an eccentricity between the upper and lower channels. At the beginning of the process, there is no eccentricity (Phi=0) (Figure 3.A) but after rotation, it gradually increases to the maximum value of 2mm when the lower die is rotated 180 degrees (Phi=180°) (Figure 3.B).

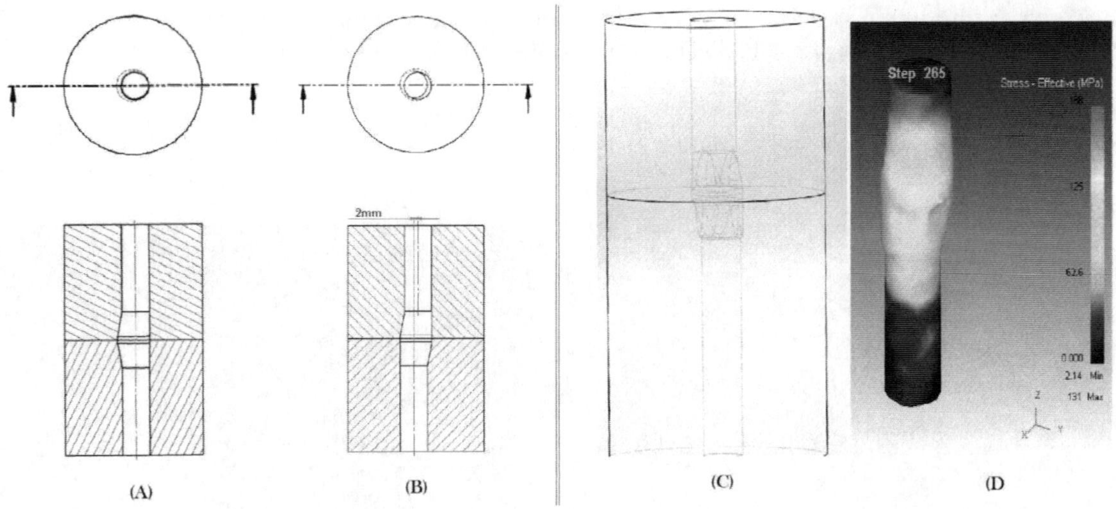

Figure 3. HPTE die with eccentric channels: (A) At the beginning of the process, both upper and lower dies are perfectly aligned and there is no eccentricity; (B) Maximum eccentricity appears after 180° of rotation; (C) Wireframe illustration of HPTE die with eccentric channels; (D) Numerical modeling of the process

3. Results and discussion:

3.1. Mechanical testing

Following figures depict results of compression test and hardness test on the annealed and HPTE processed samples. Results of compression test are demonstrated in True stress-strain diagram in which yield strength of 44MPa as the initial case reached up to 141 MPa after processing.

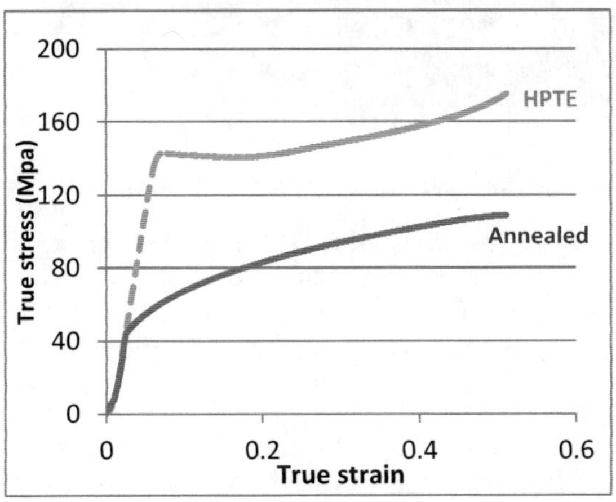

Figure 4. True stress-true strain diagram of HPTE processed and initial samples obtained by compression testing

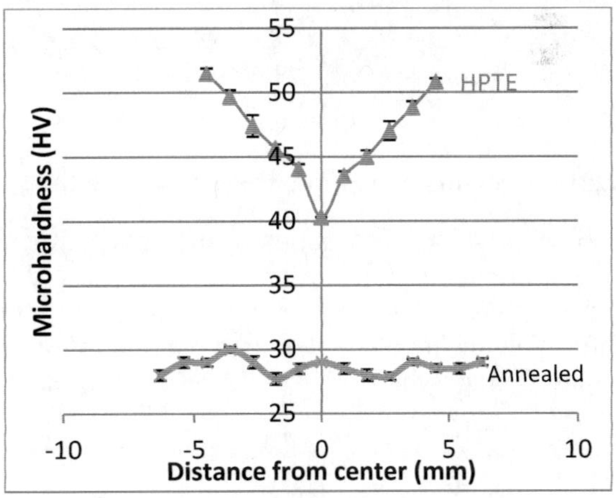

Figure 5. Results of Hardness testing

Results of hardness testing showed a sharp increase in hardness values after HPTE processing. One interesting point about this graph is that hardness distribution at the cross section of the samples is quite similar to that of HPT samples [11], that is to say lower values are located in the center and higher values are found near the periphery. The gradient of hardness varies from a minimum value of

40HV to the maximum value of 52 along the radius of the specimen, which implies a successful approach in improvement of mechanical properties of materials by means of HPTE.

3.2. Simulation results:

Simulated models were cut in the middle and then twenty nodes, evenly spaced apart, were selected from the cross section of the specimens.

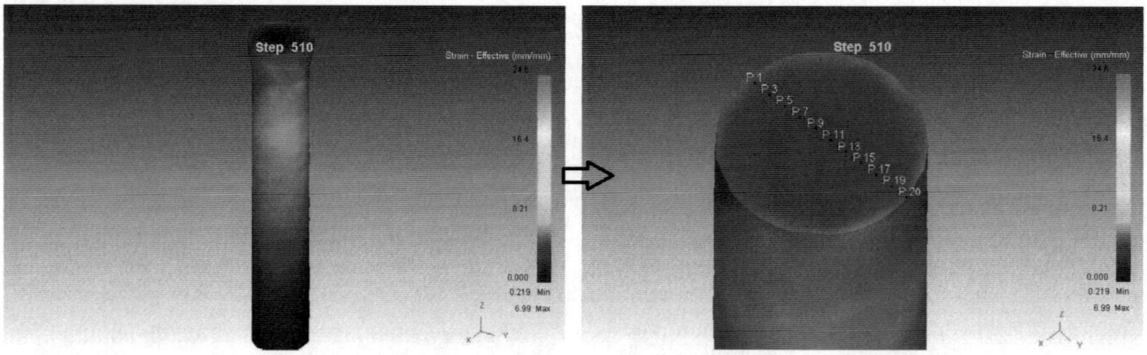

Figure 6. Selection of random nodes from the cross section of the models

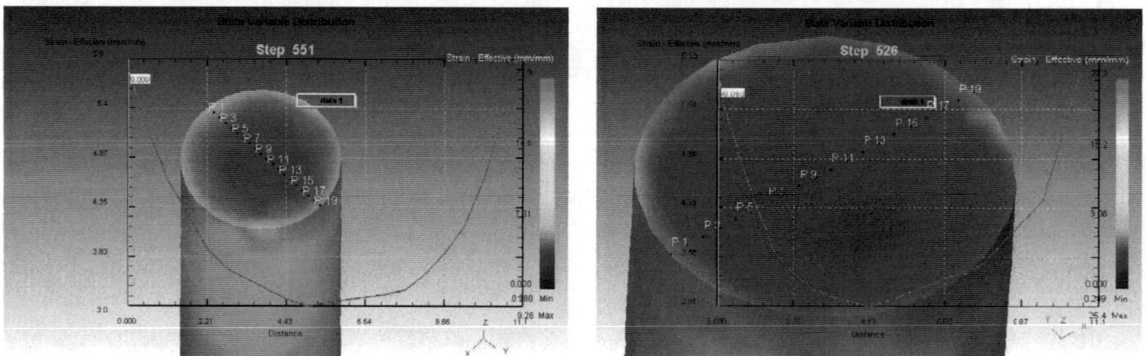

(A) Base die (B) Eccentric die

Figure 7. Strain distribution diagrams within the specimens for the base die (A) and eccentric die (B)

Figure 7 shows distribution of effective strain applied to the materials after HPTE. In this figure, strain distribution varies from 3 to 5.5 in base die, while in the eccentric die this variation changes from 2.9 to 6.1. Based on the results, there is no big difference in the minimum level of strain in the central part of both samples. But maximum values of strain in those areas which are close to the periphery, slightly increased in the presence of eccentricity.

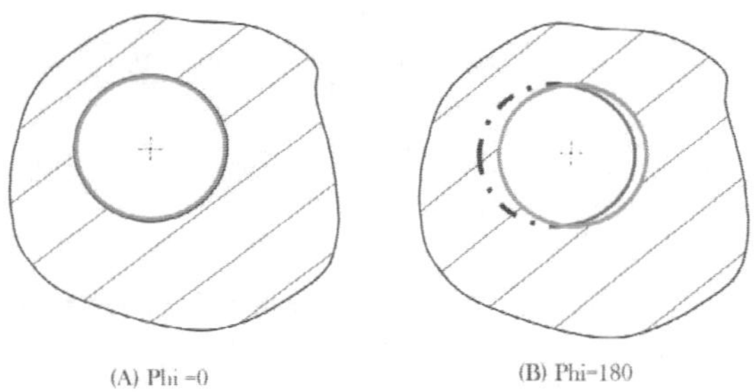

Firure 8. Cross section view of an eccentric die

(A) Both upper and lower channels are aligned; effective radius and nominal radius are equal.

(B) Eccentricity of two channels after rotation of the lower die at the phase of 180°; Effective radius in torsion is bigger that the nominal one

This phenomenon can be explained by considering the fact that strain in the central area of the specimens is mainly resulted from expansion and extrusion; and eccentricity does not really affect the strain at this area. On the other hand, at the outer parts of the specimen, when eccentricity reaches to its maximum value, effective radius at the deformation zone will be increased; and consequently strain which is caused by torsion will rise in value (Figure 8).

4. Conclusion

A new SPD process was evaluated and analyzed using Aluminium specimens. Mechanical properties of materials by two different methods of hardness testing and compression testing were studied. Moreover, some modifications in die design were proposed by implementing FEM. Based on the experimental results, after one pass of HPTE hardness values increased by 79% from an initial state of 29 HV to the final state of 52HV near the edge of the samples; and yield strength in compression test improved by 220% from an initial value of 44MPa to 141MPa.

According to the simulation results, the new die with eccentric channels showed a gradual increase in the level of strain near the periphery of the sectioned samples.

5. Acknowledgment

This project was sponsored in part by *Deutscher Akademischer Austauschdienst* (DAAD)- Germany and in part by *National Council of Science and Technology* (CONACYT)- Mexico. Technical and scientific support for running the compression test at CIDET-UANL is greatly appreciated.

References

[1] Rosochowski A, Olejnik L and Richert M 2007 *Advanced Methods in Material Forming*; Banabic D (Heidelberg: Springer); pp215-221.

[2] Wang Y, Chen M and Zhou F 2002 Nature; 419 pp 912-914.

[3] Ivanisenko Y, et al 2016 Materials science& Engineering A; 664 pp 247-256.

[4] Zhilyaev A and Langdon T 2008 Progress in Materials science; 53 pp 893-979.

[5] Pardis N, et al 2011 Materials science and Engineering A; 528 pp 7537-7540.

[6] Pardis N, et al 2015 Materials science and engineering A; 628 pp 423-432.

[7] Estrin Y and Vinogradov A 2013 Acta materialia; 61 pp 782-817.

[8] Djavanroodi F, Omranpour B and Sedighi M 2013 Materials and Manufacturing processes; 28 pp 276-281.

[9] Djavanroodi F, Omranpour B, Sedighi M, Ebrahimi M. Progress in natural science: Materials international. 2012; 22: p. 452-460.

[10] Rosochowski A and Olejnik L 2007 Journal of materials: Design and Applications; 221 pp 187-197.

[11] Sabbaghianrad S and Langdon TG 2015 Journal of materials science; 50 pp 4357-4365.

IOP Conf. Series: Materials Science and Engineering **194** (2017) 012048 doi:10.1088/1757-899X/194/1/012048

Multiscale Crystal Plasticity Modeling Considering Nucleation of Dislocations Based on Thermal Activation Process on Ultrafine-grained Aluminum

Y Aoyagi

Department of Finemechanics, Tohoku University
6-6-01 Aoba, Aramaki, Aoba-ku, Sendai, 980-8579, JAPAN

aoyagi@tohoku.ac.jp

Abstract. In this study, a crystal plasticity model expressing the behavior of the dislocation source and the mobile dislocations is proposed by considering a thermal activation process of dislocations. In order to predict the variation of critical resolved shear stress due to grain boundaries, mobile dislocations, or dislocation sources, information on these crystal defects is introduced into a hardening law of crystal plasticity. The crystal orientation and shape of ultrafine-grained (UFG) aluminum produced by accumulative roll bonding processes are measured by electron backscatter diffraction (EBSD). Mechanical properties of the UFG aluminum are estimated using tensile test and indentation test. Results obtained by EBSD are introduced into a computational model. Finite element simulation for polycrystal of aluminum investigates the effect of microstructure on mechanical properties of UFG aluminum.

1. Introduction

High-strength, light-weight metal materials are required to improve safety and reduce transportation cost. Ultrafine-grained (UFG) metals produced by severe plastic deformation have attracted interest as high-strength materials [1]-[4]. UFG metals with a grain size of less than 1μm exhibit remarkable mechanical properties (e.g., increased yield stress, decreased hardening ratio, yield point drop of FCC metal [5], temperature dependency [6], tension-compression asymmetry [7], and increased strain-rate sensitivity [8]). A computational model predicting these properties is desired in the field of materials science and engineering. Multiscale crystal plasticity models expressing a size effect on grain size have been proposed by introducing information on grain size as a material parameter [9], [10]. However, conventional theory cannot express the effect of the dislocation source and the grain boundary precisely and directly. It is assumed that such unusual mechanical properties originate from the exhaustion of dislocation sources and the enormous volume fraction of the grain boundary. Recently, the effect of the grain boundary has attracted the attention of many researchers as an important factor determining the mechanical properties of UFG metal [11]-[15].

In this study, a crystal plasticity model expressing the behavior of the dislocation source and the mobile dislocations is proposed by considering a thermal activation process of dislocations. In order to predict the variation of critical resolved shear stress due to grain boundaries, mobile dislocations, or dislocation sources, information on these crystal defects is introduced into a hardening law of crystal plasticity. The crystal orientation and shape of UFG aluminum produced by accumulative roll bonding (ARB) processes are measured by electron backscatter diffraction (EBSD). Mechanical properties of the UFG aluminum are estimated using tensile and indentation tests. Using results obtained by EBSD, information on crystal orientation and shape are introduced into a computational model for multiscale crystal plasticity simulation. Finite element simulation for polycrystal of aluminum investigates the effects of dislocation behavior and crystal orientation on mechanical properties of UFG aluminum.

2. Multiscale crystal plasticity model

In previous work, we defined the flow stress of crystal plasticity considering information on dislocations and grain boundaries, which are dislocation sources [16]. Since the work hardening caused by accumulation of dislocations and the generation of dislocations should alternatively occur, the flow stress of this model is modified as

$$g^{(\alpha)}=\max\{\tau_d^{(\alpha)},\ \min(\tau_s^{(\alpha)},\ \tau_m^{(\alpha)})\} \tag{1}$$

where $g^{(\alpha)}$ is flow stress on the slip system α, $\tau_d^{(\alpha)}$ is the deformation resistance originating in accumulated dislocations, $\tau_s^{(\alpha)}$ is that originating in the initial dislocation sources, and $\tau_m^{(\alpha)}$ is that originating in mobile dislocations. Equation (1) indicates that plastic deformation is caused by activation of the dislocation source that is most easily activated. Usually, $\tau_d^{(\alpha)}$ is given by the extended Bailey-Hirsch equation [17]. Frequency on thermal activation process a at a dislocation segment is given by

$$p_a=\frac{N_a}{N_r}v\exp\left\{\frac{-\Delta G_a(1-\tau/\tau_a)}{k_B T}\right\} \tag{2}$$

where $N_a=\rho_a/\tilde{b}$ is the number of dislocation segments, N_r is the reference value of N_a, v is the atomic frequency, ΔG_a is the thermal activation energy, τ is the applied shear stress, τ_a is the reference stress on the process a, k_B is the Boltzmann constant, and T is the temperature. When $p_a=1$, $\tau_s^{(\alpha)}$ and $\tau_m^{(\alpha)}$ are obtained as

$$\tau_s^{(\alpha)}=\tau_{FR}-\frac{k_B T}{\tilde{b}^3}\ln\frac{\rho_s^{(\alpha)}}{\tilde{b}}\frac{v}{N_r},\quad \tau_m^{(\alpha)}=\tau_{MD}^{(\alpha)}-\frac{k_B T}{\tilde{b}^3}\ln\frac{\rho_{gn}^{(\alpha)}}{\tilde{b}}\frac{v}{N_r} \tag{3}$$

where τ_{FR} is the resolved shear stress when dislocation loops are released from a Frank-Read (F-R) source, $\tilde{b}$ is the magnitude of the Burgers vector, $\rho_s^{(\alpha)}$ is the density of the dislocation source, $\tau_{MD}^{(\alpha)}$ is the resolved shear stress when the mobile dislocations move, and $\rho_{gn}^{(\alpha)}$ is the mobile dislocation density. The variables in equation (3) are given by

$$\rho_s^{(\alpha)}=\rho_{s0}^{(\alpha)}+e(\rho_d^{(\alpha)}-\rho_0^{(\alpha)}),\ \rho_{gn}^{(\alpha)}=\left\|\boldsymbol{\rho}_{gn}^{(\alpha)}\right\|=\left\|(\boldsymbol{s}^{(\alpha)}\otimes\boldsymbol{m}^{(\alpha)})\times\nabla\gamma^{(\alpha)}/\tilde{b}\right\|,$$
$$\tau_{MD}^{(\alpha)}=(\tau_{GT}+\tau_{PN})/2-(\tau_{GT}-\tau_{PN})\tanh\left\{c(\phi_g^{(\alpha)}-\phi_r)/\phi_r\right\}/2,\ \phi_g^{(\alpha)}=|\mathrm{curl}\boldsymbol{m}^{(\alpha)}| \tag{4}$$

where $\rho_{s0}^{(\alpha)}$ is the initial density of dislocation sources, e is the ratio of the dislocation density on dislocation sources and the others, $\rho_d^{(\beta)}$ is the dislocation density, $\rho_0^{(\alpha)}$ is the initial dislocation density, $\boldsymbol{\rho}_{gn}^{(\alpha)}$ is the geometrically necessary (GN) dislocation density tensor [17], $\boldsymbol{s}^{(\alpha)}$ is the slip direction vector, $\boldsymbol{m}^{(\alpha)}$ is the normal direction of the slip plane, $\gamma^{(\alpha)}$ is the slip, τ_{GT} is the resolved shear stress when the mobile dislocations across grain boundary, τ_{PN} is the critical shear stress based on the Peierls-Nabarro stress, $\phi_g^{(\alpha)}$ is the misorientation parameter, and ϕ_r is the threshold of misorientation. Equation (3) represents the local increase of CRSS in a region without a dislocation source or a mobile dislocation. It is reported that thermal activation of the F-R sources hardly affects some mechanical properties of UFG metals [18]. Under the situation assumed in this paper, the value of τ is about the same as or larger than that of τ_a. Therefore, the influence of the thermal activation process on activation of the F-R source is extremely small. Equation $(3)_1$ is modeled to express information on the decrease in the frequency of plastic deformation due to exhaustion of dislocation sources, i.e. the change of $\tau_s^{(\alpha)}$ depending on the density of dislocation source. An elasto-viscoplastic constitutive equation for a crystal plasticity model, the hardening modulus representing the interaction of slip systems, and the definition of dislocation density are same as the previous model [16]. The rate of dislocation density $\dot{\rho}_d^{(\beta)}$ is given by

IOP Conf. Series: Materials Science and Engineering **194** (2017) 012048 doi:10.1088/1757-899X/194/1/012048

$$\dot{\rho}_d^{(\beta)} = (1 - k\rho_d^{(\beta)}) r \rho_f \left|\dot{\gamma}^{(\beta)}\right| / \tilde{b}, \tag{5}$$

where k is the annihilation ratio and corresponds to the inverse of the saturated density of dislocation, r is the mean length of the remaining dislocation lines, and ρ_f is the density of forest dislocations.

3. Methods of measuring and properties of UFG material

3.1. Experiment Procedures

Ultrafine-graining using the ARB process [3] is performed for industrial pure aluminum JIS-A1050. In this study, a 2 mm-thick aluminum sheet is rolled to 1 mm thickness at room temperature. A UFG aluminum is produced by eight cycles of ARB. The material after eight cycles of ARB is not annealed. We treat unrolled specimens that are annealed for 1 h at 673 K as a starting material. The grain size and crystal orientation of specimens made from the starting and the UFG materials are estimated using EBSD from the transverse direction (TD) that is perpendicular to the normal direction (ND) and the rolling direction (RD). Mechanical properties of these materials are measured by tensile tests with a universal testing machine. The tensile direction coincides with the RD. The thickness of specimens for the starting materials is 2 mm and that for the UFG materials is 1 mm, and the gage length for the UFG material is 12mm, and the width is 5mm. Micro indentation tests using a spherical indenter are performed for the UFG specimen after 0.0%, 0.5%, 1.0%, 1.5%, 2.0%, and 2.5% tension in order to investigate the effect of pre-strain on the mechanical properties of the UFG material. The radius of the indenter is 50 μm. The indenter is forced into the surface of the specimen from the ND. The maximum load is 0.8826 N, the forcing time is 10 s, the force duration is 1 s, and the unloading time is 10 s.

3.2. Experiment Results for the Starting and UFG Materials

Figures 1 and 2 depict the distribution of crystal orientation and the <111> pole figure of the starting and UFG materials measured using EBSD. The color map depicted in these figures is hereinafter used

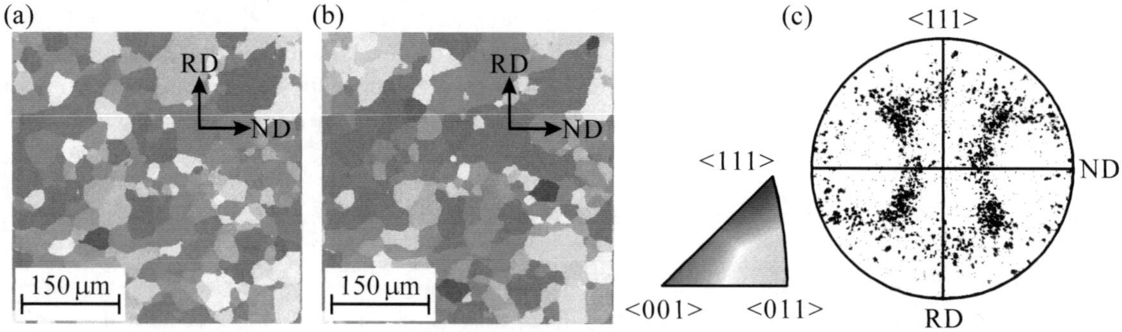

Figure 1. Crystal orientation for (a) ND and (b) RD, (c) the <111> pole figure for the starting material measured using EBSD.

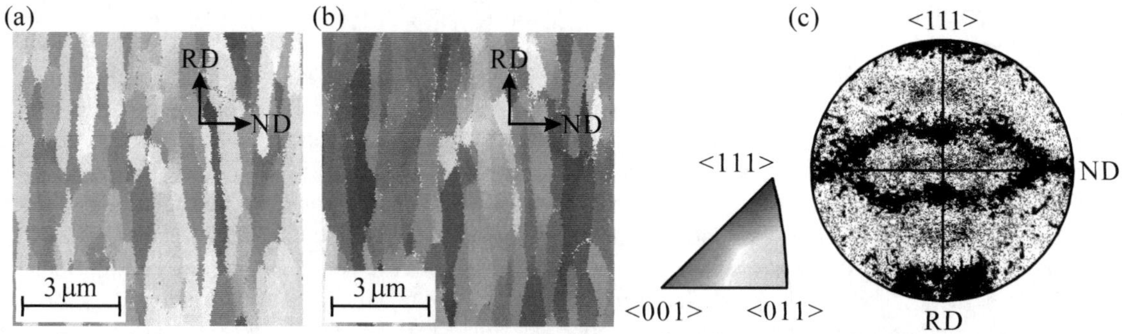

Figure 2. Crystal orientation for (a) ND and (b) RD, and (c) the <111> pole figure for the UFG material measured using EBSD.

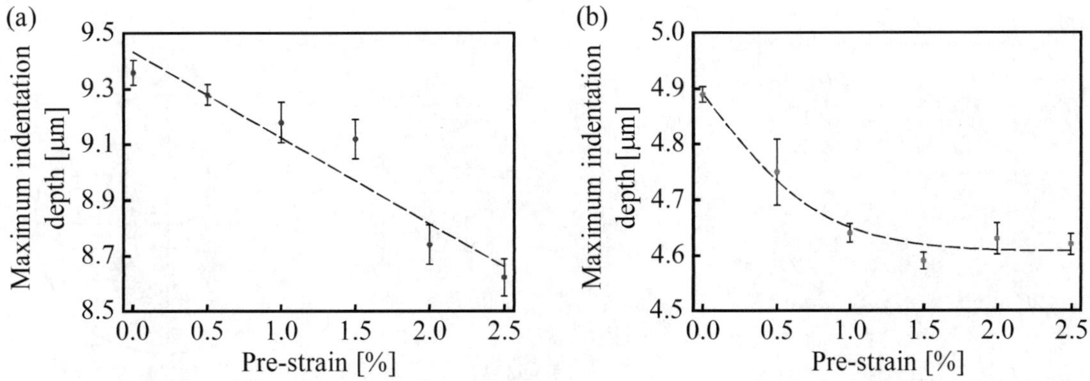

Figure 3. Maximum indentation depth for the starting (a) and UFG (b) materials.

for the color of crystal orientation. The mean grain size of the starting material along the ND is 30 μm, and that along the RD is 32 μm. In contrast, the mean grain size of the UFG material along the ND is 0.39 μm, and that along the RD is 1.8 μm. The rolling texture is developed by eight cycles of ARB. Figure 2 indicates that <111> directions tend to orient to the RD. Result obtained by the tensile tests are shown later with simulation results.

Figure 3 plots the maximum indentation depth measured by the micro indentation test for the pre-strained starting and UFG materials. The maximum indentation depth of the starting material decreases proportionally with the increase of pre-strain because of work-hardening by tensile deformation. The maximum indentation depth for the UFG material strengthened by the ARB process is less than that of the starting material and approaches the fixed value after 1.0% pre-strain. This result indicates that dislocation density is saturated after 1.0% pre-strain, and the work-hardening ratio decreases.

4. Pseudo three-dimensional simulation of aluminum

4.1. Numerical Procedures

FE method is applied for the present crystal plasticity analysis. Figures 4 and 5 depict periodic unit-cell models used in this calculation. We conduct a pseudo three-dimensional computation, which restricts the three-dimensional plastic deformation of metal to two-dimensional components on an observation plane [16]. In the present simulation, a polycrystal of aluminum under plane stress is assumed, and we control displacement at the overlapping vertexes depicted by black dots in figures 4(a) and 5(a), so that the volume mean of the stress component *xx* becomes zero, and tensile displacement in the RD is applied at the upper vertexes of the specimen. The aspect ratio of grains is 1 for the starting material or 5 for the UFG material. The initial crystal orientations of each grain, which are randomly selected from the results of EBSD, are depicted in figures 4 and 5. The red dots in figures 4(c) and 5(c) represent the <111> pole figures for the computational model.

The material constants and numerical parameters are the critical shear stress based on Peierls-Nabarro stress $\tau_{PN}=10\,\mathrm{MPa}$; the minimum shear stress when dislocations across grain boundaries $\tau_{GT}=300$ MPa ; the threshold of misorientation $\phi_r=3\times10^7$; the annihilation ratio $k=2\times10^{-15}$; and the mean length of the remaining dislocation lines $r=4$ nm . In general, the shear stress when dislocation loops are released from the F-R source are given by $\tau_{FR}=\mu b/l$, where l is the distance between slip planes. The values of the total initial dislocation density ρ_{t0} is measurable physical quantities. However, it is difficult to estimate experimentally these values. Therefore, this value is determined by fitting a simulated stress-strain curve to the experiment curve. The size of the dislocation source for the starting material is one third the size of each grain (i.e., 10 μm), and that for the UFG material is 0.3 μm , which is slightly smaller than the mean grain size along the ND.

4.2. Numerical Results and Discussion

Figure 6 plots the nominal stress versus nominal strain curves with the experiment one for the starting

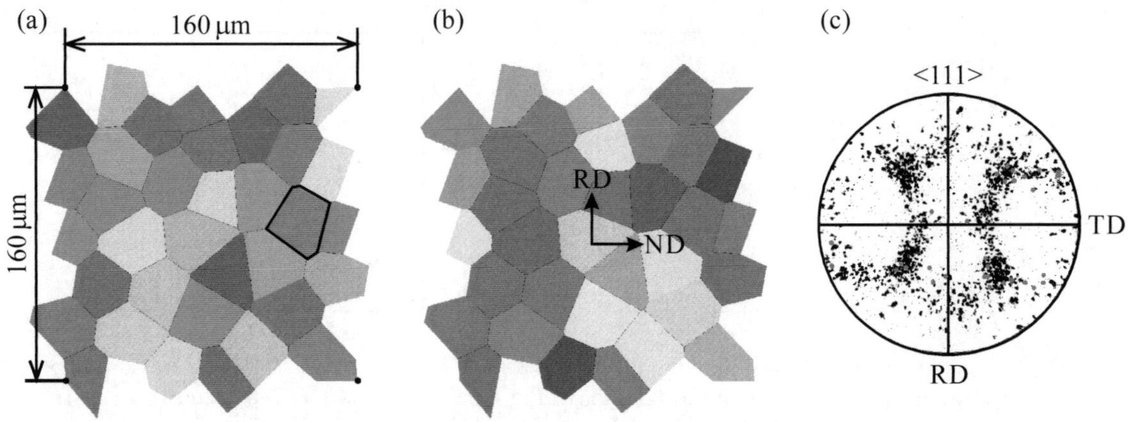

Figure 4. Crystal orientation in the ND (a), RD (b), and <111> pole figure for the starting material.

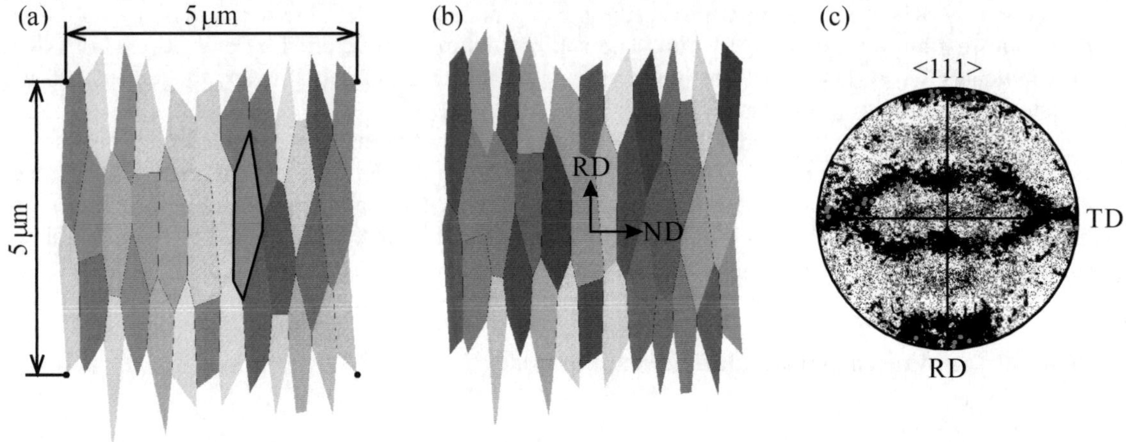

Figure 5. Crystal orientation to in ND (a), RD (b), and <111> pole figure for the UFG material (c).

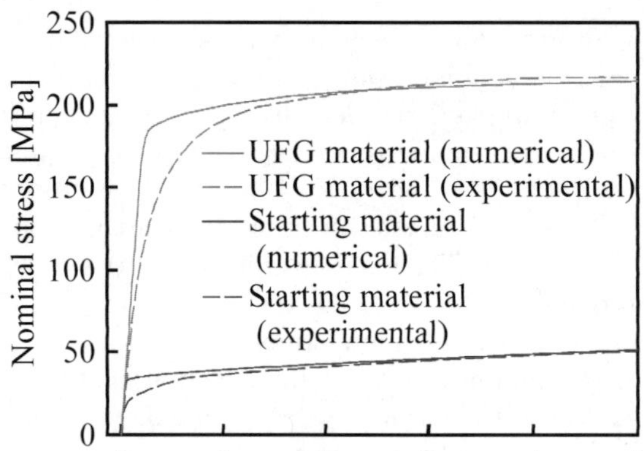

Figure 6. Nominal stress versus nominal strain curves for the starting and UFG materials calculated by the multiscale crystal plasticity model.

and UFG materials. The values of ρ_{t0} are determined as 3×10^{12} m^{-2} and 1×10^{14} m^{-2} for the starting and UFG materials by this simulation. General curves can be reproduced by adjustment of these values.

Figure 7 plots the mean dislocation density of the primary and secondary slip systems of a grain, denoted by bold lines in figures 4(a) and 5(a). The dislocation density of the starting material does not reach the saturated dislocation density at 5% elongation. The dislocation density of the primal slip

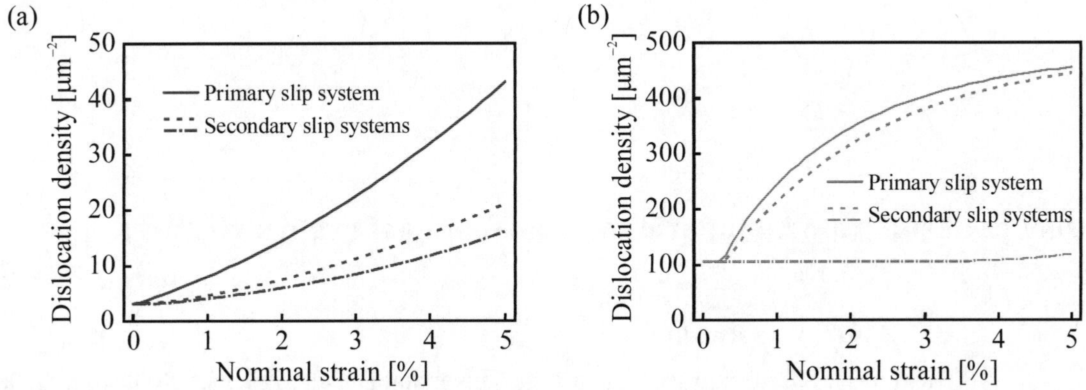

Figure 7. Dislocation density of the primary and secondary slip systems in a central grain of the specimen for the (a) starting and (b) UFG materials.

system for the UFG material approaches the saturated dislocation density at the early stage of deformation. Comparison of figures 3 and 7 indicates that the nominal strains at which the gradient of curve changes are similar. The simulation result agrees well with the experimental tendency [19] that the dislocation density of the starting material increases gradually after yielding and the increase rate of the dislocation density of the UFG material decreases with the decrease of the work-hardening ratio.

These results suggest that the work-hardening ratio becomes low when the dislocation density in the primary slip system is saturated. Since the decrease of work-hardening caused by saturation of dislocation density reduces ductility, total elongation for the UFG material becomes lower than for the starting material.

5. Conclusions

Fitting a stress-strain curve calculated by the multiscale crystal plasticity model to the experiment curve determines the initial dislocation density. The dislocation density of the UFG material is saturated in the early stage of tensile deformation, and the work-hardening ratio becomes low.

References

[1] Horita Z, Smith D J, Furukawa M, Nemoto M, Valiev R Z and Langdon T G 1996 *J. Mater. Res.* **11** 1880
[2] Belyakov A, Gao W, Miura H and Sakai T 1998 *Metall. Mater. Trans A* **29** 2957
[3] Saito Y, Tsuji N, Utsunomiya H, Sakai T and Hong R G 1998 *Scr. Mater.* **39** 1221
[4] Kim H-W, Kanga S-B, Tsuji N and Minamino Y 2005 *Acta Mater.* **53** 1737
[5] Segal V M, Ferrasse S and Alford F 2006 *Mater. Sic. Eng. A* **422** 321
[6] Duckham A, Zhang D Z, Liang D, Luzin V, Cammarata R C, Leheny R L, Chien C L and Weihs T P 2003 *Acta Mater.* **51** 4083
[7] Cheng S, Spencer J A and Milligan W 2003 *Acta Mater.* **51** 4505
[8] Sabirov I, Barnett M R, Estrin Y and Hodgson P D 2009 *Scrpta Mater.* **61** 181
[9] Ohashi T, Kawamukai M and Zbib H M 2007 *Int. J. Plasticity* **23** 897
[10] Ohno N and Okumura D 2007 *J. Mech. Phys. Solids* **55** 1879
[11] Lim H, Lee M G, Kim J H, Adams B L and Wagoner R H 2011 *Int. J. Plasticity* **27** 1328
[12] Duhamel C, Brechet Y and Champion Y 2010 *Int. J. Plasticity* **26** 747
[13] Lee M G, Lim H, Adams B L, Hirth J P and Wagoner R H 2010 *Int. J. Plasticity* **26** 925
[14] Liu Z L, Zhuang Z, Liu X M, Zhao X C and Zhang Z H 2011 *Int. J. Plasticity* **27** 201
[15] Aoyagi Y, Kobayashi R, Kaji Y and Shizawa K 2013 *Int. J. Plasticity* **47** 13
[16] Aoyagi Y, Tsuru T and Shimokawa T 2014 *Int. J. Plasticity* **55** 43
[17] Aoyagi Y and Shizawa K 2007 *Int. J. Plasticity* **23** 1022
[18] Estrin Y, Kim H S and Nabarro F R N 2007 *Acta Mater.* **55** 6401
[19] Adachi H, Miyajima Y, Sato M and Tsuji N 2014 *J. Jpn. Inst. Light Met.* **64** 463 (Japanese)

IOP Conf. Series: Materials Science and Engineering **194** (2017) 012049 doi:10.1088/1757-899X/194/1/012049

Influence of subgrain boundaries on coarsening of grain structures

D. Zöllner, W. Skrotzki

Institut für Strukturphysik, Technische Universität Dresden, Haeckelstraße 3, D-01069 Dresden, Germany

dana.zoellner@ovgu.de

Abstract. In the present work, the influence of subgrain boundaries on the coarsening kinetics of individual grains embedded in an average environment as well as within a grain structure is investigated. It is found that a specific introduction of subgrain boundaries not only influences the speed with which grains shrink or grow, but in contrast to the von Neumann-Mullins-law, a distinct manipulation of the location of the subgrain boundaries allows even grains with few edges to grow, while grains with many edges shrink. During these circumstances one fact stays the same: the area of the individual grains is a linear function of annealing time as long as the environment does not change.

1. Introduction

In the classical treatment of normal grain growth the von Neumann-Mullins-law [1-3]

$$\dot{A}(n) = -m\gamma\left(2\pi - \frac{\pi}{3}n\right) = m\gamma\frac{\pi}{3}(n-6). \qquad (1)$$

describes the temporal evolution of individual grains, according to which the area, A, of a two-dimensional grain increases or decreases linearly with time solely depending on the number of neighboring grains or edges, n. The rate of area change $\dot{A}(n)$ is independent of the shape of the grain. The materials parameters taken into account are the specific boundary mobility, m, and the boundary surface tension, γ. It follows directly from Eq. (1) that all grains of an ensemble with more than six edges will grow and all those with less than six edges will shrink making six the critical number of neighboring grains. However, this relation holds only under the assumption of ideal grain coarsening, where all boundaries are of high angle type. In such a case three grains always meet at a vertex forming 120° dihedral angles.

In contrast, processes like severe plastic deformation (SPD) introduce a large number of subgrain boundaries within the framework of grain refinement by a multi-step process [4,5]. Initially the dislocations are distributed more or less homogeneously throughout the grain structure. During deformation they start to rearrange themselves forming cells. With ongoing deformation more dislocations enter the cell walls changing them to subgrain boundaries. The misorientation along these boundaries is rather small such that they are clearly of low angle type. As deformation continues, continuous dynamic recrystallization leads to a fine steady state grain structure [6].

However, quite often during the grain refinement process the steady state grain structure is not reached such that within the grain structure a majority of the grain boundaries may very well be of low angle type, for which the classical treatment of grain coarsening as described by Eq. (1) does no longer hold. In order to understand resulting microstructural changes and therewith also property changes due to thermal treatments, the migration of low angle grain boundaries has been of great scientific interest

IOP Conf. Series: Materials Science and Engineering **194** (2017) 012049 doi:10.1088/1757-899X/194/1/012049

for decades (e.g., [7-9]). For example a combined experimental and computational study [7] has shown that the simulated shrinkage of cylindrical grains with low angle <1 0 0> boundaries results in a strongly decreasing boundary mobility with decreasing misorientation angles smaller than 10°. The shape evolution of the analyzed grains can be related directly to energy computations.

In the present work, the influence of subgrain boundaries on the coarsening kinetics of individual grains is systematically investigated under two conditions: In the first case, the grains are each embedded in an average environment neglecting structural changes around the grains focusing purely on the interplay of a grain with its associated subgrain boundaries. In the second step, it is shown how individual grains connected to subgrain boundaries behave within a polycrystalline grain ensemble.

2. Simulation method

The Monte Carlo Potts model has been in use for simulation of grain coarsening phenomena for more than 30 years. It is a lattice based method that is in its basics rather simple, but allows in its specifics complex modeling. Among others the grain boundaries of a microstructure can be manipulated individually giving them a well-defined form as well as distinct values for mobility and energy. Following the standard procedure, the grain structures are mapped in the current investigations onto a quadratic 2D lattice, where each lattice point is a Monte Carlo Unit. The time unit of the simulation is one Monte Carlo Step comprising N reorientation attempts, where N is the total number of MCUs of the lattice. In each reorientation attempt the local energy of a selected lattice point is minimized (for details refer to, e.g., [10-12]).

Particularly for the present investigations certain grain boundaries are introduced as subgrain boundaries with lower values of boundary energy, γ_{SB}, and mobility, m_{SB}, therewith treating them as low angle grain boundaries compared to the common high angle grain boundaries with γ_{GB} and m_{GB}. The influence of energy, mobility, location and number of these subgrain boundaries is examined carefully.

3. Simulation results

3.1. Subgrain boundaries in individual grains

According to Eq. (1) we know that the area change rate of an n-edged grain is proportional to the boundary energy, where a lower energy means a smaller rate of growth for $n > 6$ or rate of shrinkage for $n < 6$. For example an ideal grain with five edges embedded in an average environment should have a rate of shrinkage of $\dot{A}(n = 5) = -m\gamma\pi/3$. This is, of course, true for a grain with high angle boundaries only (compare [3]). If a number of subgrain boundaries is introduced and put solely inside the five-edged grain, they will strongly interact with the high angle boundaries and therewith influence the motion of the five edges and also influence the growth kinetics of the grain in total as it is shown in figure 1.

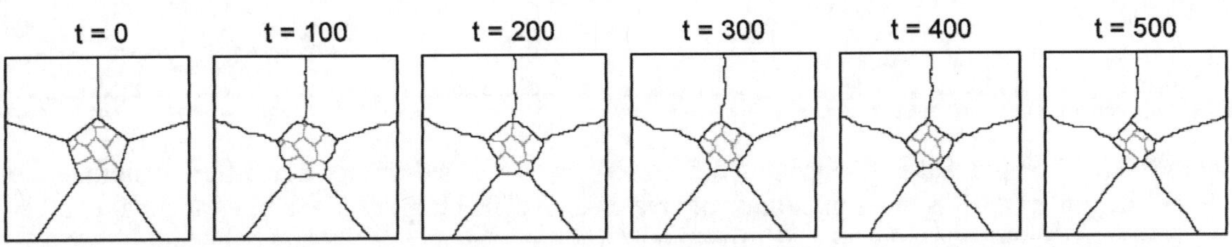

Figure 1. Temporal development of a five-edged grain with internal subgrain boundaries (in grey).

Apparently, the morphology of the grain changes because the subgrain boundaries exert a drag on the adjoining high angle boundaries. The strength of the drag strongly depends on the subgrain boundary energy as presented in figure 2a. The linear relation between grain area (which is the area enclosed by the high angle boundaries) and annealing time only holds as long as the morphology of the grain does not change. A reduction in the number of grain boundaries (high angle as well as low angle type) changes the relation $A(t)$, e.g., lower left hand in figure 2a. The limiting case for $\gamma_{SB} = \gamma_{GB}$ yields normal-like coarsening behavior with only high angle boundaries and has the highest rate of shrinkage, which equals to the fastest shrinkage. The smaller the energy of the subgrain boundaries, the slower is the grain shrinking tending to zero for $\gamma_{SB} \ll \gamma_{GB}$. A systematic parameter variation shows that the rate of area change, $\dot{A} = \frac{dA}{dt}$, is a linear function of $\ln(\gamma_{SB}:\gamma_{GB}) = \ln(\bar{\gamma})$ connecting the two limiting cases and introducing the relative energy $\bar{\gamma} = \gamma_{SB}:\gamma_{GB}$ (figure 2b). A similar behavior can be observed when the subgrain boundary mobilities are changed as presented in figure 2c.

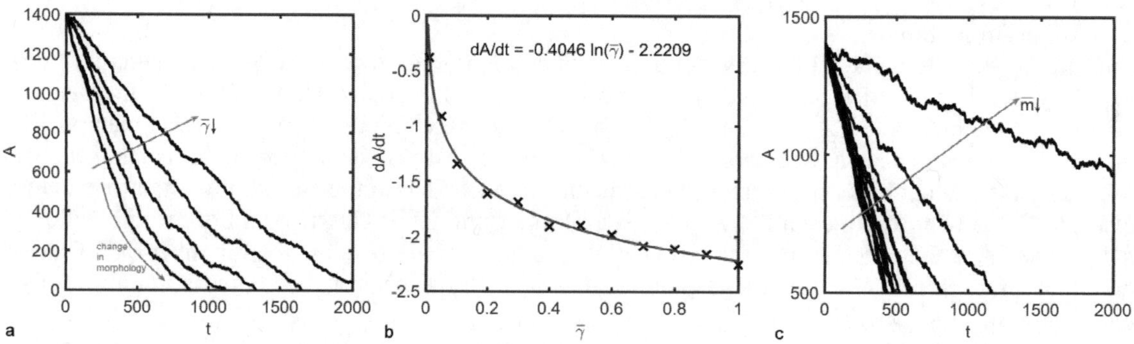

Figure 2. a – Temporal development of the area of a five-edged grain with internal subgrain boundaries (as in figure 1) changing with the relative energy; b – Resulting initial rate of area change depending on relative energy; c – Temporal development of the area of a five-edged grain with internal subgrain boundaries changing with the relative boundary mobility.

In addition, an immense influence on the growth kinetics is the number of subgrain boundaries present as shown in figures 3a and b. An increase in the number of internal subgrains, n', all with the same energy and mobility increases the rate of shrinkage of a five-edged grain such that $\dot{A} \propto n'$ for a broad range of n'. Hence, apart from the relations $\gamma_{SB}:\gamma_{GB}$ and $m_{SB}:m_{GB}$, also the number of internal subgrain boundaries influences the shrinkage of grains with fewer than six edges.

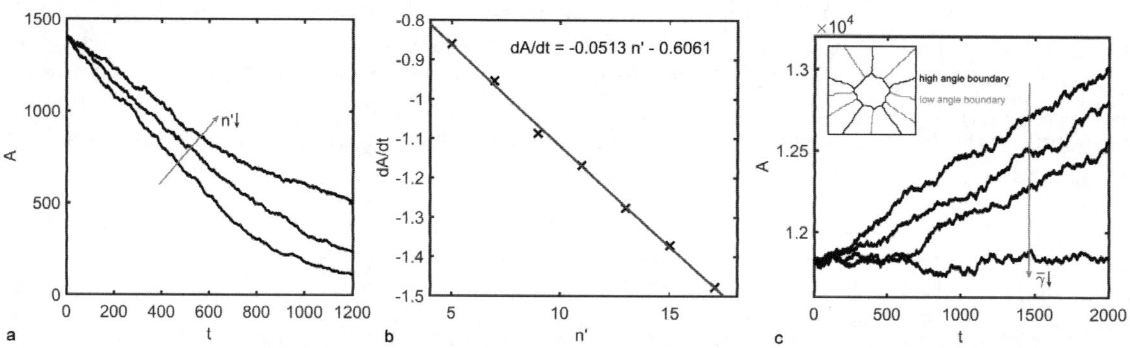

Figure 3. a – Temporal development of the area of a five-edged grain with different numbers of internal subgrains; b – Resulting initial rate of area change depending on the number of subgrains n'; c – Temporal development of the area of a five-edged grain with external subgrain boundaries.

On the other hand, a selective placement of subgrain boundaries on the outside of the central grain yields an external drag effect. The fastest growth is naturally reached for the limiting case of $\gamma_{SB} = \gamma_{GB}$, whereas a reduction of the energy of the subgrain boundaries slows the temporal development down and tends to zero for $\gamma_{SB} \ll \gamma_{GB}$ (figure 3c). A similar behavior of nearly constant size can also be achieved by introduction of internal and external subgrain boundaries of higher energy.

In contrast, grains with more than six edges should grow according to Eq. (1). However, as shown in figure 4, the internal subgrain boundaries of a nine-edged grain with $\gamma_{SB} < \gamma_{GB}$ force the grain to shrink. Initially A is a linear function of t up to a point, where the number of internal boundaries is reduced therewith also reducing the rate of shrinkage (dotted line) in agreement with figures 3a and b.

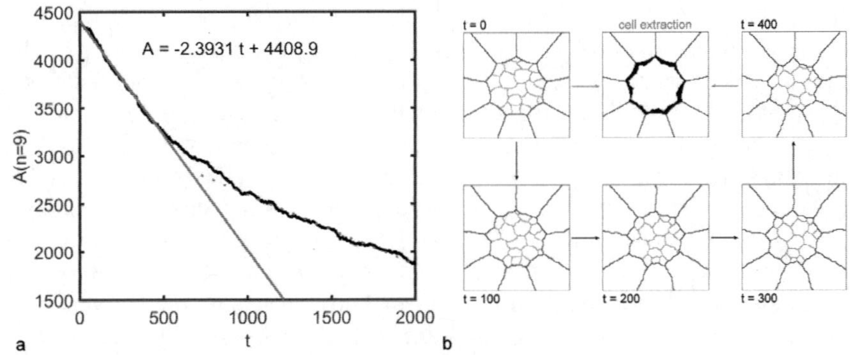

Figure 4. Temporal development of a nine-edged grain with internal subgrain boundaries that shrinks due to the drag of the subgrains in contrast to the prediction according to Eq. (1). High angle boundaries are plotted in black and low angle boundaries in grey.

3.2. *Subgrain boundaries in polycrystalline aggregates*

As discussed above, during SPD microstructures can be achieved that contain many subgrain boundaries. This is shown exemplarily in figure 5a for a grain structure produced by accumulative roll bonding (ARB) of thin aluminum sheets using commercially pure aluminum (99.5 wt.% Al). The high angle grain boundaries are marked in red and the low angle boundaries in blue [13].

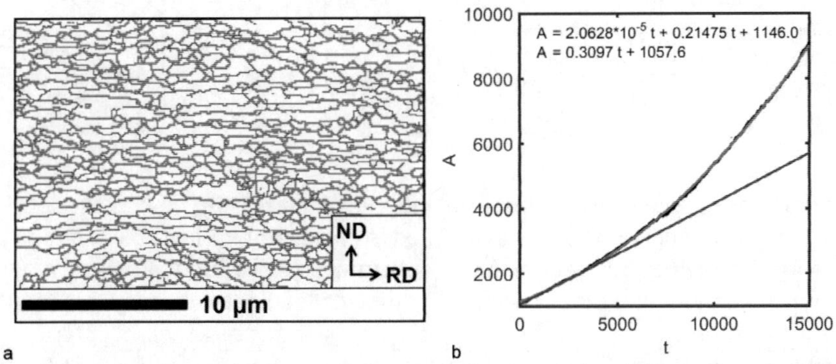

Figure 5. a – Grain boundary structure in commercially pure aluminium after 6 ARB cycles. Low angle grain boundaries (between 3° and 15°) and high angle grain boundaries (> 15°) are coloured blue and red, respectively [13]; b – Temporal development of the size of a nine-edged grain with many subgrain boundaries in neighboring grains.

How does such a neighborhood influence the growth kinetics? To model this, the considerations above are extended by embedding in a first step a nine-edged grain in a polycrystalline aggregate, where all nine neighboring grains are relatively large and contain many subgrain boundaries. The mobility and energy of the low angle subgrain boundaries is assumed to be 10% of that of the high angle boundaries. As a result it is seen in figure 5b that the speed of growth is initially constant (linear least-squares fit), but then starts to increase with increasing annealing time (quadratic least-squares fit for long annealing times). This is due to the fact that the growing grain gets into contact with more and more subgrain boundaries. Consequently, as demonstrated in figures 3a and b, a higher number of subgrains n' increases the rate of shrinkage and also in this particular case growth.

Extending this consideration to polycrystalline aggregates by inserting subgrain boundaries into 1/3 or 2/3 of the grains, where the mobility and energy of the low angle subgrain boundaries is 15% of that of the high angle boundaries, it is demonstrated in figure 6 how the average and individual growth kinetics are influenced. Particularly, in figure 6a the temporal development of the areas of individual grains is shown for the microstructure with 1/3 of the grains containing subgrains. The grains that contain no subgrains (black curves) evolve normally: large ones grow, small ones shrink, and average grains stagnate (compare [14]), while all those grains that contain subgrains (grey curves) are mostly shrinking very fast even if they are rather large. This is in agreement with figures 2 and 4, where internal subgrain boundaries force the grains to shrink even if they have more than six edges and are larger than average, since most of the grains have basically no external subgrain boundaries in case of figure 6a. As a result, after approximately 500 annealing time steps most of the grains containing subgrains are about to vanish. This in turn changes the average growth law as shown in figure 6b, where the loss of many grains between $t = 500$ and $t = 1000$ speeds up the average growth kinetics. In contrast, the microstructure, where 2/3 of the grains contain subgrain boundaries shows a slightly different behavior regarding the individual growth kinetics (figure 6c). Subgrain-free grains grow even when they are originally smaller than average, while grains that contain subgrains display a much more complex behavior. In this second case many such grains are surrounded by grains that also contain subgrain boundaries themselves, which can be shown to result in stagnation, growth, or shrinkage depending on the location and density of the subgrain boundaries.

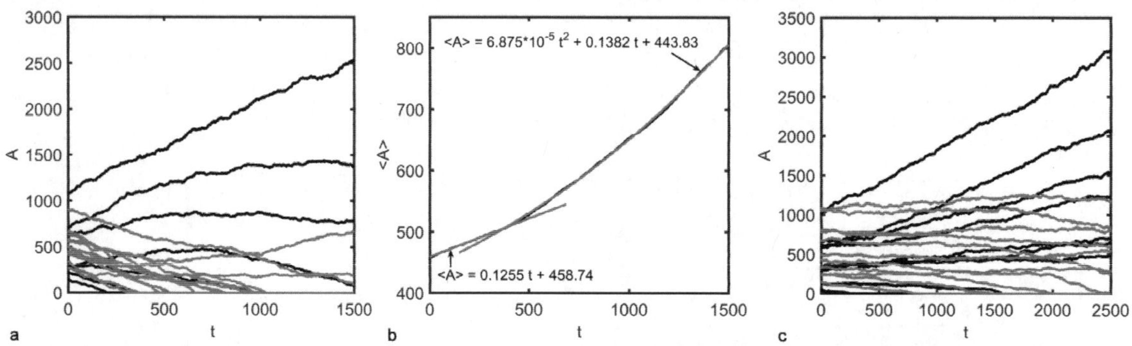

Figure 6. Temporal development of: a – area of individual grains for microstructure with 1/3 of the grains containing subgrain boundaries; b – associated average growth law; c – average growth law for microstructure with 2/3 of the grains containing subgrain boundaries. In images a and c grains that contain no subgrains are plotted in black, whereas grains that enclose subgrains are plotted in grey.

Taking a closer look at figure 6a, the area change rate as a function of the number of neighboring grains can be derived for different annealing times. Initially (see figure 7a) the grains not containing subgrains (black crosses) almost fulfill the von Neumann-Mullins-law, Eq. (1) (associated linear least-squares fit), whereas all those grains that contain subgrain boundaries (grey diamonds) form a separate cluster of data showing high shrinking rates independent of the number of edges (these data points have not been taken into account for the least-squares fit). As coarsening progresses, figures 7b and c show that the grains containing subgrain boundaries usually have a characteristically high rate of shrinkage, while for all other grains $\dot{A}$ is a linear function of n with a critical number of edges slightly deviating from 6. Particularly in the last figure it can be seen that the microstructure balances the high shrinking rates (lower right hand side) with high growth rates of neighboring grains (upper right hand side). This can easily be understood from the fact that grain boundaries form a network. If a boundary moves rather fast to the center of a certain grain, it also moves just as fast away from the center of the adjunct grain.

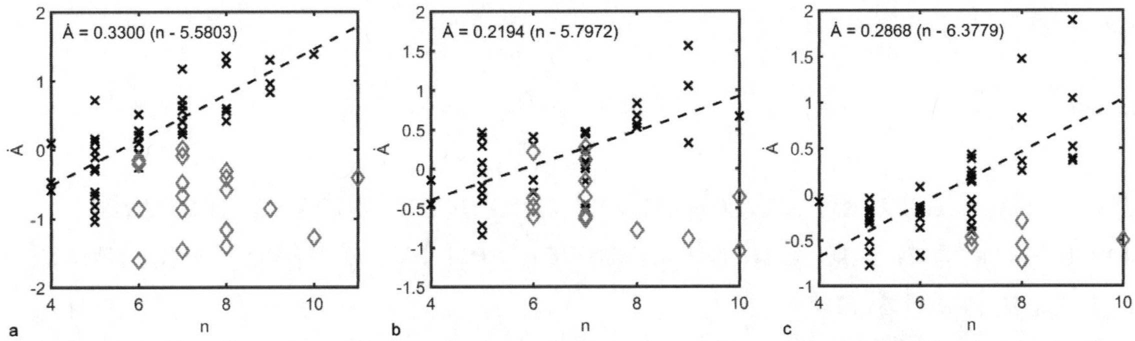

Figure 7. Area change rate of individual grains as a function of the number of neighboring grains for a microstructure with 1/3 of the grains containing subgrains. Black crosses mark all grains without internal subgrains and grey diamonds marks all those with internal subgrain boundaries: a – $t = 0$; b – $t = 500$; c – $t = 800$.

4. Summary

The present work shows the influence of subgrain boundaries on the coarsening kinetics of individual grains embedded in an average environment as well as within a polycrystalline grain ensemble. A specific introduction of subgrain boundaries not only influences the speed with which grains shrink or grow depending on their number of neighboring grains or edges, but in contrast to the von Neumann-Mullins-law a distinct manipulation of the location of the subgrain boundaries allows even grains with few edges to grow, while grains with many edges shrink. During these circumstances one fact stays the same: the area of the individual grains is always a linear function of annealing time as long as the environment does not change. However, once the number of internal or external subgrain boundaries changes clear deviations from the linear relation become visible. For polycrystalline grain aggregates produced by SPD this relation is even more complicated due to the complex interaction of area change rate, grain size respectively number of edges, and the existence or non-existence of subgrains within a grain and in its neighboring grains. Nevertheless, growth and shrinkage rates far larger and smaller, respectively, than under normal grain growth conditions have been observed.

References

[1] von Neumann J 1952 *Metal interfaces*, Cleveland OH: American Society for Metals 108-110
[2] Mullins WW 1956 *J Appl Phys* **27** 900-904
[3] Zöllner D and Rios PR 2014 *Acta Mater* **70** 290-297
[4] Azushima A, Kopp R, Korhonen A, Yang DY, Micari F, Lahoti GD, Groche P, Yanagimoto J, Tsuji N, Rosochowski A and Yanagida A 2008 *CIRP Annuals – Manu Technol* **57** 716-735
[5] Valiev RZ, Islamgaliev RK and Alexandrov IV 2000 Progr Mater Sci 45 103-189
[6] Humphreys FJ and Hatherly M 2004 Recrystallization and Related Annealing Phenomena, Oxford: Elsevier
[7] Brandenburg J-E, Barrales-Mora LA and Molodov DA 2014 *Acta Mater* **77** 294-309
[8] Winning M, Rollett AD, Gottstein G, Srolovitz DJ, Lim A and Shvindlerman LS 2010 *Phil Mag* **90** 3107-3128
[9] Lim A 2012 Migration and Mobility of Low-Angle Grain Boundaries, PhD-Thesis Princeton University
[10] Zöllner D 2011 *Comput Mater Sci* **50** 2712-2719
[11] Zöllner D 2014 *Comput Mater Sci* **86** 99-107
[12] Zöllner D 2016 *Grain growth*, Reference Module in Materials Science and Materials Engineering, Oxford: Elsevier 1-29
[13] Chekhonin P, private communication
[14] Zöllner D, Streitenberger P and Fielden I 2012 *Pract Metall* **49** 428-445

IOP Conf. Series: Materials Science and Engineering **194** (2017) 012050 doi:10.1088/1757-899X/194/1/012050

Microstructure and mechanical characteristics of gradient structured Cu and Cu alloys processed by surface mechanical attrition treatment

XZ Hu, LP Cheng, HL Chen, Z Yin, Z Zhang, BP Shu, YL Gong and XK Zhu*

Faculty of Materials Science and Engineering, Kunming University of Science and Technology, Kunming, Yunnan 650093, China

E-mail: xk_zhu@hotmail.com

Abstract: Cu-Al-Zn alloys with different stacking fault energy (SFE) were processed by surface mechanical attrition treatment (SMAT) at cryogenic temperature (CT), mechanical properties of gradient structured Cu-Al-Zn alloys were investigated in this study. Al and Zn content in alloys, which result in the decrease of SFE, can contribute to the increase in strength. Cu-4.5wt%Al-14.3wt%Zn alloy with the lower SFE shows that the strength increased, the ductility did not decrease significantly with increasing processing time, and the strength can be improved by a thicker gradient structure (GS) layer. The better combination of strength and ductility was achieved in Cu-4.5wt%Al-14.3wt%Zn alloy with lower SFE.

1 Introduction

In the traditional engineering materials, strength and plasticity are incompatible. There has been increasing interest in severe plastic deformation (SPD) processing in metals over the past decades[1, 2],such as cold rolling (CR) [3-5], equal-channel angular pressing (ECAP) [6, 7], and high-pressure torsion (HPT) [8, 9],The primary reason for this interest is previous experimental evidence that the strength of materials significantly increases with decreasing grain sizes according to the well-known Hall-Petch relation[10], An extremely high strength and hardness has been achieved in bulk ultrafine grained (UFG) materials at the expense of plasticity condition compared with coarse grained counterpart. In the previous studies, which demonstrated that the strength and ductility increase simultaneously when the grain sizes is less than 100nm [11].Our expectation is to obtain high strength and good plasticity from this new discovery. Some early reports found that strength and plasticity can be improved simultaneously in Cu–Al alloys and Cu–Zn alloys by decreasing the stacking fault energy (SFE) [12, 13]. However, it was not enough investigated yet in Cu–Al–Zn alloys with different SFE.

In this work, mechanical properties of gradient structured Cu-Al-Zn alloys with different SFE under surface mechanical attrition treatment (SMAT) was systematically investigated, which can realize gradient structure in Cu-Al-Zn alloys. Metals with gradient structure (GS) exhibits a superior mechanical properties compared with coarse grained materials[14]. In this paper, we found that the enhancement of the mechanical properties is different from that of Cu–Al alloys and Cu–Zn alloys. The possible deformation mechanisms in Cu–Al–Zn alloys were discussed accordingly.

2 Experimental procedure

The materials used in the experiments are Cu-4.5wt%Al-14.3wt%Zn alloy and Cu-1.86%Al-23.89wt%Zn alloys with different SFEs of 7 and 10 mJ/m^2[15], respectively. Cu–Al–Zn alloys were rolling to plate with a thickness of 3 mm and annealed in vacuum at 1023k for 2h. SMAT was performed on two surface of these samples. The experiment is under vacuum at cryogenic temperature (CT). Uniaxial tensile tests were carried out on a Shimadzu Universal Tester at room temperature with a maximum load of 10KN and at a constant strain rate of 1.0×10−4 s−1. Hardness measurements were carried out on a HX-1Vickers hardness testing machine. The hardness of all samples was performed on the cross–sectional of the samples with a load of 50 g and a loading time of 15 s. The detailed microstructure characteristic of GS samples were analyzed by means of electron backscattered diffraction (EBSD) observations, which was carried out on a Zeiss Auriga, with a field emission scanning electron microscope, the operating voltage is 15KV, and the original EBSD data were analyzed by the channel5 software.

3 Results and Discussion

The engineering stress–strain curves of Cu–Al–Zn sample are given in Figure 1. The mechanical properties, which including yield strength (YS) $\sigma_{0.2}$, ultimate tensile strength (UTS) σ_{UTS} and uniform elongation (UE) ε_u are shown in Figure 1. The YS and UTS of the Cu–Al–Zn samples increased with the processing time increasing. Compared with the Cu-1.86wt%Al-23.89wt%Zn alloys, the Cu-4.5wt%Al-14.3wt%Zn alloys with a lower SFE exhibits a better strength–ductility combination with different SMAT time.

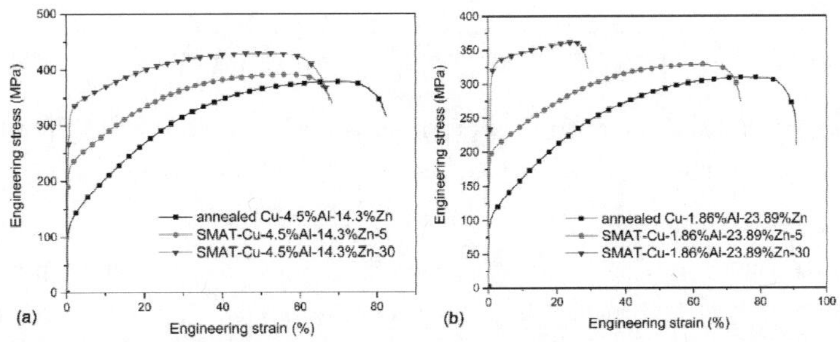

Figure.1. Engineering stress–strain curves of annealed Cu–Al–Zn samples and processed by SMAT at CT with different time, (a) Cu-4.5wt%Al-14.3wt%Zn samples, and (b) Cu-1.86wt%Al-23.89wt%Zn samples.

The normalized work hardening rate against the true strain are shown in Fig. 2. It is obvious that the annealed samples have the higher normalized work hardening rate than the GS samples. The normalized work hardening rate decreases gradually with the increasing processing time The work hardening rate behavior of the Cu-4.5wt%Al-14.3wt%Zn alloy originate from its unique microstructure, which is given and discussed in the following sections.

Fig. 3 shows the hardness profiles of Cu–Al–Zn samples. The average hardness of the annealed samples is ~1325 MPa and ~1072 MPa, respectively. Hardness has a substantial increase after the SMAT process and reaches the highest value at the topmost surface layer. The maximum hardness value is approximately 1.6 and 1.8 times higher than annealed samples. The hardness of GS samples decreased from the topmost surface layer to central matrix and remained in a constant at a certain depth. The thickness of the GS layer of the Cu-1.86wt%Al-23.89wt%Zn samples have apparent increment with the time increasing, which is unlike that the thickness of Cu-4.5wt%Al-14.3wt%Zn. The maximum value can be explained as the largest strains and strain rates occurred in the topmost surface layer during the cryogenic SMAT process [16].

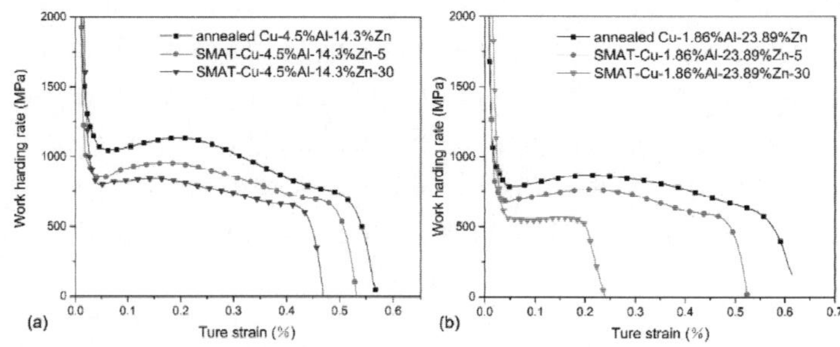

Figure.2. The normalized work hardening rate against the true strain of Cu–Al–Zn samples. (a) Cu-4.5wt%Al-14.3wt%Zn samples, and (b) Cu-1.86wt%Al-23.89wt%Zn samples.

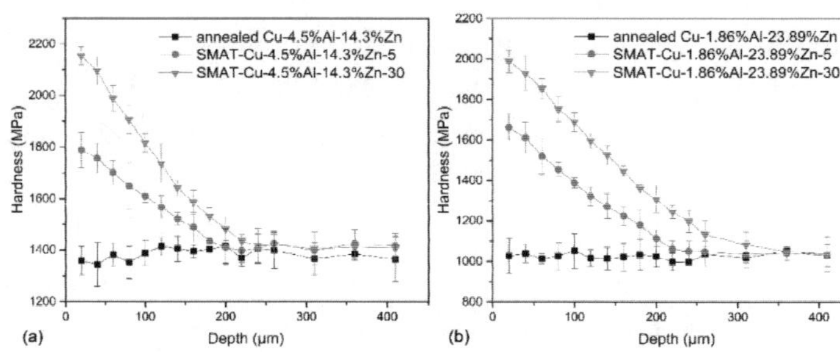

Figure. 3 The hardness profiles as a function of the depth for Cu–Al–Zn samples. (a) Cu-4.5wt%Al-14.3wt%Zn samples, and (b) Cu-1.86wt%Al-23.89wt%Zn samples.

Fig. 4 shows EBSD mappings of annealed Cu-1.86wt%Al-23.89wt%Zn samples, and the red lines in Figure 4 (b) represents twin. The significant annealing twin can be seen in annealed Cu-1.86wt%Al-23.89wt%Zn samples and the grain size of the whole sample is nearly homogeneous. EBSD mappings of Cu-1.86wt%Al-23.89wt%Zn samples (SMAT for 5, 30 min) is shown in Fig. 5. The grain size of Cu-1.86wt%Al-23.89wt%Zn alloys after SMAT increased from the surface to the center and the grain size gradually tended to be homogeneous. The thickness of the GS layer increased with the processing time from 5min to 30min, twin of surface layer decreased gradually with the time increasing.

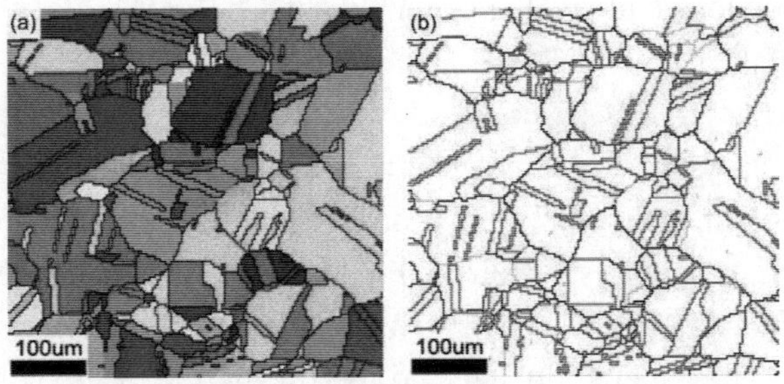

Figure. 4. EBSD images of annealed Cu-1.86wt%Al-23.89wt%Zn samples: (a) Orientation map. (b) Boundary misorientation (2° ≤ θ < 15°) map

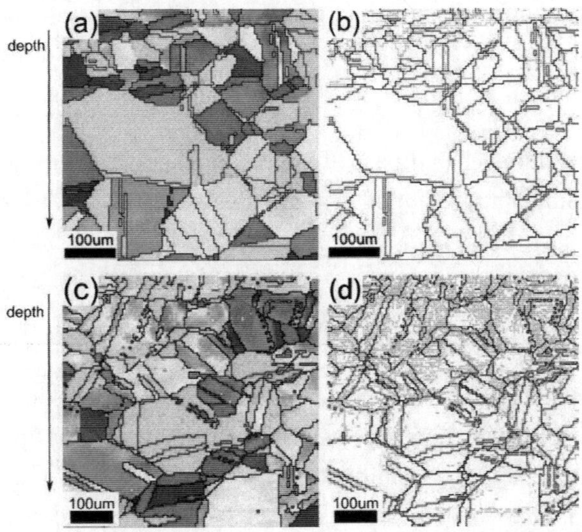

Figure. 5. EBSD images of GS samples from surface to center: (a) Orientation map, and (b) Boundary misorientation (2° ≤ θ < 15°) map of the Cu-1.86wt%Al-23.89wt%Zn SMAT-5min sample. (c) Orientation map, and (d) Boundary misorientation (2° ≤ θ < 15°) map of the Cu-1.86wt%Al-23.89wt%Zn SMAT-30 min sample.

3.1 Effect of GS layer on mechanical behaviors

The hardness has an obvious increase from the CG core region to the treated surface layer, and the thickness of the GS layer with the high SFE samples is larger than that of the lower SFE samples under the same time. The hardness of Cu-4.5wt%Al-14.3wt%Zn samples is always higher than that of Cu-1.86wt%Al-23.89wt%Zn samples at the same depth and time, Zhu et al. demonstrated that the yield strength increases with the thickness of GS layer increasing [17]. Cu-4.5wt%Al-14.3wt%Zn alloys have a higher hardness and smaller grain size on the surface layer compared with Cu-1.86wt%Al-23.89wt%Zn alloys. Therefore, thicker GS layer and smaller grain size give rise to higher yield strength.

The annealed Cu-Al-Zn sample with lower hardness is easier to deform, which can produce a thicker deformation layer when processed by SMAT. However, Cu-4.5wt%Al-14.3wt%Zn samples processed by 5min has reached a higher hardness value, which is difficult to produce further deformation. Cu-4.5wt%Al-14.3wt%Zn samples need more energy to produce further deformation, and Cu-1.86wt%Al-23.89wt%Zn samples produce greater deformation extent for the same energy. Therefore, the GS layer of the higher SFE sample is thicker than that of the low SFE under the same time. Thicker GS layer of

Cu-4.5wt%Al-14.3wt%Zn alloys and smaller grain size of Cu-4.5wt%Al-14.3wt%Zn alloys both can result in higher yield strength.

The ductility of Cu-4.5wt%Al-14.3wt%Zn alloys have no significant decrease compared to Cu-1.86wt%Al-23.89wt%Zn alloys with the process time increasing, which can attribute the gradient structure[18]. In GS samples, plastic deformation occurs first in the coarse-grain region as the tensile strain increasing, coarse-grain region is surrounded and constrained by GS layer, which stay in elastic deformation. The GS layer thickness of Cu-4.5wt%Al-14.3wt%Zn alloys almost remain unchanged with increasing processing time, and the ductility have no significant decrease. Therefore, the thicker coarse-grain region may contribute to plastic strain and improve the ductility[18]. There exists an optimum thickness of GS layer for the superior combination of strength and ductility in Cu-Al-Zn alloys with low SFE., the strength increases continually and ductility remains the same with the time increasing. The GS obtained by SMAT provides an effective way for high strength and good ductility.

3.2 Effect of SFE

SFE is a several material parameter and has influence on microstructure and mechanical properties, which can transform the dominant plastic deformation mechanism from dislocation slip to deformation twinning by lowering SFE[13, 19]. Local softening can lead to plastic instability, which is caused by dislocation annihilation through cross-slip and climb [11, 20]. Unlike the enhancement in strength, the improvement in ductility results from the increase in the strain hardening rate with decreasing SFE. As mentioned previously, restraining dynamic recovery can effectively improve the strain hardening rate, and the increase of strain hardening rate can delay the onset of necking and enhance the tensile ductility[21, 22]. With the decreasing of SFE, it is difficult to make cross-slip and climb of dislocations, which can effectively inhibit dislocation annihilation or dynamic recovery [23].

Twins density is also increased with decreasing SFE, as mentioned previously, Wang et al. [24] indicated that the increase of twins can improve the ductility by a decrease of SFE. Annealing twin is also usually found, meanwhile, there is a bump at the true strain of 0.1 nearby, which is likely to be twins, a series of previous studies demonstrated that there exists twins [25, 26]. An et al. [12] indicate that there exists an optimal combination of strength and ductility in Cu-Al alloy with the lowest SFE. Therefore, the work hardening rate of Cu-4.5wt%Al-14.3wt%Zn samples is higher than Cu-4.5wt%Al-14.3wt%Zn samples under the same processing time, which is derived from lower SFE and more twins.

4 Conclusion

Cu-Al-Zn samples with different stacking fault energy were processed by surface mechanical attrition treatment at cryogenic temperature, which have a significant increase in yield strength and a slight decrease in ductility compared with annealed Cu-Al-Zn samples. Thicker GS layer of Cu-4.5wt%Al-14.3wt%Zn alloys and smaller grain size of Cu-4.5wt%Al-14.3wt%Zn alloys both can result in higher yield strength. The work hardening rate of Cu-4.5%Al-14.3%Zn samples is higher than Cu-4.5wt%Al-14.3wt%Zn samples, which is derived from a lower SFE. There exists an optimum thickness of GS layer for the superior combination of strength and ductility in Cu-4.5wt%Al-14.3wt%Zn alloys.

Acknowledgements

The authors would like to acknowledge financial supports by the National Natural Science Foundation of

China (NSFC) under Grant No. 51561015 and No. 51664033, and the introduction of talent fund project of Kunming University of Science and Technology (KKSY201407100).

Reference

[1] Segal V M 2005 *Materials Science and Engineering: A* **406** 205-16
[2] Estrin Y and Vinogradov A 2013 *Acta Materialia* **61** 782-817
[3] Wang Y, Chen M, Zhou F and Ma E 2002 *Nature* **419** 912-5
[4] Huges D A and Hansen N 2000 *Acta Materialia* **48** 2985-3004
[5] Zhang Y, Tao N R and Lu K 2008 *Acta Materialia* **56** 2429-40
[6] Mishra A, Kad B, Gregori F and Meyers M 2007 *Acta Materialia* **55** 13-28
[7] Iwahashi Y, Horita Z, Nemoto M and Langdon T G 1998 *Acta Materialia* **46** 3317-31
[8] Xu C, Horita Z and Langdon T 2007 *Acta Materialia* **55** 203-12
[9] Edalati K, Fujioka T and Horita Z 2008 *Materials Science and Engineering: A* **497** 168-73
[10] Hall E O 1951 *Proceedings of the Physical Society* **64** 495
[11] Valiev R Z, Alexandrov I V, Zhu Y T and Lowe T C 2002 *Journal of Materials Research* **17** 5-8
[12] An X H, Lin Q Y, Wu S D, Zhang Z F, Figueiredo R B, Gao N and Langdon T G 2011 *Scripta Materialia* **64** 954-7
[13] Zhang Z J, Duan Q Q, An X H, Wu S D, Yang G and Zhang Z F 2011 *Materials Science and Engineering: A* **528** 4259-67
[14] Youssef K M, Scattergood R O, Murty K L, Horton J A and Koch C C 2005 *Applied Physics Letters* **87** 091904
[15] Denanot M F and Villain J P 2010 *Physica Status Solidi* **8** K125-K7
[16] Wang K, Tao N R, Liu G, Lu J and Lu K 2006 *Acta Materialia* **54** 5281-91
[17] Wu X L, Jiang P, Chen L, Zhang J F, Yuan F P and Zhu Y T 2014 *Materials Research Letters* **2** 185-91
[18] Wu X L, Yang M, Yuan F, Wu G, Wei Y, Huang X and Zhu Y 2015 *Proceedings of the National Academy of Sciences of the United States of America* **112** 14501-5
[19] Zhao Y H, Horita Z, Langdon T G and Zhu Y T 2008 *Materials Science and Engineering: A* **474** 342-7
[20] Kolobov Y R, Grabovetskaya G P, Ivanov M B, Zhilyaev A P and Valiev R Z 2001 *Scripta Materialia* **44** 873-8
[21] An X H, Han W Z, Huang C X, Zhang P, Yang G, Wu S D and Zhang Z F 2008 *Applied Physics Letters* **92** 201915
[22] Lu L, Zhu T, Shen Y, Dao M, Lu K and Suresh S 2009 *Acta Materialia* **57** 5165-73
[23] Zhao Y H, Zhu Y T, Liao X Z, Horita Z and Langdon T G 2006 *Applied Physics Letters* **89** 121906
[24] Wang Z W, Wang Y B, Liao X Z, Zhao Y H, Lavernia E J, Zhu Y T, Horita Z and Langdon T G 2009 *Scripta Materialia* **60** 52-5
[25] Hamdi F and Asgari S 2010 *Scripta Materialia* **62** 693-6
[26] Rohatgi A and Vecchio K S 2001 *Metallurgical and Materials Transactions A* **32** 135-45

IOP Conf. Series: Materials Science and Engineering **194** (2017) 012051 doi:10.1088/1757-899X/194/1/012051

Microstructural characteristics of nano-structured Fe-28.5Ni steel by means of severe plastic deformation

S H Mousavi Anijdan[1]*, H R Jafarian[2], A Bahrami[3]

[1]*Department of Materials Engineering, Science and Research Branch, Islamic Azad University, Tehran, Iran
[2]School of Metallurgy and Materials Engineering, Iran University of Science and Technology (IUST), Tehran, Iran
[3]Department of Materials Engineering, Isfahan University of Technology, Isfahan 84156-83111, Iran

Emails: hashemmousavi@gmail.com, hashemmousavi@srbiau.ac.ir

Abstract. Microstructural evolution together with changes in mechanical properties of an high nickel content steel processed by various cycles of accumulative roll bonding (ARB) is explored. It is shown by Electron Backscatter Diffraction (EBSD) analysis that after successive roll bonding processes a stabilized nano-structure is developed containing sufficient amount of ductility under severe plastic deformation. A mean grain size of few hundred nano meter was obtained after 6-cycle of ARB process meaning that the successive ARB cycles made the structure quite refined. The starting material was mainly coming from the transformation of martensite to retained austenite, particularly under high temperature and in high cycle of ARB process. This is an early indication of the stabilization of retained austenite during the ARB process through grain refinement phenomenon. Uniaxial tensile test demonstrated that yield strength significantly improves by only one cycle of ARB process. Successive cycles of ARB process gradually increased the yield and the ultimate tensile strengths at the expense of ductility. The main cause of such a substantial increase in yield strength is discussed. Remarkably high amount of ductility was still observed in a very high amount of deformation that was applied in the 6-cycle ARB process.

1. Introduction

In the past two decades scientists have extensively studied bulk nano/ultrafine-grained (UFG) structure materials that had grain size below 1 μm. The studies were mostly driven by excellent mechanical properties of these materials and their relative ease of manufacturing [1-3]. Among the manufacturing processes to achieve nano grain size range materials, several severe plastic deformation (SPD) methods have been employed [4, 5]. These methods include accumulative roll bonding (ARB) [6, 7], equal channel angular pressure (ECAP) [8], hot pressure torsion (HPT) [9], repetitive-upsetting extrusion (RUE) [10, 11], and twist extrusion [12] etc. Of the several SPD methods used, ARB has singularly been the subject of a large number of researches, for both steels and aluminum alloys, due mainly to the relative ease of application [13]. The low ductility of SPD materials is mainly attributed to the early necking and the lack of strain hardenability of the UFG materials. Small amount of total elongation were partially overcome by a mixed of UFG microstructure and transformation induced plasticity (TRIP) effect. The total elongation is also highly dependent on the stabilization of austenite. Such stabilization, in addition to engineering the chemical composition, can be achieved through SPD [14, 15]. Kitahara et al found that the martensite transformation temperature decreases proportionally by increasing SPD amount [15].

The transformation behavior of high nickel, low carbon, content alloys has not been fully understood. Therefore, it was the subject of this research to elaborate the structural changes, and its resultant mechanical properties, that the steel would undergo during several ARB processes.

2. Materials and Experimental procedure

An Fe-28.5Ni-0.05C-0.003Si-0.015Mn-0.005P (wt.%) sheet steel with the thickness of 1 mm was used as the starting material. Apparently, having such a high Ni content would make the steel corrosion resistance and suitable for harsh industrial applications [14]. Initially, the sheets were subjected to annealing process at 900 °C for one hour. Cutting was performed by guillotine subsequent to this process to the dimensions of 5 × 14 cm followed by cleaning with acetone. Then, the wire brushed sheets were stacked and heated up to 500 °C in a box furnace and maintained at this temperature for 10 min. Roll bonding followed after the heat treating process. The specimen after each rolling process was quenched in water. A two-high mill with the rolls diameter of 310 mm were used to conduct ARB processes. The sheet was cut by half at the end of each rolling process and further rolling were conducted by the same procedure as explained above. The ARB processes were performed from one cycle to 4-cycle and for 6-cycle as well.

Characterization was performed by a Filed-Emission Scanning Microscopy (SEM/Philips XL30S-FEG). Specimens were prepared by usual preparation method prior to the SEM analyses. The SEM machine was equipped with an Electron Backscatter Diffraction (EBSD) detector. EBSD raw data were subsequently analyzed by the TSL-OIM analysis software. Mechanical properties were also measured based on ASTM 8 standard for the preparation of the samples. Room temperature uniaxial tensile tests were conducted on materials. The tensile test machine was SHIMADZU AG-100 kN model and the strain rate of 8.3×10^{-4} s^{-1} was used for all the tests. Visual inspection and extensometer measurements were used for these analyses.

3. Results and discussion

3.1. Microstructure analysis

3.1.1. Initial material. The microstructure of the starting material is shown in figure 1. Here, the structure mainly contains of retained austenite together with second phase martensite embedded in it. The former phase is shown in green color and the latter in pink. The grain boundaries map is also shown in figure 1b. The backscattered micrograph of figure 1a shows a somehow random grains structure. Moreover, the grains, to some degree, show a bimodal size distribution. As can be seen in figure 1b, some low angle grain boundaries (LAGBs) can be seen inside the martensite phase. Commonly, the LAGBs are indicative of dislocations in the microstructure. The low amount of LAGBs is associated with the low

IOP Conf. Series: Materials Science and Engineering **194** (2017) 012051 doi:10.1088/1757-899X/194/1/012051

amount of lath martensite in the structure. The martensite blocks are not differentiated by HAGBs and do not form a continuous phase in the structure.

a)

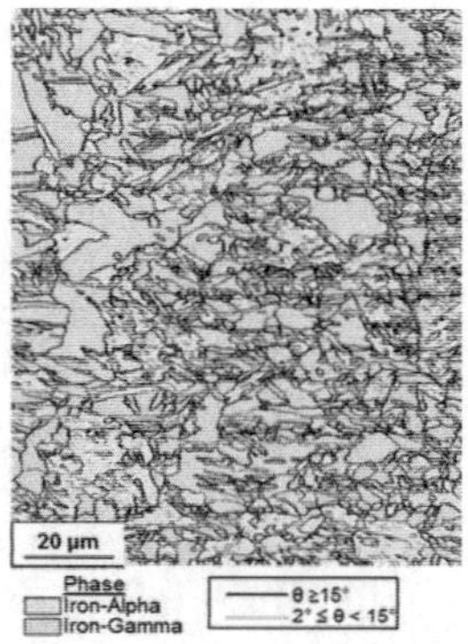

b)

Fig. 1. In initial material, (a) SEM micrograph, (b) Grain boundary and phase map.

3.1.2. Microstructure development during ARB process. Two specimens of 1-cycle, and 6-cycle ARB process, as the representatives of low and high amount of plastic deformation, were analyzed by EBSD method. The grain boundary map of the 1-cycle ARB processed is shown in figure 2a. It is clear from this figure that elongated grains are developed in this situation. This is a structure very similar to typical rolling microstructure. The grain boundaries map shows the elongated cell structures. It also shows the LAGBs developed along the grains. Another aspect of this deformation path is the substantial reduction of martensite in the system. A reversion of martensite to austenite is quite plausible in this situation. Figure 2b shows the orientation color map of this condition. Unlike the starting material, here a clear grain orientation preference is observable.

Compared to the 1-cycle microstructure, in the case of 6-cycle microstructure, shown in figure 4, ultrafine grains are being developed. The grains in this case are encircled by semi-lamellar boundaries. Also, they are mostly HAGBs with large misorientation angles and with very low amount of LAGBs. The average grain size in this case is in the range of few hundred nano meter, even in some cases less than 100 nm grains, which are similar to the literature in single phase materials [6, 7]. In this case, in line with the reduction of martensite, almost no discernible amount of martensite is found. In the 6-cycle ARB processed specimen, the grains were subdivided down to nano sizes by grain subdivision mechanism. The Ms temperature is reduced by the application of severe plastic deformation in the 6-cycle ARB processed sample, as also found in a study on a similar steel [15]. The initial austenite grain size also reduces during severe plastic deformation leading to the reduction of the Ms temperature. The formation of a large amount of HAGBs is an outcome of the reduction of Ms temperature. Consequently, retained austenite would be stabilized at the expense of martensite.

The grain size continuously decreases as the number of ARB cycle increase. This is shown in the grain size profiles of figure 4 for the starting specimen as well as for the 1-cycle, and the 6-cycle ARB processed specimens. Apparently, the grain size varies from a few micron to nano meter size, even in some areas below 100 nm, as the ARB cycle increases up to 6-cycle.

IOP Conf. Series: Materials Science and Engineering **194** (2017) 012051 doi:10.1088/1757-899X/194/1/012051

a) b)

Fig. 2. In 1-cycle ARB processed sample (a) Grain boundary and phase map. (b) Orientation color map.

a) b)

Fig. 3. In 6-cycle ARB processed sample (a) Grain boundary and phase map. (b) Orientation color map.

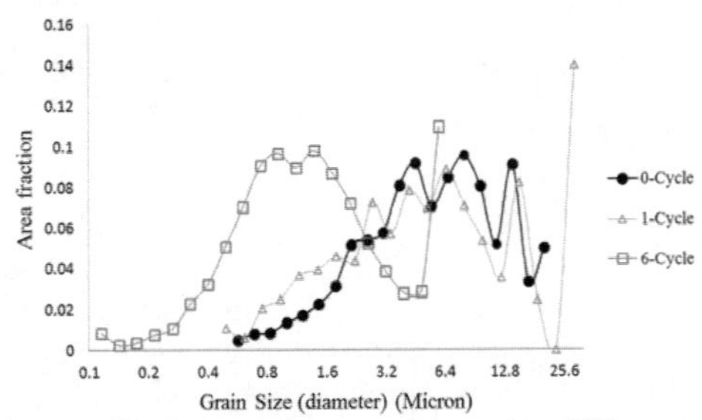

Fig. 4. Grain size distribution profile for specimens subjected to different number of ARB cycles.

3.2. Mechanical properties

The engineering stress-strain curves for the starting material, 1-cycle, 2-cycle, 3-cycle, 4-cycle, and 6-cycle ARB processed specimens are shown in figure 5. Here, a typical stress-strain curve is observed in the case of starting specimen. The yield strength is in the range of 300 MPa and a large strain hardening follows, together with a continuous yielding, up to an ultimate strength of about 470 MPa. As well, the ductility of the starting material is quite significant and stands at about 44%. Performing the ARB process at a relatively high temperature of 500 ºC would make martensitic transformation through deformation-induced martensite mechanism possible by reducing martensite transformation. In fact, the specimen in the 6-cycle situation has substantial amount of ductility after this severe plastic deformation condition.

There is also a rapid increase of yield strength up to 498 MPa after only 1-cycle of ARB processing. Such an increase in the yield strength level after 1-cycle of ARB process have been reported only in single phase materials, particularly for Al alloys [7]. Grain refinement and substantial increase in dislocation density were proposed as the main causes of such an increase in the yield strength level.

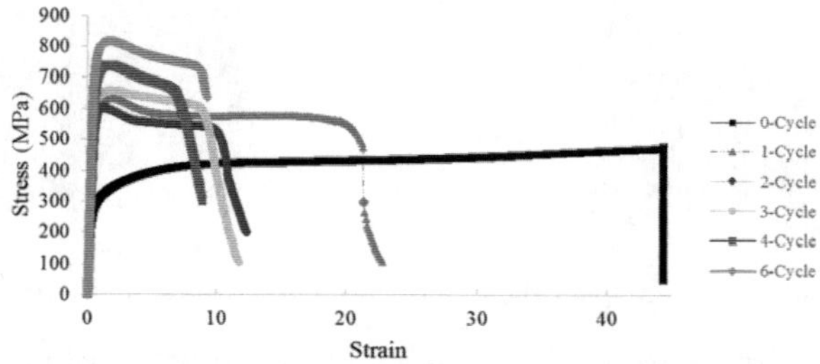

Fig. 5. Engineering stress-strain curves of different cycles of ARB processed specimens.

The extracted yield strength and ultimate tensile strength for the starting material together with the different ARB processes are given in figure 6a. Figure 6b also shows the changes of ductility under the above conditions. It can be seen in this table that the strength values increase quite rapidly in the early stages of ARB process. Ductility, on the other hand, reduces in such conditions. The pace of strength increase slows down after the first cycle and up to 4-cycle after which the rate of strength increase goes up again. The second increase in strength level is smaller than the first increment.

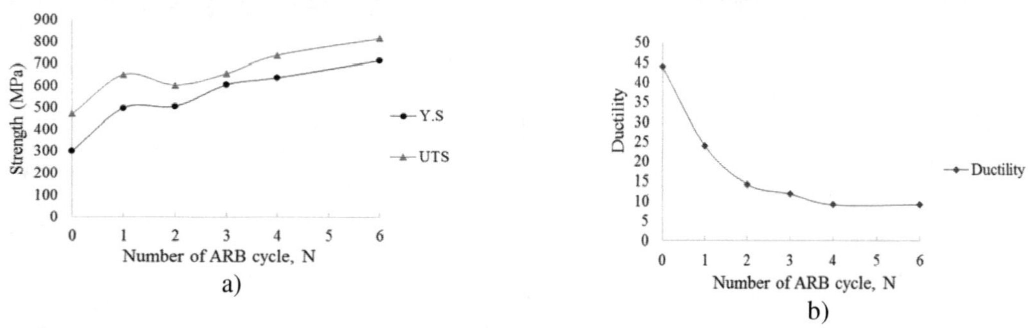

Fig. 6. Relationship between the number of ARB cycle, a) yield strength and ultimate tensile strength, (b) ductility.

4. Concluding remarks

The following results were obtained by conducting this research:

1- The starting material contained retained austenite together with martensite phase.
2- The amount of martensite was substantially reduced during the ARB cycles. This was related to reversed austenite mechanism and the stabilization of austenite by lowering martensite start temperature due to severe plastic deformation.
3- Although dislocation density increased after only 1-cycle of ARB process, it was almost diminished after 6-cycle ARB process due to dynamic restoration phenomenon.
4- A substantial increase in flow strength was obtained after one cycle of ARB process. The pace of increment was greatly reduced in further cycles. The ductility on the other hand, although decreased gradually, was still significant, i.e. 6-cycle.

5. References

[1] Shibata A, Jafarian H, Tsuji N 2012 Microstructure and crystallographic features of martensite transformed from ultrafine-grained austenite in Fe-24Ni-0.3 C alloy Mater. Trans. 53 8186

[2] Jafarian H, Habibi-Livar J, Razavi SH 2015 Microstructure evolution and mechanical properties in ultrafine grained Al/TiC composite fabricated by accumulative roll bonding Compos. Part B Eng. 77 8492

[3] Jafarian H, Eivani A 2014 Texture development and microstructure evolution in metastable austenitic steel processed by accumulative roll bonding and subsequent annealing J. Mater. Sci. 49 65706578

[4] Valiev R Z, Islamgaliev R K, Alexandrov IV 2000 Bulk nanostructured materials from severe plastic deformation Prog. Mater. Sci. 45 103189

[5] Valiev R 2004 Nanostructuring of metals by severe plastic deformation for advanced properties Nat. Mater. 3 511516

[6] Saito Y, Tsuji N, Utsunomiya H, Sakai T, Hong R G 1998 Ultra-fine grained bulk aluminum produced by accumulative roll-bonding (ARB) process Scr. Mater. 39 12211227

[7] Saito Y, Utsunomiya H, Tsuji N, Sakai T 1999 Novel ultra-high straining process for bulk materials—development of the accumulative roll-bonding (ARB) process Acta Mater. 47 579583

[8] Valiev R Z, Langdon T G 2006 Principles of equal-channel angular pressing as a processing tool for grain refinement Prog. Mater. Sci. 51 881981

[9] Zhilyaev A P, Langdon T G 2008 Using high-pressure torsion for metal processing: fundamentals and applications Prog. Mater. Sci. 53 893979

[10] Fatemi-Varzaneh S M, Zarei-Hanzaki A 2009 Accumulative back extrusion (ABE) processing as a novel bulk deformation method Mater. Sci. Eng. A 504 104

[11] Aizawa T, Tokimutu K 1999 Bulk Mechanical alloying for productive processing of functional alloys Mater. Sci. Foru. 312-314 1322

[12] Beygelzimer Y, Varyukhin V, Synkov S, Orlov D 2009 Useful properties of twist extrusion Mater. Sci. Eng. A 503 1417.

[13] Borhani E, Jafarian H, Terada D, Adachi H, Tsuji N 2012 Microstructural evolution during ARB Process of Al-0.2 mass% Sc Alloy Containing Al3Sc Precipitates in Initial Structures Mater. Trans. 53 7280

[14] Lavvafi H, Lewandowski J R, and Lewandowski J J 2014 Flex bending fatigue testing of wires, foils, and ribbons Mater. Sci. and Eng. A 601 123130

[15] Kitahara H, Tsuji N, Minamino Y 2006 Martensite transformation from ultrafine grained austenite in Fe–28.5 at.% Ni Mater. Sci. and Eng. A 438–440 233236

[16] Hughes D, Hansen N 1997 High angle boundaries formed by grain subdivision mechanisms Acta Mater. 45 38713886

IOP Conf. Series: Materials Science and Engineering **194** (2017) 012052 doi:10.1088/1757-899X/194/1/012052

Structural stability of ultra-fine grained magnesium alloys processed by equal channel angular pressing

M Janeček, T Krajňák, P Minárik, J Čížek, J Stráská, J Stráský

Charles University, Faculty of Mathematics and Physics, Ke Karlovu 3, 121 16 Praha 2, Czech Republic

E-mail: janecek@met.mff.cuni.cz

Abstract. Structural stability of two magnesium alloys AZ31 (MgAlZn) and AX41 (MgAlCa) processed by equal channel angular pressing is investigated. The mechanical properties, microstructure evolution and dislocation density were studied by microhardness, electron back scatter diffraction and positron annihilation spectroscopy, respectively. The loss of microstructure stability at high temperatures and the coarsening of the ultrafine-grained structure as a result of isochronal annealing is accompanied by the sharp decrease of microhardness and the decrease of dislocation density. The differences in thermal stability of both alloys are related to different conditions of ECAP pressing and the phase stability, namely the presence of stable Al_2Ca precipitates in AX41 alloy. Microscopic mechanisms controlling the structure stability of both alloys are discussed.

1. Introduction

Magnesium and its alloys are light-weight metallic materials which belong to prospective materials that have extensive application potential as structural components in many industrial branches, e.g. in automobile and aircraft industry. Due to the low density and relatively good mechanical properties magnesium-based materials have been increasingly demanded in weight-critical applications where they replace aluminum, steel and other traditional materials. The interest in magnesium-based materials have recently been revived primarily due to their gradually decreasing costs and the effort of scientists, researchers and engineers to reduce energy consumption and greenhouse gas emissions [1].

The mechanical and other physical properties determining the application of magnesium alloys may be significantly improved by refining the grain size down to submicrometer ($d<1\mu m$) or even nanometer ($d<100$ nm) level by severe plastic deformation (SPD) [2].

A wide variety of special techniques have been proposed and successfully employed for the production of bulk ultra-fine grained (UFG) materials, e.g. equal channel angular pressing (ECAP) [3], high pressure torsion (HPT) [4], accumulative roll-bonding (ARB) [5], twist extrusion [6], multidirectional forging [7], etc.

Among these techniques, which introduce the severe plastic deformation in the material and result in exceptional grain refinement, ECAP, for the first time reported by Segal [8] already more than 40 years ago, became a very popular and extensively used technique due to its versatility, scalability and namely the effectiveness in producing UFG structure in a wide variety of materials with different crystal structures [9,10].

However, the application of UFG materials is limited due to the structure stability at elevated temperatures and concomitant processes of recovery and recrystallization which significantly limit the application potential of these materials. Thermal stability depends on many microstructural parameters, e.g. the phase composition, the stacking fault energy, conditions of material preparation, properties of grain boundaries, etc. Investigation of thermal stability is therefore a challenge to obtain a full portrait of properties of materials after SPD. Unfortunately, this has been done only rarely [11]. Poor thermal stability was found in pure FCC metals Cu, Al [12, 13]. Enhanced thermal stability was reported in Cu- and Al-alloys and attributed to the grain boundary stabilization role of alloying elements [14,15]. However, detail studies describing the positive effect of alloying elements on grain growth suppression in magnesium alloys are only sporadic [16, 17].

The objective of this work is the investigation of the structure stability of two commercial magnesium alloys by a detail characterization of the microstructure, dislocation density development and the loss of stability of UFG structure as a result of isochronal annealing and to correlate these properties with microhardness evolution.

2. Experimental procedures

Two commercial cast alloys were used in this investigation, namely the AX41 (Mg - 4 wt.% Al - 1 wt.% Ca) and the AZ31(Mg - 3 wt.% Al - 1 wt.% Zn). The alloy AZ31 was subsequently extruded with the extrusion ratio of 22:1 at 350 °C and a speed of 60 mm/min. Homogeneous ultrafine-grained structure of both materials consisting of fine equiaxed grains was produced by ECAP following route B_C [18] at the temperature of 180 °C and the velocity of 50 mm/min. by 4 passes, and 220 °C, 20 mm/min. by 8 passes in AZ31 and AX41 alloy, respectively. These different processing procedures were employed purposely to obtain approximately the same grain sizes in both alloys (d=1.4±0.1 μm).

Thermal stability of the ultrafine-grained microstructure was investigated by isochronal annealing of the respective samples in the temperature range of 150-500 °C for 1 h followed by a water-quench. Characterization of the microstructure prior and after annealing was performed by scanning electron microscopy, electron diffraction and positron annihilation spectroscopy (PAS) [19]. The evolution of mechanical properties with the temperature of annealing was determined by Vickers microhardness. Both the microstructure investigations and the microhardness measurements were performed on the cross section of the billets lying perpendicular to the pressing direction (plane X) [20]. The samples for SEM observations and microhardness measurements were first mechanically grinded and polished using diamond suspensions (particle size down to 250 nm) to the mirror like surface quality. Specimens for PAS measurements were additionally etched for 30 s in the mixture of nitric acid, water and ethylene glycol. For EBSD investigations, the mechanically polished surface was subsequently ion milled using a Leica RES ion mill at 5 kV and an incidence angle of 5°.

The microstructure was investigated by a Zeiss Auriga Compact scanning electron microscope operated at 10 kV, equipped with the BSE detector and EDAX EBSD camera. Measured EBSD data were cleaned and evaluated by software TSL OIM Analysis 7.

The density of dislocations was evaluated from PAS data using a simple two-state trapping model [19] and employing the specific trapping rate of positrons to dislocations in Mg $\nu_D = 10^{-5}\ s^{-1}m^2$. The details of PAS data evaluation and the experimental device are described in detail elsewhere [21-23]. Vickers microhardness measurements were carried out employing an applied load of 100 g for 10 s, using automatic microhardness tester Qness Q10a. The hardness values were determined as the average of 36 separate measurements.

3. Experimental results and discussion

3.1. Microstructure evolution

EBSD inverse pole figures maps, captured the evolution of the microstructure of AX41 alloy at selected temperatures of isochronal annealing, are shown in Figure 1. The microstructure of the original ECAP processed AX41 alloy (in non-annealed condition) is homogenous, formed by equiaxed grains with the average grain size of 1.6 μm, see Figure 1 (RT). After isochronal annealing in the temperature range of 160-240 °C the microstructure remained almost unchanged, only a slight grain coarsening was observed.

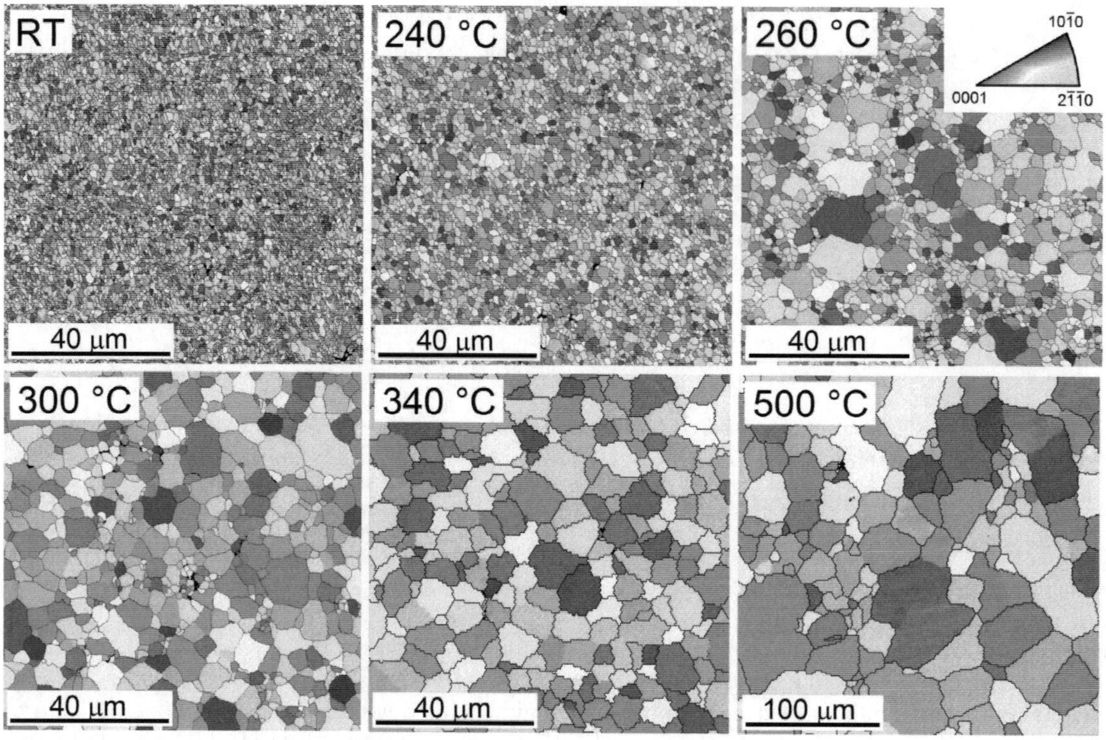

Figure 1 - EBSD inverse pole figure maps of the non-annealed (labeled as RT) and annealed AX41 magnesium alloy at various temperatures for 1h

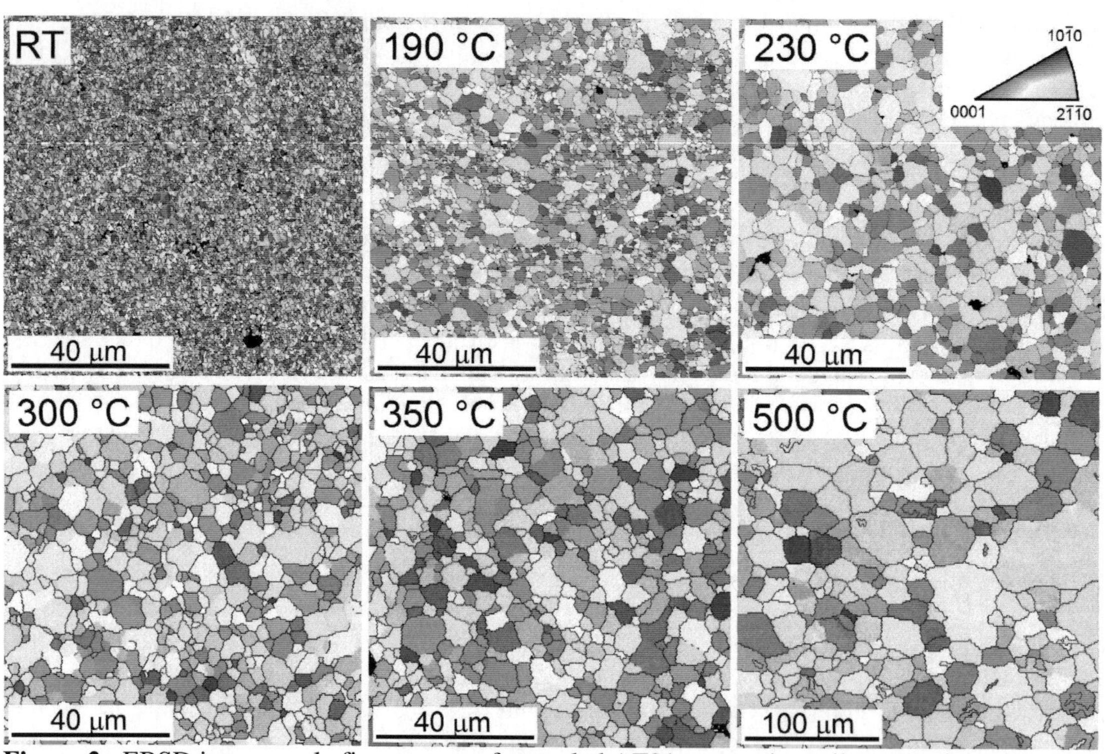

Figure 2 - EBSD inverse pole figure maps of annealed AZ31 magnesium alloy at various temperatures for 1h

Starting from the temperature of 260 °C, inhomogeneous grain growth, preferentially in the areas of microstructure without Al_2Ca phase (see below) was observed. During further annealing the fraction of coarse grains gradually increases and original small grains disappeared after annealing at 340 °C and homogeneous microstructure coarsening up to the final annealing temperature of 500 °C occurred.

The microstructural evolution in AZ31 magnesium alloy during isochronal annealing is shown in Figure 2. The original microstructure with the average grains size of 1.3 μm starts to grow already around the temperature of 190 °C. A bimodal microstructure was observed at this temperature consisting of coarse grains and the original ultrafine grains. Further annealing results in the gradual and homogeneous coarsening of the microstructure in the whole volume. Above the temperature of 230 °C no original ultrafine grains were observed.

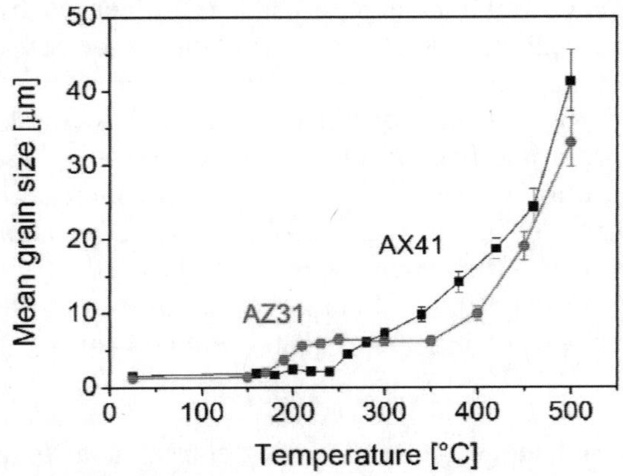

Figure 3 - The mean grain size of AZ31 and AX41 magnesium alloys after 1h of isochronal annealing

The mean grain size calculated from EBSD maps by software TSL OIM Analysis 7 using the area fraction statistics is plotted in Figure 3 as a function of the temperature of annealing for both AX41 and AZ31 alloys. The area fraction statistics was selected instead of the number fraction statistics, as the former is more sensitive to the inhomogeneous grain growth. Thus, the coarsening of even a small number of grains will affect the average grain size. Whereas in AZ31 alloy the grain size increases already from the temperature of 190 °C, in AX41 alloy the grain size starts to increase only above 240 °C. Nevertheless, the rate of the coarsening is much higher in AX41 alloy than in AZ31 resulting in higher average grain size in AX41 than in AZ31 above the temperature of 300 °C.

Improved thermal stability of AX41 alloy in the range of 180 – 260°C can be attributed to higher temperature and lower strain rate of ECAP processing of AX41 alloy as compared to AZ31 alloy. The achieved dynamic equilibrium grain size during ECAP at 220°C in AX41 alloy remains constant upon static annealing up to 240°C. Moreover, it is important to note that despite AX41 alloy was processed at higher temperature and lower strain rate, the achieved grain size is comparable to that of AZ31 alloy. This can be attributed to thermally stable secondary Al_2Ca particles (melting point 1079 °C, unsolvable in Mg matrix [24]), that are present during ECAP processing of the AX41 alloy, which contrasts to thermally instable $Mg_{17}Al_{12}$ particles in AZ31 alloy as shown in SEM micrographs in Figure 4. Increased grain growth of AX41 in the temperature range of 300 – 400°C is not fully resolved yet. It might be also caused by the absence of dissolved zinc in AX41 in contrast to AZ31 alloy.

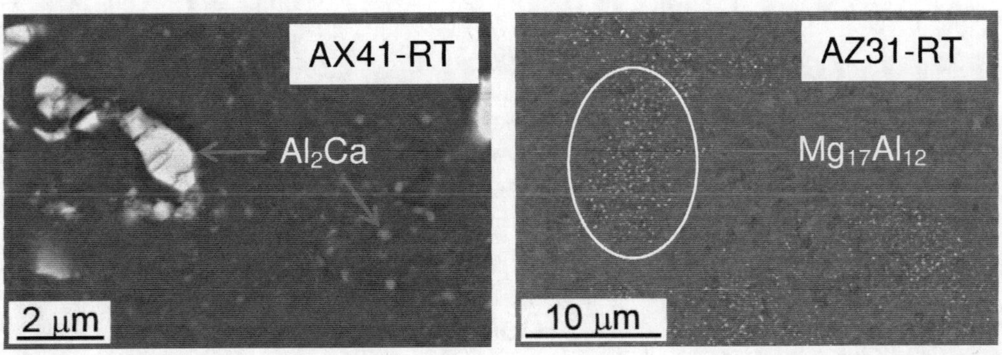

Figure 4 - SEM micrographs (BSE) of secondary phases in AX41 and AZ31 magnesium alloys at RT

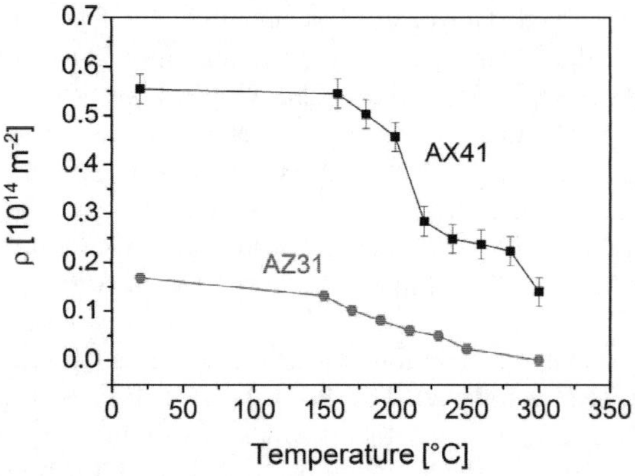

Figure 5 - The evolution of dislocation density with temperature of isochronal annealing in AX41 and AZ31 alloy

The values of dislocation density (ρ), obtained from PAS measurements are plotted in Figure 5 as a function of the annealing temperature. The dislocation density in both alloys starts to decrease at the temperature of ≈ 150 °C. After annealing at 300 °C, ρ in AZ31 drops below the detectable limit of PAS technique ($\rho < 5 \times 10^{12}$ m^{-2}). Note, that AX41 in the initial condition exhibits almost 4 times higher ρ than AZ31. The reason of this difference is rather complex and is influenced by many parameters, e.g. different processing conditions (temperature and number of passes) and the microstructure differences (thermally stable Al_2Ca particles which are present already in the cast AX41 alloy vs. rather instable $Mg_{17}Al_{12}$ which were probably formed in AZ31 only during cooling after ECAP), etc. The detail discussion of these effects is beyond the scope of this paper. On the other hand, the thermal stability and mechanical properties are determined by the temperature dependence of dislocation density rather than by its absolute value.

3.2. Microhardness evolution

Microhardness values of the ECAP processed samples and samples after 1 h of isochronal annealing are plotted in Figure 6. The first data points in both curves correspond to the initial UFG conditions after ECAP. A slightly higher microhardness (HV) was found in AZ31 alloy as compared to the AX41 alloy which corresponds to slightly smaller average grain size in AZ31 than in AX41, cf. Figures 3 and 6. HV values of both alloys do not differ significantly up to the temperature of 190 °C. For higher temperatures (T>190 °C) HV values of AZ31 decline abruptly up to 250 °C and continue to decrease up to 500 °C. On the other hand, AX41 alloy exhibits apparently higher thermal stability. HV values of AX41 decrease only slightly up to 240 °C, while a sharp drop in the HV can be observed only in the temperature range of 250-330 °C, which continues up to the temperature of 500 °C. After the annealing at 500 °C both alloys exhibit approximately the same values of HV. The evolution of HV is consistent with the evolution of grain size, cf. Figures 3 and 6.

In order to assess the role of grain size and the dislocation density in strengthening of both alloys the Hall-Petch (H-P) plot relating the microhardness with the grain size ($d^{-1/2}$) as determined from EBSD

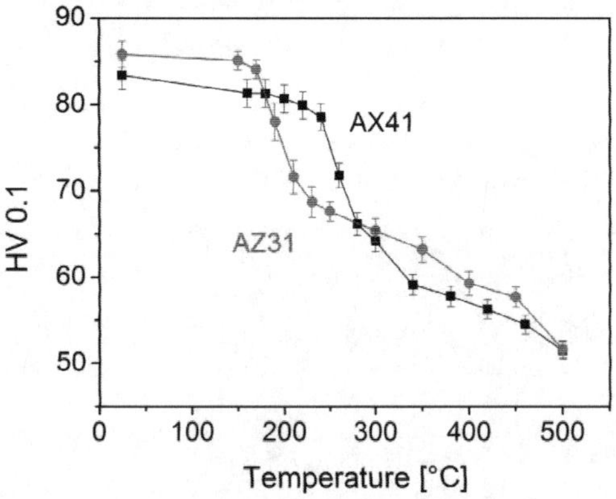

Figure 6 - The evolution of Vickers microhardness the with temperature of isochronal annealing

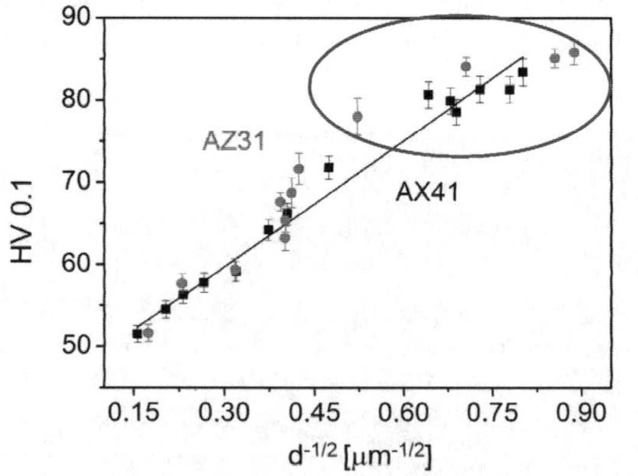

Figure 7 - The Hall-Petch plot for both alloys

analysis is displayed in Figure 7. It is clearly seen that the data of AX41 alloy obey the H-P law in the whole temperature range confirming the controlling role of the grain size on mechanical properties while the role of dislocations is minor. On the other hand, the data of AZ31 fit well the H-P relation only for high annealing temperatures (T>250 °C). At low temperatures (T< 250 °C) the data deviate from the linear relationship indicating that both the grain size and the dislocation density contribute to the strength of the material.

Conclusions

The following conclusions may be drawn from this investigation:
- the thermal stability of UFG AX41 alloy is superior to the thermal stability of AZ31 of the same initial grain size,
- enhanced thermal stability of the UFG structure of AX41 in the range of 180-260°C is caused by the conditions of ECAP processing and the presence of thermally stable Al_2Ca precipitates which stabilize the microstructure during ECAP,
- mechanical properties of AX41 are predominantly controlled by the grain size in the whole range of annealing temperatures. On the other hand, in AZ31 alloy the grain size affects the materials strength at high temperatures while at low temperatures (T< 250 °C) both the grain size and the dislocation density contribute to the strengthening.

Acknowledgement

This work was financially supported by the Czech Science Foundation under the project GB 14-36566G and by ERDF under the project “Nanomaterials centre for advanced applications”, project No. CZ.02.1.01/0.0/0.0/15_003/0000485. J.C acknowledges financial support by the Czech Science Foundation project no. 16-12828S.

References

[1] Gupta M and Sharon N M L 2011 *Magnesium, Magnesium Alloy, and Magnesium Composites*, (New Jersey, John Wiley &Sons, Inc.)
[2] Koch C C 2009 *Bulk nanostructured materials* (Wiley, Weinheim) 9
[3] Valiev R Z and Langdon T G 2006 *Progress in Materials Science* **51** 881
[4] Zhilyaev A P and Langdon T G 2008 *Progress in Materials Science* **53** 893
[5] Saito Y et al 1999 *Acta Materialia* **47** 579
[6] Varyutkhin N V et al 2006 *Mater. Sci. Forum* **503-504** 335
[7] Yang X et al 2012 *J. Mater. Sci.* **47** 2823
[8] Segal et al 1981 *Russ. Metall.* **1** 99
[9] Estrin Y and Vinogradov A 2013 *Acta Materialia* **61** 782
[10] Langon T G 2013 *Acta Materialia* **61** 1237
[11] Hellmig R J et al 2008 *Mater. Trans.* **49** *31*
[12] Kužel et al 2008 *Z. Krystallogr. Suppl.* **27** 73
[13] Azushima et al 2008 *Manufacturing Technology* **57** 716
[14] Vinogradov A et al 2005 *Acta Materialia* **53** 2181
[15] Sitdikov 2008 *Acta Materialia* **56** *821*
[16] Strásková J et al 2014 *Mater. Char.* **94** 69
[17] Čížek J et al 2006 *Phys. Stat. Sol.* **203** 466
[18] Furukawa M, Horita Z, Nemoto M and Langdon T G 2001 *J. Mater. Sci.* **36** 2835
[19] West R N 1979 *Positrons in Solids*. (Springer Berlin, Heidelberg) 89
[20] Iwahashi Y, Wang J, Horita Z, Nemoto M and Langdon T G 1998 *Mater. Sci. Eng.* **A 257** 328
[21] Janeček M, Čížek J, Gubicza J and Vrátná J 2012 *J. Mater. Sci.* **47** 7860
[22] Bečvář F, Čížek J, Procházka I and Janotová J 2005 *Nucl. Instrum. Methods* A **539** 372
[23] Procházka I, Novotný I and Bečvář F 1997 *Mater. Sci. Forum* **255-257** 772
[24] Massalski T B, Okamoto H *Binary alloy phase diagram* (Mater. Information Society, Ohio)